Medicinal Plants in the Asia Pacific for Zoonotic Pandemics

Medicinal Plants in the Asia Pacific for Zoonotic Pandemics Series

Series Editor:
Christophe Wiart

This series provides an unprecedented comprehensive overview of the antimicrobial activity of more than 100 medicinal plants in Asia and the Pacific. It discusses their actions against viruses, bacteria, and fungi representing a threat of zoonotic pandemics, with an emphasis on the molecular basis and cellular pathways. Scientific names, botanical classifications and descriptions, medicinal uses and chemical constituents are presented, as well as chemical structures, and a vast selection of bibliographical references. This series is a critical reference for anyone involved in the discovery of leads for the treatment of pandemic viral, bacterial, and fungal infections.

Medicinal Plants in the Asia Pacific for Zoonotic Pandemics,
Volume 1: Family Amborellaceae to Vitaceae, Christophe Wiart

Medicinal Plants in the Asia Pacific for Zoonotic Pandemics,
Volume 2: Family Zygophyllaceae to Muntingiaceae, Christophe Wiart

Medicinal Plants in the Asia Pacific for Zoonotic Pandemics,
Volume 3: Family Thymelaeaceae to Portulacaceae, Christophe Wiart

Medicinal Plants in the Asia Pacific for Zoonotic Pandemics,
Volume 4: Family Alangiaceae to Araliaceae, Christophe Wiart

Medicinal Plants in the Asia Pacific for Zoonotic Pandemics

Family Alangiaceae to Araliaceae

Volume 4

Christophe Wiart

CRC Press
Taylor & Francis Group
Boca Raton London New York

CRC Press is an imprint of the
Taylor & Francis Group, an **informa** business

First edition published 2022
by CRC Press
6000 Broken Sound Parkway NW, Suite 300, Boca Raton, FL 33487-2742

and by CRC Press
4 Park Square, Milton Park, Abingdon, Oxon, OX14 4RN

CRC Press is an imprint of Taylor & Francis Group, LLC

ISBN: 978-1-032-01850-8 (hbk)
ISBN: 978-1-032-01034-2 (pbk)
ISBN: 978-1-003-18034-0 (ebk)

DOI: 10.1201/9781003180340

Typeset in Times
by codeMantra

Contents

Foreword

Natural products are compounds produced by organisms that often possess biological activity. For thousands of years, humans have harnessed this biological activity to help treat a variety of diseases. Initially, people harnessed herbal plant preparations, but also explored enriching and purifying these active natural products. Important natural product-derived drugs from plants currently in medicine are morphine for pain relief, artemisinin as an antimalarial, and acetyldigitoxin in heart failure, as well as paclitaxel, vinblastine, maytansine, and ingenol mebutate for cancer (Butler,2004; Butler et al., 2014).

Zoonotic diseases, which are caused by pathogens that can be transferred between vertebrate animals and humans (Rees et al., 2021), have been a major source of human infections for thousands of years. In the last few decades, the number of new zoonotic infections has increased, which has been linked with animal habit destruction and increased animal contact. Examples of zoonotic diseases that have caused considerable deaths include Avian influenza, the Plague, Ebola, and more recently SARS-CoV2. Increased population mobility also facilitates the dissemination of these diseases, which can turn regional outbreaks into national and global outbreaks. The most dramatic recent example is the rapid spread of a zoonotic virus, SARS-CoV2, that went from a few isolated cases in December 2019 to a worldwide pandemic in a few months. COVID-19, which is the disease caused by SARS-CoV-2, is still causing immense health issues all over the world, and it will be many years before its effects are diminished.

Rapid deforestation of tropical rainforests has also been highlighted as a major contributor to the emergence of new zoonotic diseases, both now and in future outbreaks (Keesing and Ostfeld, 2021). Once natural ecosystems have been perturbed or destroyed, the interactions between the rainforest animals, endemic or migratory, and humans and their livestock will increase. In addition, intact rainforest can also help to contain zoonotic diseases and help limit their escape. As demonstrated with COVID-19, it is far better for both health and economic outcomes to control the spread of SARS-CoV2, rather than be wholly reliant upon medical treatment. Therefore, in addition to the benefits of the rainforest toward the discovery of new medicines, biodiversity, climate change, and wellness, they will have a key role in helping to reduce the future zoonotic diseases. With so many positive benefits from rainforests, it is imperative that Governments and people protect these vital treasures for future generations. Once they are gone, these ancient and complex ecosystems cannot be easily regenerated and will almost certainly never return to their original state.

In the series "Medicinal Plants in the Asia Pacific for Zoonotic Pandemics," Dr Christophe Wiart has meticulously prepared Volume 4 that details important information about the biological activities of Asian rainforest plants from the families Cornaceae to Pittosporaceae. This book will be an important resource for scientists, as well as people passionate about helping to save these critical resources. Perhaps a new drug lead will be identified from at least one of these plants.

Dr Mark S. Butler
MSBChem Consulting, Brisbane, Australia

References

Butler, M.S., 2004. The role of natural product chemistry in drug discovery. *Journal of Natural Products*, *67*, pp. 2141–2153.

Butler, M.S., Robertson, A.A.B. and Cooper, M.A., 2014. Natural product and natural product derived drugs in clinical trials. *Natural Product Reports*, *31*, pp. 1612–1661.

Keesing, F. and Ostfeld, R.S. 2021. Impacts of biodiversity and biodiversity loss on zoonotic diseases. *Proceedings of the National Academy of Sciences*, *118*, p. e2023540118.

Rees, E.M., Minter, A., Edmunds, W.J., Lau, C.L., Kucharski, A.J. and Lowe, R., 2021. Transmission modelling of environmentally persistent zoonotic diseases: a systematic review. *The Lancet Planetary Health*, *5*, pp. e466–e478.

Preface

The recent emergence of life-threatening waves of viral zoonosis of increasing severity, culminating today with the COVID-19 pandemic, marks the beginning of an era that might be the last before the ultimate disappearance of human beings. Throughout history, plagues threatened human survival. The Roman Empire was struck from about 166 to 189 A.D. by a smallpox pandemic known as "the Antonine plague" claiming the lives of half of the population of the Empire including the emperor Marcus Aurelius and contributing to the socioeconomic decline of the Western Roman Empire until its fall in 476 A.D. The zoonotic bacterium *Yersinia pestis* accounted for two major plagues in the Middle Ages: "the Justinian plague" (6th century) and the "Black Death" (14th century) claiming the lives of millions. The Spanish flu that followed World War I due to a zoonotic Influenza A (H1N1) virus made more than 20 million victims. For the last about 50 years, novel life-threatening zoonotic viruses have emerged and spread globally including the human immunodeficiency virus, the Ebola virus, the Zika virus, the Middle East respiratory syndrome coronavirus, and today the ignominious severe acute respiratory syndrome coronavirus 2 (SARS-CoV-2) which according to the WHO has infected so far 413 million people, caused 5.83 million deaths, and paralyzes the whole world as never seen before. Among the reasons for the emergence of zoonotic pandemics is the increasing contact of humans with animal reservoirs due to deforestation and industrial animal farming. According to Wolfe et al. (2005), the richness of microbes in a region, environmental changes, and increased human and animal contact with wildlife result in the emergence of new zoonotic pathogenic microorganisms. Thus, one could envisage the increasing waves of zoonosis to be biological "negative feedback loops" meant to compel humans to put an end to primary rainforest burning, carbon dioxide emission, global warming, as well as intensive poultry, swine, and other animal farming. The pharmaceutical industry has managed to produce vaccines offering some levels of protection against SARS-CoV-2. However, if pressures on wildlife and nature continue, other new zoonotic pandemics will come with the possibility of the emergence of a completely untreatable fast killing virus, bacteria, or even fungi. Furthermore, between the time of appearance of a new virus and the time of production of a vaccine, people are left vulnerable. The intention behind the writing of this book is to offer researchers, academics, and students a comprehensive and interrelated set of botanical, ethnopharmacological, and pharmacological evidence to facilitate the discovery of natural products or even herbal remedies for the treatment or prevention of viral, bacterial, or fungal zoonotic pandemics. It is also to offer an intellectual rationale or tool to understand and thus cogently approach the fascinating realm of drug discovery from medicinal plants. There is a law for the rational selection and use of plants for the treatment of diseases. This fourth volume covers the medicinal plants traditionally used for the treatment of microbial infection in Asia and the Pacific classified in the Clade Asterids, Lamids, and Campanulids. Within these Clades, plants are presented according to their respective orders and families. Scientific and common names are given as well as synonyms, habitat, distribution, personal botanical observation, pharmacology, commentaries, personally handmade botanical plates, and carefully selected references. Rabbi Nachman of Breslov (1772–1810) said "The day you were born is the day G-D decided the world could not exist without you" and these 4 Volumes are meant to assist those who will find cures to eradicate COVID-19 and other global pandemics from the medicinal plants of Asia and the Pacific.

Christophe Wiart
Kota Kinabalu, Sabah, Malaysia
January 24, 2022

Author

Christophe Wiart, PharmD, PhD, is a French Ethnopharmacologist attached to the Institute for Tropical Biology and Conservation, Universiti Malaysia Sabah. His fields of expertise are Asian ethnopharmacology, chemotaxonomy, and ethnobotany. He has collected, identified, classified, and examined the therapeutic potentials of several hundred species of medicinal plants from Asia and the Pacific.

He appeared on HBO's *Vice* (TV series) in season 3, episode 6 (episode 28 of the series), titled "The Post-Antibiotic World and Indonesia's Palm Bomb," April 17, 2015. It highlighted the need to find new treatments for infections that were previously treatable with antibiotics but are now resistant to multiple drugs. He was guest at Al Jazzera inside Story "Rise of the Superbug: is resistance futile?" in February 2016 and was on BBC radio Crowdscience in February 2020.

9 Medicinal Plants in Asia-Pacific for Microbial Infections

The Clade Asterids

9.1 ORDER CORNALES LINK (1829)

9.1.1 Family Alangiaceae A.P. de Candolle (1828)

The family Alangiaceae consists of the single genus *Alangium* Lam.

9.1.1.1 *Alangium chinense* (Lour.) Harms

Synonyms: *Alangium begoniifolium* (Roxb.) Baill.; *Diacicarpium rotundifolium* Hassk.; *Diacicarpium tomentosum* Blume; *Marlea begoniifolia* Roxb.; *Marlea tomentosa* (Blume) Endl. ex Hassk.; *Stylidium chinense* Lour.; *Stylis chinensis* (Lour.) Poir.

Common names: Chinese Alangium; sa nhan (Cambodia; Laos); ba jiao feng (China); belkel phang (India); timangan (Indonesia); bagaloan (the Philippines); khao yen (Thailand); cây thôi chanh (Vietnam)

Habitat: Forests margins

Distribution: India, Bhutan, Nepal, China

Botanical observation: This sinister shrub grows up to 5 m tall. The stems are zig-zag-shaped and hairy at apex when young. The leaves are simple, alternate, and exstipulate. The petiole is 4–6 cm long and somewhat curved. The blade is dull light green, broadly lanceolate to lobed, cordate and asymmetrical at base, 8–20 cm × 5–12 cm, hairy below, marked with about 5 pairs of secondary nerves, the first pair arising from the base, and acuminate at apex. The cymes are axillary, lax, and slender. The calyx is tubular and minutely 4–7-lobed. The perianth is very characteristic with six petals which are pure white, recurved, 1–1.5 cm long, and oblong. A disc is present. The androecium is made of 6–8 stamens with the anthers merged in an oblong (somewhat Solanaceous) mass from which emerges the stigma. The drupes are ellipsoid, 5–7 mm, and blue-berry colored, and dreadfully poisonous.

Medicinal uses: Fever, tuberculosis (China)

Strong antiviral (non-enveloped monopartite linear single-stranded (+) RNA) amphiphilic cadinane-type sesquiterpenes: 2,7-Dihydroxycadalene (LogD 3.8 at pH 7.4; molecular mass = 230.3 g/mol), 3,7-dihydroxycadalene, 7-hydroxycadalene, 2,7-dihydroxycadalene, and (1*S*,4*R*)-7,8-dihydroxycalamenene isolated from the roots inhibited the replication of the Coxsackie virus B3 in Vero cells with the IC_{50} values of 1.4, 11.3, 5.3, 10.1, and 7.4 µM, respectively (Zhang et al., 2013).

DOI: 10.1201/9781003180340-1

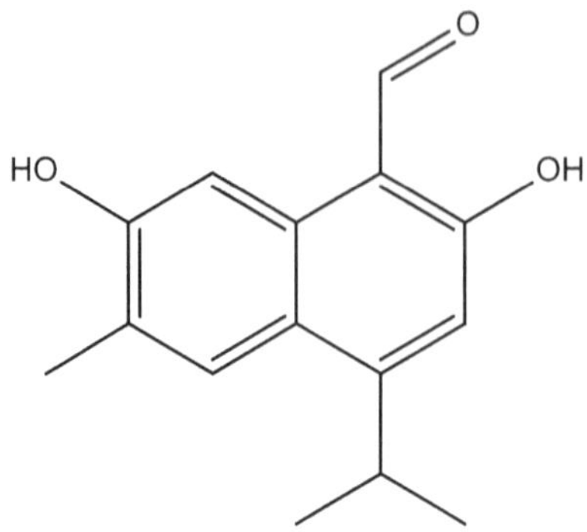

2,7-Dihydroxycadalene

Strong antiviral (non-enveloped monopartite linear single-stranded (+) RNA) amphiphilic cadinane sesquiterpenes: Mansonone E and mansonone H isolated from the roots inhibited the replication of the Coxsackie virus B3 in Vero cells with the IC_{50} values of 4.7 and 3.1 µM, respectively (Zhang et al., 2013).

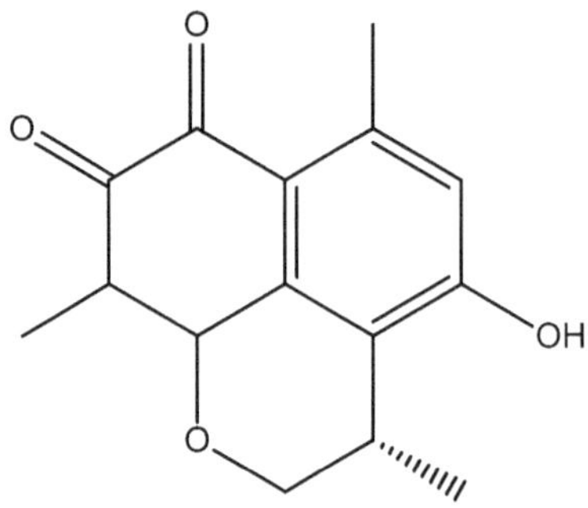

Mansonone

Strong antiviral (non-enveloped monopartite linear single-stranded (+) RNA) amphiphilic β-carboline alkaloid: 4,5-Dimethoxycanthin-6-one (also known as methyl nigakinone; LogD 2.4 at pH 7.4; molecular mass=280.2 g/mol) isolated from the roots inhibited the replication of the Coxsackie virus B3 in Vero cells with the IC_{50} value of 7.4 µM (Zhang et al., 2013).

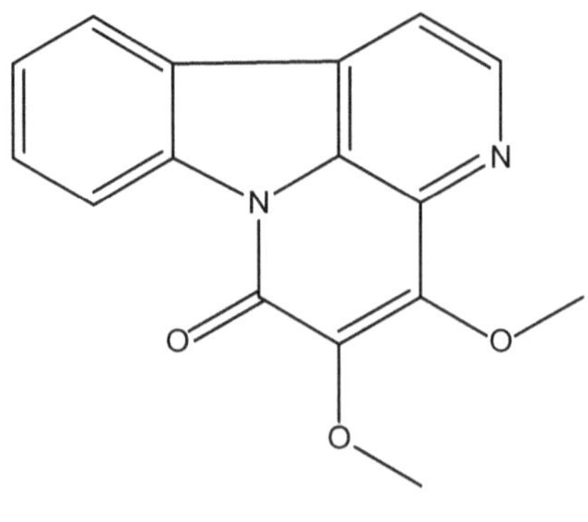

4,5-Dimethoxycanthin-6-one

Antiviral (non-enveloped monopartite linear single-stranded (+) RNA) norditerpenes: Norditerpenes inhibited the replication of the Coxsackie virus B3 (Zhang et al., 2015).

Antiviral (non-enveloped monopartite linear single-stranded (+) RNA) polar extract: Ethanol extract (10 mg/mL) of leaves inhibited the replication of the Tobacco mosaic virus (strain U1) on *Nicotiana glutinosa* L. by 48.6% (Jing et al., 2012)

Commentaries

i. The plant produces phenolic glycosides (Zhang et al., 2017), pyridine alkaloids (Wei et al., 2020) of which as the dreadfully poisonous pyridine alkaloid anabasine (Sims et al., 1999; Wang & Yaling, 2016).

ii. Cadinane-type sesquiterpenes and sesquiterpene quinones from this plant are most probably antibacterial and antifungal. Being active against Coxsackie virus which is an (+)-RNA, sesquiterpenes in this plant might have activity against coronaviruses including the severe acute respiratory syndrome-associated coronavirus 2 virus and vaccination-resistant mutants. Cadinane sesquiterpenes often inhibit the replication of the human immunodeficiency virus and as such could be of value against betacoronaviruses.

iii. *Alangium salviifolium* (L. f.) Wangerin (synonym: *Alangium lamarckii* Thwaites) is used in China (local name: *Lang gia*) for the treatment of external injuries and in India (local name: *Ankola*) for bite of rabid animals, diarrhea, dysentery, fever, leprosy, syphilis, and tuberculosis. *Alangium salviifolium* (L. f.) Wangerin produces a series of tetrahydroisoquinoline-monoterpene and iridoid glycosides. Iridoids are monoterpene derivatives, which compared with the previous Clades, are common in the Clades Lamiids and Asterids (Sherden et al., 2018), and often display antiviral properties *in vitro.*

iv. Consider that indole alkaloids of therapeutic interest are particularly produced by members of the order Gentianales in the Clade Lamiids.

References

Jing, B., Ma, Z., Feng, J., Liang, H., Li, C. and Zhang, X., 2012. Evaluation of the antiviral activity of extracts from plants grown in the qinling region of China against infection by Tobacco mosaic virus (TMV). *Journal of Phytopathology*, *160*(4), pp. 181–186.

Sherden, N.H., Lichman, B., Caputi, L., Zhao, D., Kamileen, M.O., Buell, C.R. and O'Connor, S.E., 2018. Identification of iridoid synthases from Nepeta species: Iridoid cyclization does not determine nepetalactone stereochemistry. *Phytochemistry*, *145*, pp. 48–56.

Sims, D.N., James, R. and Christensen, T., 1999. Another death due to ingestion of Nicotiana glauca. *Journal of Forensic Science*, *44*(2), pp. 447–449.

Wang, Q. and Yaling, X.U., 2016. Content determination of L (–)-Anabasine in different medicinal parts of Alangium chinense by HPLC. *China Pharmacy*, *27*(27), pp. 3877–3879.

Wei, X., Yang, J., Dai, Z., Yu, H.F., Ding, C.F., Khan, A., Zhao, Y.L., Liu, Y.P. and Luo, X.D., 2020. Antitumor pyridine alkaloids hybrid with diverse units from Alangium chinense. *Tetrahedron Letters*, *61*(7), p. 151502.

Zhang, Y., Liu, Y.B., Li, Y., Ma, S.G., Li, L., Qu, J., Zhang, D., Chen, X.G., Jiang, J.D. and Yu, S.S., 2013. Sesquiterpenes and alkaloids from the roots of Alangium chinense. *Journal of Natural Products*, *76*(6), pp. 1058–1063.

Zhang, Y., Liu, Y.B., Li, Y., Li, L., Ma, S.G., Qu, J., Jiang, J.D., Chen, X.G., Zhang, D. and Yu, S.S., 2015. Terpenoids from the roots of Alangium chinense. *Journal of Asian Natural Products Research*, *17*(11), pp. 1025–1038.

Zhang, Y., Liu, Y.B., Li, Y., Ma, S.G., Li, L., Qu, J., Zhang, D., Jiang, J.D. and Yu, S.S., 2017. Phenolic constituents from the roots of Alangium chinense. *Chinese Chemical Letters*, *28*(1), pp. 32–36.

9.1.2 Family Cornaceae Bercht. & J. Presl (1825)

The family Cornaceae consists of the single genus *Cornus* L.

9.1.2.1 *Cornus officinalis* Sieb. & Zucc.

Synonym: *Macrocarpium officinale* (Siebold & Zucc.) Nakai

Common names: Japanese cornel; shan zhu yu (China); sanshuyu (Japan)

Habitat: Forests, cultivated

Distribution: China, Korea, Japan

Botanical observation: This tree grows up to 10 m tall. The leaves are simple, opposite, and exstipulate. The blade is lanceolate, 5.5–10 cm×2.5–4.5 cm, hairy below, and marked with 7–11 pairs of curved secondary nerves sunken above and with a characteristic arrangement (angle of insertion with midrib decreases the closest it is to the apex of the blade). The inflorescences are cauliflorous fascicles. The calyx is minute, tubular, and 4-lobed. The four petals are yellow, lanceolate, about 3 mm long. A disc is present. The androecium includes four stamens with the same color with the petals and about 1 cm long. The ovary is hairy. The drupes are red, glossy, narrowly oblong, and up to about 2 cm long, cherry-like, and marked at apex.

Medicinal uses: Fever (China)

Broad-spectrum antibacterial polar extract: Ethanol extract (4 mg/6 mm disc) inhibited the growth of *Bacillus cereus* (KCTC 1012), *Escherichia coli*, *Escherichia coli* O157:H7 (ATCC 43894), *Listeria monocytogenes*, *Pseudomonas aeruginosa* (KCTC 1750), *Staphylococcus aureus* (KCTC 1916), and *Salmonella typhimurium* (KCTC 2515) with the inhibition zone diameters of 12, 18.3, 16.3, 18.2, 9.5, 10.3, and 8.5 mm, respectively (Choi et al., 2004).

Antifungal polar extract: Methanol extract (300 mg of fresh plant parts/mL) inhibited the growth of *Puccinia recondita* by 67%.

Strong antiviral (non-enveloped monopartite linear single-stranded (+) RNA) extract: Extract (solvent not specified) inhibited the replication of the Coxsackie virus A group 16 strain (CVA16) in Vero cells with the IC_{50} value of 39.2 µg/mL (CC_{50} value > 50 µg/mL) (Song et al., 2015).

Viral enzyme inhibition by gallotannins: 1, 2, 3, 6-Tetragalloyl-β-D-glucopyranose and 1,2,3,4,6-pentagollayl-β-D glucopyranose inhibited the Hepatitis C virus NS3 protease with IC_{50} values of 6.9 and 5.1 µM (Yue et al., 2006).

Viral enzyme inhibition by ellagitannins: Tellimagrandin I and tillimagrandin II inhibited the Hepatitis C virus NS3 protease with IC_{50} values of 7 and 4.8 µM (Yue et al., 2006).

Commentaries

i. The plant is used in China for COVID-19 (Zeng et al., 2020). Although the active principles are not known yet, one could contemplate the role of iridoid glycosides which abound in this tree. One such iridoid glycoside is morroniside (20 mg/Kg intraperitoneally) which protected rats against Coxsackie virus B3 infections (Li et al., 2019). One other main iridoid glycoside in this plant is loganin which impeded the entry of the Hepatitis C virus in host cells by about 25% at the concentration of 20 µg/mL (Zhang et al., 2009). Cornuside from this plant inhibited severe acute respiratory syndrome-associated coronavirus RNA synthesis (Zhang et al., 2021). *Cornus officinalis* Sieb. & Zucc. phytomedication could be of value against COVID-19 and could be added (as well as its iridoid glycosides) to the list of first-line medicinal plant products to be tested for the coming viral pandemics.

ii. The plant builds a series of cadinane-type sesquiterpene glycosides (He et al., 2019), and as seen earlier, cadinane-type sesquiterpenes are antimicrobials, and accumulate iridoid glycosides such as morroniside, loganin, and cornuside (Lee et al., 2012).

iii. Plants in the genus *Cornus* L. produce tannins, explaining at least in part, why methanol extract of *Cornus kousa* Bürger ex Hance stem and leaves at the concentration of 100 µg/mL inhibited RNA-dependent DNA polymerase and ribonuclease H activities of human immunodeficiency virus type-1 reverse transcriptase and human immunodeficiency virus type-1 protease by 98.9%, 17.3%, and 8.5%, respectively (Min et al., 2001) and why ethanol extract of aerial parts of *Cornus walteri* Wangerin inhibited the human immunodeficiency virus type-1 protease (Park, 2003). Tannins are of no therapeutic value for the systemic treatment or prevention of infectious diseases. They are however having potentials as disinfectants and antiseptics.

iv. In Turkey, *Cornus mas* L. (local name: *Kizilcik*) is used to treat cold, cough, diarrhea, flu, and fungal infections. In China, *Cornus chinensis* Wangerin (local name: *Hu tui tz*) is used to treat cough.

References

Choi, G.J., Jang, K.S., Kim, J.S., Lee, S.W., Cho, J.Y., Cho, K.Y. and Kim, J.C., 2004. In vivo antifungal activities of 57 plant extracts against six plant pathogenic fungi. *Plant Pathology Journal*, *20*, pp. 184–191.

He, J., Xu, J.K., Pan, X.G., Ye, X.S., Gao, P.Y., Yan, Y., Xu, C.Y., Qiang, G.F., Du, G.H., Cheng, Y.C. and Zhang, W.K., 2019. Unusual cadinane-type sesquiterpene glycosides with α-glucosidase inhibitory activities from the fruit of Cornus officinalis Sieb. et Zuuc. *Bioorganic Chemistry*, *82*, pp. 1–5.

Li, W., Chen, M., Xu, L., Lv, Z., Chen, L., Li, Y. and He, W., 2019. Morroniside alleviates coxsackievirus B3-induced myocardial damage apoptosis via restraining NLRP3 inflammasome activation. *RSC Advances*, *9*(3), pp. 1222–1229.

Lee, A.Y., Kim, H.S., Jo, J.E., Kang, B.K., Moon, B.C., Chun, J.M., Ji, Y. and Kim, H.K., 2012. Optimization of extraction condition for major iridoid components in fruit of corni (Cornus officinalis) by UPLC-PDA using response surface methodology. *Food Science and Biotechnology*, *21*(4), pp. 1023–1029.

Min, B.S., Kim, Y.H., Tomiyama, M., Nakamura, N., Miyashiro, H., Otake, T. and Hattori, M., 2001. Inhibitory effects of Korean plants on HIV-1 activities. *Phytotherapy Research*, *15*(6), pp. 481–486.

Park, J.C., 2003. Inhibitory effects of Korean plant resources on human immunodeficiency virus type 1 protease activity. *Oriental Pharmacy and Experimental Medicine*, *3*(1), pp. 1–7.

Song, J.H., Park, K., Shim, A., Kwon, B.E., Ahn, J.H., Choi, Y.J., Kim, J.K., Yeo, S.G., Yoon, K. and Ko, H.J., 2015. Complete sequence analysis and antiviral screening of medicinal plants for human coxsackievirus a16 isolated in Korea. *Osong Public Health and Research Perspectives*, *6*(1), pp. 52–58.

Yue, W., Zhengquan, L., Lirong, C. and Xiaojie, X., 2006. Antiviral compounds and one new iridoid glycoside from Cornus officinalis. *Progress in Natural Science*, *16*(2), pp. 142–146.

Zeng, Y., Du, J., Pu, X., Yang, X., Li, X. and Yang, J., 2020. Remarkable efficacy of COVID-19 treatment by traditional Chinese medicine. *Military Medical Research*, *7*, Article number: 4

Zhang, H., Rothwangl, K., Mesecar, A.D., Sabahi, A., Rong, L. and Fong, H.H., 2009. Lamiridosins, hepatitis C virus entry inhibitors from Lamium album. *Journal of Natural Products*, *72*(12), pp. 2158–2162.

Zhang, Q.Y., Deng, C.L., Liu, J., Li, J.Q., Zhang, H.Q., Li, N., Zhang, Y.N., Li, X.D., Zhang, B., Xu, Y. and Ye, H.Q., 2021. SARS-CoV-2 replicon for high-throughput antiviral screening. *Journal of General Virology*, *102*(5), p. 001583.

9.1.3 Family Hydrangeaceae Dumortier (1829)

The family Hydrangeaceae consists of about 20 genera and 250 species of graceful yet poisonous herbs, shrubs, trees, or climbers. The leaves are mainly opposite, simple, and exstipulate. The blade is often serrate. The inflorescences are terminal or axillary, cymes, panicles, racemes, or corymbs, or solitary. The calyx and corolla present 4–12 sepals and 4–12 petals, respectively. The androecium consists of up to 200 stamens. The gynoecium includes 2–12-carpels fused into a 1–12 locular ovary, each locule sheltering 1–50 ovules growing on axile or parietal placentas. The fruits are dehiscent capsules or berries.

9.1.3.1 *Hydrangea heteromalla* D. Don

Synonyms: *Hydrangea khasiana* Hook. f. & Thomson; *Hydrangea mandarinorum* Diels; *Hydrangea vestita* Wall.

Common names: Fang han mi, wei rong xiu qiu (China); qidcug (Myanmar)

Habitat: Forest margins, mountains, ornamental

Distribution: India, Bhutan, Nepal, Sikkim, China

Botanical observation: This magnificent shrub grows up to 5 m tall. The leaves are simple, opposite, and exstipulate. The petiole is purplish red, 2–4 cm long, and hairy. The blade is elliptic, broadly lanceolate, 6–15 cm × 3–8 cm, hairy below, marked with 7–9 pairs of secondary nerves, cuneate at base, serrate, and acute to acuminate at apex. The cymes are up to about 30 cm across. The calyx is tubular, minute, and 4-lobed. The 4 petals are white or yellowish, oblong-ovate, and about 2 mm long. The 10 stamens are of unequal length. The gynoecium includes 3 or 4 styles. Sterile flowers develop four showy sepals, somewhat bluish to white, growing up to about 1.5 cm long. The capsules are ovoid to subglobose, about 3 mm across, and contain a few tiny and winged seeds.

Medicinal use: Diarrhea, wounds (China)

Commentaries

i. The plant has apparently not been studied for antimicrobial properties. Plants in the genus *Hydrangea* L. produce 1,4-benzoquinones (Harasawa et al., 1994) as well as quinazolinone alkaloids (Patnam et al., 2001). Benzoquinones are broad-spectrum antimicrobial agents (see earlier) and the quinazolinone moiety is a broad-spectrum antibacterial, antifungal, and antiviral pharmacophore (Hassanzadeh et al., 2012; Wang et al., 2012). Natural products with broad-spectrum antimicrobial effects often target microbial genetic material. The quinazolinone moiety is a cytotoxic pharmacophore in mammalian cells via DNA intercalation (Rakesh et al., 2019). Further, quinazolinones interact with the angiotensin-converting enzyme (Hodges, 1994) and as such could be of value for the fight against COVID-19 and other pandemic coronaviruses.

ii. In Taiwan, *Hydrangea umbellata* Rehder is used to treat gonorrhea.

References

Hassanzadeh, F., Jafari, E., Hakimelahi, G.H., Khajouei, M.R., Jalali, M. and Khodarahmi, G.A., 2012. Antibacterial, antifungal and cytotoxic evaluation of some new quinazolinone derivatives. *Research in Pharmaceutical Sciences*, *7*(2), p. 87.

Harasawa, A. and Tagashira, A., 1994. Isolation of 2,6-dimethoxy 1,4- benzoquinone from Hydrangea (Hydrangea macrophylla Seringe var otaksa Makino) and its deodorant activity against methyl mercaptan. *Bioscience, Biotechnology, and Biochemistry*, *58*, pp. 2073–2074.

Hodges, J.C., 1994. Overview Cardiovascular & Renal: Angiotensin antagonists: Agents that interact with the AT2 receptor and their potential therapeutic utilities. *Expert Opinion on Therapeutic Patents*, *4*(11), pp. 1325–1338.

Patnam, R., Chang, F.R., Chen, C.Y., Kuo, R.Y., Lee, Y.H. and Wu, Y.C., 2001. Hydrachine A, a novel alkaloid from the roots of hydrangea c hinensis. *Journal of Natural Products*, *64*(7), pp. 948–949.

Wang, Z., Wang, M., Yao, X., Li, Y., Tan, J., Wang, L., Qiao, W., Geng, Y., Liu, Y. and Wang, Q., 2012. Design, synthesis and antiviral activity of novel quinazolinones. *European Journal of Medicinal Chemistry*, *53*, pp. 275–282.

9.2 ORDER ERICALES BERCHT. & J. PRESL (1820)

9.2.1 Family Balsaminaceae Berchtold and J. Presl (1820)

The family Balsaminaceae consists of 2 genera and about 1000 species of herbs. The leaves are simple, opposite, verticillate or spiral, and exstipulate. The blade is not uncommonly serrate. The inflorescence is a raceme or a fascicle. The flowers are zygomorphic and colorful. The calyx comprises 3–5 sepals which are free or form a tube. The corolla consists of five petals which are membranous. The androecium consists of five stamens. The 4–5 carpels are united to form a 4–5-celled ovary containing numerous ovules. The fruits are berries or explosively dehiscent capsules containing numerous tiny seeds. Members of the family produce antibacterial and antifungal naphthoquinones.

9.2.1.1 *Impatiens balsamina* L.

Synonyms: *Balsamina hortensis* Desp.; *Impatiens eriocarpa* Launert; *Impatiens stapfiana* Gilg

Common names: Garden balsam dopati (Bangladesh); chu jia hua, feng xian hua (China); gul mendhi, kasitumbai, teera (India); inai ayam; pacar air (Indonesia); tsuri-fune-so (Japan); bong-seonhwa (Korea); bunga embung (Malaysia); dandalet (Myanmar); saponga, solonga, camantigui, solonga, suranga, (the Philippines); kudalu (Sri Lanka); thian baan (Thailand)

Habitat: Cultivated

Distribution: Tropical Asia

Botanical observation: This beautiful herb grows to about 1 m tall. The stem is fleshy, light green, and longitudinally striated. The leaves are simple, spiral, and exstipulate. The petiole is 1–3 cm long and channeled. The blade is elliptic to lanceolate, 4–12 cm × 1.5–3 cm, with 4–7 pairs of secondary nerves, tapering at base, serrate, and acuminate at apex. The inflorescences are axillary fascicles of showy and white-pink flowers. The flower pedicels are 2-2.5 cm long and pubescent. Lateral sepals 2, ovate and minute. Lower sepal navicular, up to 2 cm long, and narrowed into an incurved spur which is 1–2.5 cm long. The upper petal is orbicular. Lateral united petals are clawed, 2-lobed, and up to 2.5 cm long. The androecium consists of five stamens with linear filaments. The ovary is minute, fusiform, and pubescent. The capsules are fusiform, explosively dehiscent, 1–2 cm long, hairy, and contain numerous tiny seeds.

Medicinal uses: Boils (India); burns (Bangladesh, India, Sri Lanka); fungal infections (India)

Strong antibacterial (Gram-positive) polar extract: Ethanol extract inhibited the growth of clinical isolates of methicillin-resistant *Staphylococcus aureus* and *Staphylococcus aureus* with MIC/MBC values of 6.3/25 and 6.3 /25 μg/mL, respectively (Voravuthikunchai & Kitpipit, 2005). Methanol extract of leaves inhibited the growth of *Staphylococcus aureus* (ATCC 25923) with MIC value of 2.5 $\times 10^3$ μg/mL (Niyomkam et al., 2010). Methanol/acetone/water (3/1/1) extract of aerial parts inhibited the growth of *Staphylococcus aureus* (ATCC 25293), *Staphylococcus epidermidis* (ATCC 1228), *Micrococcus luteus* (ATCC 10240), *Bacillus subtilis* (ATCC 6633), and *Bacillus cereus* (ATCC 10876) with MIC of 500, 125, 250, 500, and 1000 μg/mL, respectively (Szewczyk et al., 2016).

Strong broad-spectrum antibacterial amphiphilic naphthoquinones: 2-Methoxy-1,4-naphthoquinone (also known as lawsone methyl ether; LogD 1.6 at pH 7.4; molecular mass = 188.1 g/mol) from aerial parts inhibited the growth of *Staphylococcus aureus* (236), *Staphylococcus aureus* Cowan, *Bacillus cereus*, *Bacillus megaterium*, *Bacillus subtilis* (168), *Aeromonas salmonicida* (A449), and *Aquaspirillum serpens* with MIC of 16, 16, 64, 4, 8, 2, and 16 μg/mL, respectively (Yang et al., 2001) and inhibited the growth of methicillin-resistant *Staphylococcus aureus* with MIC value of 15.6 μg/mL (Panichayupakaranant et al., 2019).

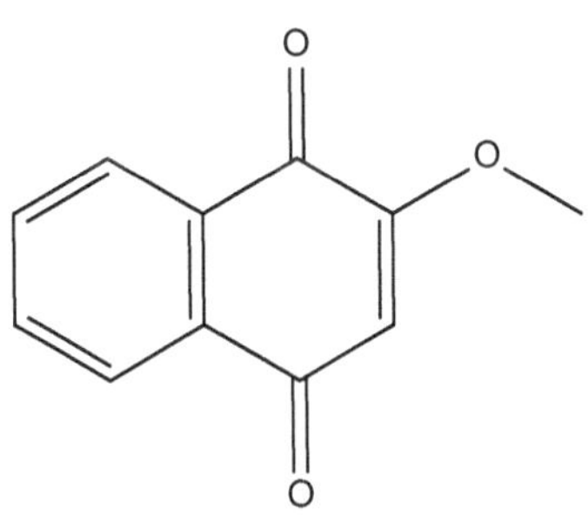

2-Methoxy-1,4-naphthoquinone

Strong antibacterial (Gram-positive) hydrophilic naphthoquinone: 5-Hydroxy-1,4-naphthoquinone (also known as juglone; LogD 0.9 at pH 7.4; molecular mass = 174.1 g/mol) inhibited the growth of *Streptococcus pyogenes* with MIC of 1.5 μg/mL and MBC of 100 μg/mL (Macé et al., 2017).

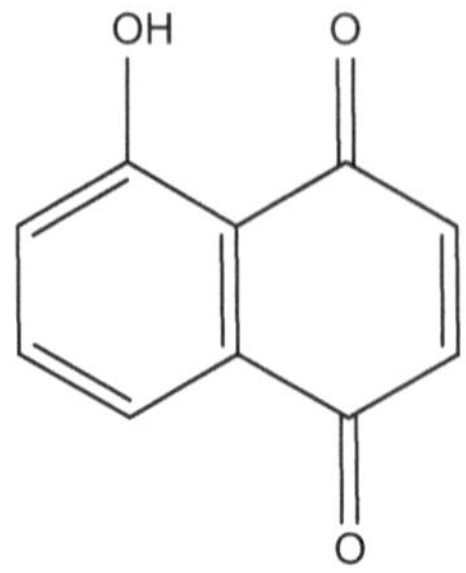

5-Hydroxy-1,4-naphthoquinone

Moderate antibacterial (Gram-positive) naphthoquinone: Methylene-3,3′-bilawsone from the leaves inhibited the growth of *Bacillus subtilis* with MIC of 46.9 µg/mL and MBC of 93.8 µg/mL (Sakunphueak & Panichayupakaranant, 2012).

Antibiotic potentiator naphthoquinone: Juglone at the concentration of 182 µg/mL inhibited by 50% ethidium bromide efflux in *Staphylococcus aureus* (ATCC 25923) (Zmantar et al., 2016).

Strong broad-spectrum antifungal amphiphilic naphthoquinone: 2-Methoxy-1,4-naphthoquinone inhibited the growth of *Candida albicans* (CN1A), *Fusarium oxysporum*, *Aspergillus fumigatus*, *Microsporum gypseum*, and *Trichophyton mentagrophytes* (Yang et al., 2001) with the MIC values of 0.6, 1.2, 0.3, 0.6, and 1.2 µg/mL (Yang et al., 2001). Lawsone methyl ether from leaves inhibited the growth of *Trichophyton rubrum* and *Candida albicans* with the MIC values of 7.8 and 3.9 µg/mL, respectively (Panichayupakaranant et al., 2019).

Antifungal potentiator naphthoquinone: Lawsone methyl ether increased the sensitivity of *Trichophyton rubrum* and *Candida albicans* to clotrimazole (Panichayupakaranant et al., 2019).

Commentaries

i. In Nepal, *Impatiens urticifolia* Wall. (local name: *Tiuri*) is used to treat urinary tract infections and burns. In Pakistan, *Impatiens edgeworthii* Hook. f. (local name: *Chatty char*) is used to treat burns.
ii. Juglone inhibits calcium channels in mammalian cells (Neuhaus-Carlisle et al., 1997) providing further evidence to support the contention that natural products with calcium channel inhibitory activity tend to inhibit bacterial (and fungal) efflux pumps.

References

Macé, S., Truelstrup Hansen, L. and Rupasinghe, H.P., 2017. Anti-bacterial activity of phenolic compounds against streptococcus pyogenes. *Medicines*, *4*(2), p. 25.

Neuhaus-Carlisle, K., Vierling, W. and Wagner, H., 1997. Screening of plant extracts and plant constituents for calcium-channel blocking activity. *Phytomedicine*, *4*(1), pp. 67–71.

Niyomkam, P., Kaewbumrung, S., Kaewnpparat, S. and Panichayupakaranant, P., 2010. Antibacterial activity of Thai herbal extracts on acne involved microorganism. *Pharmaceutical Biology*, *48*(4), pp. 375–380.

Panichayupakaranant, P., Septama, A.W. and Sinviratpong, A., 2019. Synergistic activity of lawsone methyl ether in combination with some antibiotics and artocarpin against methicillin-resistant Staphylococcus aureus, Candida albicans, and Trychophyton rubrum. *Chinese Herbal Medicines*, *11*, pp. 321–325.

Sakunphueak, A. and Panichayupakaranant, P., 2012. Comparison of antimicrobial activities of naphthoquinones from Impatiens balsamina. *Natural Product Research*, *26*(12), pp. 1119–1124.

Szewczyk, K., Zidorn, C., Biernasiuk, A., Komsta, Ł. and Granica, S., 2016. Polyphenols from Impatiens (Balsaminaceae) and their antioxidant and antimicrobial activities. *Industrial Crops and Products*, *86*, pp. 262–272.

Voravuthikunchai, S.P. and Kitpipit, L., 2005. Activity of medicinal plant extracts against hospital isolates of methicillin-resistant Staphylococcus aureus. *Clinical Microbiology and Infection*, *11*(6), pp. 510–512.

Yang, X., Summerhurst, D.K., Koval, S.F., Ficker, C., Smith, M.L. and Bernards, M.A., 2001. Isolation of an antimicrobial compound from Impatiens balsamina L. using bioassay-guided fractionation. *Phytotherapy Research*, *15*(8), pp. 676–680.

Zmantar, T., Miladi, H., Kouidhi, B., Chaabouni, Y., Slama, R.B., Bakhrouf, A., Mahdouani, K. and Chaieb, K., 2016. Use of juglone as antibacterial and potential efflux pump inhibitors in Staphylococcus aureus isolated from the oral cavity. *Microbial Pathogenesis*, *101*, pp. 44–49.

9.2.1.2 *Impatiens trilobata* Colebr.

Synonym: *Impatiens flavida* Hook. f. & Thomson

Common names: basalm; haturi nola kher (Bangladesh); jongli dopati (India)

Distribution: India, Bangladesh, Myanmar

Botanical observation: This herb grows up to about 70 cm tall. The stem is fleshy, light green, and longitudinally striated. The leaves are simple, spiral, and exstipulate. The petiole is up to 5 cm long. The blade is elliptic to lanceolate, 2–8.5 cm × 1.5–3.5 cm, with 4–7 pairs of secondary nerves, tapering at base, serrate, and acuminate at apex. The inflorescences are axillary fascicles of showy and purple to red to yellowish flowers. The flower pedicels are up to 1.5 cm long. The calyx includes 5 sepals. The corolla includes an upper petal which is obcordate and lateral united petals which are 2-lobed and up to 8 mm long. The androecium consists of five stamens with linear filaments. The ovary is minute. The capsules are up to about 1.5 cm long and contain numerous tiny seeds.

Medicinal uses: Boils (Bangladesh)

Commentary: The plant has apparently not been subjected to antimicrobial study. Antimicrobial naphthoquinones are probably present. Consider that the naphthoquinone framework is an antiviral pharmacophore for the (+) RNA viruses: Zika virus (Gonzaga et al., 2019), Dengue virus (Roa-Linares et al., 2019), and naphthoquinones from Balsaminaceae (and other plant families) could be assessed for their anti-coronavirus properties. The global vaccination for COVID-19 might offer some levels of protection against the severe acute respiratory syndrome-associated coronavirus, but one ought to grasp that if pressures on the environment continue other viruses (including coronaviruses) are bound to emerge. Deforestation and increased contact of humans with mice, rats, domestic cats, raccoons, wild pigs, civet cats, and bats (Quarles, 2020) will result in continuous waves of pandemic zoonosis.

References

Gonzaga, D.T., Gomes, R.S., Marra, R.K., Silva, F.C.D., Gomes, M.W., Ferreira, D.F., Santos, R., Pinto, A., Ratcliffe, N.A., Cirne-Santos, C.C. and Barros, C.S., 2019. Inhibition of Zika Virus replication by synthetic bis-naphthoquinones. *Journal of the Brazilian Chemical Society*, *30*(8), pp. 1697–1706.

Quarles, W. (2020) Coronavirus and Ecology. The IPM Practitioner. Volume XXXVII, Number ½; pp. 1–7.

Roa-Linares, V.C., Miranda-Brand, Y., Tangarife-Castaño, V., Ochoa, R., García, P.A., Castro, M., Betancur-Galvis, L. and San Feliciano, A., 2019. Anti-herpetic, anti-dengue and antineoplastic activities of simple and heterocycle-fused derivatives of terpenyl-1, 4-naphthoquinone and 1, 4-anthraquinone. *Molecules*, *24*(7), p. 1279.

9.2.2 Family Ebenaceae Gürke (1891)

The family Ebenaceae consists of 3 genera and about 500 species of tree, often majestic, or shrubs. The wood is often dark, very durable, dense, and the inner bark often turns bluish on light exposure. The leaves are simple, alternate, or opposite, and exstipulate. The flowers are actinomorphic. The inflorescences are cymes or solitary. The calyx is tubular and 3–7-lobed. The corolla is tubular and 3–7-lobed. The androecium includes up to about 30 stamens. The ovary is 2–16-loculed, with 1 or 2 per ovules per locules, 2–8 styles, and entire or 2-lobed stigma. The berries present remnant calyx at base. Plants in this family are known to produce antimicrobial naphthoquinones, triterpenes, and condensed tannins.

9.2.2.1 *Diospyros malabarica* (Desr.) Kostel.

Synonyms: *Diospyros embryopteris* Pers., *Diospyros glutinosa* Koening, *Diospyros siamensis* Ridl. (non Hoch.)

Common names: Gaub ebony, Indian persimmon, mountain ebony, riber ebony; gap (Bangladesh); pei shih (China); kledung (Indonesia); gaub, kandu (India); lang dam (Laos); kayu arang (Malaysia); bok pyin, yengan bok (Myanmar); phlap (Thailand); thi dầu heo (Vietnam)

Habitat: Forests

Distribution: India, Sri Lanka, Myanmar, Thailand, Malaysia Cambodia, Laos, Vietnam, Indonesia, the Philippines.

Botanical observation: This magnificent timber tree grows to a height of 35 m tall and a girth of 2 m. The bole is straight, and the wood is hard and dense. The bark is black, smooth and the inner bark turns bluish on light exposure. The leaves are simple and exstipulate. The petiole is up to about 1.2 cm long. The blade is elliptic or ovate, reddish at first, pointed or rounded at base, and shows a midrib sunken above and 4–8 pairs of secondary nerves. The inflorescences are 3–5-flowered axillary cymes. The calyx is tubular, about 5 mm long, and 4-lobed. The corolla is tubular, whitish yellow, up to about 1.2 cm long, and 4-lobed. The androecium includes about 40 stamens. The gynoecium includes a globose ovary, 4 styles, and an 8-celled ovary. The berries are globose, red, velvety, up to 6 cm in diameter and seated on persistent calyx.

Medicinal uses: Diarrhea (India)

Broad-spectrum antibacterial halo elicited by mid-polar extract: Chloroform extract of young fruits (400 µg/mL/paper disc) inhibited the growth of *Bacillus cereus*, *Staphylococcus aureus*, *Escherichia coli*, *Pseudomonas aeruginosa*, and *Shigella dysenteriae* with the inhibition zone diameters of 32, 25, 27, 24, and 27 mm, respectively (Moniruzzaman et al., 2019).

Broad-spectrum antifungal halo elicited by mid-polar extract: Chloroform extract of young fruits (400 µg/mL/paper disc) inhibited the growth of *Aspergillus niger*, *Candida albicans*, and *Saccharomyces cerevisiae* with the inhibition zone diameters of 22, 30, and 20 mm, respectively (Moniruzzaman et al., 2019).

Commentaries

i. The antibacterial and antifungal constituents in this plant are apparently unknown.
ii. Members of the genus *Diospyros* L. and *Euclea* L. assemble a series of naphthoquinones and bisnaphthoquinones with strong broad-spectrum antibacterial and antimycobacterial properties *in vitro*: 2-Methoxy-7-methyl juglone halved the growth of *Micrococcus luteus*, *Staphylococcus aureus*, *Mycobacterium smegmatis*, and *Mycobacterium avium* at a concentration of 50, 200, 50, and 50 µg/mL, respectively (Gu et al., 2004), and inhibited *Mycobacterium tuberculosis* (H37Rv) with the MIC value of 0.5 µg/mL and a selectivity index of 30.2 (Mahapatra et al., 2007).

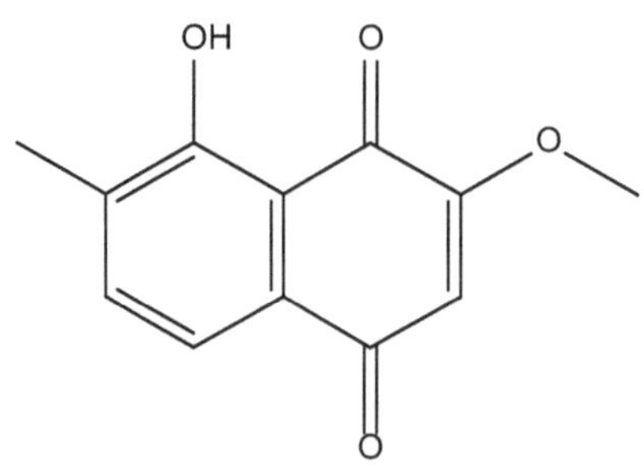

2-Methoxy-7-methyl juglone

7-Methyl juglone (also known as ramentaceone; LogD 1.3 at pH 7.4; molecular mass = 188.1 g/mol) inhibited the growth of *Mycobacterium tuberculosis* H37Rv (ATCC 27294 strain), *Mycobacterium tuberculosis* (clinical strain), and *Mycobacterium tuberculosis* (clinical strain) with MIC value of 0.5 µg/mL (Bapela et al., 2006). This compound had synergistic activity with isoniazid (Bapela et al., 2006) via probable efflux pump inhibition. In a subsequent study, 7-methyljuglone inhibited the growth of *Mycobacterium smegmatis* (ATCC 1441), *Mycobacterium fortuitum* (ATCC 6841), *Mycobacterium bovis* (Pasteur strain P1172), and *Mycobacterium bovis* (ATCC 19210) with MIC/MBC of 1.5/15.6, 22.1/>250, 11.7/>39.0, and 1.5/39.0 µg/mL, respectively (McGaw et al., 2008).

Plumbagin (LogD 1.4 at pH 7.4; molecular mass= 188.1 g/mol) evoked 50% inhibition of growth of *Micrococcus luteus*, *Staphylococcus aureus*, *Mycobacterium smegmatis*, and *Mycobacterium avium* at a concentration of 80, 100, 300, and 600 µg/mL, respectively (Gu et al., 2004). In a subsequent study, plumbagin inhibited the growth of *Mycobacterium smegmatis* and *Mycobacterium tuberculosis* (H37Rv (ATCC 27294) and *Neisseria gonorrhoeae* (ATCC 49226) (Kuete et al., 2009). Plumbagin is planar and intercalate into DNA (Demma et al., 2009). The question arises as why are natural products with quinone moiety often antimycobacterial *in vitro*?

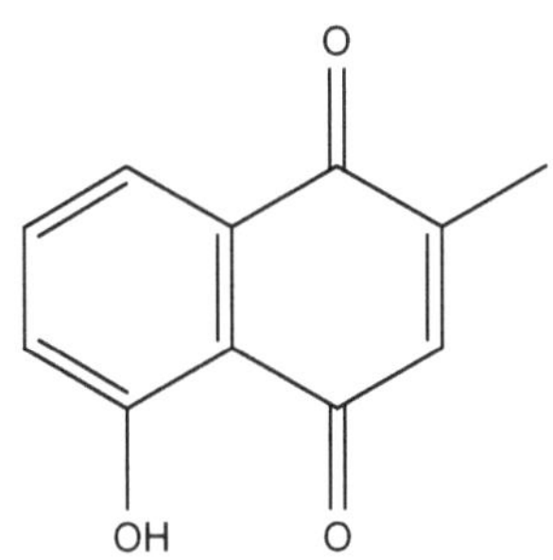

Plumbagin

The furanonaphthoquinone crassiflorone (molecular mass = 360.3 g/mol) inhibited the growth of *Mycobacterium smegmatis* and *Mycobacterium tuberculosis* (H37Rv (ATCC 27294) with the MIC/MBC values of 9.7/19.5 and 4.8/4.8 µg/mL and *Neisseria gonorrhoeae* (ATCC 49226) with the MIC value of 4.8 µg/mL (Kuete et al., 2009).

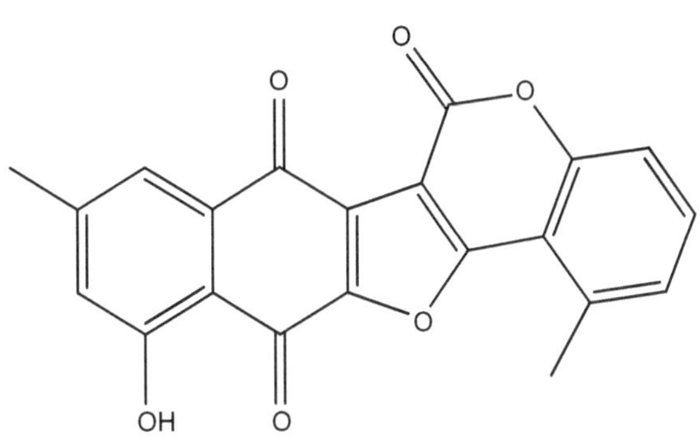

Crassiflorone

Menadione (also known as 2-methyl-1,4-naphthoquinone; LogD 2.0 at pH 7.4; molecular mass = 172.1 g/mol) inhibited the growth of *Staphylococcus aureus*, *Streptococcus pneumoniae*, *Streptococcus pyogenes*, multidrug-resistant clinical strain of *Staphylococcus aureus*, *Staphylococcus saprophyticus*, *Staphylococcus epidermidis*, *Streptococcus pyogenes*, *Streptococcus agalactiae*, *Bacillus subtilis*, *Bacillus cereus*, *Escherichia coli*, multidrug-resistant *Klebsiella pneumoniae*, *Pseudomonas aeruginosa*, *Proteus vulgaris*, *Klebsiella oxytoca*, multidrug-resistant *Proteus mirabilis*, *Serratia marcescens*, multidrug-resistant *Proteus mirabilis*, *Acinetobacter baumannii*, *Moraxella catarrhalis*, *Vibrio cholerae*, *Salmonella typhi*, *Salmonella paratyphi* A, *Shigella flexneri*, *Shigella sonnei*, *Burkholderia cepacia*, and *Stenotrophomonas maltophilia* (Dey et al., 2014).

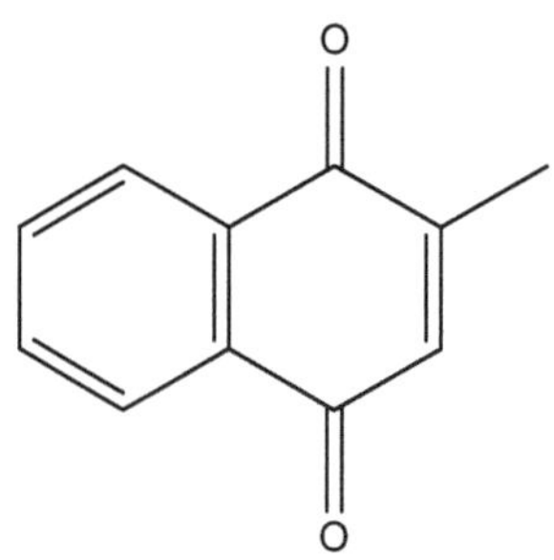

Menadione

Shinanolone inhibited the growth of *Mycobacterium smegmatis* (ATCC 1441), *Mycobacterium fortuitum* (ATCC 6841), *Mycobacterium bovis* (Pasteur strain P1172), and *Mycobacterium bovis* (ATCC 19210) (McGaw et al., 2008).

The bisnaphthoquinone maritinone (also known as 8,8'-biplumbagin; molecular mass = 374.3 g/mol) evoked 50% inhibition of growth of *Staphylococcus aureus* at the concentration of 9 µg/mL (Gu et al., 2004) and inhibited the growth of *Mycobacterium tuberculosis* (H37Rv ATCC 27294, which is sensitive to all first-line antituberculosis drugs) and *Mycobacterium tuberculosis* (clinical isolate CIBIN 99, resistant to all first-line antituberculosis drugs) with MIC value of 3.1 µg/mL (Uc-Cachón et al., 2014).

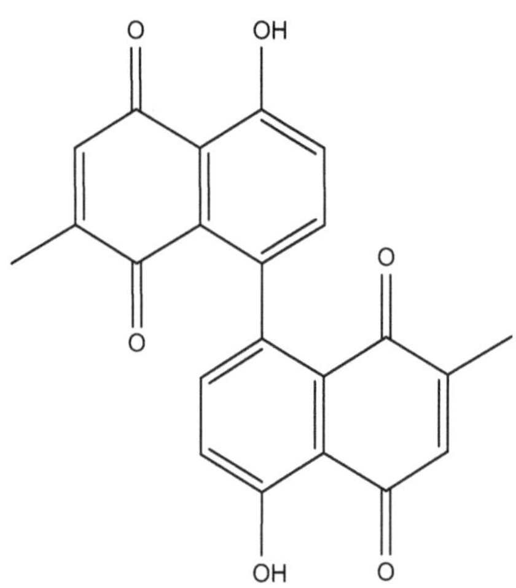

Maritinone

The bisnaphthoquinone chitranone (LogD 0.8 at pH 7.4; molecular mass=374.3 g/mol) evoked 50% inhibition of growth of *Micrococcus luteus*, *Staphylococcus aureus*, *Mycobacterium smegmatis*, and *Mycobacterium avium* at a concentration of 10, 90, 6, and 5 µg/mL, respectively (Gu et al., 2004). Chitranone inhibited the growth of *Mycobacterium tuberculosis* (H37Rv ATCC 27294) and *Mycobacterium tuberculosis* (clinical isolate CIBIN 99, resistant to all first-line antituberculosis drugs) with MIC values of 3.1 and 3.1 µg/mL, respectively (Uc-Cachón et al., 2014).

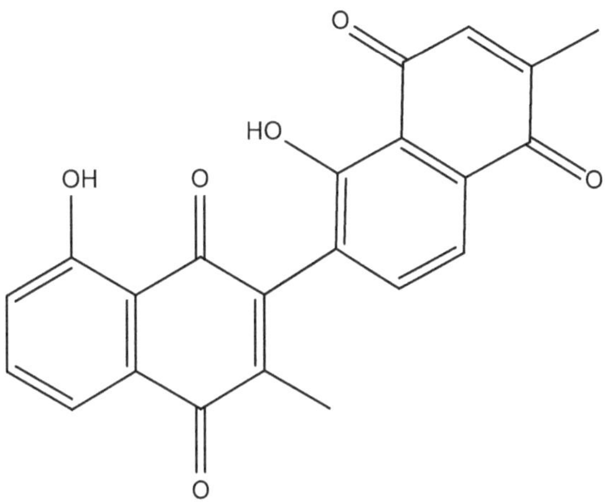

Chitranone

Diospyrone inhibited the growth of *Mycobacterium smegmatis* and *Mycobacterium tuberculosis* (H37Rv (ATCC 27294) with the MIC/MBC values of 1.2/2.4 and 2.4/2.4 µg/mL and *Neisseria gonorrhoeae* (ATCC 49226) with the MIC value of 1.2 µg/mL (Kuete et al., 2009).

The bisnaphthoquinone diospyrone (molecular mass=722.6 g/mol) inhibited the growth of *Escherichia coli* (ATCC 10536), *Escherichia coli* (AG100A), *Enterobacter aerogenes* (ATCC 13048), *Enterobacter aerogenes* (EA298), *Klebsiella pneumoniae* (ATCC 11296), *Klebsiella pneumoniae* (KP63 MDR clinical isolates), *Pseudomonas aeruginosa* (PA124, multidrug-resistant clinical isolates), *Enterobacter cloacae* (Ec1194 clinical isolates) (Kuete et al., 2010), *Mycobacterium smegmatis* (ATCC 700084), *Mycobacterium tuberculosis* H37Rv (ATCC 27294), *Mycobacterium tuberculosis* (MTCS2), *Neisseria gonorrhoeae* (ATCC 49226), and *Neisseria gonorrhoeae* (NGCS7 β-lactamase positive) (Kuete et al., 2009). According to Benz and Bauer (1988), the exclusion limits of the porins of *Escherichia coli* and of *Salmonella typhimurium* are between 550 and 800 g/mol which would be consistent with diameters around 1.1–1.3 nm. Bisnaphthoquinones often inhibit bacterial efflux pumps, diospyrone is an example (Kuete et al., 2010)

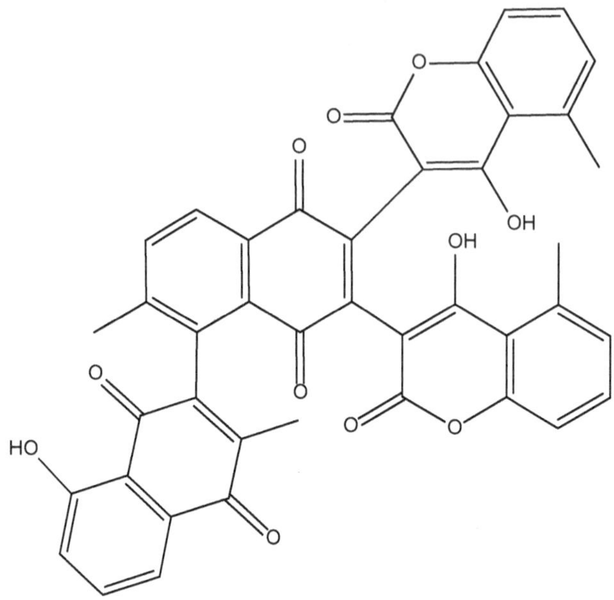

Diospyrone

Zeylanone (molecular mass=374.3 g/mol) inhibited the growth of *Mycobacterium tuberculosis* (H37Rv ATCC 27294) and *Mycobacterium tuberculosis* (clinical isolate CIBIN 99) with the MIC values of 25 and 12.5 µg/mL, respectively (Uc-Cachón et al., 2014). Zeylanone evoked 50% inhibition of growth of *Micrococcus luteus, Staphylococcus aureus, Mycobacterium smegmatis*, and *Mycobacterium avium* at a concentration of 50, 6, 50, and 200 µg/mL, respectively (Gu et al., 2004).

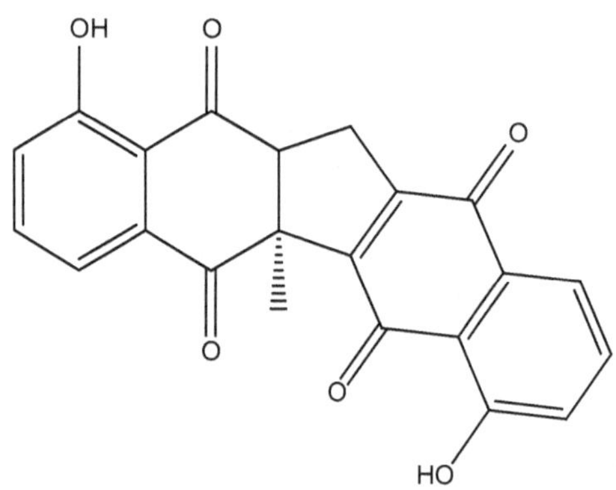

Zeylanone

Diospyrin (LogD 0.5 at pH 7.4; molecular mass=374.3 g/mol) inhibited the growth of *Staphylococcus aureus, Staphylococcus saprophyticus, Staphylococcus epidermidis, Streptococcus agalactiae, Bacillus subtilis, Bacillus cereus, Enterococcus faecalis*, and *Moraxella catarrhalis* with MIC/MBC of 64 /128, 64 /128, 128 /256, 64 /128, 64 /128, 64 /128, 128 /256, and 128 /256 µg/mL, respectively (Dey et al., 2014). Diospyrin inhibited the growth of *Staphylococcus aureus* (ATCC 6538), *Bacillus subtilis* (ATCC 6633), *Escherichia coli* (ATCC 10536), *Pseudomonas aeruginosa* (ATCC 15443), *Corynebacterium diphtheriae*

(ATCC 27010), *Listeria monocytogenes* (ATCC 15313), *Mycobacterium smegmatis* (ATCC 607), *Mycobacterium chelonae* (ATCC 19977), *Micrococcus luteus*, *Salmonella typhi* (ATCC 6539), *Streptococcus pyogenes* (ATCC 12344), and *Streptococcus pneumoniae* (ATCC 3340) with the MIC values of 12.5, 50, 100, 100, 3.1, 50, 25, 25, 50, 1.5, and 3.2 µg/mL, respectively (Adeniyi et al., 2000). Diospyrin inhibited the growth of *Mycobacterium smegmatis* (ATCC 1441), *Mycobacterium fortuitum* (ATCC 6841), *Mycobacterium bovis* (Pasteur strain P1172), and *Mycobacterium bovis* (ATCC 19210) with the MIC/MBC of 2.4/31.2, 41.6/250, 33.6/78.1, and 1.7/39.0 µg/mL, respectively (McGaw et al., 2008).

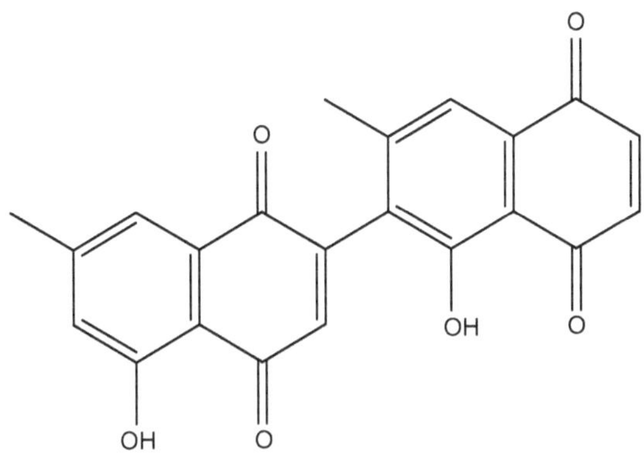

Diospyrin

The bisnaphthoquinone isodiospyrin (LogD 0.5 at pH 7.4; molecular mass = 374.3 g/mol) inhibited the growth of *Staphylococcus aureus* (ATCC 6538), *Bacillus subtilis* (ATCC 6633), *Escherichia coli* (ATCC 10536), *Pseudomonas aeruginosa* (ATCC 15443), *Corynebacterium diphtheriae* (ATCC 27010), *Listeria monocytogenes* (ATCC 15313), *Mycobacterium chelonae* (ATCC 19977), *Micrococcus luteus* (ATCC 9341), *Salmonella typhi* (ATCC 6539), *Streptococcus pyogenes* (ATCC 12344), *Streptococcus pneumoniae* (ATCC 3340), and *Streptococcus faecalis* (ATCC 14506) with the MIC values of 12.5, 6.2, 100, 100, 1.6, 6.2, 6.2, 6.2, 50, 0.7, 0.7, and 12.5 µg/mL, respectively (McGaw et al., 2008).

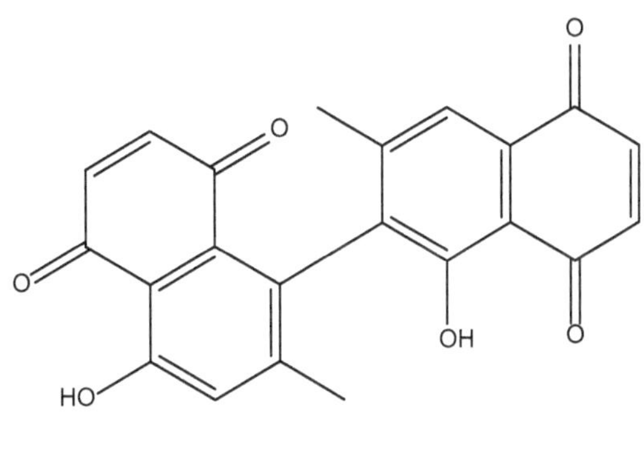

Isodiospyrin

The bisnaphthoquinone 3,3′-biplumbagin (molecular mass = 374.3 g/mol) inhibited the growth of *Mycobacterium tuberculosis* (H37Rv ATCC 27294) and *Mycobacterium tuberculosis* (clinical isolate CIBIN 99) with the MIC values of 3.1 and 3.1 µg/mL, respectively (Uc-Cachón et al., 2014).

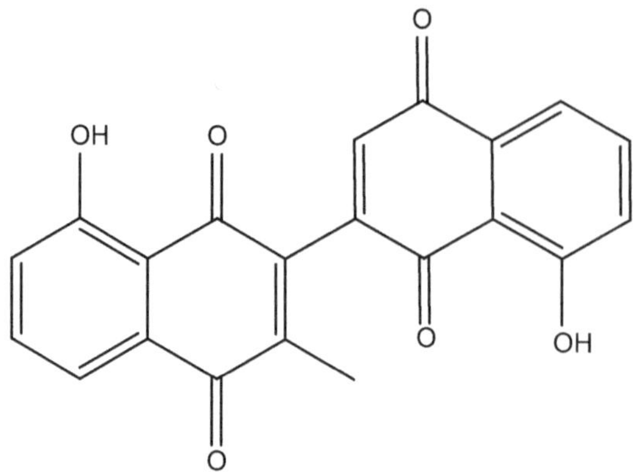

3,3'-Biplumbagin

iii. Condensed tannins (proanthocyanidins) in fruits of members of the genus *Diospyros* L. are antibacterial (Fu et al., 2016).

iv. Plants in the genus *Diospyros* L. are often used to treat microbial infections in Asia. In India, *Diospyros cordifolia* Roxb. (local name: *Rajaan*) is used to treat abscesses and tooth infections; *Diospyros embryopteris* Pers. (local name: *Gaub*, *Tinduka*) is used for diarrhea and dysentery; *Diospyros melanoxylon* Roxb. (local names: *Tendu*, *Tereldare*) is used to treat diarrhea, pimples, and jaundice; *Diospyros peregrina* (Gaertn.) Gürke. is used to treat wounds, ulcers, sore throats, fever, diarrhea, and gangrene; *Diospyros tomentosa* Poir. (local name: *Viralaa*) is used to treat pneumonia. Ethanol extract of leaves of *Diospyros peregrina* (Gaertn.) Gürke. inhibited the growth of *Staphylococcus aureus*, *Shigella* dysenteriae, *Escherichia coli*, and *Pseudomonas aeruginosa* with zones of inhibition of 8, 9, 7, and 8 mm, respectively (0.5 mg/disc) (Uddin et al., 2008).

In Nepal, *Diospyros lancifolia* Roxb. (local name: *Khallu*) is used for sore throat. *Diospyros blancoi* A. DC. is used in Bangladesh to treat diarrhea, and in the Philippines, this plant is locally known as *Mabolo*. In Korea, *Diospyros kaki* Thunb. (local name: *Gamnamu*) is used to treat cholera, and in India it is used to treat cough. *Diospyros marmorata* R. Parker (local name: *Kala-lakdi*) is used as antiseptic in the Andaman Islands. In Thailand, *Diospyros montana* Roxb. (local name: *Tan dam*) is used for the treatment of ulcers and urinary tract infections, and in India it is used for jaundice and very high fever, pneumonia, and boils. In the Philippines, *Diospyros multiflora* Blanco (local name: *Kanumai*) is used for fever, herpes, and ringworms. In Malaysia, *Diospyros wallichii* King & Gamble is applied to yaws. *Diospyros malabarica* (Desr.) Kostel. is used in Indonesia (local name: *Makusi*) to treat ulcers, wounds, and sore throat; in Myanmar (local name: *Bok-pyin*) it is used for diarrhea, dysentery, and vaginal discharges, and in India it is used for aphthae, fever, dysentery, and wounds. In Papua New Guinea, *Diospyros lolin* Bakh. (local name: *Palasoup*) is used to heal ulcers.

v. The phenolic usnic acid (Log −0.1 at pH 7.4; molecular mass = 344.3 g/mol) inhibited the replication of the Dengue virus by 90% at the concentration of 5 µg/mL and inhibited Dengue virus RNA-dependent RNA polymerase with IC_{50} of 4.7 µM (Peyrat et al., 2016). Is usnic active against coronaviruses including the severe acute respiratory syndrome-associated coronavirus?

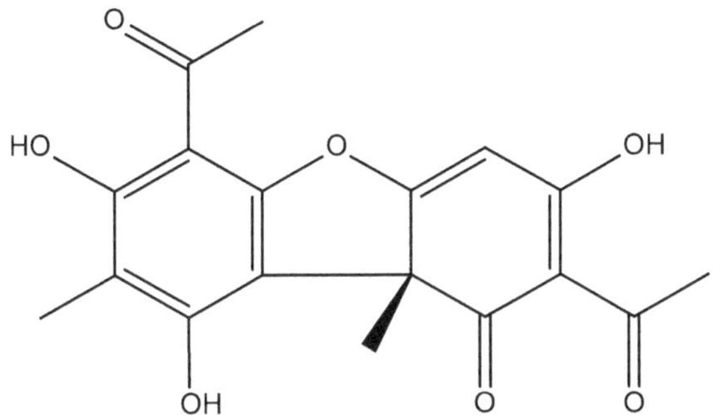

Usnic acid

The triterpenes betulinic acid (LogD 5.1 at pH 7.4; molecular mass=456.7 g/mol), (3β)-3.23-dihydroxylup-20(29)-en-28-oic acid, (3β)-3-(acetyloxy)-urs-12-en-28-oic acid, betulinic aldehyde, and (3β)-3.23-dihydroxylup-12.20(29)-dien-28-oic acid at the concentration of 10 μg/mL inhibited the replication of the Dengue virus by 57%, 45%, 44%, 55%, and 52%, respectively (Peyrat et al., 2016).

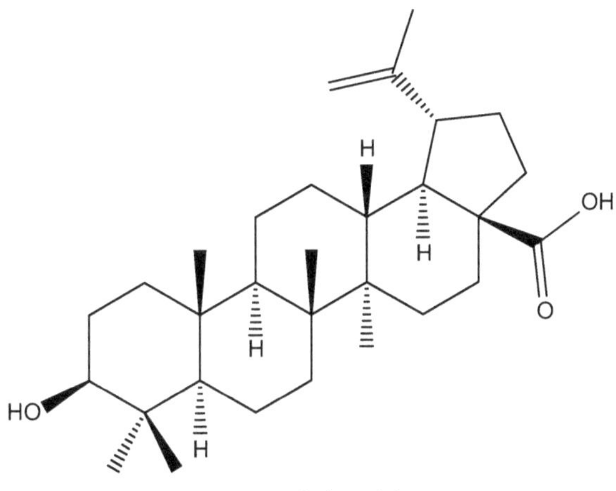

Betulinic acid

References

Adeniyi, B.A., Fong, H.H.S., Pezzuto, J.M., Luyengi, L. and Odelola, H.A., 2000. Antibacterial activity of diospyrin, isodiospyrin and bisisodiospyrin from the root of Diospyros piscatoria (Gurke)(Ebenaceae). *Phytotherapy Research*, *14*(2), pp. 112–117.

Bapela, N.B., Lall, N., Fourie, P.B., Franzblau, S.G. and Van Rensburg, C.E.J., 2006. Activity of 7-methyljuglone in combination with antituberculous drugs against Mycobacterium tuberculosis. *Phytomedicine*, *13*(9–10), pp. 630–635.

Benz, R. and Bauer, K., 1988. Permeation of hydrophilic molecules through the outer membrane of gram-negativ bacteria: Review of becterial porins. *European Journal of Biochemistry*, *176*(1), pp. 1–19.

Demma, J., Hallberg, K. and Hellman, B., 2009. Genotoxicity of plumbagin and its effects on catechol and NQNO-induced DNA damage in mouse lymphoma cells. *Toxicology in Vitro*, *23*(2), pp. 266–271.

Dey, D., Ray, R. and Hazra, B., 2014. Antitubercular and antibacterial activity of quinonoid natural products against multi-drug resistant clinical isolates. *Phytotherapy Research*, *28*(7), pp. 1014–1021.

Fu, L., Lu, W. and Zhou, X., 2016. Phenolic compounds and in vitro antibacterial and antioxidant activities of three tropic fruits: Persimmon, guava, and sweetsop. *BioMed Research International, 2016*. doi: 10.1155/2016/4287461.

Gu, J.Q., Graf, T.N., Lee, D., Chai, H.B., Mi, Q., Kardono, L.B., Setyowati, F.M., Ismail, R., Riswan, S., Farnsworth, N.R. and Cordell, G.A., 2004. Cytotoxic and antimicrobial constituents of the bark of diospyros ma ritima collected in two geographical locations in Indonesia. *Journal of Natural Products, 67*(7), pp. 1156–1161.

Kuete, V., Tangmouo, J.G., Meyer, J.M. and Lall, N., 2009. Diospyrone, crassiflorone and plumbagin: three antimycobacterial and antigonorrhoeal naphthoquinones from two Diospyros spp. *International Journal of Antimicrobial Agents, 34*(4), pp. 322–325.

Kuete, V., Ngameni, B., Tangmouo, J.G., Bolla, J.M., Alibert-Franco, S., Ngadjui, B.T. and Pagès, J.M., 2010. Efflux pumps are involved in the defense of Gram-negative bacteria against the natural products isobavachalcone and diospyrone. *Antimicrobial Agents and Chemotherapy, 54*(5), pp. 1749–1752.

McGaw, L.J., Lall, N., Hlokwe, T.M., Michel, A.L., Meyer, J.J.M. and Eloff, J.N., 2008. Purified compounds and extracts from Euclea species with antimycobacterial activity against Mycobacterium bovis and fast-growing mycobacteria. *Biological and Pharmaceutical Bulletin, 31*(7), pp. 1429–1433.

Mahapatra, A., Mativandlela, S.P., Binneman, B., Fourie, P.B., Hamilton, C.J., Meyer, J.J.M., Van der Kooy, F., Houghton, P. and Lall, N., 2007. Activity of 7-methyljuglone derivatives against Mycobacterium tuberculosis and as subversive substrates for mycothiol disulfide reductase. *Bioorganic & Medicinal Chemistry, 15*(24), pp. 7638–7646.

Moniruzzaman, M., Kuddus, M.R., Chowdhury, A.S. and Rashid, M.A., 2019. Antioxidant, antimicrobial, anti-diarrheal and analgesic activities of diospyros malabarica (Desr.) Kostel. *Bangladesh Pharmaceutical Journal, 22*(1), pp. 27–33.

Peyrat, L.A., Eparvier, V., Eydoux, C., Guillemot, J.C., Stien, D. and Litaudon, M., 2016. Chemical diversity and antiviral potential in the pantropical Diospyros genus. *Fitoterapia, 112*, pp. 9–15.

Uc-Cachón, A.H., Borges-Argáez, R., Said-Fernández, S., Vargas-Villarreal, J., González-Salazar, F., Méndez-González, M., Cáceres-Farfán, M. and Molina-Salinas, G.M., 2014. Naphthoquinones isolated from Diospyros anisandra exhibit potent activity against pan-resistant first-line drugs Mycobacterium tuberculosis strains. *Pulmonary Pharmacology & Therapeutics, 27*(1), pp. 114–120.

Uddin, S.J., Rouf, R., Shilpi, J.A., Alamgir, M., Nahar, L. and Sarker, S.D., 2008. Screening of some Bangladeshi medicinal plants for in vitro antibacterial activity. *Oriental Pharmacy and Experimental Medicine, 8*(3), pp. 316–321.

9.2.3 Family Ericaceae A.L. de Jussieu (1789)

The family Ericaceae consists of 128 genera and about 400 species of trees, shrubs, and herbs often growing in high-altitude Tropical Asia. The leaves are simple, spiral or decussate, and exstipulate. The blade is often coriaceous, serrate, and dotted with tiny glands. The inflorescence is an axillary cyme or solitary. The flowers are actinomorphic. The calyx comprises mostly 5 sepals. The corolla is often characteristically urceolate and 5-lobed. The androecium comprises 10 stamens. The gynoecium is multilocular, with numerous ovules growing on axile placentas. The fruits are capsules or berries. Plants in this family generate a fascinating array of phenolic glycosides.

9.2.3.1 *Arbutus unedo* L.

Common names: Strawberry tree; dağ çileği, davulga, kocayemiş yaprağı, yemis (Turkey)

Habitat: Sun-exposed rocks, boulders, cultivated

Distribution: Turkey

Botanical observation: This magnificent tree grows to about 10 m tall. The bark is reddish brown. The stems are somewhat reddish and smooth. The leaves are simple, alternate, and exstipulate. The petiole is up to 6 mm long. The blade is elliptic, oblong, or obovate, up to 10 cm long, smooth, somewhat glossy, cuneate at base, sharply serrate, and acute at apex. The inflorescences are terminal and panicles. The calyx is tubular and 5-lobed. The corolla is urceolate, whitish, 5 lobed, and up to 7 mm long. The androecium comprises 10 stamens with appendages. The ovary comprises 5 carpels fused into a 5-locular ovary and a peltate stigma. The berries are warty, globose, edible, of a mesmerizing array of gorgeous red, orange, yellow, edible, and up to about 2 cm across.

Medicinal uses: Diarrhea (Turkey)

Broad-spectrum antibacterial halo elicited by polar extract: Aqueous extract of leaves (5 mm diameter agar well, 85 µL/well of a 5 µg/mL solution) inhibited the growth of *Staphylococcus aureus*, *Staphylococcus epidermidis*, *Pseudomonas aeruginosa*, and *Escherichia coli* with inhibition zone diameters of about 7 mm (Malheiro et al., 2012).

Antibacterial (Gram-positive) halo developed by simple phenol glycoside: Arbutin (LogD −1.1 at pH 7.4; molecular mass=272.2 g/mol) 50 µg/ 6mm disc) inhibited the growth of *Staphylococcus aureus*, methicillin-resistant *Staphylococcus aureus*, and extended-spectrum beta-lactamase-*Staphylococcus aureus* with inhibition zone diameter of 10, 7, and 9 mm, respectively (Ma et al., 2019).

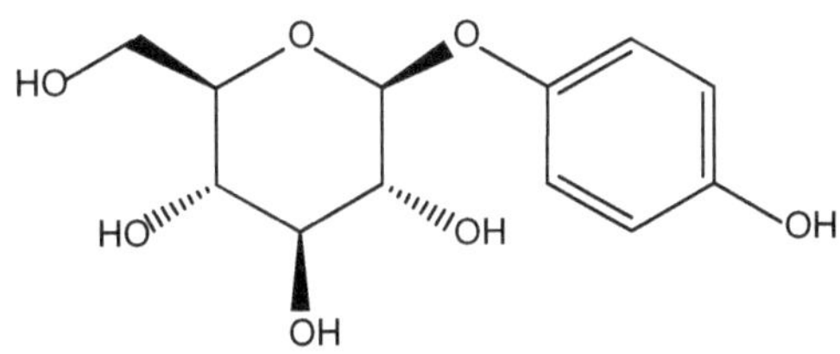

Arbutin

Moderate antibacterial (Gram-positive) polar extract: Acetone extract of aerial part inhibited the growth of *Pseudomonas aeruginosa*, *Staphylococcus aureus*, *Enterococcus faecalis*, *Staphylococcus epidermidis*, and *Staphylococcus saprophyticus* with the MIC values of 1200, 300, 1200, 300, and 300 µg/mL, respectively (Sassi et al., 2007).

Moderate antibacterial (Gram-negative) extract: Extract of stems inhibited the growth of *Klebsiella pneumoniae* with the MIC values of 110 µg/mL (Dib et al., 2010).

Moderate antifungal (yeast) extract: Extract of stems inhibited the growth of *Candida albicans* with the MIC value of 110 µg/mL (Dib et al., 2010).

Strong (Gram-positive) antibacterial amphiphilic simple phenol: Methyl gallate (LogD 1.1 at pH 7.4; molecular mass=184.1 g/mol) inhibited the growth of *Pseudomonas aeruginosa* (Z61), *Escherichia coli* (DC2), and *Staphylococcus aureus* (RN4220) with the MIC values of 12.5, 25, and 200 µg/mL, (Saxena et al., 1994).

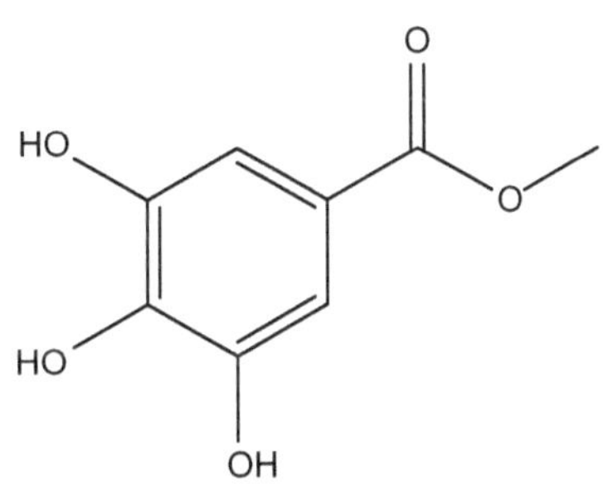

Methyl gallate

Moderate broad-spectrum antibacterial amphiphilic simple phenols: Gallic acid inhibited the growth of methicillin-resistant *Staphylococcus aureus* (clinical strain) with the MIC of 128 µg mL (Hatano et al., 2003). Gallic acid inhibited the growth of methicillin-sensitive *Staphylococcus aureus* (209P) with the MIC of 128 µg/mL (Hatano et al., 2003). In a subsequent study, gallic acid inhibited the growth of *Staphylococcus aureus* (CMCC (B) 26003) with the MIC of 232 µg/mL (Xu et al., 2015) and inhibited the growth of *Bacillus subtilis* (CMCC (B) 63501) with MIC of 464 μg/mL (Xu et al., 2015). Ethyl gallate inhibited the growth *of Klebsiella pneumoniae* (ATCC 700603) with MIC/MBC values of 100/300 μg/mL (Li et al., 2016).

Moderate antibacterial (Gram-positive) simple phenol glycoside: Arbutin inhibited the growth of *Staphylococcus aureus* (ATCC 6358) and *Streptococcus faecalis* (ATCC 29212) with the MIC value of 100 μg/mL (González-Cortazar et al., 2019). In a parallel study, arbutin inhibited the growth of *Staphylococcus aureus* with the MIC value of 2500 μg/mL (Ma et al., 2019).

Antibiotic potentiator simple phenolic: Gallic acid decreased the resistance of methicillin-resistant *Staphylococcus aureus* to oxacillin, penicillin G, streptomycin, and ampicillin (Hatano et al., 2003).

Strong antiviral enveloped segmented linear single-stranded (-) RNA) amphiphilic simple phenol: Gallic acid (LogD 1 at pH 7.4; molecular mass = 170.1 g/mol) inhibited the replication of the Influenza A (H1N1) (A/California/07/2009) virus in Madin-Darby canine kidney with the IC_{50} value of 2.6 μg/mL (You et al., 2018).

Moderate antiviral (enveloped monopartite linear single-stranded (+)-RNA) simple phenol: Gallic acid inhibited the replication of the Human coronavirus NL63 in LLC-MK2 cells with the IC_{50} value of 71.4 μM (Weng et al., 2019).

In vivo antiviral (enveloped segmented linear single-stranded (-) RNA) amphiphilic simple phenolic: Protocatechuic acid (LogD 1 at pH 7.4; molecular mass = 154.1 g/mol) given orally at the dose of 10 mg/kg (3 doses given post-infection) protected mice against the Influenza virus A (H9N2) (A/Chicken/Hebei/4/2008) virus (Ou et al., 2014).

Commentary:

i. The plant produces mainly simple phenolics such as gallic acid, protocatechuic acid, ethyl gallate; the phenolic glycoside arbutin; flavonoids such as catechin, gallocatechin glycosides of quercetin and myricetin; and condensed tannins such as epicatechin-4,8-epicatechin-4,8-epicatechin. All these phenolics work synergistically for the astringency of this plant. Is gallic acid of value against COVID-19?

ii. Fu et al. (2016) reported that gallic acid inhibited *Staphylococcus aureus, Bacillus cereus, Staphylococcus epidermidis, Escherichia coli, Salmonella typhimurium, Salmonella flexneri*, and *Pseudomonas aeruginosa* with MIC/MBC of 600/600, 2500/2500, 600/600, 2500/5000, 2500/5000, 2500/5000, and 5000/5000 μg/mL, respectively. This is different from Hatano et al., 2003. In this case, the difference of MIC for gallic acid can be explained by the use of 96 wells broth dilution technique in the first study and agar well in the second. The method used for MIC influences the MIC values.

iii. *Arctostaphylos uva-ursi* (L.) Spreng was already known of our peer the French physician Charles de l' Écluse (1526–1609) as a remedy for urinary tract infections. One of its active constituents is arbutin which is hydrolyzed into hydroquinone by intestinal bacteria and excreted in urine mainly as glucuronide or sulphate which are taken up by urinary tract bacteria to liberate hydroquinone which is antibacterial (Siegers et al., 2003). Hydroquinone damages the bacterial cell wall and membrane, increases permeability, and induces leakage of intracellular substance. Hydroquinone inhibited the growth of *Staphylococcus aureus*, methicillin-resistant *Staphylococcus*, and extended spectrum beta-lactamase-*Staphylococcus aureus* with the MIC values of 300, 1200, and 600 μg/mL, respectively (Ma et al., 2019).

iv. *Arbutus andrachne* L. is used in Turkey (local name: *Sandal ağacı*) as antiseptic.

References

Dib, M.A., Paolini, J., Bendahou, M., Varesi, L., Allali, H., Desjobert, J.M., Tabti, B. and Costa, J., 2010. Chemical composition of fatty acid and unsaponifiable fractions of leaves, stems and roots of Arbutus unedo and in vitro antimicrobial activity of unsaponifiable extracts. *Natural Product Communications*, 5(7), p. 1085.

Fu, L., Lu, W. and Zhou, X., 2016. Phenolic compounds and in vitro antibacterial and antioxidant activities of three tropic fruits: Persimmon, guava, and sweetsop. *BioMed Research International, 2016*, pp. 1–9.
González-Cortazar, M., López-Gayou, V., Tortoriello, J., Domínguez-Mendoza, B.E., Ríos-Cortes, A.M., Delgado-Macuil, R., Hernández-Beteta, E.E., Blé-González, E.A. and Zamilpa, A., 2019. Antimicrobial gastrodin derivatives isolated from Bacopa procumbens. *Phytochemistry Letters, 31*, pp. 33–38.
Hatano, T., Kusuda, M., Hori, M., Shiota, S., Tsuchiya, T. and Yoshida, T., 2003. Theasinensin A, a tea polyphenol formed from Epigallocatechin gallate, suppresses antibiotic resistance of methicillin-resistant Staphylococcus aureus. *Planta Medica, 69*(11), pp. 984–989.
Li, K., Lin, Y., Li, B., Pan, T., Wang, F., Yuan, R., Ji, J., Diao, Y. and Wang, S., 2016. Antibacterial constituents of Fructus Chebulae Immaturus and their mechanisms of action. *BMC Complementary and Alternative Medicine, 16*(1), p. 183.
Ma, C., He, N., Zhao, Y., Xia, D., Wei, J. and Kang, W., 2019. Antimicrobial mechanism of hydroquinone. *Applied Biochemistry and Biotechnology, 189*(4), pp. 1291–1303.
Malheiro, R., Sá, O., Pereira, E., Aguiar, C., Baptista, P. and Pereira, J.A., 2012. Arbutus unedo L. leaves as source of phytochemicals with bioactive properties. *Industrial Crops and Products, 37*(1), pp. 473–478.
Ou, C., Shi, N., Yang, Q., Zhang, Y., Wu, Z., Wang, B., Compans, R.W. and He, C., 2014. Protocatechuic acid, a novel active substance against avian Influenza virus H9N2 infection. *PloS One, 9*(10), p. e111004.
Sassi, A.B., Harzallah-Skhiri, F. and Aouni, M., 2007. Investigation of some medicinal plants from Tunisia for antimicrobial activities. *Pharmaceutical Biology, 45*(5), pp. 421–428.
Saxena, G., McCutcheon, A.R., Farmer, S., Towers, G.H.N. and Hancock, R.E.W., 1994. Antimicrobial constituents of Rhus glabra. *Journal of Ethnopharmacology, 42*(2), pp. 95–99.
Siegers, C., Bodinet, C., Ali, S.S. and Siegers, C.P., 2003. Bacterial deconjugation of arbutin by Escherichia coli. *Phytomedicine, 10*, pp. 58–60.
Weng, J.R., Lin, C.S., Lai, H.C., Lin, Y.P., Wang, C.Y., Tsai, Y.C., Wu, K.C., Huang, S.H. and Lin, C.W., 2019. Antiviral activity of Sambucus FormosanaNakai ethanol extract and related phenolic acid constituents against human coronavirus NL63. *Virus Research, 273*, p. 197767.
Xu, T., Wang, Z., Lei, T., Lv, C., Wang, J. and Lu, J., 2015. New flavonoid glycosides from Sedum aizoon L. *Fitoterapia, 101*, pp. 125–132.
You, H.L., Huang, C.C., Chen, C.J., Chang, C.C., Liao, P.L. and Huang, S.T., 2018. Anti-pandemic influenza A (H1N1) virus potential of catechin and gallic acid. Journal of the Chinese Medical Association, 81(5), pp. 458–468.

9.2.3.2 *Calluna vulgaris* (L.) Hull

Synonym: *Erica vulgaris* L.; *Ericoides vulgaris* (L.) Merino

Common names: Scotch heather; piren (Turkey)

Habitat: Rocks, hills, dry moorlands

Distribution: Turkey

Botanical observation: This beautiful lignose and diffuse herb grows to about 50 cm tall. The plant like many members of the Ericaceae has an aura of dryness. It can be said that plant accumulating phenolics have often dry auras. The leaves are simple, sessile, opposite, and exstipulate. The blade is coriaceous, ovate-lanceolate, and 2.5–3.5 mm×0.5–0.7 mm. The inflorescences are terminal and axillary racemes of fragrant and minute flowers. The calyx comprises 4 sepals which are about 4 mm long. The 4 petals are purplish to pink and 3 mm long. The androecium comprises 8 stamens. The ovary is 5-locular. The capsules are minute.

Medicinal uses: Urinary infection (Turkey)

Moderate antibacterial (Gram-positive) polar extract: Methanol extract inhibited the growth of *Staphylococcus aureus* (10788) and *Staphylococcus hominis* (11320) with MIC value of 100 µg/mL (Kumarasamy et al., 2002). Aqueous extract of aerial parts inhibited multidrug-resistant *Staphylococcus aureus* and *Staphylococcus epidermidis* with the MIC values of 562 and 281 µg/mL, respectively (Braghiroli et al., 1996).

Weak antibacterial (Gram-negative) polar extract: Aqueous extract inhibited the growth of *Proteus vulgaris* with the MIC/MBC of 2500/5000 µg/mL (Vučić et al., 2014). Aqueous extract of inflorescences inhibited the growth of *Escherichia coli, Klebsiella pneumoniae, Morganella morganii, Neisseria gonorrhoeae, Listeria monocytogenes,* and methicillin-resistant *Staphylococcus aureus* (Mandim et al., 2019). Aqueous extract of leaves and flowers inhibited the growth of clinical urinary tract *Escherichia coli, Enterococcus faecalis,* and *Proteus vulgaris* with the MIC/MBC values of 20/40, 20/40, and 2.5/5 μg/mL, respectively (Vučić et al., 2014).

Moderate antifungal (yeast) polar extract: Aqueous extract inhibited the growth of *Candida albicans* and *Cryptococcus neoformans* with the MIC values of 562 μg/mL (Braghiroli et al., 1996).

Commentaries

i. Water is used as a solvent for the extraction of hydrophilic principles such as phenolic glycosides. Consider that aqueous extract of *Calluna vulgaris* (L.) Hull concentrates the antibacterial (Gram-positive) activity suggesting the involvement of glycosides. The plant produces simple phenolics such as orcinol and phenolic glycosides such as orcinol β-*D*-glucoside and arbutin (Simon et al., 1993), kaempferol glycosides such as kaempferol-3-*O*-β-D-galactoside (Orhan et al., 2007), and quercetin glycosides (Simon et al., 1993) which may be, at least partially, accounting for the antibacterial properties.

ii. Aqueous extracts of Ericaceae are often antiviral *in vitro.* This is exemplified with aqueous extract of leaves of *Rhododendron ferrugineum* L.(containing 2*R*,3*R*-taxifolin-3-*O*-β-L-arabinoside) which inhibited the replication of the Herpes simplex virus type-1 (strain 17 syn$^+$) with the IC_{50} of 7.4 μg/mL and a selectivity index of 64 via inhibition of host cell penetration (Gescher et al., 2011). One can observe that flavonoids glycosides often hamper the entry of viruses into host cells.

References

Braghiroli, L., Mazzanti, G., Manganaro, M., Mascellino, M.T. and Vespertilli, T., 1996. Antimicrobial activity of Calluna vulgaris. *Phytotherapy Research, 10*(1), p. S86.

Gescher, K., Kühn, J., Hafezi, W., Louis, A., Derksen, A., Deters, A., Lorentzen, E. and Hensel, A., 2011. Inhibition of viral adsorption and penetration by an aqueous extract from Rhododendron ferrugineum L. as antiviral principle against herpes simplex virus type-1. *Fitoterapia, 82*(3), pp. 408–413.

Kumarasamy, Y., Cox, P.J., Jaspars, M., Nahar, L. and Sarker, S.D., 2002. Screening seeds of Scottish plants for antibacterial activity. *Journal of Ethnopharmacology, 83*(1–2), pp. 73–77.

Mandim, F., Barros, L., Calhelha, R.C., Abreu, R.M., Pinela, J., Alves, M.J., Heleno, S., Santos, P.F. and Ferreira, I.C., 2019. Calluna vulgaris (L.) Hull: Chemical characterization, evaluation of its bioactive properties and effect on the vaginal microbiota. *Food & Function, 10*(1), pp. 78–89.

Orhan, I., Küpeli, E., Terzioğlu, S., Yesilada, E. 2007. Bioassay-guided isolation of kaempferol-3-O-b-d-galactoside with anti-inflammatory and antinociceptive activity from the aerial part of Calluna vulgaris L. *Journal of Ethnopharmacology, 114*, pp. 32–37.

Simon, A., Chulia, A.J., Kaouadji, M., Allais, D.P. and Delage, C., 1993. Further flavonoid glycosides from Calluna vulgaris. *Phytochemistry, 32*(4), pp. 1045–1049.

Vučić, D.M., Petković, M.R., Rodić-Grabovac, B.B., Stefanović, O.D., Vasić, S.M. and Čomić, L.R., 2014. In vitro activity of heather [Calluna vulgaris (L.) Hull] extracts on selected urinary tract pathogens. *Bosnian Journal of Basic Medical Sciences, 14*(4), p. 234.

9.2.3.3 *Pyrola japonica* Klenze ex Alef.

Synonyms: *Pyrola subaphylla* Maxim.

Common names: Shinleaf; ri ben lu ti cao (China); nok je cho (Korea); ichiyaku so (Japan)

Habitat: Forests

Distribution: China, Korea, Japan, Taiwan

Botanical observation: This beautiful rhizomatous herb grows up to about 30 cm tall. It has at first glance an Orchidaceous look. The leaves are simple, spiral, basal, and exstipulate. The petiole is 3–8 cm long. The blade is dark green, obovate, 4–7 cm×2.5–4.5 cm, somewhat coriaceous, serrulate, and rounded at apex. The scape is erect, 15–30 cm tall, and sustains a few flowered raceme of nodding flowers. The calyx develops 5 broadly lanceolate to ovate lobes which are up to 6 mm long. The 5 petals are ovate, about 5 mm long, and pure white. The 10 stamens have pure white and somewhat fleshy filaments and orangish anthers. The style is exserted, curved, up to 1.3 cm long, and terminated with a capitate stigma. The capsules are somewhat globose, 8 mm across, and nodding.

Medicinal uses: Cuts (Korea)

Antibacterial halo (Gram-positive) developed by polar extracts: Ethanol extract (50 μg/disc) inhibited the growth of *Bacillus subtilis* with the inhibition zone diameter of about 20 mm (Park et al., 2006).

Strong antimycobacterial amphiphilic naphthoquinone: Chimaphilin (LogD 2.3 at pH 7.4; molecular mass = 186.2 g/mol) inhibited the growth of *Mycobacterium tuberculosis* (H37Ra) with the IC_{50} value of 5.4 μg/mL (Li et al., 2018).

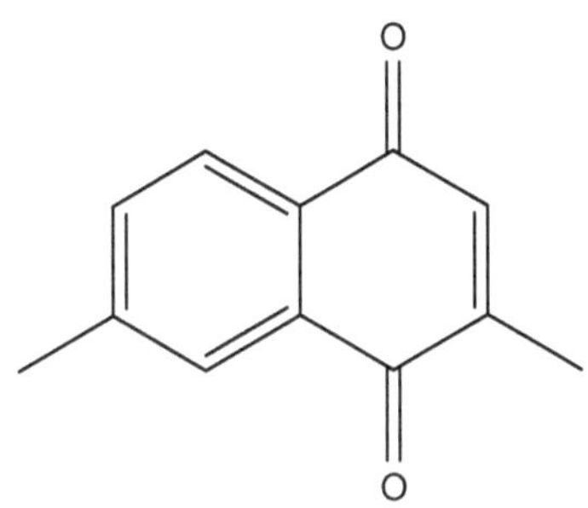

Chimaphilin

Strong antibacterial (Gram-positive) amphiphilic naphthoquinone: Chimaphilin inhibited the growth of *Staphylococcus aureus* (SAP00017) and *Bacillus subtilis* (Vernon) with the MIC values of 25 and 50μg/mL, respectively (Saxena et al., 1996).

Moderate anticandidal amphiphilic naphthoquinone: Chimaphilin inhibited the growth of *Candida albicans* with the MIC values of 100 μg/mL (Saxena et al., 1996).

Commentaries

i. To have a plant which looks like an orchid in the Ericaceae is very strange. In fact, it can be observed that in flowering plant families, some species have first glance features of other families. Even in the mineral world, some stones are very much like the shape of plants and even proteins (!). Some plant inflorescences and flowers sometimes display shapes of stones, viruses, and even birds. The human mind is not yet able to decipher the meaning of these signs. What is Nature telling us?

ii. *Pyrola japonica* Klenze ex Alef. is partially mycoheterotrophic meaning that it gets its nutrients from fungi (Merckx et al., 2009). This plant produces a series of naphthoquinones (are these of fungal origin?) of which 5, 8-dihydrochimaphilin which upon exposure to air turns into chimaphilin (Lee et al., 2001) exemplifying the fact that a number of natural products isolated are artifacts.

iii. Methanol extract of roots *Lycopus coreanus* H. Lév. at the concentration of 100 μg/mL inhibited RNA-dependent DNA polymerase activity of the human immunodeficiency virus type-1 reverse transcriptase and human immunodeficiency virus type-1 protease by 66.9 and 24.2%, respectively (Min et al., 2001).

iv. Ethyl acetate extract of *Pyrola calliantha* Andres inhibited neuraminidase with the IC_{50} value of 34.4 μg/mL (Yang et al., 2016).

References

Li, H., Bos, A., Jean, S., Webster, D., Robichaud, G.A., Johnson, J.A. and Gray, C.A., 2018. Antimycobacterial 1, 4-napthoquinone natural products from Moneses uniflora. *Phytochemistry Letters*, *27*, pp. 229–233.

Merckx, V., Bidartondo, M.I. and Hynson, N.A., 2009. Myco-heterotrophy: When fungi host plants. *Annals of Botany*, *104*(7), pp. 1255–1261.

Min, B.S., Kim, Y.H., Tomiyama, M., Nakamura, N., Miyashiro, H., Otake, T. and Hattori, M., 2001. Inhibitory effects of Korean plants on HIV-1 activities. *Phytotherapy Research*, *15*(6), pp. 481–486.

Park, H.G., Cha, M.R., Hwang, J.H., Kim, J.Y., Park, M.S., Choi, S.U., Park, H.R. and Hwang, Y.I., 2006. Antimicrobial activity of the extract from Pyrola japonica against Bacillus subtilis. *Journal of Life Science*, *16*(6), pp. 989–993.

Yang, X.Y., Liu, A.L., Liu, S.J., Xu, X.W. and Huang, L.F., 2016. Screening for neuraminidase inhibitory activity in traditional Chinese medicines used to treat influenza. *Molecules*, *21*(9), p. 1138.

9.2.3.4 *Vaccinium vitis-idaea* L.

Synonyms: *Rhodococcum vitis-idaea* Avrorin; *Vaccinium jesoense* Miq.

Common names: Red whortleberry, cowberry; yue ju (China)

Habitat: Shady wet forests, bog margins, dry lands

Distribution: Temperate Asia

Botanical observation: This handsome herb grows to about 30 cm tall and has an aura of vitality. The leaves are simple, spiral, and exstipulate. The petiole is minute. The blade is coriaceous, dark green, glossy, ovate-elliptic, and 0.7–2 cm×4–8 mm, with 5–6 pairs of secondary nerves, and rounded at apex. The inflorescences are terminal racemes. The calyx comprises 4 sepals which are about 1 mm long. The corolla is campanulate, 5 mm long, and 4-lobed. A disc is present. The androecium comprises about 4 brownish stamens. The ovary is multilocular. The berries are up to 1 cm long, bright red, glossy, long, and edible.

Medicinal uses: Cystitis, genitourinary infection, urethritis (Russia)

Weak antiviral condensed tannin: Proanthocyanidin A-1 inhibited the replication of the Herpes simplex type-2 (strain 196) in Vero cells with the IC_{50} of 73.3 μM and a selectivity index value of 3.8 and inhibited plaque formation with IC_{50} of 41.9 μM and a selectivity index of 6.7 (Cheng et al., 2005). Proanthocyanidin A-1 inhibited the attachment and penetration of virus to host cells (Cheng et al., 2005).

Commentaries

i. The plant generates a fascinating array of proathocyanidins, phenolic and flavonoid glycosides, as well as triterpenes (Shamilov et al., 2020).
ii. In Turkey, *Vaccinium myrtillus* L. (local name: *Çoban üzümü*) is used as antiseptic.

References

Cheng, H.Y., Lin, T.C., Yang, C.M., Shieh, D.E. and Lin, C.C., 2005. In vitro anti-HSV-2 activity and mechanism of action of proanthocyanidin A-1 from Vaccinium vitis-idaea. *Journal of the Science of Food and Agriculture*, *85*(1), pp. 10–15.

Shamilov, A.A., Bubenchikova, V.N., Chemikov, M.V., Pozdnyakov, D.I. and Garsiya, E.R., 2020. Vaccinium vitis-idaea L.: Chemical contents, pharmacological activities. *Pharmaceutical Sciences*, *26*(4), pp. 344–362.

9.2.4 Family Lecythidaceae A. Rich. (1825)

The family Lecythidaceae consists of about 15 genera and 325 species of trees or shrubs. The leaves simple, alternate, spiral, exstipulate, or stipulated. The blades are often coriaceous, glossy, dark green, and obovate. The inflorescences are solitary or racemes or spikes. The flowers are not uncommonly showy and somewhat fleshy. The calyx is tubular and produces 4–6 lobes. The corolla comprises 4–6 petals which are free or partially fused in a tube. A disc is present. The androecium is often massive (very much like as in some members of the family Myrtaceae), and often comprises numerous linear stamens. The gynoecium includes a 2-locular ovary, each locule sheltering a single or several ovules growing from axile placentas and a slender style. The fruits are berries or capsule, with persistent style and calyx. Capsules in this family are often somewhat characteristically urn-shaped and large, hence the name of the family which derives from the Greek *lekythos*= a narrow-necked oil flask. Plants in this family produce triterpene saponins.

9.2.4.1 *Barringtonia acutangula* (L.) Gaertn.

Synonyms: *Barringtonia luzonensis* (Presl) Vidal.; *Eugenia acutangula* L.; *Michelia acutangula* (L.) Kuntze; *Stravadium acutangulum* (L.) Sweet

Common names: Cut nut, Indian putat, hijla daru (Bangladesh); hijjala, nichula, sumundur phal, samandarpha, sjeria-samstravadi (India); alakang (Indonesia); putat (Indonesia, the Philippines); gadon nam (Laos); jempalang, khyaia tha, langkong (Malaysia); kyi (Myanmar); ko (Papua New Guinea); tong (Thailand); chiê'c mung (Vietnam)

Habitat: Coastal wetlands, riverbanks, fresh-water mangroves

Distribution: From Afghanistan to Asia-Pacific

Botanical observation: This tree grows up to about 8 m tall. The bark is grayish and cracked longitudinally. The wood is whitish and soft. The leaves are simple, spiral, without stipules, and crowded at apex of stems. The petiole is 6 mm–1 cm long and flattened above. The blade is spathulate to obovate, 1.5–3.2 cm×4–7 cm, glossy, soft, serrate, and marked with 7–9 pairs of sunken secondary nerves. The tertiary nerves are visible and raised above the blade. The inflorescences are terminal, flexuous, slender, pendulous, and up to 50 cm long racemes. The flowers are very 6 mm long with a quadrangular receptacle. The calyx includes 4 rounds and about 3-mm long sepals. The corolla presents 4 petals which are 1.2 cm long, plain red, and elliptic. A disc is present. The androecium is showy and includes numerous stamens. The style is up to about 2 cm long and slender. The capsules are ovoid, quadrangular, up to about 2.5 cm×6 cm, and contain a single up to 4 cm long ridged seed.

Medicinal uses: Cholera, wound, cold with pain in chest, influenza, dermatitis, gingivitis, toothache in wounds (Bangladesh); fever (India; Vietnam); cough (India); eye infection (Myanmar); diarrhea (Bangladesh; Myanmar; India); dysentery (Bangladesh; Myanmar); fever (Laos)

Strong broad-spectrum antibacterial non-polar extract: Petroleum ether extract of stem bark inhibited the growth of *Staphylococcus aureus* (NCTC1088), *Bacillus subtilis*, *Pseudomonas aeruginosa*, and *Escherichia coli* (NCTC9001) with the MIC values of 100, 25, 50, and 50 µg/mL, respectively (Rahman et al., 2005).

Moderate broad-spectrum antibacterial extract: Ethanol extract of seeds inhibited the growth of urinary tract infection isolates of *Staphylococcus aureus*, *Pseudomonas aeruginosa*, *Klebsiella pneumoniae*, *Enterococcus faecalis*, and *Escherichia coli* with the MIC values of 62.5, 125, 250, 62.5, and 125 µg/mL, respectively (Sahoo et al., 2008).

Strong anticandidal polar extract: Methanol extract of leaves inhibited the growth of *Candida albicans* and *Candida tropicalis* with the MIC values of 31.2 and 62.5 µg/mL, respectively (Mishra et al., 2009).

Commentary: From the bark were isolated epigallocatechin, gallocatechin 4′-*O*-methyl ether, and gallocatechin 4′-*O*-methyl ether 5-*O*-β-D-glucopyranoside (Thi et al., 2014) as well as flavonol glycosides (Van et al., 2017), and triterpene saponins (Pal et al., 1994) which most probably contribute, at least in part, to the antimicrobial properties. One could endeavor to isolate antibacterial principles from the petroleum ether extract of stem bark. Are antibacterial triterpene present in this plant?

References

Mishra, S., Sahoo, S., Mishra, S.K., Rout, K.K., Nayak, S.K., Panda, P.K. and Dhal, N.K., 2009. Antimicrobial investigation of leaves of Barringtonia acutangula Linn. *Medicinal and Aromatic Plant Science and Biotechnology*, *3*(1), pp. 55–58.

Pal, B.C., Chaudhuri, T., Yoshikawa, K. and Arihara, S., 1994. Saponins from Barringtonia acutangula. *Phytochemistry*, *35*(5), pp. 1315–1318.

Rahman, M.M., Polfreman, D., MacGeachan, J. and Gray, A.I., 2005. Antimicrobial activities of *Barringtonia acutangula*. *Phytotherapy Research*, *19*(6), pp. 543–545.

Sahoo, S., Panda, P.K., Mishra, S.R., Parida, R.K., Ellaiah, P. and Dash, S.K., 2008. Antibacterial activity of Barringtonia acutangula against selected urinary tract pathogens. *Indian Journal of Pharmaceutical Sciences*, *70*(5), p. 677.

Thi, L.N., Huy, T.T., Tuan, A.H., Van Huong, B., Hoai, N.N. and Tien, D.N., 2014. Flavan-3-ols from the barks of Barringtonia acutangula. *Biochemical Systematics and Ecology*, *55*, pp. 219–221.

Van, Q.T.T., Hanh, T.T.H., thanh Huong, P.T., Thuy, N.T.K., Cuong, N.T., Dang, N.H., Van Thanh, N., Cuong, N.X., Nam, N.H., Van Kiem, P. and Van Minh, C., 2017. Flavonoid glycosides from Barringtonia acutangula. *Bioorganic & Medicinal Chemistry Letters*, *27*(16), pp. 3776–3781.

9.2.4.2 *Barringtonia racemosa* (L.) Spreng

Synonyms: *Barringtonia ceylanica* (Miers) Gardner ex C.B. Clarke; *Barringtonia elongata* Korth.; *Barringtonia timorensis* Blume; *Butonica apiculata* Miers; *Butonica ceylanica* Miers; *Butonica inclyta* Miers; *Butonica terrestris* Miers; *Eugenia racemosa* L.

Common names: Fish-killer tree; dedaowi, moha shomudro gach (Bangladesh); yu rui (China); mohasomudra, samutra pullum, sam stravadi (India); putat sungai (Indonesia); putat ayam (Malaysia); kyi (Myanmar); koragas (Palau); pao pao (Papua New Guinea); putat (the Philippines); chik ban, jik suan (Thailand); tim lang (Vietnam)

Habitat: Seashores

Distribution: India to Pacific Islands

Botanical observation: This sinister tree native to India grows to 6 m tall. The leaves are simple, spiral, and exstipulate. The petiole is winged and up to 1.5 cm long. The blade is obovate, membranous, glossy, 20–35 cm×6–14 cm, base cuneate, margin serrate, edible (Indonesia), apex acute, and presents about 15 pairs of secondary nerves. The inflorescences are pendulous racemes terminal or axillary. The calyx is 2–4-lobed, and about 1.2 cm in diameter. The 4 petals are green or tinged red or yellow, oblong, and up to about 1.5 cm long. The stamens are many, showy, white or pink, and about 3–3.5 cm long. The ovary develops a slender style which is 4–6 cm long. The capsules are 4-angled, 5–9 cm×3–4 cm, coriaceous, and contain a single seed which is ovoid and 2–4 cm long.

Medicinal uses: Fever, leucorrhea, sore throat (India); otitis media (Palau)

Broad-spectrum antibacterial halo elicited by neo-clerodane diterpene: Nasimalun A (also known as 14-*neo*-clerodatrien-18,19-olide-17-carboxylic acid; 200 µg/disc) inhibited the growth of *Staphylococcus aureus*, *Bacillus subtilis*, *Shigella flexneri*, *Salmonella paratyphi* B, and *Vibrio mimicus* with the zones of inhibitions of 9, 10, 7, 10, and 11 mm, respectively (Khan et al., 2001).

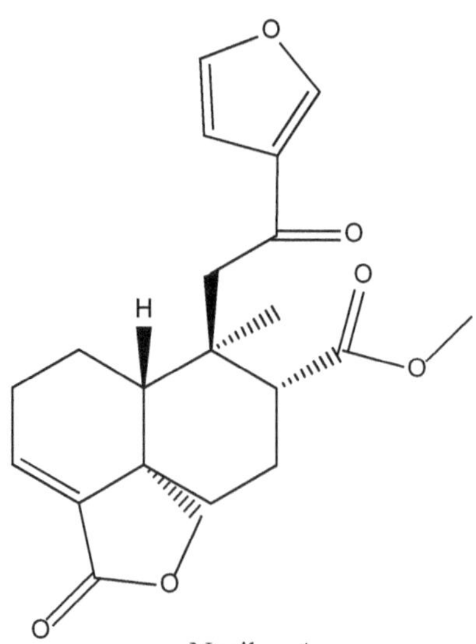

Nasilun A

Moderate antibacterial (Gram-positive) lupane-type triterpene: Lupeol inhibited *Enterococcus faecalis* (ATCC 29212) with MIC_{50} of 62.5 µg/mL (Al Bonvicini et al., 2017).

Antifungal (filamentous) polar extract: Methanol extract of leaves (10 mg/solid agar) inhibited the mycelial growth of *Aspergillus* sp., *Fusarium* sp., *Ganoderma lucidum*, *Ganoderma tropicum*, *Penicillium* sp., *Rhizopus* sp., and *Trichoderma* (Hussin et al., 2009).

Strong antiviral (enveloped monopartite linear double-stranded DNA) polar extract: Aqueous extract inhibited the replication of the Herpes simplex virus type-1 with the IC_{50} value of 23 µg/mL (Pefile, 2001).

Commentaries

i. The plant constructs triterpenes such as betulinic acid, lupeol, taraxerol, barringtogenol C, and bartogenic acid as well as the neo-clerodane-type diterpenoids nasimaluns A and B (Gowri et al., 2009).
ii. In the Philippines, *Barringtonia asiatica* (L.) Kurz (local name: *Vutun*) is used to repel evil spirits (!), and in Papua New Guinea this dreadfully poisonous tree affords a remedy for bronchitis, cough, diarrhea, influenza, skin infections, sores, and sore throat.

References

Al Bonvicini, F., Antognoni, F., Mandrone, M., Protti, M., Mercolini, L., Lianza, M., Gentilomi, G.A. and Poli, F., 2017. Phytochemical analysis and antibacterial activity towards methicillin-resistant Staphylococcus aureus of leaf extracts from Argania spinosa (L.) Skeels. *Plant Biosystems-An International Journal Dealing with all Aspects of Plant Biology*, *151*(4), pp. 649–656.

Gowri, P.M., Radhakrishnan, S.V.S., Basha, S.J., Sarma, A.V.S. and Rao, J.M., 2009. Oleanane-type isomeric triterpenoids from Barringtonia racemosa. *Journal of Natural Products*, *72*(4), pp. 791–795.

Hussin, N.M., Muse, R., Ahmad, S., Ramli, J., Mahmood, M., Sulaiman, M.R., Shukor, M.Y.A., Rahman, M.F.A. and Aziz, K.N.K., 2009. Antifungal activity of extracts and phenolic compounds from Barringtonia racemosa L. (Lecythidaceae). *African Journal of Biotechnology*, *8*(12), p. 2835.

Khan, S., Jabbar, A., Hasan, C.M. and Rashid, M.A., 2001. Antibacterial activity of Barringtonia racemosa. *Fitoterapia*, *72*(2), pp. 162–164.

Pefile, S.C., 2001. *A study of the antiherpes simplex virus type 1 properties of barringtonia racemosa* (Doctoral dissertation, University of Cape Town).

9.2.4.3 *Careya arborea* Roxb.

Synonyms: *Barringtonia arborea* (Roxb.) F. Muell.; *Barringtonia careya* F. Muell.; *Planchonia careya* (F. Muell.) R. Knuth

Common names: Cockatoo apple, cocky apple, patana oak, wild guava; putat kedun (Malaysia); baubwai, hou no, mai pingo, sangaw-gmawt, thelaw (Myanmar); kumbhi (India; Nepal); kyanum (India); kumh (Thailand)

Habitat: Forests

Distribution: From Afghanistan to Myanmar to Australia

Habitat: Forests

Distribution: Tropical Asia, Australia

Botanical observation: This handsome tree grows up to about 15 m tall. The bark is thick and suberose. The stems are striate and lenticelled. The leaves are simple, spiral, and exstipulate. The petiole is 4 cm long and winged. The blade is obovate, glossy, light green above, serrate, 10–15 cm × 15–25 cm, marked with 8–10 pairs of secondary nerves, and rounded, notched, or acuminate at apex. The spikes are terminal and up to 15 cm long. The calyx is 2 cm long and develops 4 round lobes which are about 8 mm long. The 4 petals are obovate-oblong, about 3.5 cm long, light yellow, and recurved. A disc is present. The stamens are 3–5 cm long, numerous, fused and reddish at base. The style is slender, somewhat straight, and about 4 cm long. The berries are ellipsoid, smooth, up to about 9 cm × 5 cm, marked at apex by persistent disk, elongated style, and sepals. The seeds are numerous and about 5 mm long. The berries somewhat look like some sort of guava fruit from which they differ by a dull light green color, smooth surface, and long vestigial style.

Medicinal uses: Cold, cough (India); smallpox, chickenpox (Cambodia, Laos, Vietnam); ulcers (Myanmar); wounds (Australia); fever (India; Nepal)

Broad-spectrum antibacterial halo developed by polar extract: Methanol extract of stem bark (100 µg/mL/ 6 mm disc) inhibited the growth of *Pseudomonas aeruginosa*, *Escherichia coli*, *Salmonella typhi*, *Vibrio cholerae*, *Staphylococcus aureus*, *Streptococcus pneumoniae*, *Micrococcus luteus*, and *Staphylococcus epidermidis* with the inhibition zone diameters ranging from 9 to 15 mm, respectively (Sambath et al., 2006).

Antibacterial halo developed by fatty acid: α-Dimorphecolic acid (100 µg/disc) inhibited the growth of *Bacillus subtilis* (SBUG 14); *Micrococcus flavus* (SBUG 16); *Staphylococcus aureus* (SBUG11), and *Staphylococcus aureus* (ATCC 25923) (Mundt, et al., 2003).

Moderate antibacterial (Gram-positive) fatty acid: α-Dimorphecolic acid inhibited the growth of *Staphylococcus aureus* (SBUG 11) with the MIC value of 100 µg/mL (Mundt, et al., 2003).

Weak antibacterial (Gram-positive) oleanane-type triterpene: Hyptatic acid A (also known as 4,2α,3β,24-trihydroxyolean-12-en-28-oic acid) inhibited the growth of methicillin-resistant *Staphylococcus aureus* (M99320) and vancomycin-resistant *Enterococcus* (193272) with the MIC values of 800 and 400 µg/mL, respectively (McRae et al., 2008).

Very weak antibacterial (Gram-positive) fatty acid: α-Dimorphecolic acid inhibited the growth of methicillin-resistant *Staphylococcus aureus* (M99320), vancomycin-resistant *Enterococcus* (193272), *Staphylococcus aureus* (ATCC 25923), and *Mycobacterium smegmatis* with the MIC values of 1000, 1000, 500, and 2000 µg/mL, respectively (McRae et al., 2008).

Very weak antimycobacterial (Gram-positive) fatty acid: α-Dimorphecolic acid inhibited the growth of *Mycobacterium smegmatis* with the MIC values of 2000 µg/mL (McRae et al., 2008).

Broad-spectrum antifungal halo developed by polar extract: Methanol extract of stem bark (100 µg/mL/ 6 mm disc) inhibited the growth of *Candida albicans*, *Aspergillus niger*, *Aspergillus flavus*, and *Alternaria solani* with the inhibition zone diameters of 10, 7, 7, and 10 mm, respectively (Sambath et al., 2006).

Strong antiviral (enveloped monopartite linear double-stranded DNA) polar extract: Methanol extract of bark inhibited the replication of the Herpes simplex virus type-1 with the IC_{50} value of 10 µg/mL in Vero cells (CC_{50} value: 49 µg/mL) (Rajbhandari et al., 2001).

Commentaries

i. Extract of *Careya herbacea* Roxb. at the concentration of 12.5 µg/mL inhibited the growth of the Vaccinia virus (enveloped linear double-stranded DNA virus) (Babbar et al., 1982). In Bali, *Planchonia valida* Bl. (local name: *Kutat kedis*) is used to treat viral infections, suggesting that antiviral principles might await discovery in the genus *Planchonia* Bl. and *Careya* Roxb. Further, in the Lecythidaceae, ethanol extract of *Bertholletia excelsa* Bonpl. inhibited the replication with the Vesicular stomatitis virus in Vero cells with the MIC value of 0.005 µg/mL (Hamidi et al., 1996).

ii. Methanol extract of this plant demonstrated central nervous depressant activities (Kumar et al., 2008) and as such might contain some bacterial and/or fungal efflux pumps.

iii. The plant produces anti-leishmanial triterpene saponins such as 3-*O*-β-D-glucopyranosyl (1→2)-β-D-glucopyranosyl-2α,3β-dihydroxy-taraxast-20-en-28-oic acid (Mandal et al., 2006). Consider that molecules with antileishmanial properties are often antifungal and *vice versa*.

iv. *Careya arborea* Roxb. and *Planchonia careya* (F. Muell.) R. Knuth are not uncommonly treated separately.

v. The plant produces (+)-gallocatechin and gallocatechin-(4α → 8)-gallocatechin (McRae et al., 2008).

vi. Consider that α-Dimorphecolic acid could be produced by symbiotic cyanobacteria growing on the plant. Cyanobacteria are known to produce and live on plants from the genus *Careya* Roxb. (Ambika & Krishnamurthy, 2019). Is α-dimorphecolic acid antifungal?

References

Ambika, H.D. and Krishnamurthy, S.R., 2019. Diversity of subaerial algae and cyanobacteria on tree bark in tropical mountain habitats. *Algological Studies*, *155*, pp. 15–27.
Babbar, O.P., Joshi, M.N. and Madan, A.R., 1982. Evaluation of plants for antiviral activity. *Indian Journal of Medical Research*, *76*, pp. 54–65.
Hamidi, J.A., Ismaili, N.H., Ahmadi, F.B. and Lajisi, N.H., 1996. Antiviral and cytotoxic activities of some plants used in Malaysian indigenous medicine. *Pertanika Journal of Tropical Agricultural Science*, *19*(2/3), pp. 129–136.
Kumar, R.S., Sundram, R.S., Sivakumar, P., Nethaji, R., Senthil, V., Murthy, N.V. and Kanagasabi, R., 2008. CNS activity of the methanol extracts of Careya arborea in experimental animal model. *Bangladesh Journal of Pharmacology*, *3*(1), pp. 36–43.
Mandal, D., Panda, N., Kumar, S., Banerjee, S., Mandal, N.B. and Sahu, N.P., 2006. A triterpenoid saponin possessing antileishmanial activity from the leaves of Careya arborea. *Phytochemistry*, *67*(2), pp. 183–190.
Mundt, S., Kreitlow, S. and Jansen, R., 2003. Fatty acids with antibacterial activity from the cyanobacterium Oscillatoria redekei HUB 051. *Journal of Applied Phycology*, *15*(2), pp. 263–267.
McRae, J.M., Yang, Q., Crawford, R.J. and Palombo, E.A., 2008. Antibacterial compounds from *Planchonia careya* leaf extracts. *Journal of Ethnopharmacology*, *116*(3), pp. 554–560.
Rajbhandari, M., Wegner, U., Jülich, M., Schoepke, T. and Mentel, R., 2001. Screening of Nepalese medicinal plants for antiviral activity. *Journal of Ethnopharmacology*, *74*(3), pp. 251–255.
Sambath, K.R., Sivakumar, T., Sunderam, R., Sivakumar, P., Nethaji, R., Gupta, M. and Mazumdar, U.K., 2006. Antimicrobial and antioxidant activities of Careya arborea Roxb. stem bark.

9.2.4.4 *Chydenanthus excelsus* Miers

Common names: Brosol, puyung puyung (Indonesia)

Habitat: Forests

Distribution: Andaman Islands, Myanmar, Indonesia, Papua New Guinea

Botanical observation: The handsome tree grows up to a height of 30 m. The bark is smooth and light brown. The leaves are simple, spiral, and stipulate. The petiole is up to about 2 cm long. The blade is elliptical, 7–25 cm×3–10 cm, papery, dark green and glossy above, marked with about 10–15 pairs of secondary nerves, and acuminate at apex. The panicles are stout, hairy, and up to about 25 cm long and bear showy flowers opening at night. The calyx is tubular, about 2 cm long, and 4-lobed. The corolla includes 4 tepals which are about 3 cm long and whitish. A disc is present. The androecium includes numerous stamens which are about 4 cm long and fused at base. The ovary develops and slender style. The drupe is ovoid and about 12 cm long, quadrangular, warty, and marked at apex with vestigial calyx.

Medicinal uses: Diarrhea (Indonesia)

Commentary: This endangered tree has apparently not been examined for antimicrobial properties.

9.2.4.5 *Couroupita guianensis* Aubl.

Common name: Cannon ball tree; nagalingamaram (India)

Habitat: Ornamental

Distribution: Tropical Asia, Pacific

Botanical observation: This massive tree is native to South America and grows to about 45 m tall. It is sacred in India and grown in Temples. The bark is rough and dark brown. The leaves are simple, alternate, and exstipulate. The blade is oblong, obovate, elliptic or broadly lanceolate, glossy, 10–16 cm × 4–6 cm, and acute at apex. The inflorescences are cauliflorous, woody, stout, and pendulous racemes which can reach about 70 cm long. The calyx is tubular and 6-lobed. The corolla includes 6 spongy petals which are about 3–5 cm × 2–3 cm, oblong-ovate, wavy, and reddish to pink. The androecium is fleshy and includes numerous stamens and (has somewhat a look of sea anemone). The gynoecium includes 5–7-locular ovary with several ovules in each locule. The capsules are somewhat globose, woody, heavy, indehiscent, can reach up to about 20 cm across, and shelter numerous seeds embedded in a pulp.

Medicinal uses: Skin diseases (India)

Broad-spectrum antibacterial halo by mid-polar extract: Chloroform extract of fruits (5 mg/disc) inhibited the growth of *Staphylococcus aureus*, *Enterococcus aerogenes*, *Micrococcus luteus*, and *Bacillus subtilis* with the inhibition zone diameters of 26, 19, 18, and 14 mm, respectively (Al-Dhabi et al., 2012). This extract inhibited the growth of *Shigella flexneri*, *Proteus vulgaris*, *Pseudomonas aeruginosa*, and *Klebsiella pneumoniae* with the inhibition zone diameters of 20, 12, 8, and 18 mm, respectively (Al-Dhabi et al., 2012).

Moderate antibacterial (Gram-positive) indole alkaloid: Tryptanthrin inhibited the growth of methicillin-resistant *Staphylococcus aureus* with the MIC of 62.5 μg/mL with alterations in the bacterial structure (Costa et al., 2017).

Antifungal (yeast) halo by mid-polar extract: Chloroform extract of fruits (5 mg/disc) inhibited the growth of *Candida albicans* and *Malassezia pachydermatis* with the inhibition zone diameters of 8 and 16 mm, respectively (Al-Dhabi et al., 2012).

Moderate broad-spectrum antifungal indole alkaloid: Tryptanthrin at the concentration of 50 μg/mL inhibited the growth of *Fusarium oxysporum* f. sp. *cucumeris*, *Cercospora arachidicola* Hori, *Physalospora piricola*, *Rhizoctonia cerealis*, *Bipolaris maydis*, *Fusarium moniliforme*, *Alternaria solani*, *Fusarium graminearum*, *Phytophthora infestans*, *Phytophthora capsica*, *Sclerotinia sclerotiorum*, *Botrytis cinerea*; and *Rhizoctonia solani* by (Hao et al., 2020).

Strong antiviral (enveloped monopartite linear single-stranded (+) RNA) amphiphilic indole alkaloid: Tryptanthrin (LogD 2.4 at pH 7.4; molecular mass = 248.2 g/mol) inhibited the replication of the human coronavirus NL63 with the IC_{50} values of 0.06 μg/mL and a selectivity index above 6600 (Tsai et al., 2020).

Weak antiviral (non-enveloped linear monopartite single-stranded (+) RNA) amphiphilic indole alkaloid: Tryptanthrin at the concentration of 500 μg/mL inhibited the replication of the Tobacco mosaic virus by about 37% (Hao et al., 2020).

Commentary: Tryptanthrin is planar and heterocyclic alkaloid and as such, most probably interferes with the RNA replication of human coronavirus. Natural products which are both antiviral and antifungal are often targeting genetic material. In mammalian cells, tryptanthrin inhibit the expression of P-glycoprotein efflux pumps (Yu et al., 2007) and one could look into its effect on the expression of efflux pumps in bacteria and fungi.

References

Al-Dhabi, N.A., Balachandran, C., Raj, M.K., Duraipandiyan, V., Muthukumar, C., Ignacimuthu, S., Khan, I.A. and Rajput, V.S., 2012. Antimicrobial, antimycobacterial and antibiofilm properties of Couroupita guianensis Aubl. fruit extract. *BMC Complementary and Alternative Medicine*, *12*(1), pp. 1–8.

Costa, D.C.M., Azevedo, M.M.B.D., Silva, D.O.E., Romanos, M.T.V., Souto-Padrón, T.C.B.S., Alviano, C.S. and Alviano, D.S., 2017. In vitro anti-MRSA activity of Couroupita guianensis extract and its component Tryptanthrin. *Natural Product Research*, *31*(17), pp. 2077–2080.

Tsai, Y.C., Lee, C.L., Yen, H.R., Chang, Y.S., Lin, Y.P., Huang, S.H. and Lin, C.W., 2020. Antiviral action of tryptanthrin isolated from strobilanthes cusia leaf against human coronavirus NL63. *Biomolecules*, *10*(3), p. 366.

Yu, S.T., Chen, T.M., Tseng, S.Y. and Chen, Y.H., 2007. Tryptanthrin inhibits MDR1 and reverses doxorubicin resistance in breast cancer cells. Biochemical and Biophysical Research Communications, 358(1), pp. 79–84.

9.2.5 Family Primulaceae Batsch ex Borkhausen (1797)

The family Primulaceae consists of about 25 genera and 1000 species of herbs. The leaves are simple, often in a rosette, and exstipulate. The inflorescences are panicles, racemes, spikes, fascicles, or heads. The flowers are actinomorphic. The calyx is tubular. The corolla is tubular and develops 5 lobes. The androecium consists of 5 stamens. The gynoecium consists of 5 carpels fused into a unilocular ovary with one to many ovules growing on basal or free central placenta, a single style and a capitate stigma. The capsules are dehiscent and contain numerous minute seeds.

9.2.5.1 *Aegiceras corniculatum* (L.) Blanco

Synonyms: *Aegiceras majus* Gaertn.; *Rhizophora corniculata* L.; *Umbraculum corniculatum* (L.) Kuntze

Common names: Khalsi, khalshe, narikandam (India); saka mata (Malaysia)

Habitat: Mangroves, sea muds

Distribution: Pakistan, Southeast Asia, Solomons, to Northeast Australia

Botanical observation: This tree grows up to about 5 m tall. The leaves are simple, spiral, and exstipulate. The petiole is stout and about 5 mm long. The blade is obovate, coriaceous, dull green, smooth, 4–8 cm×2.5–4 cm, glaucous below, and round at apex. The inflorescences are terminal umbels. The flower pedicels are slender, straight, and about 3 cm long. The calyx is tubular, about 7 mm long, and develops 5 lobes. The corolla is beautiful, pure white, tubular, about 1.5 cm long, and develops 5 lanceolate lobes which are recurved. The 5 stamens are pure white and exserted. The style is pure white, straight, pointed, longer than the corolla, and shorter than stamens. The capsules are falcate, coriaceous, somewhat bean-shaped (whence the local name *Kacang kacang* meaning bean in Malay), cylindrical, about 5 cm long, terete, pointed at apex, light green to reddish or purplish-brown, and seated in persistent calyx.

Medicinal uses: Ear infection, wounds (India)

Broad-spectrum antibacterial halo developed by polar extract: Ethanol extract of bark (500 μg/disc) inhibited the growth of *Staphylococcus epidermidis*, *Streptococcus pyogenes*, *Shigella sonnei*, *Salmonella typhi*, *Escherichia coli*, *Vibrio cholerae*, and *Enterobacter aerogenes* with zones of inhibition of 18, 14, 20, 14, 13, 12, and 13 mm, respectively (Uddin et al., 2008).

Strong antimycobacterial non-polar extract: Hexane extract of leaves inhibited the growth of *Mycobacterium tuberculosis* ($H_{37}R_v$) with the MIC value of 50 μg/mL (Janmanchi et al., 2017).

Broad-spectrum antifungal oleanane-type triterpene saponin: Acornine 2 isolated from the bark inhibited the growth of *Candida albicans*, *Trichophyton mentagrophytes*, *Trichophyton rubrum*, and *Termitomyces clypeatus* with the minimum amount of 50, 50, 100, and 150 μg/disc, respectively (Gupta et al., 2014).

Commentaries

i. The plant produces the oleanane-type triterpenes 16α-hydroxy-13,28-epoxyoleanan-3-one, protoprimulagenin, aegicerin, embelinone as well as syringic acid, gallic acid, and isorhamnetin 3-*O*-α-l-rhamnopyranosyl-(1→6)-D-glucopyranoside (Zhang et al., 2005). In India, the plant is a source of fish poison. Note that plants used for fish poison often abound with saponins, and this is the case for this plant (Vinh et al., 2019).
ii. Other constituents are benzoquinones such as 2-hydroxy-5-ethoxy-3-nonyl-1,4-benzoquinone, 5-*O*-butyl-embelin, and 2,5-dihydroxy-6-methyl-3-pentadecyl-1,4-benzoquinone (Li et al., 2020) which are most probably antimicrobial.
iii. This mangrove tree is a host for microorganisms producing antimicrobial principles (Bin et al., 2014). Mangrove trees are often home to bacteria and fungi producing antimicrobial principles.

References

Bin, G., Yanping, C., Hong, Z., Zheng, X., Yanqiu, Z., Huaiyi, F., Qiupin, Z. and Chenxiao, Z., 2014. Isolation, characterization and anti-multiple drug resistant (MDR) bacterial activity of endophytic fungi isolated from the mangrove plant, Aegiceras corniculatum. *Tropical Journal of Pharmaceutical Research*, *13*(4), pp. 593–599.

Gupta, V.K., Mukherjee, K. and Roy, A., 2014. Two novel antifungals, acornine 1 and acornine 2, from the bark of mangrove plant Aegiceras corniculatum (Linn.) Blanco from Sundarban Estuary. *Pharmacognosy Magazine*, *10*(Suppl 2), p. S342.

Janmanchi, H., Raju, A., Degani, M.S., Ray, M.K. and Rajan, M.G.R., 2017. Antituberculosis, antibacterial and antioxidant activities of Aegiceras corniculatum, a mangrove plant and effect of various extraction processes on its phytoconstituents and bioactivity. *South African Journal of Botany*, *113*, pp. 421–427.

Li, Y., Dong, C., Xu, M.J. and Lin, W.H., 2020. New alkylated benzoquinones from mangrove plant Aegiceras corniculatum with anticancer activity. *Journal of Asian Natural Products Research*, *22*(2), pp. 121–130.

Uddin, S.J., Rouf, R., Shilpi, J.A., Alamgir, M., Nahar, L. and Sarker, S.D., 2008. Screening of some Bangladeshi medicinal plants for in vitro antibacterial activity. *Oriental Pharmacy and Experimental Medicine*, *8*(3), pp. 316–321.

Vinh, L.B., Nguyet, N.T.M., Yang, S.Y., Kim, J.H., Thanh, N.V., Cuong, N.X., Nam, N.H., Minh, C.V., Hwang, I. and Kim, Y.H., 2019. Cytotoxic triterpene saponins from the mangrove Aegiceras corniculatum. *Natural Product Research*, *33*(5), pp. 628–634.

Zhang, D., Wu, J., Zhang, S. and Huang, J., 2005. Oleanane triterpenes from Aegiceras corniculatum. *Fitoterapia*, *76*(1), pp. 131–133.

9.2.5.2 *Ardisia crispa* A. DC.

Synonyms: *Ardisia hortorum* Maxim., *Ardisia crispa* var. *angusta* Clarke, *Ardisia crenata* Sims.

Common names: Village ardisia; ping chap (Cambodia); hen's eye; bai liang jin (China); mata ayam, mata pelanduk (Malaysia); tagpo (the Philippines); chamkhruea (Thailand); trojng duka

Habitat: Roadsides, forests

Distribution: India, Sri Lanka, Myanmar, Vietnam, Thailand, Malaysia, China, Korea, the Philippines, Taiwan, Japan

Botanical observation: This shrub grows to a height of 3 m. The leaves are simple, alternate, and exstipulate. The blade is about 5–15 cm × 2–4 cm, narrowly elliptic, dotted beneath, tapering at apex and base, dark green, glossy, succulent, marked with 12–18 inconspicuous pairs of secondary nerves and tiny dots and the margin laxly crenate. The inflorescences are terminal umbel-like fascicles of whitish-pink flowers. The calyx consists of 5 free sepals which are oblong and minute. The corolla is tubular, short, and consists of 5 spreading lobes which are about 5 mm long and white. The androecium comprises 5 short pediceled stamens with large anthers, the whole structure forming a conspicuous cone around the style. The berries are red, glossy, globose, 5 mm across, and look like somewhat some bird or animal eyes whence the Malay local names: *mata ayam* (chicken eye) or *mata pelanduk* (deer eye).

Medicinal uses: Cough, diarrhea (India; Malaysia); fever (Malaysia)

Weak antifungal (filamentous) oleanane-type triterpene saponins: Ardisicrenoside K and ardisicrenoside L inhibited the growth of *Pyricularia oryzae* (Liu et al., 2007)

Commentaries

i. Asian traditional systems of medicines often use members of the genus *Ardisia* Sw. for the treatment of microbial infections: In Malaysia, *Ardisia ridleyi* King & Gamble is used to treat fever and as post-partum remedy while *Ardisia sanguinolenta* Wall ex. (local name: *Kayu mata buah*) is used to treat cough, diarrhea, used as post-partum remedy and *Ardisia villosa* Roxb. is used for cough, fever. In India, *Ardisia solanacea* Roxb. (local name: *Junglikhari*) is used to treat fever and diarrhea, and in Bangladesh (local name: *Barochulliya*) it is used for boils. In India, *Ardisia paniculata* Roxb. (local name:

Naunuar) is used for jaundice. In China, *Ardisia humilis* Vahl (local name: *Jiu jie*) is used to treat orchitis, *Ardisia villosa* Roxb. (local name: *Niu shan xue dan*) is used to treat diarrhea, and *Ardisia chinensis* Benth. (local name: *Huang dan cao*) is used for tuberculosis. In Japan, *Ardisia japonica* (Hornst.) Blume. (local name: *Yama tachibana*) is used for fever. Ethylacetate extract of *Ardisia chinensis* Benth. inhibited the replication of the Coxsackie virus B3 in HeLa cells with the IC_{50} value of 62.5 μg/mL (Su et al., 2006). In Thailand, *Ardisia elliptica* Thunb. (Local name: *Philungkasa*) is used to treat gonorrhea. In Korea, *Ardisia japonica* (Thunb.) Bl. is used to treat COVID-19 (Lee et al., 2021), and from this plant, norbergenin (LogD – 0.04 at pH 7.4; molecular mass=324.2 g/mol) inhibited the replication of the human immunodeficiency virus in C8166 cells with the IC_{50} value of 20 μg/mL and a selectivity index above 25 (Piacente et al., 1996).

ii. The traditional anti-infectious use of plants in the genus *Ardisia* Sw. (as with plants in the genus *Diospyros* L.) in several countries point to the presence of strong antimicrobial principles. *Ardisia crispa* A. DC. produces 2-methoxy-6-tridecyl-1,4-benzoquinone (Kang et al., 2001) which is probably antibacterial. Alkylphenols and benzoquinones are the main antibacterial principles in members of the genus *Ardisia* Sw. The alkyl phenols methoxy-2-methyl-5-pentylphenol, 1,3-dimethoxy-2-methyl-5-pentylbenzene, 1-(3,5-dihydroxyphenyl)nonan-1′-one, and belamcandol isolated from the roots of *Ardisia cornudentata* Mez inhibited the growth of *Mycobacterium tuberculosis* (H37Rv strain) with MIC values of ≤2.5, ≤2.5, 6.0, and 33.8 μg/mL, respectively (Chang et al., 2011). 2,5-Dihydroxy-3-heptyl-2,5-cyclohexadiene-1,4-dione from a member of the family Myrsinaceae inhibited the growth of *Escherichia coli* (ATCC 8739), *Escherichia coli* (AG100), *Escherichia coli* (AG102), *Escherichia coli* (AG100Atet), *Klebsiella pneumoniae* (ATCC 11296), *Klebsiella pneumoniae* (KP55), and *Klebsiella pneumoniae* (KP63) with the Mic Value of 32, 4, 4, 8, 32, 8, and 4 μg/mL, respectively (Omosa et al., 2016). 2,5-Dihydroxy-3-butyl-2,5-cyclohexadiene-1,4-dione inhibited *Pseudomonas aeruginosa* (PA01), *Pseudomonas aeruginosa* (PA124), *Enterobacter aerogenes* (ATCC 13048), *Enterobacter aerogenes* (ATCC 13048), *Enterobacter aerogenes* (AM-CM64), *Enterobacter aerogenes* (EA3), and *Enterobacter aerogenes* (EA27) (Omosa et al., 2016). Ardisiaquinone B from a member of the genus *Ardisia* Sw. inhibited *Escherichia coli* (ATCC 8739), *Escherichia coli* (AG100), *Escherichia coli* (AG102), *Escherichia coli* (AG100Atet), *Klebsiella pneumoniae* (ATCC 11296), *Klebsiella pneumoniae* (KP55), and *Klebsiella pneumoniae* (KP63) (Omosa et al., 2016). Ardisiaquinone B inhibited the growth of *Pseudomonas aeruginosa* (PA124), *Enterobacter aerogenes* (ATCC 13048), *Enterobacter aerogenes* (AM-CM64), *Enterobacter aerogenes* (EA3), *Enterobacter aerogenes* (EA27) (Omosa et al., 2016).

iii. The benzoic acid derivative syringic acid isolated from the fruits of *Ardisia elliptica* Thunb. inhibited *Salmonella enteritidis* (DMST 8536) and *Salmonella typhimurium* (ATCC 13311) with MIC values of 125 and 62.5 μg/mL, respectively (Phadungkit & Luanratana, 2006). The flavonol isorhamnetin isolated from the fruits of *Ardisia elliptica* inhibited *Salmonella enteritidis* (DMST 8536) and *Salmonella typhimurium* (ATCC 13311) with MIC values of 31.3 and 15.6 μg/mL, respectively (Phadungkit & Luanratana, 2006). Quercetin isolated from the fruits of *Ardisia elliptica* Thunb. inhibited *Salmonella enteritidis* (DMST 8536) and *Salmonella typhimurium* (ATCC 13311) with MIC values of 15.6 and 15.6 μg/mL, respectively (Phadungkit & Luanratana, 2006).

iv. Aqueous extract of a member of the genus *Ardisia* Sw. inhibited Adenovirus-8 with EC_{50} of 96.7 μg/mL (selectivity index: 17.6) (Chiang et al., 2003).

v. The flavone glycosides myricitrin, desmanthin-1, and the ellagitannin corilagin isolated from the leaves of *Ardisia splendens* Pit. inhibited Coxsackie viruses A16 with IC_{50} values of 40.1, 32.2, and 30.5 μM, respectively (Van Nguyen et al., 2014).

References

Chang, C.P., Chang, H.S., Peng, C.F., Lee, S.J. and Chen, I.S., 2011. Antitubercular resorcinol analogs and benzenoid C-glucoside from the roots of Ardisia cornudentata. *Planta Medica*, *77*(01), pp. 60–65.

Chiang, L.C., Cheng, H.Y., Liu, M.C., Chiang, W. and Lin, C.C., 2003. In vitro anti-herpes simplex viruses and anti-adenoviruses activity of twelve traditionally used medicinal plants in Taiwan. *Biological and Pharmaceutical Bulletin*, *26*(11), pp. 1600–1604.

Kang, Y.H., Kim, W.H., Park, M.K. and Han, B.H., 2001. Antimetastatic and antitumor effects of benzoquinonoid AC7-1 from Ardisia crispa. *International Journal of Cancer*, *93*(5), pp. 736–740.

Lee, K., Jeong, S., Jeong, M., Choi, Y., Song, M. and Jang, I., 2021. Review on herbal medicine treatment for late complications of COVID-19 patients. *The Journal of Internal Korean Medicine*, *42*(1), pp. 53–66.

Liu, D.L., Wang, N.L., Zhang, X., Gao, H. and Yao, X.S., 2007. Two new triterpenoid saponins from Ardisia crenata. *Journal of Asian Natural Products Research*, *9*(2), pp. 119–127.

Omosa, L.K., Midiwo, J.O., Mbaveng, A.T., Tankeo, S.B., Seukep, J.A., Voukeng, I.K., Dzotam, J.K., Isemeki, J., Derese, S., Omolle, R.A. and Efferth, T., 2016. Antibacterial activities and structure–activity relationships of a panel of 48 compounds from Kenyan plants against multidrug resistant phenotypes. *SpringerPlus*, *5*(1), p. 901.

Phadungkit, M. and Luanratana, O., 2006. Anti-Salmonella activity of constituents of Ardisia elliptica Thunb. *Natural Product Research*, *20*(7), pp. 693–696.

Piacente, S., Pizza, C., De Tommasi, N. and Mahmood, N., 1996. Constituents of Ardisia japonica and their in vitro anti-HIV activity. *Journal of Natural Products*, *59*(6), pp. 565–569.

Su, M., Li, Y., Leung, K.T., Cen, Y., Li, T., Chen, R. and Ooi, V.E., 2006. Antiviral activity and constituent of Ardisia chinensis benth against coxsackie B3 virus. *Phytotherapy Research*, *20*(8), pp. 634–639.

Van Nguyen, T.H., Vien, T.A., Nhiem, N.X., Van Kiem, P., Van Minh, C., Long, P.Q., Anh, L.T., Cuong, N.M., Song, J.H., Ko, H.J. and Kim, N., 2014. Chemical components of Ardisia splendens leaves and their activity against coxsackie A16 viruses. *Natural Product Communications*, *9*(5), pp. 643–645.

9.2.5.3 *Embelia ribes* Burm.f.

Common names: False black pepper; baiberung, long dhama shak (Bangladesh); bai hua suan teng guo (China); baberang, vidanga (India); kacembang (Indonesia); barang kaboli (Iran); akar sulur kerang (Malaysia); som kung (Thailand)

Habitat: Rainforest

Distribution: India to Papua New Guinea

Botanical observation: This rare climber grows to a length of 15 m long. The stems are angled and hairy at apex. The leaves are simple, alternate, and exstipulate. The petiole is up to about 1 cm long. The blade is elliptic, oblong, to obovate, dull dark green, 5–8 cm×3–3.5 cm, coriaceous, with about 15 pairs of secondary nerves, cuneate at base, and acuminate at apex. The inflorescence is a showy panicle of minute flowers which grows up to 30 cm long. The calyx includes 5 sepals. The corolla comprises 5 petals which are lanceolate and white. The androecium includes 5 stamens. The drupes are globose, bluish, up to 5 mm diameter which when dry look like somewhat pepper.

Medicinal use: Carbuncle, cough, flu, gonorrhea, urinary tract infection (India); jaundice (Bangladesh; Iran), diarrhea (India; Iran)

Antibacterial halo elicited by benzoquinone: Embelin (100 µg/disc) inhibited the growth of *Staphylococcus aureus*, *Streptococcus faecalis*, *Streptococcus pyogenes*, *Escherichia coli*, *Salmonella typhi*, *Shigella flexneri*, *Shigella boydii*, *Shigella sonnei*, *Proteus mirabilis*, *Vibrio cholera*, *Pseudomonas aeruginosa*, and *Klebsiella pneumonia* with inhibition zone diameters of 23, 28, 23, 22, 16, 24, 23, 23, 16, 29, 24 and 23 mm, respectively (Chitra et al., 2003).

Strong broad-spectrum hydrophilic benzoquinone: Embelin (LogD 0.5 at pH 7.4; molecular mass=294.3 g/mol) inhibited the growth of *Bacillus cereus* (ATCC 10876), *Bacillus subtilis* (ATCC 11774), *Staphylococcus aureus* (ATCC 6538), *Micrococcus luteus* (ATCC 4698), *Escherichia coli* (ATCC 4157), *Shigella flexneri* (ATCC 9199), *Klebsiella pneumoniae* (ATCC 4352), *Pseudomonas aeruginosa* (ATCC 9027), and *Proteus mirabilis* (ATCC 7002) with the MIC/MBC values of 20/75, 20/75, 20/75, 25/100, 45/325, 30/175, 50/400, 25/125, and 25/150 µg/mL, respectively (Radhakrishnan et al., 2011).

Moderate broad-spectrum antifungal benzoquinone: Embelin inhibited the growth of *Epidermophyton floccosum*, *Microsporum canis*, *Trichophyton mentagrophytes*, and *Trichophyton rubrum* with the MIC values of 50 µg/mL, and *Microsporum gypseum* with the IC_{50} value of 100 µg/mL (Feresin et al., 2003). It also inhibited the growth of *Candida albicans*, *Candida tropicalis*, *Cryptococcus neoformans*, and *Saccharomyces cerevisiae* with the MIC value of 125 µg/mL (Feresin et al., 2003).

Strong antiviral (enveloped segmented linear single-stranded (-) RNA) hydrophilic benzoquinone: Embelin inhibited the replication of the Influenza virus A/Puerto Rico/8/34 (H1N1) an IC_{50} of 0.3 µM and a selectivity index of 10 was most effective when added at early stages of the viral life cycle (Hossan et al., 2018).

Commentaries

i. Ethanol extract of fruits inhibited angiotensin-converting enzyme with the IC_{50} value of 761.5 µg/mL (Khan & Kumar, 2019) suggesting the presence of principle with possible activity against the severe acute respiratory syndrome-associated coronavirus 2 and other zoonotic betacoronaviruses. In Iran, the plant is considered as hot and dry as per the humoral classification of Hippocrates. A formulation containing the plant evoked reduction of viral load in COVID-19 patients (Patankar et al., 2021). The Sanskrit name of this plant is *Vrishanasana* which is said to mean "destroyer of enemy. Consider that despite global vaccination to control COVID-19, some variants seem to affect vaccinated people (Carneiro et al., 2021; Moore, 2021) suggesting that there is a need to look into medicinal plants and natural products as means to participate against the pandemic. Do the fruits of *Embelia ribes* Burm.f. hold potentials to fight COVID-19 and other coming betacoronaviruses? Note that the fruits were given in India as a mean of protection against the Spanish flu that followed the first world war. There is the urgent need to develop phytomedication for pandemic zoonotic viral infections. We argue that some plants in Asia and the Pacific hold the key to solve global viral pandemics.

ii. Embelin has central nervous system activities (Kundap et al., 2017) and as such could be an antibiotic or antifungal potentiator by means of inhibition of efflux pumps.

iii. *Embelia robusta* Roxb. is used in India (local name: *Bayabirang*) to treat dysentery.

References

Carneiro, D.C., Sousa, J.D. and Monteiro-Cunha, J.P., 2021. The COVID-19 vaccine development: A pandemic paradigm. *Virus Research*, *301*, p. 198454.

Chitra, M., Shyamala Devi, C.S. and Sukumar, E., 2003. Antibacterial activity of embelin. *Fitoterapia*, *74*(4), pp. 401–403.

Feresin, G.E., Tapia, A., Sortino, M., Zacchino, S., Rojas de Arias, A., Inchausti, A., Yaluff, G., Rodriguez, J., Theoduloz, C. and SchmedaHirschmann, G. 2003. Bioactive alkyl phenols and embelin from Oxalis erythrorhiza. *Journal of Ethnopharmacology*, 88(2–3), pp. 241–247.

Hossan, M.S., Fatima, A., Rahmatullah, M., Khoo, T.J., Nissapatorn, V., Galochkina, A.V., Slita, A.V., Shtro, A.A., Nikolaeva, Y., Zarubaev, V.V. and Wiart, C., 2018. Antiviral activity of Embelia ribes Burm. f. against influenza virus in vitro. *Archives of Virology*, *163*(8), pp. 2121–2131.

Kundap, U.P., Bhuvanendran, S., Kumari, Y., Othman, I. and Shaikh, M., 2017. Plant derived phytocompound, embelin in CNS disorders: a systematic review. *Frontiers in Pharmacology*, *8*, p. 76.

Moore, J.P., 2021. Approaches for optimal use of different COVID-19 vaccines: issues of viral variants and vaccine efficacy. *JAMA*, *325*(13), pp. 1251–1252.

Patankar, S.B., Rangnekar, H., Joshi, K., Suryawanshi, K., Soni, P., Gorde, A., Shah, T., Patankar, S., Jha, D. and Raje, R., 2021. Efficacy and Safety of Polyherbal formulation as an add-on to the standard of care in mild to moderate COVID-19: A randomized, double-blind, placebo-controlled trial. *medRxiv*. doi: 10.1101/2021.05.14.21256900.

Radhakrishnan, N., Gnanamani, A. and Mandal, A.B., 2011. A potential antibacterial agent Embelin, a natural benzoquinone extracted from Embelia ribes. *Biology and Medicine*, *3*(2), pp. 1–7.

9.2.5.4 *Labisia pumila* (Bl.) F-Vill. & Naves

Synonyms: *Ardisia pumila* Bl., *Angiopetalum punctatum* Reinw., *Ardisia pumila* var. *alata* Scheff., *Labisia pothoina* Lindl., *Labisia pumila* (Bl.) F-Vill. & Naves, *Labisia pumila* var. *alata* (Scheff.) Mez, *Labisia punctata* (Reinw.) Airy-Shaw.

Common names: Common Labisia; selusoh Fatimah, kacip Fatimah (Malay)

Botanical observation: This is an understorey herb of the lowland and hill rainforests of Asia-Pacific, which grows up to a height of about 30 cm. The leaves are simple, spiral, and exstipulate. The petiole is short and thick. The blade is lanceolate, fleshy, light green above and purplish below, and slightly toothed. The inflorescences are 12–20 cm long racemes of small and white campanulate flowers. The berries are black, glossy, and about 8 mm across.

Medicinal uses: Cut, dysentery, post-partum, wounds (Malaysia)

Broad-spectrum antibacterial polar extract: Aqueous extract of leaves (300 µg/disc) inhibited the growth of *Staphylococcus aureus* (S1431), *Escherichia coli* (E256), *Pseudomonas aeruginosa* (PI96), *Micrococcus luteus*, *Klebsiella pneumoniae* (K36), *Bacillus subtilis* (B145), *Bacillus cereus* (B43) and *Enterococcus aeruginosa* (Karimi et al., 2012).

Commentaries: Note that alkylphenols are assembled by this plant like, for instance, demethylbelamcandaquinone B, fatimahol, and dexyloprimulanin (Ali & Khan, 2011). Alkylphenols in this plant are most probably antibacterial.

References

Ali, Z. and Khan, I.A., 2011. Alkyl phenols and saponins from the roots of Labisia pumila (Kacip Fatimah). *Phytochemistry*, *72*(16), pp. 2075–2080.

Karimi, E., Jaafar, H.Z. and Ahmad, S., 2013. Antifungal, anti-inflammatory and cytotoxicity activities of three varieties of labisia pumila benth: From microwave obtained extracts. *BMC Complementary and Alternative Medicine*, *13*(1), pp. 1–10.

9.2.5.5 *Myrsine seguinii* H. Lév.

Synonyms: *Athruphyllum neriifolium* (Siebold & Zucc.) H. Hara; *Athruphyllum seguinii* (H. Lév.) Nakai; *Athruphyllum taiwanianum* Nakai; *Athruphyllum yunnanensis* (Mez) Nakai; *Myrsine neriifolia* Siebold & Zucc.; *Rapanea neriifolia* Mez

Common names: Mi hua shu (China); irahaja, shima daimintachibana, taimin-tachibana (Japan)

Habitat Forests

Distribution: Myanmar, Vietnam, China, Japan

Botanical observation: This shrub grows up to about 6 m tall and has an aura of dryness. The leaves are simple, spiral, exstipulate, and at apex of stems. The petiole is up to 1.5 cm long. The blade is narrowly elliptic, glossy, fleshy, smooth, without apparent secondary nerves, 7–17 cm × 1.3–6 cm, coriaceous, cuneate at base, and acute at apex. The inflorescences are numerous axillary fascicles of up to about 10 flowers. The calyx is tubular and develops 5 tiny lobes. The 5 petals are ovate, fused at based, elliptic, white, recurved, and up to 4 mm long. The androecium comprises 5 stamens which are opposite to the petals. The ovary is light green, bottle-shaped, and with an asymmetric stigma. The drupes are globose, grayish, coriaceous, and up to 5 mm in diameter.

Medicinal use: Infections (Myanmar)

Strong broad-spectrum amphiphilic benzoquinone: Rapanone (also known as 2,5-dihydroxy-3-tridecyl-2,5-cyclohexadiene-1,4-dione; LogD 1.5 at pH 7.4; molecular mass = 322.4 g/mol) inhibited the growth of methicillin-resistant *Staphylococcus* aureus, *Escherichia coli* (AG100ATet), *Klebsiella pneumoniae* (KP55), and *Klebsiella pneumoniae* (KP63) with the MIC values of 4, 16, 32, and 16 µg/mL, respectively (Omosa et al., 2016).

Commentaries

i. The plant produces the prenylated benzoic acid derivative myrsinoic acid B (Ito et al., 2008) which is probably antibacterial since a hydrogenated derivative of myrsinoic acid A inhibited the growth of *Bacillus subtilis* (ATTC 23858), *Staphylococcus aureus* (ATCC 6538P), *Pseudomonas aeruginosa* (ATCC 15442), and *Escherichia coli* (ATCC 25922) with MIC values below 20µg/mL (Burger et al., 2015). Myrsinoic acid B might have some levels of antifungal activity since lanceaefolic acid methyl ester isolated from a member of the genus *Piper* L. inhibited the growth of *Candida albicans* with the MIC value of 100 µg/mL (López et al., 2002). Note that the plant produces series of *p*-hydroquinone diglucosides: seguinosides G – J (Xu et al., 2015), which could be examined for their antimicrobial properties.

ii. In India, *Myrsine africana* L. (local name: *Kakhum*) is used to treat ringworms. In Bhutan, *Myrsine semiserrata* Wall. (local name: *Byi tang ga*) is used for microbial infections.
iii. The plant contains rapanone as well as megastigmane glycosides (Otsuka et al., 2001). The alkylated benzoquinone rapanone inhibited angiotensin-converting enzyme with an IC_{50} value of 36 µM (Lund et al., 1998) and as such might have potential for COVID-19 and other betacoronavirus.
iv. Rapanone inhibits oxidative phosphorylation in mitochondria (Makawiti et al., 1990) suggesting that it might inhibit bacterial respiratory chain and as such inhibit efflux pumps.

References

Burger, M.C.D.M., Terezan, A.P., Cunha, G.S.D.O., Fernandes, J.B., Silva, M.F.D.G.F., Vieira, P.C. and Menezes, A.C.S., 2015. Antimicrobial activity of the myrsinoic acid A from Myrsine coriacea and the semi-synthetic derivatives. *Revista Brasileira de Farmacognosia*, *25*(5), pp. 451–454.

Ito, S., Narise, A. and Shimura, S., 2008. Identification of a methioninase inhibitor, myrsinoic acid B, from Myrsine seguinii Lev., and its inhibitory activities. *Bioscience, Biotechnology, and Biochemistry*, *72*(9), pp. 2411–2414.

López, A., Ming, D.S. and Towers, G.N., 2002. Antifungal activity of benzoic acid derivatives from Piper l anceaefolium. *Journal of Natural Products*, *65*(1), pp. 62–64.

Lund, A.K., Adsersen, A. and Nyman, U., 1998. Biological activities of benzoquinones from Badula barthesia and Embelia angustifolia. *Phytomedicine*, *5*(3), pp. 199–203.

Makawiti, D.W., Konji, V.N. and Olowookere, J.O., 1990. Interaction of benzoquinones with mitochondria interferes with oxidative phosphorylation characteristics. *FEBS Letters*, *266*(1–2), pp. 26–28.

Omosa, L.K., Midiwo, J.O., Mbaveng, A.T., Tankeo, S.B., Seukep, J.A., Voukeng, I.K., Dzotam, J.K., Isemeki, J., Derese, S., Omolle, R.A. and Efferth, T., 2016. Antibacterial activities and structure–activity relationships of a panel of 48 compounds from Kenyan plants against multidrug resistant phenotypes. *SpringerPlus*, *5*(1), pp. 1–15.

Otsuka, H., Zhong, X.N., Hirata, E., Shinzato, T. and Takeda, Y., 2001. Myrsinionosides AE: Megastigmane glycosides from the leaves of Myrsine seguinii Lev. *Chemical and Pharmaceutical Bulletin*, *49*(9), pp. 1093–1097.

Xu, W.H., Liang, Q., Zhang, Y.J. and Zhao, P., 2015. Naturally occurring arbutin derivatives and their bioactivities. *Chemistry & Biodiversity*, *12*(1), pp. 54–81.

9.2.6 Family Sapotaceae A.L. de Jussieu (1789)

The family Sapotaceae consists of about 75 genera and 800 species of laticiferous trees, shrubs, or climbers. The leaves are simple, spiral, and exstipulate or not. The petiole is often straight and swollen at apex. The blade is often coriaceous and spathulate. The flowers are solitary or in axillary cymes and actinomorphic. The calyx comprises 4–6 sepals. The corolla is tubular and develops 4–16 lobes. The androecium comprises up to 12 stamens which are often fused at base. The gynoecium includes a 1-to-many-loculed ovary with one or more ovules in each locule, and a straight style. The berries are elongated and contain 1–6 smooth and often glossy seeds.

9.2.6.1 *Madhuca indica* J.F. Gmel.

Synonyms: *Bassia longifolia* J. Koenig ex L.; *Madhuca longifolia* (J. Koenig ex L.) J.F. Macbr.

Common names: Indian butter tree, madhuca; mahua; mahul; mohua; mool (Bangladesh); mahua, gudapushpa, ipa chetu (India)

Habitat: Roadsides, wastelands

Distribution: Pakistan, India, Sri Lanka, Myanmar

Botanical observation: This tree grows up to about 15 m tall. The stems are hairy when young and somewhat reddish. The leaves are simple, spiral, and exstipulate. The blade is 7–25 cm×5–11 cm, hairy when young, broadly elliptic-lanceolate, and with 12–15 pairs of secondary nerves. The calyx is tubular and develops 4 lobes which are ovate, up to 1.5 cm long, and hairy. The corolla is somewhat fleshy, up to 1.7 cm long, and develops 8 lobes which are cream white and ovate-acuminate. The androecium comprises numerous stamens. The berries are ovoid to somewhat bullet-shaped, hairy, up to 3.5 cm long, with persistent style, and seated on persistent calyx.

Medicinal uses: Boils, fever, cough, tuberculosis, small pox, cholera (Bangladesh); dysentery, skin disease, phthisis, tonsilitis (India)

Antibacterial (Gram-positive) halo elicited by polar extract: Methanol extract of flowers (10 mg/disc) inhibited the growth of *Clostridium perfringens* (ATCC 13124) with the inhibition zone diameter of above 20 mm (Park et al., 1998).

Strong antiviral polar extract: Methanol extract of bark inhibited the replication of the human immunodeficiency virus type-1 (NL4.3 strain) virus in human CD4+ T cell line CEM-GFP by 39% with a maximum non-cytotoxic concentration of 15 μg/mL (Sabde et al., 2011).

Commentaries

i. The bark accumulates ellagitannins (Rajak et al., 2017) which are, at least partially, involved in the traditional uses. Other secondary metabolites in this plant are oleanane-type triterpene saponins (Yoshikawa et al., 2000).
ii. In Vietnam, *Madhuca pasquieri* (Dubard) H.J. Lam (local name: *Sê´n*) is used to treat burns. *Madhuca neriifolia* (Moon) H. is used in India (local name: *Attilipa*) for the treatment of bronchitis, cough, and tonsilitis.

References

Park, S.J., Choi, D.H., Cho, H.C., Hiremath, I.G. and Ahn, Y.J., 1998. Growth-inhibitory responses of human intestinal bacteria to extracts from Indian and African plants. *Applied Biological Chemistry*, *41*(1), pp. 104–109.

Rajak, R.C., Singh, A. and Banerjee, R., 2017. Biotransformation of hydrolysable tannin to ellagic acid by tannase from Aspergillus awamori. *Biocatalysis and Biotransformation*, *35*(1), pp. 27–34.

Sabde, S., Bodiwala, H.S., Karmase, A., Deshpande, P.J., Kaur, A., Ahmed, N., Chauthe, S.K., Brahmbhatt, K.G., Phadke, R.U., Mitra, D. and Bhutani, K.K., 2011. Anti-HIV activity of Indian medicinal plants. *Journal of Natural Medicines*, *65*(3–4), pp. 662–669.

Yoshikawa, K., Tanaka, M., Arihara, S., Pal, B.C., Roy, S.K., Matsumura, E. and Katayama, S., 2000. New oleanene triterpenoid saponins from Madhuca l ongifolia. *Journal of Natural Products*, *63*(12), pp. 1679–1681.

9.2.6.2 *Manilkara zapota* (L.) P. Royen

Synonyms: *Achras zapota* L.; *Nispero achras* (Miller) Aubréville; *Pouteria mammosa* (L.) Cronquist

Common names: Chicle tree; naseberry, sapodilla plum; sofeda (Bangladesh); sawo Manila (Indonesia); lamud (Laos); ciku (Malaysia); thagya (Myanmar); vhiko (the Philippines); lamut (Thailand); tam luc (Vietnam)

Habitat: Cultivated

Distribution: Native to Mexico, brought by Spaniards to the Philippines, Tropical Asia and Pacific

Botanical observation: This tree grows to a height of about 15 m and produces a pure white latex. The leaves are simple, spiral, and exstipulate. The petiole is up to about 3 cm long. The blade is coriaceous, elliptic to oblong-elliptic, 6–14 cm×2–5 cm, cuneate at base, and acute to acuminate at apex. The inflorescence is solitary and axillary. Flower pedicel and flower bud are light brown and hairy. The calyx includes 6 ovate sepals, which are up to about 1 cm long and hairy. The 6 petals are pure white and up to about 1 cm long. The androecium comprises 6 stamens. The berries are globose to ovoid, up to 8 cm across, light brown, rough, and contain up to 10 compressed seeds in an edible flesh.

Medicinal uses: Diarrhea (Bangladesh); fever (Myanmar)

Broad-spectrum antibacterial halo elicited by a polar extract: Methanol extract of fruit peel (2 mg/ 8.5 mm agar well) inhibited *Bacillus subtilis* (ATCC 6633), *Staphylococcus aureus* (ATCC 25293), *Staphylococcus epidermidis* (ATCC 12228), *Citrobacter freundii* (NCIM2489), *Enterobacter aerogenes* (ATCC 13048), *Klebsiella pneumoniae* (NCIM2719), and *Salmonella typhimurium* (ATCC 2364) (Rakholiya et al., 2014).

Moderate broad-spectrum antibacterial polar extract: Acetone extract of seeds inhibited *Vibrio cholerae* (MTCC 3906) and *Pseudomonas oleovorans* (MTCC 617) with MIC/MBC of 216.5/216.5 and 323/323 μg/mL, respectively (Kothari and Seshadri, 2010). Methanol of leaves inhibited the growth of *Escherichia coli* (AG102), *Enterococcus aerogenes* (ATCC 13048), *Klebsiella pneumoniae* (ATCC 11296), *Providencia stuartii* (PS2636), *Pseudomonas aeruginosa* (PA01), and methicillin-resistant *Staphylococcus aureus* with the MIC values of 512, 512, 512, 512, 1024, and 256 μg/mL, respectively (Ngongang et al., 2020).

Commentaries

i. The fruits abound with phenolics including ethyl 4-*O*-galloylchlorogenic acid, methyl chlorogenic acid, gallic acid, as well as the flavonoids dihydromyricetin, quercitrin, myricitrin, (+)-catechin, (−)-epicatechin, and (+)-gallocatechin (Ma et al., 2003) which are probably, at least in part, synergistically involved on the medicinal property.
ii. *Manilkara kauki* (L.) Dubard. (also known as *Mimusops kauki* L.) is used in India (local names: *Khirn*, *Manilkara*) to treat diarrhea, fever, eye infection, and leprosy.
iii. Consider that plants accumulating phenolics are often used to treat fever that would classify phenolics under a cold category in the Hippocratic system of humors.

References

Kothari, V. and Seshadri, S., 2010. In vitro antibacterial activity in seed extracts of Manilkara zapota, Anona squamosa, and Tamarindus indica. *Biological Research*, *43*(2), pp. 165–168.

Ma, J., Luo, X.D., Protiva, P., Yang, H., Ma, C., Basile, M.J., Weinstein, I.B. and Kennelly, E.J., 2003. Bioactive novel polyphenols from the fruit of Manilkara zapota (Sapodilla). *Journal of Natural Products*, *66*(7), pp. 983–986.

Ngongang, F.C., Fankam, A.G., Mbaveng, A.T., Wamba, B.E., Nayim, P., Beng, V.P. and Kuete, V., 2020. Methanol extracts from Manilkara zapota with moderate antibacterial activity displayed strong antibiotic-modulating effects against multidrug-resistant phenotypes. *Pharmacology*, *3*(1), p. 37.

Rakholiya, K., Kaneria, M. and Chanda, S., 2014. Inhibition of microbial pathogens using fruit and vegetable peel extracts. *International Journal of Food Sciences and Nutrition*, *65*(6), pp. 733–739.

9.2.6.3 *Mimusops elengi* L.

Synonym: *Kaukenia elengi* (L.) Kuntze

Common names: Bullet wood, elengee, spanish cherry; bakul, bohur (Bangladesh); tanjung (Indonesia); mahila, maghidam, sthirapushpa (India); buah tanjung (Indonesia); bitis, mengkular (Malaysia); khayay pin (Myanmar); bansalagon, betis (the Philippines); kun (Thailand)

Habitat: Cultivated

Distribution: From Pakistan to Hawaii

Botanical observation: This handsome and laticiferous tree grows up to about 10 m tall. It has a positive aura of life. The leaves are simple, spiral, and stipulate. The stipules are minute and caducous. The petiole is about 3 cm long. The blade is coriaceous, glossy, 3–5 cm×7–12.5 cm, acute at base, shortly acuminate at apex, glossy, and marked with numerous faint pairs of secondary nerves. The flowers are slightly fragrant, axillary, solitary on a 2 cm long peduncle or in small fascicles. The calyx comprises 8 lobes. The corolla is white, ephemeral, 8-lobed, the lobes deeply incised. The androecium includes 8 stamens joined in a cone. The ovary is minute. The fruits are ovoid, bullet-shaped, dull red-orange, about 3 cm long, and contain a few tear-shaped glossy brown seeds.

Medicinal uses: Cough, cuts, gingivitis, gonorrhea, trush (Myanmar); dysentery, fever (India); skin infection, syphilis, tuberculosis (Bangladesh); tooth infections (Bangladesh; India, Myanmar); fever, dysentery, wounds (India; Myanmar); diarrhea (India; Myanmar)

Viral enzyme inhibition by polar extract: Ethanol extract of leaves inhibited human immunodifficiency virus type-1 integrase activity with IC_{50} value of 62.1 μg/mL (Suramin IC_{50} value of 3.4 μg/mL) (Chen et al., 2005).

Viral enzyme inhibition by flavanols: From this extract, a mixture of gallocatechin and epigallocatechin inhibited human immunodifficiency virus type-1 integrase with IC_{50} value of 35 μM (Suedee et al., 2014). Epigallocatechin inhibited the severe acute respiratory syndrome-associated coronavirus 3C-like protease with the IC_{50} values of 100 μM (Chen et al., 2005).

Commentaries

i. The astringency of the plant is a suggestion that it could produce tannins or phenolic compounds, and polar extracts of such plants almost always inhibit enzyme activity *in vitro.*
ii. Flowers produce an essential oil (Wong & Tenf, 1994).
iii. Use of the plant in several Asian countries point to the presence of antimicrobial compounds yet to be examined fully.
iv. The saponin arganine C isolated from a member of the genus *Tieghemella* Pierre abrogated the entry of the human immunodeficiency virus in host cells entry at the concentration of 20 μM (Gosse et al., 2002). One can observe that triterpene saponins often viral entry in host cells.

References

Chen, C.N., Lin, C.P., Huang, K.K., Chen, W.C., Hsieh, H.P., Liang, P.H. and Hsu, J.T.A., 2005. Inhibition of SARS-CoV 3C-like protease activity by theaflavin-3, 3'-digallate (TF3). *Evidence-Based Complementary and Alternative Medicine*, *2*(2), pp. 209–215.

Gosse, B., Gnabre, J., Bates, R.B., Dicus, C.W., Nakkiew, P. and Huang, R.C.C., 2002. Antiviral Saponins from Tieghemella h eckelii. *Journal of Natural Products*, *65*(12), pp. 1942–1944.

Suedee, A., Tewtrakul, S. and Panichayupakaranant, P., 2014. Anti-HIV-1 integrase activity of Mimusops elengi leaf extracts. *Pharmaceutical Biology*, *52*(1), pp. 58–61.

Wong, K.C. and Teng, Y.E., 1994. Volatile components of Mimusops elengi L. flowers. *Journal of Essential Oil Research*, *6*(5), pp. 453–458.

9.2.7 Family Theaceae Mirbel (1816)

The family Theaceae consists of about 19 genera and 600 species of handsome shrubs or trees. The leaves are simple, spiral, and exstipulate. The blade is not uncommonly glossy, dark green, serrate, and coriaceous. The inflorescence is a raceme or solitary. The flowers are actinomorphic. The calyx presents 5 sepals. The 5 petals are free. The androecium comprises numerous showy stamens. The gynoecium is made up of 3–5 carpels which are free and shelter 2–5 ovules on axile placentas. The fruits are drupes or capsules.

9.2.7.1 *Camellia japonica* L.

Synonym: *Thea japonica* (L.) Baill.

Common names: Camellia, tea rose; shan cha (China); tsubaki (Japan)

Habitat: Forests

Distribution: China, Korea, Japan

Botanical observation: This magnificent tree grows up to 6 m tall. The stems are terete. The leaves are simple, spiral, and exstipulate. The petiole is up to about 1 cm long. The blade is elliptic, 5–10 cm×2.5–6 cm, smooth, coriaceous, cuneate at base, glossy, serrate, with up to about 9 pairs of secondary nerves, and acuminate at apex. The flowers are axillary, solitary, cauliflorous, and showy. The calyx comprises 9 sepals which are up to about 2 cm long. The 7 petals are pink, magnificent, obovate, and up to about 4.5 cm long. The androecium comprises numerous stamens which are up to about 3.5 cm long. The ovary develops a 3-lobed stigma. The capsules are globose, glossy, reddish, dehiscent, up to about 4.5 cm in diameter, and contain few 2 cm long seeds.

Medicinal use: Hair and skin care (Japan; Korea)

Moderate broad-spectrum polar extract: Aqueous extract inhibited the growth of *Bacillus subtilis* (ATCC 6633), *Clostridium perfringens* (ATCC 13124), *Staphylococcus aureus* (ATCC 29213), *Vibrio parahaemolyticus* (ATCC 17802), *Vibrio fulvinicus* (ATCC 27562), *Escherichia coli* (ATCC 25922), *Yersinia enterocolitica* (ATCC 9610), *Pseudomonas aeruginosa* (ATCC 27853), and *Klebsiella pneumoniae* (ATCC 13883) (Taguri et al., 2006).

Antifungal oleanane-type triterpene saponin: Camellidin I and II inhibited the growth of *Gloeosporium theae-sinensis*, *Pestalotia longiseta*, *Botrytis cinerea*, *Pyricularia oryzae*, and *Cochliobolus miyabeanus* (Nagata et al., 1985; Hamaya et al., 1984).

Strong antiviral (enveloped monopartite linear single-stranded (+)-RNA) lipophilic oleanane-triterpenes: Primulagenin A (also known as (3β,16α)-olean-12-en-3,16,28-triol; LogD 7.2 at pH 7.4; molecular mass=458.7 g/mol), 3β-hydroxy-28-noroleana-12,17-dien-16-one, and schimperinone isolated from the flowers inhibited the growth of porcine epidemic diarrhea virus with IC_{50} values of 0.06, 0.2, and 0.06 µM, respectively, and selectivity indexes of 12.9, 32.7, and 44.5, respectively (Yang et al., 2015).

Viral enzyme inhibition by ellagitannins: Camelliatannin A, F, and H from the pericarps inhibited the human immunodeficiency virus type-1 protease with IC_{50} of 77, 67, and 0.9 µM, respectively (Park et al., 2002).

Commentaries

i. Tannins inhibit a broad spectrum of enzymes, and because they are decomposed in the guts, they can't, per se, be of any use for systemic treatment of bacterial, fungal, or viral infection. Such compounds can be however of interest as antiseptic or preservatives.
ii. Porcine epidemic diarrhea virus belongs to the Coronaviridae, genus *Alphacoronavirus*, thus primulagenin A might have some activity against the severe acute respiratory syndrome-associated coronavirus 2 and could be added to the list of natural products to test for the coming zoonotic coronavirus pandemics or vaccine-resistant strains of COVID-19. What is the antiviral mode of action of oleanane-type triterpenes? 3β-*O*-[2, 4-di-*O*-(α-L-rhamnopyranosyl)-β-D-glucopyranosyl]-olean-12-en-28-oic acid benzyl ester inhibited the replication of severe acute respiratory syndrome-associated coronavirus 2 with an IC_{50} value of 16.7 µM (Li et al., 2021). Series of oleanane-type triterpenes inhibited the replication of severe acute respiratory syndrome-associated coronavirus 2 in 293T-ACE2 cells via, at least in part, inhibition of membrane fusion of mediated by virion S protein (Li et al., 2021). Glycyrrhizic acid is a known to block the interaction between S protein and the angiotensin-converting enzyme type 2 (Yu et al., 2021). Won Keun Oh et al. isolated a series of oleanane-type triterpenes with strong activity against the porcine epidemic diarrhea virus (Yang et al., 2016).

References

Hamaya, E., Tsushida, T., Nagata, T., Nishino, C., Enoki, N. and Manabe, S., 1984. Antifungal components of Camellia plants. *Japanese Journal of Phytopathology*, *50*(5), pp. 628–636.

Li, C., Li, Z., Zou, Y., Wicht, O., van Kuppeveld, F.J., Rottier, P.J. and Bosch, B.J., 2013. Manipulation of the porcine epidemic diarrhea virus genome using targeted RNA recombination. *PLoS One*, *8*(8), p. e69997.

Li, H., Cheng, C., Li, S., Wu, Y., Liu, Z., Liu, M., Chen, J., Zhong, Q., Zhang, X., Liu, S. and Song, G., 2021. Discovery and structural optimization of 3-O-β-chacotriosyl oleanane-type triterpenoids as potent entry inhibitors of SARS-CoV-2 virus infections. *European Journal of Medicinal Chemistry*, *215*, p. 113242.

Nagata, T., Tsushida, T., Hamaya, E., Enoki, N., Manabe, S. and Nishino, C., 1985. Camellidins, antifungal saponins isolated from Camellia japonica. *Agricultural and Biological Chemistry*, *49*(4), pp. 1181–1186.

Park, J.C., Hur, J.M., Park, J.G., Hatano, T., Yoshida, T., Miyashiro, H., Min, B.S. and Hattori, M., 2002. Inhibitory effects of Korean medicinal plants and camelliatannin H from Camellia japonica on human immunodeficiency virus type 1 protease. *Phytotherapy Research*, *16*(5), pp. 422–426.

Taguri, T., Tanaka, T. and Kouno, I., 2006. Antibacterial spectrum of plant polyphenols and extracts depending upon hydroxyphenyl structure. *Biological and Pharmaceutical Bulletin*, *29*(11), pp. 2226–2235.

Yang, J.L., Ha, T.K.Q., Dhodary, B., Pyo, E., Nguyen, N.H., Cho, H., Kim, E. and Oh, W.K., 2015. Oleanane triterpenes from the flowers of Camellia japonica inhibit porcine epidemic diarrhea virus (PEDV) replication. *Journal of Medicinal Chemistry*, *58*(3), pp. 1268–1280.

Yang, J.L., Ha, T.K.Q. and Oh, W.K., 2016. Discovery of inhibitory materials against PEDV corona virus from medicinal plants. *Japanese Journal of Veterinary Research*, *64*(Supplement 1), pp. S53–S63.

Yu, S., Zhu, Y., Xu, J., Yao, G., Zhang, P., Wang, M., Zhao, Y., Lin, G., Chen, H., Chen, L. and Zhang, J., 2021. Glycyrrhizic acid exerts inhibitory activity against the spike protein of SARS-CoV-2. *Phytomedicine*, *85*, p. 153364.

9.2.7.2 *Camellia oleifera* Abel

Synonyms: *Thea biflora* Hayata; *Thea oleifera* (Abel) Rehder & E.H. Wilson; *Thea podogyna* H. Lév.

Common names: Tea-oil camellia; shan you (China)

Habitat: Forests, thickets, cultivated

Distribution: Myanmar; Laos, Vietnam, China

Botanical observation: This gorgeous tree grows up to 5 m tall. The stems are terete and somewhat hairy at apex. The leaves are simple, spiral, and exstipulate. The petiole is up to about 1 cm long. The blade is elliptic, 3–10 cm×2–4 cm, coriaceous, cuneate at base, glossy, dark green serrate, with up to about 8 pairs of secondary nerves, and acute to acuminate at apex. The flowers are solitary, axillary, terminal, and showy. The 5–7 sepals are up to about 1.2 cm long and hairy. The 5–7 petals are pure white, membranous, oblong, up to about 2 cm long, and bifid at apex or not. The androecium is showy and includes numerous stamens growing up to 1.5 cm long and fused at base. The gynoecium is globose, hairy, comprises a 3-locular ovary developing a slender style, and a 3-lobed stigma. The capsules are globose, brownish, dehiscent into 3 valves (Euphorbiaceous-like), up to about 4 cm in diameter, and contain 6 smooth, dull-brown, oily seeds which are about 2 cm long.

Medicinal uses: Dandruff (China)

Broad-spectrum antibacterial halo developed by fixed oil: Oil of seeds (100 μL/disc) inhibited the growth of *Escherichia coli* and *Bacillus subtilis* (Zhou et al., 2014a).

Strong antibacterial saponin fraction: Saponin fraction of seeds inhibited the growth of *Staphylococcus aureus*, *Escherichia coli*, and *Bacillus subtills* with the MIC of 31.3, 31.3, and 62.5 μg/mL, respectively (Hu et al., 2012). This fraction demonstrated MBC of 62.5, 62.5, and 125 μg/mL for *Staphylococcus aureus*, *Escherichia coli*, and *Bacillus subtilis*, respectively (Hu et al., 2012).

Broad-spectrum antifungal halo developed by fixed oil: Oil of seeds (100 μL/disc) inhibited the growth of *Aspergillus niger* and *Saccharomyces cerevisiae* (Zhou et al., 2014b).

Commentary: The fruits abound with oleanane-type saponins (Zhou et al., 2014a) which may have some antimicrobial properties. Note that saponins are not or poorly absorbed orally and as such are not theoretically of possible use as a source of systemic antimicrobial agents. Saponins are however good candidates as antiseptics or disinfectants if not toxic.

References

Hu, J.L., Nie, S.P., Huang, D.F., Li, C., Xie, M.Y. and Wan, Y., 2012. Antimicrobial activity of saponin-rich fraction from Camellia oleifera cake and its effect on cell viability of mouse macrophage RAW 264.7. *Journal of the Science of Food and Agriculture*, *92*(12), pp. 2443–2449.

Zhou, H., Wang, C.Z., Ye, J.Z. and Chen, H.X., 2014a. New triterpene saponins from the seed cake of Camellia oleifera and their cytotoxic activity. *Phytochemistry Letters*, *8*, pp. 46–51.

Zhou, Q.F., Jia, X.J., Li, Q.Q., Yang, R.W., Zhang, L., Zhou, Y.H. and Ding, C.B., 2014b. Antioxidant and antimicrobial activities of camellia oleifera seed oils. *Journal of Applied Biological Chemistry*, *57*(2), pp. 123–129.

9.2.7.3 *Camellia sinensis* (L.) Kuntze

Synonyms: *Camellia thea* Link; *Thea assamica* J.W. Mast.; *Thea sinensis* L.

Common names: Tea; tae (Cambodia); cha (Bangladesh); cha (China); chai, syamaparni, thayilail (India); teh (Indonesia, Malaysia); chiya (Nepal); chay (Pakistan); miang, tsa (the Philippines); cay che (Vietnam)

Habitat: Forests, thickets

Distribution: Pakistan, India, Bangladesh, Myanmar, Thailand, Vietnam, China, Korea, Japan

Botanical observation: This handsome tree grows up to 5 m tall and has a positive aura. The stems are terete. The leaves are simple, spiral, and exstipulate and the source of the healthiest drink on Earth after water: tea. The petiole is up to about 7 mm long. The blade is elliptic, 5–13.4 cm×2.5–7.5 cm, slightly fragrant, smooth, coriaceous, cuneate at base, glossy, serrate, marked with up to about 9 pairs of secondary nerves, and acute at apex. The flowers are axillary, in clusters or solitary, and showy. The 5 sepals are up to about 5 mm long. The corolla comprises up to 8 petals which are pure white, membranous, round, up to about 2 cm long and broadly ovate. The androecium comprises numerous showy stamens which are up to about 1 cm long. The ovary develops slender style. The capsules are somewhat 3-lobed, up to about 3 cm long, and contain a few globose seeds which are about 1.5 cm in diameter.

Medicinal uses: Conjunctivitis, wounds (Bangladesh); diarrhea, dysentery (China); fever (Pakistan)

Antibacterial (Gram-negative) halo elicited by flavans: Epigallocatechin-3-gallate (100 μL of a of 2.5 mg/mL solution/well) inhibited the growth of highly multidrug-resistant *Acinetobacter baumannii* with the inhibition zone diameter of 7 mm (Osterburg et al., 2009). Epigallocatechin-3-gallate (200 μg/6 mm disc) inhibited the growth of *Stenotrophomonas maltophilia* (ATCC 10258) with the inhibition zone of 25 mm (Gordon & Wareham, 2010).

Weak broad-spectrum antibacterial polar extract: Methanol extract of leaves inhibited methicillin-resistant *Staphylococcus aureus* (clinical), *Escherichia coli* (clinical), *Salmonella enterica* serovar *typhi* (multidrug-resistant, clinical), *Klebsiella pneumoniae* (multidrug-resistant; clinical), *Citrobacter freundii* (multidrug-resistant; clinical), *Enterobacter cloacae* (multidrug-resistant; clinical), *Bacillus subtilis*, *Streptococcus pneumoniae* (clinical), *Micrococcus* sp., *Salmonella paratyphi* A, (Clinical) *Campylobacter jejuni* (Clinical), *Staphylococcus aureus* (ATCC 25923), *Salmonella typhi* (ATCC 13311), and *Shigella flexneri* (ATCC 9199) (Farooqui et al., 2015).

Strong broad-spectrum antibacterial phenolic fraction: Phenolic fraction inhibited the growth of *Staphylococcus aureus*, *Bacillus subtilis*, *Enterococcus faecalis*, *Listeria monocytogenes*, *Escherichia coli*, *Pseudomonas aeruginosa*, *Salmonella typhi*, and *Shigella dysenteriae* with MIC of 31.2, 15.6, 15.6, 31.2, 31.2, 62.5, 62.5, and 62.5 μg/mL, respectively (Aman et al., 2014). Polyphenols demonstrated MIC ranging from 50 to 180 μg/mL towards strains of *Staphylococcus aureus* from patients treated with oxacillin (Cho et al., 2008).

Strong antibacterial (Gram-positive) amphiphilic flavan: Epigallocatechin gallate (LogD 1.8 at pH 7.4; molecular mass=458.3 g/mol) inhibited the growth of methicillin-resistant *Staphylococcus aureus* (clinical strain) with the MIC of 32 μg/mL (Hatano et al., 2003). Epigallocatechin gallate inhibited the growth of methicillin-sensitive *Staphylococcus aureus* (209P) with MIC of 32 μg/mL (Hatano et al., 2003).

Strong antibacterial (Gram-positive) antibacterial amphiphilic proanthocyanidins: Theasinensin A and theasinensin D inhibited the growth of methicillin-resistant *Staphylococcus aureus* (clinical strain) with MIC of 64 and 64 μg/mL, respectively (Hatano et al., 2003). Theasinensin A (LogD 2.8 at pH 7.4; molecular mass=914.7 g/mol) and D inhibited the growth of methicillin-sensitive *Staphylococcus aureus* (209P) with MIC of 32 μg/mL and 64 μg/mL, respectively (Hatano et al., 2003).

Moderate antibacterial (Gram-positive) simple phenolic: Gallic acid inhibited the growth of methicillin-sensitive *Staphylococcus aureus* (209P) with the MIC of 64 μg/mL, respectively (Hatano et al., 2003).

Weak antibacterial (Gram-positive) hydrolysable tannin: Strictinin inhibited the growth of *Propionibacterium acnes* with MIC of 250 μM (Hsieh et al., 2016). This ellagitannin was synergistic with gentamicin (FICI: 0.375) and cefmetazole (FICI: 0.5) (Hsieh et al., 2016).

Weak antibacterial (Gram-positive) flavan: Epigallocatechin gallate inhibited the growth of *Acinetobacter baumannii* with an MIC of about 625 μg/mL (Osterburg et al., 2009). At 2×MIC, this flavonoid evoked a 3-log reduction in CFU/mL after 5 hours evidencing bactericidal effect (Osterburg et al., 2009).

Antibiotic potentiator polar extract: Aqueous extract of leaves (5 g in 100 mL boiling water) combined ampicillin with decreased MIC of ampicillin from 0.2 to 0.06 μg/mL towards methicillin-sensitive *Staphylococcus aureus* (ATCC 25923) (Peng et al., 2010). Combined with cefazolin it decreased MIC of cefazolin from 1 to 0.2 μg/mL towards methicillin-sensitive *Staphylococcus aureus* (ATCC 25923) (Peng et al., 2010). Combined with amoxicillin it decreased MIC of amoxicillin from 0.5 to 0.1 towards methicillin-sensitive *Staphylococcus aureus* (ATCC 25923) (Peng et al., 2010). Combined with oxacillin it decreased MIC of oxacillin from 2 to 0.2 μg/mL towards methicillin-sensitive *Staphylococcus aureus* (ATCC 25923) (Peng et al., 2010). Methanol extract of leaves was synergistic with nalidixic acid with FICI of 0.3 for *Salmonella enterica* serovar *typhi* and synergistic with oxacillin with FICI of 0.14 for methicillin-resistant *Staphylococcus aureus* (Farooqui et al., 2015).

Antibiotic potentiator phenolics: Polyphenols at≤0.5 ×MIC were synergistic with oxacillin for methicillin-resistant *Staphylococcus aureus* strains (Cho et al., 2008). At 0.2×MIC of oxacillin+0.5×MIC of polyphenols, there was a fall of 3 logs CFU/mL at 8 hours (Cho et al., 2008). These polyphenols downregulated the expression of protein (Cho et al., 2008).

Antibiotic potentiator proanthocyanidin: Upon the addition of theasinensin A [3.5×10^{-5} M (32 μg/mL) half of the MIC of theasinensin A], the MICs of oxacillin decreased to 4 μg/mL for methicillin-resistant *Staphylococcus aureus* (Hatano et al., 2003).

Antiviral (enveloped segmented linear single-stranded (−) RNA virus) polar extracts: Aqueous decoction (2 g/100 mL) evoked after 10 mins pretreatment a reduction in Log10 titer of the Influenza virus A H5N1 virus (Chicken/yamaguchi/7/04) in Madin-Darby canine kidney cells (Baatartsogt et al., 2016). Aqueous decoction (2 g/100 mL) evoked after 10 minutes treatment a reduction in Log10 titer of Influenza A Avian/Japan/9UO0036/09, Avian/Japan/11OG1083/11, chicken/Ibaraki/8/05, Whistlingswan/Shimane/499/83, and Duck/Hong Kong/820/80 by 2.3, ≥3.7, ≥4.5, 1.5, and ≥2.0 in Madin-Darby canine kidney cells (Baatartsogt et al., 2016).

Strong antiviral (enveloped segmented linear single-stranded (−) RNA virus) amphiphilic flavan: Epigallocatechin gallate inhibited the replication of the Influenza A virus (Oh7 strain) with EC_{50} of 8 μM (Kim et al., 2010).

Strong (enveloped double stranded circular DNA) amphiphilic flavan: Epigallocatechin-3-gallate at 0.4 μM inhibited the secretion of the Hepatitis B surface antigen and Hepatitis B e-antigen by 76% and 78%, respectively, by HepG2.2.15 cells and inhibited extracellular Hepatitis B DNA by 52% (Pang et al., 2014).

Strong antiviral (enveloped segmented linear single-stranded (−) RNA virus) hydrophilic polyphenols: Theaflavin-3,3'-digallate (LogD 0.3 at pH 7.4; molecular mass=868.7 g/mol) inhibited

A/PR/8/34, H1N1 with IC_{50} of 0.7 µg/mL in Madin-Darby canine kidney cells (Zu et al., 2012). Theaflavin-3'-gallate (molecular mass=704.6 g/mol) inhibited A/PR/8/34, H1N1 with IC_{50} of 1.7 µg/mL in Madin-Darby canine kidney cells (Zu et al., 2012). In this experiment, ribavirin and oseltamivir inhibited A/PR/8/34, H1N1 with IC_{50} values of 5 and 1.8 µg/mL, respectively (Zu et al., 2012).

Strong antiviral (non-enveloped segmented linear double-stranded RNA) hydrophilic polyphenol: Theaflavin-3'-gallate inhibited the replication of the Bovine rotavirus (NCDV-Lincoln strain) with the IC_{50} value of 5.5 µg/mL (Clark et al., 1998).

Strong antiviral (enveloped monopartite linear double-stranded DNA) polyphenolics: A fraction of theaflavins at 1.4 µM decreased Herpes simplex virus type-1 viral titer from 8.0×10^6 to 6.0×10^4 plaque-forming unit/mL in Vero cells after 1 h (Cantatore et al., 2013). This extract prevented virus attachment and penetration in host cells (Cantatore et al., 2013). Theaflavin at 50 µM caused a 77.2% reduction in Herpes simplex virus type-1 plaque-forming units by 91% in A549 cells (de Oliveira et al., 2015). Theaflavin monogallate at 50 µM caused a 89% reduction in Herpes simplex virus type-1 plaque-forming units in Vero cells by 99.3% in A549 cells (de Oliveira et al., 2015). Theaflavin-3,3-digallate at 50 µM caused a 93.5% reduction in Herpes simplex virus type-1 plaque-forming units in Vero cells and more than 99% in A549 cells (de Oliveira et al., 2015). Theaflavin-3,3-digallate at 75 µM inhibited the binding of Herpes simplex virus type-1 in Vero cells (de Oliveira et al., 2015).

Moderate antiviral (enveloped double-stranded circular DNA) polyphenol: Theaflavins reduced the secretion of Hepatitis B e-antigen with EC_{50} of 70.3 µg/mL, from Hep2 2.2.15 (Pei et al., 2011).

Viral enzyme inhibition by polar extract: Aqueous extract of green tea and black tea inhibited severe acute respiratory syndrome-associated coronavirus 3C-like protease with the IC_{50} values of 125 and 70 µg/mL, respectively (Chen et al., 2005).

Viral enzyme inhibition by polyphenols: From black tea, theaflavin, theaflavin-3-gallate, and theaflavin-3,3′-digallate inhibited the severe acute respiratory syndrome-associated coronavirus 3C-like protease with the IC_{50} values of 56, 43, and 9.5 µM (Chen et al., 2005). Theaflavin-3,3′-digallate from black tea inhibited neuraminidase of the Influenza virus A/PR/8/34(H1N1), A/Sydney/5/97 (H3N2) and B/Jiangsu/10/2003 with IC_{50} values of 26.2, 10.6, and 42 µM, respectively (Zu et al., 2012). Theaflavin-3-gallate inhibited hemagglutinin A of A/PR/8/34(H1N1), A/Sydney/5/97 (H3N2), and B/Jiangsu/10/2003 with minimum effective concentrations of 4.69, 3.125, and 25 µg/mL, respectively (Zu et al., 2012). Theaflavin-3′-gallate inhibited hemagglutinin A of A/PR/8/34(H1N1), A/Sydney/5/97 (H3N2), and B/Jiangsu/10/2003 with minimum effective concentrations of 13.8, 8.33, and >50 µg/mL, respectively (Zu et al., 2012). Theaflavin-3,3-digallate inhibited hemagglutinin A of A/PR/8/34(H1N1), A/Sydney/5/97 (H3N2), and B/Jiangsu/10/2003 with minimum effective concentrations of 13.15, >135.42, and>50 µg/mL, respectively (Zu et al., 2012).

Viral enzyme inhibition by flavans: From green tea, catechin, epigallocatechin, epigallocatechin gallate, epicatechin, and epicatechin gallate inhibited severe acute respiratory syndrome-associated coronavirus 3C-like protease with the IC_{50} values of 100 µM (Chen et al., 2005). In a subsequent study, epigallocatechin gallate and gallocatechin gallate inhibited the severe acute respiratory syndrome-associated coronavirus 3C-like protease with the IC_{50} value of 73 and 47 µM, respectively (Nguyen et al., 2012).

Viral enzyme inhibition by xanthine alkaloids: Caffeine and theophylline inhibited severe acute respiratory syndrome-associated coronavirus 3C-like protease with the IC_{50} values of 100 µM (Chen et al., 2005).

Commentaries

i. Epigallocatechin gallate inhibits mammalian topoisomerase (Bandele & Osheroff, 2008) and might inhibit bacterial topoisomerases. The antibacterial, antifungal, and antiviral activities of phenolics *in vitro* depend on their concentrations. High concentration of phenolics coagulates proteins.
ii. Theaflavins are formed during the fermentation process of tea leaves and are absent in fresh fresh leaves (green tea prepared by heating fresh leaves prevents fermentation).

This illustrates the fact that the drying process of plant material is of critical importance when looking for antimicrobial compounds. Fresh plant samples often shelter compounds attached to sugars (glycosides) to probably facilitate their transport in aqueous plant tissues. Plugging plant samples and boiling them in water immediately, a method often used in traditional medicine or even eating fresh plant part, will result in absorption of glycosides in the guts. Should these glycosides be antimicrobial, they cannot be isolated by phytochemists working on dried plant samples. When leaves or bark are removed, plant enzymes are activated to liberate aglycones.

iii. In Nepal, the plant is used to prevent COVID-19 and one could ask if drinking black tea could afford some protection against COVID-19. A study by Ohgitani et al. (2021) demonstrates that drinking black tea affords some levels of protection against the severe acute respiratory-associated coronavirus 2 and other zoonotic coronaviruses.

iv. Black tea (without milk) could be of value against Influenza (Iwata et al., 1997). According to Harrington et al. (2021), there have been 4 Influenza pandemics since the beginning of the 20th century: H1N1 in 1918, H2N2 in 1957, H3N2 in 1968, and H1N1 in 2009. The H7N9 appeared in China in 2013 with a concerning mortality rate of about 30% (Qiu et al., 2018; Harrington et al., 2021). Is the mortality rate of pandemic zoonotic viruses increasing?

References

Aman, S., Naim, A., Siddiqi, R. and Naz, S., 2014. Antimicrobial polyphenols from small tropical fruits, tea and spice oilseeds. *Revista de Agaroquimica y Tecnologia de Alimentos*, *20*(4), pp. 241–251.

Bandele, O.J. and Osheroff, N., 2008. (–)-Epigallocatechin gallate, a major constituent of green tea, poisons human type II topoisomerases. *Chemical Research in Toxicology*, *21*(4), pp. 936–943.

Baatartsogt, T., Bui, V.N., Trinh, D.Q., Yamaguchi, E., Gronsang, D., Thampaisarn, R., Ogawa, H. and Imai, K., 2016. High antiviral effects of hibiscus tea extract on the H5 subtypes of low and highly pathogenic avian influenza viruses. *Journal of Veterinary Medical Science*, *78*(9), pp. 1405–1411.

Cantatore, A., Randall, S.D., Traum, D. and Adams, S.D., 2013. Effect of black tea extract on herpes simplex virus-1 infection of cultured cells. *BMC Complementary and Alternative Medicine*, *13*(1), p. 139.

Chen, C.N., Lin, C.P., Huang, K.K., Chen, W.C., Hsieh, H.P., Liang, P.H. and Hsu, J.T.A., 2005. Inhibition of SARS-CoV 3C-like protease activity by theaflavin-3, 3'-digallate (TF3). *Evidence-Based Complementary and Alternative Medicine*, *2*(2), pp. 209–215.

Cho, Y.S., Schiller, N.L. and Oh, K.H., 2008. Antibacterial effects of green tea polyphenols on clinical isolates of methicillin-resistant Staphylococcus aureus. *Current Microbiology*, *57*(6), pp. 542–546.

Clark, K.J., Grant, P.G., Sarr, A.B., Belakere, J.R., Swaggerty, C.L., Phillips, T.D. and Woode, G.N., 1998. An in vitro study of theaflavins extracted from black tea to neutralize bovine rotavirus and bovine coronavirus infections. *Veterinary Microbiology*, *63*(2–4), pp. 147–157.

de Oliveira, A., Prince, D., Lo, C.Y., Lee, L.H. and Chu, T.C., 2015. Antiviral activity of theaflavin digallate against herpes simplex virus type 1. *Antiviral Research*, *118*, pp. 56–67.

Farooqui, A., Khan, A., Borghetto, I., Kazmi, S.U., Rubino, S. and Paglietti, B., 2015. Synergistic antimicrobial activity of Camellia sinensis and Juglans regia against multidrug-resistant bacteria. PloS one, 10(2), p. e0118431.

Fu, L., Lu, W. and Zhou, X., 2016. Phenolic compounds and in vitro antibacterial and antioxidant activities of three tropic fruits: Persimmon, guava, and sweetsop. *BioMed Research International*, *2016*. doi: 10.1155/2016/4287461.

Gordon, N.C. and Wareham, D.W., 2010. Antimicrobial activity of the green tea polyphenol (–)-epigallocatechin-3-gallate (EGCG) against clinical isolates of Stenotrophomonas maltophilia. *International Journal of Antimicrobial Agents*, *36*(2), pp. 129–131.

Harrington, W.N., Kackos, C.M. and Webby, R.J., 2021. The evolution and future of influenza pandemic preparedness. *Experimental & Molecular Medicine*, *53*, pp. 1–13.

Hatano, T., Kusuda, M., Hori, M., Shiota, S., Tsuchiya, T. and Yoshida, T., 2003. Theasinensin A, a tea polyphenol formed from Epigallocatechin gallate, suppresses antibiotic resistance of methicillin-resistant Staphylococcus aureus. *Planta Medica*, *69*(11), pp. 984–989.

Hsieh, S.K., Xu, J.R., Lin, N.H., Li, Y.C., Chen, G.H., Kuo, P.C., Chen, W.Y. and Tzen, J.T., 2016. Antibacterial and laxative activities of strictinin isolated from Pu'er tea (Camellia sinensis). *Journal of Food and Drug Analysis*, *24*(4), pp. 722–729.

Iwata, M., Toda, M., Nakayama, M., Tsujiyama, H., Endo, W., Takahashi, O., Hara, Y. and Shimamura, T., 1997. Prophylactic effect of black tea extract as gargle against influenza. *Kansenshogaku zasshi. The Journal of the Japanese Association for Infectious Diseases*, *71*(6), pp. 487–494.

Kim, Y., Narayanan, S. and Chang, K.O., 2010. Inhibition of influenza virus replication by plant-derived isoquercetin. *Antiviral Research*, *88*(2), pp. 227–235.

Nguyen, T.T.H., Woo, H.J., Kang, H.K., Kim, Y.M., Kim, D.W., Ahn, S.A., Xia, Y. and Kim, D., 2012. Flavonoid-mediated inhibition of SARS coronavirus 3C-like protease expressed in Pichia pastoris. *Biotechnology Letters*, *34*(5), pp. 831–838.

Ohgitani, E., Shin-Ya, M., Ichitani, M., Kobayashi, M., Takihara, T., Kawamoto, M., Kinugasa, H. and Mazda, O., 2021. Rapid inactivation in vitro of SARS-CoV-2 in saliva by black tea and green tea. *Pathogens*, *10*(6), p. 721.

Osterburg, A., Gardner, J., Hyon, S.H., Neely, A. and Babcock, G., 2009. Highly antibiotic-resistant Acinetobacter baumannii clinical isolates are killed by the green tea polyphenol (–)-epigallocatechin-3-gallate (EGCG). *Clinical Microbiology and Infection*, *15*(4), pp. 341–346.

Pang, J.Y., Zhao, K.J., Wang, J.B., Ma, Z.J. and Xiao, X.H., 2014. Green tea polyphenol, epigallocatechin-3-gallate, possesses the antiviral activity necessary to fight against the hepatitis B virus replication in vitro. *Journal of Zhejiang University Science B*, *15*(6), pp. 533–539.

Pei, S., Zhang, Y., Xu, H., Chen, X. and Chen, S., 2011. Inhibition of the replication of hepatitis B virus in vitro by pu-erh tea extracts. *Journal of Agricultural and Food Chemistry*, *59*(18), pp. 9927–9934.

Peng, Q., Huang, Y., Hou, B., Hua, D., Yao, F. and Qian, Y., 2010. Green tea extract weakens the antibacterial effect of amoxicillin in methicillin-resistant Staphylococcus aureus infected mice. *Phytotherapy Research*, *24*(1), pp. 141–145.

Qiu, W., Chu, C., Mao, A. and Wu, J., 2018. The impacts on health, society, and economy of SARS and H7N9 outbreaks in China: A case comparison study. *Journal of Environmental and Public Health*, *2018*, pp. 1–7.

Zu, M., Yang, F., Zhou, W., Liu, A., Du, G. and Zheng, L., 2012. *In vitro* anti-influenza virus and anti-inflammatory activities of theaflavin derivatives. *Antiviral Research*, *94*(3), pp. 217–224.

9.2.7.4 *Eurya acuminata* DC.

Synonyms: *Eurya arisanensis* Hayata; *Eurya euprista* Korth.; *Eurya multiflora* DC.; *Eurya phyllanthoides* Bl.; *Eurya wallichiana* Steud.

Common names: Tapering leaf Eurya; lopat, monraicha, sagolerbori (Bangladesh); jian ye ling (China); deura (India); jirak padang, sala (Indonesia); jerak merah (Malaysia); bilaune, chonsi, jhuinsin (Nepal); halinghingon (the Philippines); jheen rangkai (Thailand)

Habitat: Open places

Distribution: India, Sri Lanka, Bangladesh, Cambodia, Laos, Vietnams, Thailand, Malaysia, Indonesia, China, Korea, Japan, and the Philippines

Botanical observation: This tree grows up to 6 m tall. The stems are terete and somewhat hairy at apex. The leaves are simple, alternate, and exstipulate. The petiole is up to about 5 mm long. The blade is oblong to lanceolate, 6–12 cm × 1–2 cm, coriaceous, dark green, glossy, obtuse at base, sharply serrulate, with up to about 10 pairs of secondary nerves sunken above and looping at margin, and acuminate to obtuse at apex. The flowers are cauliflorous. The 5 sepals are 2 mm long, lanceolate, and somewhat hairy. The 5 petals are white, up to about 2 cm long, membranous, fused at base, oblong, and minutely incised at apex. The androecium comprises numerous orangish stamens. The ovary is ovoid, hairy, and develops a slender style a a trifid stigma. The berries are globose, up to about 5 mm in diameter and with remnant style and calyx, and eaten by yellow-vented bulbul birds.

Medicinal use: Diarrhea (Bangladesh; Indonesia; Nepal); fever (Indonesia)

Antibacterial (Gram-positive) halo developed by mid- polar extract: Dichloromethane extract inhibited the growth of *Staphylococcus aureus* (Nugraha & Keller, 2011).

Anticandidal polar extract: Methanol extract of stem at the concentration of 1 mg/mL inhibited the growth of *Candida albicans* (NUM961) by 51% (Choi et al., 2017).

Commentaries

i. Early reports mention the presence of tannins (Roberts & Sarma, 1940) which, being astringent, most probably account for the traditional antidiarroeal use.
ii. Methanol extract of leaves of *Eurya ciliata* Merr. inhibited the replication of the human immunodeficiency virus type-1 with IC_{50} values of 12.1 µg/mL (CC_{50} =15.9 µg/mL) and at 50 µg/mL inhibited the Human immunodeficiency virus ribonuclease H by 96% (Tai et al., 2011).

References

Choi, H.A., Cheong, D.E., Lim, H.D., Kim, W.H., Ham, M.H., Oh, M.H., Wu, Y., Shin, H.J. and Kim, G.J., 2017. Antimicrobial and anti-biofilm activities of the methanol extracts of medicinal plants against dental pathogens Streptococcus mutans and Candida albicans. *Journal of Microbiology and Biotechnology*, *27*(7), pp. 1242–1248.

Nugraha, A.S. and Keller, P.A., 2011. Revealing indigenous Indonesian traditional medicine: anti-infective agents. *Natural Product Communications*, *6*(12), pp. 1953–1966.

Roberts, E.H. and Sarma, S.N., 1940. Tannins as hydrogen carriers in biological oxidation. *Biochemical Journal*, *34*(12), p. 1517.

Tai, B.H., Nhut, N.D., Nhiem, N.X., Quang, T.H., Thanh Ngan, N.T., Thuy Luyen, B.T., Huong, T.T., Wilson, J., Beutler, J.A., Ban, N.K. and Cuong, N.M., 2011. An evaluation of the RNase H inhibitory effects of Vietnamese medicinal plant extracts and natural compounds. *Pharmaceutical Biology*, *49*(10), pp. 1046–1051.

9.2.7.5 *Eurya chinensis* R. Br.

Common names: Chinese Eurya; mi sui hua (China)

Habitat: Margins of forests

Distribution: China, Taiwan

Botanical observation: This shrub grows up to 3 m tall. The stems are terete and ribbed. The leaves are simple, spiral, and exstipulate. The petiole is up to about 3 mm long. The blade is obovate to elliptic, 2–5.5 cm × 1–2 cm, coriaceous, dark green, glossy, tapering at base, serrate, with up to about 8 pairs of secondary nerves, and acuminate at apex. The flowers are axillary, cauliflorous, and about 8 mm long clusters. The calyx comprises five sepals which are minute. The five petals are whitish and about 3.5 mm long. The androecium comprises about 15 stamens which are orangish. The gynoecium comprises a conical ovary developing a 3-lobed stigma. The berries are globose, purplish black, up to 4 mm in diameter, and with remnant style and seated on persistent calyx. The seeds are minute, glossy, and kidney-shaped.

Medicinal uses: Fever, jaundice, influenza (China)

Weak antibacterial (Gram-positive) coumarin: Euryacoumarin A inhibited the growth of *Bacillus subtilis* and *Bacillus cereus* with MIC values of 50.5 and 35.4 µM, respectively (Song et al., 2017).

Commentary: The plant produces *ent*-kaurane diterpenes (Song et al., 2018) which could be examined for activity against (+)-RNA viruses.

References

Song, J.L., Yuan, Y., Tan, H.B., Huang, R.M., Liu, H.X., Xu, Z.F. and Qiu, S.X., 2017. Anti-inflammatory and antimicrobial coumarins from the stems of Eurya chinensis. *Journal of Asian Natural Products Research*, *19*(3), pp. 222–228.

Song, J.L., Yuan, Y., Nie, L.H., Li, B.L., Qin, X.B., Li, Y., Wu, J.W. and Qiu, S.X., 2018. A new ent-kaurane diterpene derivative from the stems of Eurya chinensis R. Br. *Natural Product Research*, *32*(2), pp. 182–188.

9.2.7.6 *Schima wallichii* (DC) Korth

Synonyms: *Gordonia chilaunia* Buch.-Ham. ex D. Don; *Gordonia wallichii* DC.; *Schima brevipes* Craib

Common names: Chinese guger tree, needle wood; makrisal (Bangladesh); hong mu he (China); chilauni (India); buah cangkok (Indonesia); laukya (Myanmar); gatal gatal (Malaysia); alue-chilauni, chilaume, kyosin, hyansing (Nepal); khai so (Vietnam)

Habitat: Forests

Distribution: India, Bangladesh, Bhutan, Myanmar, Laos, Cambodia, Vietnam, Malaysia, Indonesia, China, the Philippines

Botanical observation: This magnificent tree grows up to about 50 m tall. The bark is light grayish and cracked and the inner bark pinkish to red, fibrous, the fibers irritating. A whitish sap exudes from the bark after incision. The stems are terete, lenticellate, and somewhat hairy at apex. The leaves are simple, spiral, and exstipulate. The petiole is up to 3 cm long. The blade is elliptic to lanceolate, cuneate at base, 6–20 cm × 2–9.5 cm, glossy, dark green, glossy, coriaceous, with about 7–10 pairs of secondary nerves, laxly serrate, and acuminate at apex. The flowers are solitary, axillary, on about 2.5 cm long peduncles (consider that flowers of Theaceae are often somewhat Clusiaceous-look). The calyx comprises five broad orbicular sepals which are about 4 mm long. The five petals are obovate to spathulate, spoon-shaped, pure white, and to up about 2 cm long. The androecium comprises numerous stamens which are showy and up to about 1 cm long. The ovary is globose and develops a slender style which is about 5 mm long and has a broad stigma. The capsules are woody, globose, about 1–3 cm in diameter, opening with 5 valves, with a persistent calyx, and contain 8 mm long seeds which are winged and reniform.

Medicinal use: Cuts, wounds (Nepal); fungal infections, gonorrhea (India)

Strong broad-spectrum antibacterial polar extract: Ethanol extract of bark inhibited the growth of *Staphylococcus aureus* (29737), *Escherichia coli* (ATCC 10536), *Shigella dysenteriae*, *Shigella sonnei*, *Shigella boydii*, *Pseudomonas aeruginosa* (ATCC 25619), *Vibrio cholerae* (1037), and *Bacillus subtilis* (1037) with MIC/MBC values of 100/100, 25/25, 50/100, 50/75, 50/50, 25/50, 50/100, and 200/>200 µg/mL, respectively (Dewanjee et al., 2008).

Weak broad-spectrum antifungal polar extract: Ethanol extract of bark inhibited the growth of *Candida albicans* (ATCC 10231), *Aspergillus niger* (ATCC 6275), *Penicillium notatum* (ATCC 11625), and *Penicillium funiculosum* (NCTC 287) with MIC/MFC values of 800/1000, 1000/1200, 1000/1200, and 1000/1200 µg/mL, respectively (Dewanjee et al., 2008).

Strong antiviral (enveloped monopartite linear double-stranded DNA) polar extract: Ethanol extract of leaves at the concentration of 100 µg/mL inhibited the formation of plaques by the Herpes simplex virus type-1 by about 70% (Lipipun et al., 2003).

Moderate antiviral (enveloped monopartite linear single-stranded (-)-RNA) polar extract: Ethanol extract of leaves at the concentration of 100 µg/mL inhibited the formation of plaques by the Measles virus by about 30% (Lipipun et al., 2003).

Commentaries

i. The plant produces kaempferol-3-*O*-rhamnoside (Diantini et al., 2012) as well as ellagitannins and gallotannins (Yoshida et al., 1991), which by virtue of their astringency probably account for the medicinal uses.
ii. A plant extract which shows more antibacterial effect against Gram-negative than Gram-positive bacteria is interesting because Gram-negative bacteria are more resistant to antibacterial agents owed to their outer membrane and porins.

References

Dewanjee, S., Maiti, A., Majumdar, R., Majumdar, A. and Mandal, S.C., 2008. Evaluation of antimicrobial activity of hydroalcoholic extract Schima wallichii bark. *Pharmcolology Online*, *1*, pp. 523–528.

Diantini, A., Subarnas, A., Lestari, K., Halimah, E., Susilawati, Y., Supriyatna, S., Julaeha, E., Achmad, T.H., Suradji, E.W., Yamazaki, C. and Kobayashi, K., 2012. Kaempferol-3-O-rhamnoside isolated from the leaves of Schima wallichii Korth. inhibits MCF-7 breast cancer cell proliferation through activation of the caspase cascade pathway. *Oncology Letters*, *3*(5), pp. 1069–1072.

Lipipun, V., Kurokawa, M., Suttisri, R., Taweechotipatr, P., Pramyothin, P., Hattori, M. and Shiraki, K., 2003. Efficacy of Thai medicinal plant extracts against herpes simplex virus type 1 infection in vitro and in vivo. *Antiviral Research*, *60*(3), pp. 175–180.

Yoshida, T., Chou, T., Nitta, A. and Okuda, T., 1991. Tannins and related polyphenols of theaceous plants. IV. Monomeric and dimeric hydrolyzable tannins having a dilactonized valoneoyl group from Schima wallichii KORTH. *Chemical and Pharmaceutical Bulletin 39*(9), pp. 2247–2251.

10 Medicinal Plants in Asia-Pacific for Microbial Infections

The Clade Lamiids

10.1 ORDER BORAGINALES JUSS. EX BERCHT. & J. PRESL (1820)

10.1.1 Family Boraginaceae A.L de Jussieu (1789)

The family Boraginaceae consists of about 150 genera and 1500 species of beautiful and sun-loving herbs which often have an aura of dryness. Stems, leaves, and sepals are often covered with minute rigid hairs. The leaves are simple, spiral, and exstipulate. The inflorescences are cymes. The flowers are actinomorphic. The calyx is tubular and 5-lobed. The corolla is tubular, membranous, 5-lobed, and often with a unique type of blue. The androecium consists of five stamens. The gynoecium includes two carpels fused in a 2 or 4 locular ovary, each locule sheltering 1–2 ovules. The fruits are berries or nutlets. Plants in this family are known to produce poisonous pyrrolizidine alkaloids.

10.1.1.1 *Alkanna tinctoria* Tausch.

Synonyms: *Anchusa tinctoria* (L.) L.; *Lithospermum tinctorium* Ruiz & Pav.; *Plagiobothrys myosotoides* (Lehm.) Brand

Common names: Dyer's alkanet, Spanish buglot; dineshavali (India); ratanjot (India; Pakistan); abukhalsa shankar (Iran); anik, havaciva otu (Turkey)

Habitat: Deserts, meadows, seashores

Distribution: Turkey, Iran

Botanical observation: This creeping to diffuse herb grows up to about 60 cm long. The stems are terete and hairy. The leaves are simple, decussate, sessile, and exstipulate. The blade is triangular, hairy, and about 1 cm long. The cymes are terminal and bout 3 cm long. The calyx is hairy and develops five lanceolate lobes. The corolla is membranous, tubular, up to about 8 mm long, blue, darker or yellowish at throat, and develops 5 oblong round lobes. The androecium includes 5 stamens, The ovary is 4-lobed and develops a slender style and a forked stigma. The nutlets are minute.

Medicinal uses: Burn (Turkey); smallpox (Cambodia, Laos, Vietnam); wounds (India; Iran; Turkey)

Broad-spectrum antibacterial halo developed by naphthoquinones: Alkannin and alkannin acetate inhibited the growth of *Staphylococcus aureus* (ATCC 25923), *Staphylococcus epidermidis* (ATCC 12228), *Escherichia coli* (ATCC 25922), *Enterobacter cloacae* (ATCC 13047), *Klebsiella pneumoniae* (ATCC 13883), and *Pseudomonas aeruginosa* (ATCC 227853) (Damianakos et al., 2012).

Strong antibacterial (Gram-positive) amphiphilic naphthoquinone: Alkannin (LogD 3.2 at pH 7.4; molecular mass = 288.2 g/mol) inhibited the growth of *Staphylococcus aureus*, methicillin-resistant *Staphylococcus aureus*, *Enterococcus faecalis*, vancomycin-resistant *Enterococcus* (F935, VanA), and vancomycin-resistant *Enterococcus* (CKU-18, VanB) with the MIC/MBC values of 6.2/12.5, 6.2/12.5, 50/50, 50/50, and 25/50 μg/mL, respectively (Shen et al., 2002).

Anticandidal halo developed by naphthoquinones: Alkannin and alkannin acetate inhibited the growth of *Candida albicans* (ATCC 10231), *Candida tropicalis* (ATCC 13801), and *Candida glabrata* (ATCC 28838) (Damianakos et al., 2012).

DOI: 10.1201/9781003180340-2

Commentaries

i. The roots produce a series of naphthoquinones including alkannins, alkannin isovalerate, alkannin acetate, and alkannin angelate (Papageorgiou & Digenis, 1980).

ii. The exact antibacterial mode of action of alkannin and alkannin acetate is yet apparently unknown, but Plyta et al. (1998) observed the inhibition of topoisomerase I by alkannin and alkannin acetate via at least in part, complexation with Zn^{2+} since these naphthoquinones have adjacent ketone and hydroxy groups. One could observe the planar structure of naphthoquinones that favors DNA intercalation. Bacterial topoisomerase I is a zinc metalloprotein (Tse-Dinh and Beran-Steed, 1988). Consider that viral enzymes are not uncommonly zinc metalloproteins. RNA-dependent DNA polymerase from the avian myeloblastosis virus is a zinc-dependent enzyme and the severe acute respiratory syndrome-associated coronavirus and the severe acute respiratory syndrome-associated coronavirus 2 bind to the zinc metalloenzyme angiotensin-converting enzyme to penetrate the host cell (Doboszewska et al., 2020). Thus, natural products complexing or chelating zinc could be of value against COVID-19 and other coronaviruses.

iii. Alkannin and derivatives in this plant are most probably antiviral for (+)-RNA viruses (including coronaviruses), since shikonin (also known as (+)-alkannin) inhibited the replication of Japanese encephalitis virus in TE-671 cells with the IC_{50} value of 2.1 µg/mL and a selectivity index value of 835.2 (Lin et al., 2008).

iv. Unani practitioners in Iran consider the plant as hot and dry and as such it might have antiviral potentials

v. In Turkey, *Alkanna orientalis* (L.) Boiss. (local name: *Dikensiz kizilot*) and *Alkanna megacarpa* A. DC (local name: *Havaciva*) are used to heal wounds.

References

Damianakos, H., Kretschmer, N., Sykłowska-Baranek, K., Pietrosiuk, A., Bauer, R. and Chinou, I., 2012. Antimicrobial and cytotoxic isohexenylnaphthazarins from Arnebia euchroma (Royle) Jonst. (Boraginaceae) callus and cell suspension culture. *Molecules*, *17*(12), pp. 14310–14322.

Doboszewska, U., Wlaź, P., Nowak, G. and Młyniec, K., 2020. Targeting zinc metalloenzymes in coronavirus disease 2019. *British Journal of Pharmacology*, *177*(21), pp. 4887–4898.

Lin, C.W., Wu, C.F., Hsiao, N.W., Chang, C.Y., Li, S.W., Wan, L., Lin, Y.J. and Lin, W.Y., 2008. Aloe-emodin is an interferon-inducing agent with antiviral activity against Japanese encephalitis virus and enterovirus 71. *International Journal of Antimicrobial Agents*, *32*(4), pp. 355–359.

Papageorgiou, V.P. and Digenis, G.A., 1980. Isolation of two new alkannin esters from Alkanna tinctoria. *Planta Medica*, *39*(05), pp. 81–84.

Plyta, Z.F., Li, T., Papageorgiou, V.P., Mellidis, A.S., Assimopoulou, A.N., Pitsinos, E.N. and Couladouros, E.A., 1998. Inhibition of topoisomerase I by naphthoquinone derivatives. *Bioorganic & Medicinal Chemistry Letters*, *8*(23), pp. 3385–3390.

Shen, C.C., Syu, W.J., Li, S.Y., Lin, C.H., Lee, G.H. and Sun, C.M., 2002. Antimicrobial activities of naphthazarins from arnebia e uchroma. *Journal of Natural Products*, *65*(12), pp. 1857–1862.

Tse-Dinh, Y.C. and Beran-Steed, R.K., 1988. Escherichia coli DNA topoisomerase I is a zinc metalloprotein with three repetitive zinc-binding domains. *Journal of Biological Chemistry*, *263*(31), pp. 15857–15859.

10.1.1.2 *Anchusa hispida* Forssk.

Synonym: *Gastrocotyle hispida* (Forssk.) Bunge

Common names: Fu qi cao (China); gul i gav zaban (Iran); khatol (Pakistan)

Habitat: Deserts

Distribution: Iran, Afghanistan, Pakistan, India

Botanical observation: This discrete and diffuse herb grows up to about 35 cm tall. The stems are terete and hairy. The leaves are simple, sessile, spiral, and exstipulate. The blade is narrowly spathulate or oblanceolate, hairy, and up to about 0.4–1 cm × 3–10 cm. The flowers are axillary and solitary. The calyx is tubular, 3 mm long, hairy, and develops 5 lanceolate and hairy lobes.

The corolla is blue, about 4 mm long, and develops five ovate lobes around a throat marked with five tiny appendages. The androecium includes five stamens adnate to the corolla tube. The ovary is 4-lobed. The style is slender, minute, and forms a capitate stigma. The nutlets are about 4 mm across.

Medicinal use: Dysentery, urinary tract infection (Pakistan)

Commentaries

i. The plant builds a series of poisonous pyrrolizidine alkaloids such as retrocenine, trachelanthamine, supinine, intermedine, and lycopsamine (El-Shazly et al., 2017). The antimicrobial mode of action of pyrrolizidine alkaloids has apparently not been studied, but they are known to be cytotoxic by inducing DNA alkylation (Wink, 2012).

ii. In Turkey, *Anchusa azurea* Miller (local name: *Sigirdili*), *Anchusa leptophylla* Roem. & Schult. (local name: *Sigirdili*), and *Anchusa strigosa* Banks & Sol. (local name: *Guriz*) are used to heal wounds, whereas *Anchusa azurea* Mill. (local name: *Mljmjokasin*) is used for pimples. Essential oil of flowers of *Anchusa strigosa* Banks & Sol. (100 µg/mL) inhibited the growth of *Staphylococcus aureus*, *Staphylococcus epidermidis, Bacillus subtilis, Escherichia coli,* and *Klebsiella pneumoniae* (Al-Salihi et al., 2009). Aqueous extract of *Anchusa strigosa* Banks & Sol. (15 µg/mL) inhibited the mycelial growth of *Microsporum canis*, *Trichophyton mentagrophytes*, and *Trichophyton violaceum* by 50.1%, 6.7%, and 71.7%, respectively (Ali-Shtayeh & Abu Ghdeib, 1999).

iii. *Anchusa italica* Retz. (local name: *Govihjvika*) is used to treat a cold, sore throat, and fever and in Iran (local name: *Gavzaban*) and alcoholic extract of this plant inhibited the replication of the Influenza virus (Ketabchi et al., 2011).

References

Ali-Shtayeh, M.S. and Abu Ghdeib, S.I., 1999. Antifungal activity of plant extracts against dermatophytes. *Mycoses*, *42*(11–12), pp. 665–672.

Al-Salihi, F., Yasseen, A.I. and Al-Salihi, S.F., 2009. Antimicrobial activity of volatile oil and fixed oil extracted from Anchusa strigosa Lab. *Tikrit Journal of Pure Science*, *14*(2), pp. 21–24.

El-Shazly, A., El-Domiaty, M., Witte, L. and Wink, M., 1998. Pyrrolizidine alkaloids in members of the Boraginaceae from Sinai (Egypt). *Biochemical Systematics and Ecology*, *26*(6), pp. 619–636.

El-Shazly, A. and Wink, M., 2014. Diversity of pyrrolizidine alkaloids in the Boraginaceae structures, distribution, and biological properties. *Diversity*, *6*(2), pp. 188–282.

Ketabchi, S., Moatari, A., Shadram, M. and Rostami, Y., 2011. The anti influenza virus activity of Anchusa italica. *Asian Journal of Experimental Biological Sciences*, *2*(4), pp. 758–761.

Wink, M., 2012. Medicinal plants: A source of anti-parasitic secondary metabolites. *Molecules*, *17*(11), pp. 12771–12791.

10.1.1.3 *Carmona microphylla* (Lam.) G. Don

Synonyms: *Carmona heterophylla* Cav.; *Carmona retusa* (Vahl) Masam.; *Cordia retusa* Vahl; *Ehretia buxifolia* Roxb.; *Ehretia dentata* Courchet; *Ehretia microphylla* Lam.

Common names: Cai mo, ji ji shu shu (China); kuruvigi, thel chedi (India); cay trui (Vietnam)

Habitat: Scrub forests, cultivated

Distribution: India, Vietnam, China, Taiwan, the Philippines, Japan, Australia

Botanical observation: This ornamental hedge shrub grows up to 3 m tall. The stems are terete and glabrous. The leaves are simple, sessile, fascicled, and exstipulate. The blade is coriaceous, obovate to spathulate, 1.5–3.5 cm × 1–2 cm, hairy, dentate, dark green and glossy, and rounded to acute at apex. The inflorescences are axillary cymes and about 1.5 cm long. The calyx is 6 mm long and presents 5 linear sepals that are hairy and fused at base. The corolla is tubular, pure white to reddish, 5-lobed, and with appendages at throat. The 5 stamens are adnate to the base of the corolla tube. The ovary develops a slender and about 6 mm style and a bifid stigma. The drupes are red, somewhat glossy, globose, marked at apex with vestige of style, 4 mm across, and seated on calyx.

Medicinal uses: Cold (China); fever (the Philippines); diarrhea, syphilis (India)

Commentaries

i. The plant produces the triterpenes bauerenol and ursolic acid as well as the prenylated benzoquinones microphyllone and ehretianone (Selvanayagam et al., 1996). The benzoquinone moieties and prenylated moieties of microphyllone and ehretianone suggest these substances to be strongly antiviral. This could be examined.

ii. In the Philippines, *Ehretia philippinensis* A. DC (local name: *Alibungog*) is used to heal wounds and affords a remedy to treat toothache, diarrhea, and dysentery. In the Philippines and Taiwan, *Ehretia resinosa* Hance is used for fever. In India, *Ehretia microphylla* Lam. (synonym: *Carmonia retusa* (Vahl) Masam.) (local name: *Seethevi thalai*) is used for toothaches. In the Philippines, *Ehretia microphylla* Lam. (Philippine tea; local name: *Tsaang gubat*) is used to stop diarrhea, for toothache, and disinfectant at childbirth, and aqueous extract of this plant inhibited the replication of the Zika virus (strain MR766) in Vero cells with the EC_{50} value of 1254 μg/mL (Vista et al., 2020).

iii. In India, *Ehretia laevis* Roxb. (local name: *Khanduch akka*) is used for wounds.

References

Selvanayagam, Z.E., Gnanavendhan, S.G., Balakrishna, K., Rao, R.B., Sivaraman, J., Subramanian, K., Puri, R. and Puri, R.K., 1996. Ehretianone, a novel quinonoid xanthene from Ehretia buxifolia with antisnake venom activity. *Journal of Natural Products*, *59*(7), pp. 664–667.

Vista, F.E.S., Dalmacio, L.M.M., Corales, L.G.M., Salem, G.M., Galula, J.U. and Chao, D.Y., 2020. Antiviral effect of crude aqueous extracts from ten philippine medicinal plants against zika virus. *Acta Medica Philippina*, *54*(2). doi: 10.47895/amp.v54i2.1501.

10.1.1.4 *Coldenia procumbens* L.

Common names: Trailing coldenia; shuang zhu zi cao (China); seru padi, tripakshi (India); oreganon lalaki (the Philippines); serapadi (Sri Lanka); yaa teentukae (Thailand); casp diefn (Vietnam)

Habitat: Sun-exposed soils, dry rivers, roadsides, rice fields, wastelands

Distribution: Tropical Asia, Australia

Botanical observation: This discrete and creeping herb grows up to a length of 50 cm from a taproot. The stems are terete and hairy. The leaves are simple, alternate, and exstipulate. The petiole is up to 1 cm long. The blade is obovate to spathulate, somewhat vaguely stone-like, asymmetrical, hairy, dentate, marked with about six pairs of secondary nerves conspicuously and characteristically sunken above, 0.5–2 × 1–4 cm, and somewhat wavy. The flowers are solitary and axillary. The calyx is 4-lobed and hairy. The corolla is pure white, 2.5 mm long, and 4-lobed, the lobes rounded. The 4 stamens are adnate to the corolla tube. The ovary is ovoid and develops a bifid style. The capsules are 4-lobed and about 4 mm across.

Medicinal uses: Abscesses (Philippines); boils (India)

Strong antifungal (filamentous) polar extract: Ethanol extract of leaves inhibited the growth of *Trichophyton mentagrophytes* (clinical isolate 1), *Trichophyton simii* (clinical isolate), *Trichophyton rubrum*, *Trichophyton mentagrophytes*, *Microsporum gypseum*, *Trichophyton tonsurans*, and *Trichophyton rubrum* (Clinical isolate 2) with the MIC values of 62.5/250, 250/500, 500/1000, 125/500, 125/250, 125/500, and 500/2000 μg/mL, respectively, via a mechanism involving, at least in part, ergosterol synthesis inhibition (Chellappandian et al., 2018).

Strong antiviral (enveloped linear monopartite dimeric single-stranded (+) RNA) polar extract: Aqueous extract of leaves inhibited the replication of the human immunodeficiency virus type-1 (strain IIIB) and type-2 (Strain LAV-2ROD) in MT-4 cells with the IC_{50} values of 32.1 and 41.6 μg/mL, respectively (Venkatachalam, 2018).

Commentary: Apparently, secondary metabolites have not been characterized yet from this interesting herb.

References

Chellappandian, M., Saravanan, M., Pandikumar, P., Harikrishnan, P., Thirugnanasambantham, K., Subramanian, S., Hairul-Islam, V.I. and Ignacimuthu, S., 2018. Traditionally practiced medicinal plant extracts inhibit the ergosterol biosynthesis of clinically isolated dermatophytic pathogens. *Journal de Mycologie Medicale*, *28*(1), pp. 143–149.

Venkatachalam, D., 2018. Studies of anti-viral activity and Cytotoxicity of Sphaeranthus indicus and Coldenia procumbens. *International Journal of Medical Research and Pharmaceutical Sciences*, *5*, pp. 1–5.

10.1.1.5 *Cordia myxa* L.

Synonyms: *Argyreia arborea* Lour.; *Cornus obliqua* Raf.

Common names: Assyrian plum, clammy cherry, Indian cherry, sebestan pum; bahuri (Bangladesh); lasoora, lesora (Pakistan); bukampadaruk shleshmaataka (India); kendla (Indonesia); sepistan (Iran); thanat (Myanmar)

Habitat: Seashore, sun-exposed drylands, cultivated

Distribution: Pakistan, India, Sri Lanka, Indonesia

Botanical observation: This tree grows up to a height of 5 m. The stems are hairy. The leaves are simple, spiral, and exstipulate. The petiole is up to 5 cm long. The blade is elliptic to oblong, 6–12.5 cm × 4.5–8 cm, dull green, with about 4 pairs of secondary nerves, somewhat coriaceous, cuneate to rounded at base, acute or obtuse at apex, and crenate or dentate. The calyx is tubular, about 5–9 mm long, and 3–4-lobed. The corolla is tubular, about 1 cm long, whitish, and develops 4–6 lanceolate lobes. The 3–4 stamens are adnate to the corolla tube. The ovary develops a slender style which reaches 9 mm long and 4 stigmas. The drupe is up to 3.5 cm long, ovoid, glossy, smooth, of a very peculiar light orangish-brown, sticky within, palatable, and on persistent calyx.

Medicinal uses: Cold, fever, urinary tract infection, ringworm (India); dysentery (Pakistan); gonorrhea (Indonesia); cough (India; Iran; Pakistan)

Commentaries

i. Our peer the Persian physician Avicenna knew this plant as *Lebuck* and Unani practitioners suggest the plant to be of value for COVID-19. The plant was called *Arbor glutinosa* in the *Herbarium amboinense* of our peer the German botanist Georg Eberhard Rumphius (1627–1702).

ii. Members of the genus *Cordia* L. are often used in Asia to treat microbial infections. *Cordia dichotoma* Forst. f. is used for boils in the Philippines, in Nepal for fever, in Myanmar (local name: *Hpak mong*) for cough, in Bangladesh for vaginitis (Local name: *Bongol gash*), and in India for cough, cold, urinary tract infection, and fever. In India, *Cordia obliqua* Willd. (local name: *Naruvallai pulum*) is used for cough, diseases of the lung, and sore throat, and in Indonesia (local name: *Sekendai*) it affords a remedy for fever. In India, *Cordia wallichii* Don (local name: *Bahuvaraka*) is used for bronchitis, stomatitis, gonorrhea, eye infection, cough, fever, and sore throat. In India, *Cordia rothii* Roem. & Schult. (local name: *Nariviri*) is used to gargle, *Cordia vestita* Hook. f. & Thomson (local name: *Latora*) is used for cough, and *Cordia macleodii* Hook. f. & Thomson (local name: *Palandeku*) is used for jaundice.

10.1.1.6 *Cynoglossum amabile* Stapf & J.R. Drumm.

Common names: Chinese forget-me-not; dao ti hu (China)

Habitat: Riverbank, roadsides, open forests, cultivated

Distribution: China

Botanical observation: This magnificent herb grows up to about 60 cm tall. The stems are hairy. The leaves are simple, basal or cauline, and exstipulate. The petiole is none to 1 cm long. The blade is oblong-lanceolate to lanceolate, somewhat fleshy and dull, 5–20 cm × 2–4 cm, hairy, tapering at base, and acute at apex. The Inflorescences are terminal cymes. The calyx is about 3 mm long, hairy, and comprises 5 broadly lanceolate sepals. The corolla is tubular, marked at throat by appendages, of a very peculiar blue, and develops 5 lobes which are about 3 mm long and orbicular. The androecium presents 5 stamens adnate to the corolla tube. The nutlets are ovoid, about 4 mm across, and hairy.

Medicinal uses: Cough, tuberculosis, wounds (China)

Commentary:

i. Plants in the genus *Cynoglossum* L. are used in Asia for the external treatment of microbial infections. In Papua New Guinea, *Cynoglossum hellwigii* Brand (local name: *Semongokina*) is used to heal ulcers. In India, *Cynoglossum meeboldi* Brand (local name: *Khuriattai*) is used for oils (India). In Pakistan, *Cynoglossum lanceolatum* Forssk. (local name: *Lainda*) is used to treat eye infection. In Nepal, *Cynoglossum zeylanicum* (Vahl ex Hornem.) Thunb. ex Lehm. is used to treat cuts and wounds, and in India (local name: *Vandugadi chedi*) it is used to treat fungal infections. *Cynoglossum officinale* L. has been used by the Celts of the British Isles to heal wounds as recorded in the 14th century Red book of Hergest and in Italy for syphilis (Guidi et al., 2010). The use of plants in the genus *Cynoglossum* L. is probably owed, at least in part, to the presence of allantoin (Dresler et al., 2017). Allantoin has beneficial effects on the healing of ulcers (Kaplan, 1937). This hydrophilic imidazole alkaloid (LogD –5.9 at pH 7.4; molecular mass = 158.1 g/mol) inhibited the growth of *Bacillus subtilis*, *Staphylococcus aureus*, *Escherichia coli*, and *Klebsiella pneumoniae* with the MIC values of 4, 8, 8, and 8 µg/mL, respectively (Lakshmanan et al., 2019).

References

Dresler, S., Szymczak, A.G. and Wójcik, M., 2017. Comparison of some secondary metabolite content in the seventeen species of the Boraginaceae family. *Pharmaceutical Biology*, *55*(1), pp. 691–695.

Guidi, E., Angelini, L., Mares, D., Contini, C. and Vicentini, C.B., 2010. The treatment of syphilis in Ferrara (Italy) in the 19th century: The example of the Ferrarese Pharmacopoeia. *Pharmaceutical Historian*, *40*(1), pp. 2–9.

Kaplan, T., 1937. The allantoin treatment of ulcers. *Journal of the American Medical Association*, *108*(12), pp. 968–969.

Lakshmanan, G., Sivaraj, C., Ammar, A., Gopinath, S., Saravanan, K., Gunasekaran, K. and Murugesan, K., 2019. Isolation and structural elucidation of allantoin a bioactive compound from cleome viscosa L.: A combined experimental and computational investigation. *Pharmacognosy Journal*, *11*(6s), pp. 1391–1400.

10.1.1.7 *Heliotropium indicum* L.

Synonym: *Tiaridium indicum* (L.) Lehm

Common names: Indian heliotrope; Indian turnsole; hattii shura (Bangladesh); kantui damrey (Cambodia); da wei yao (China); bhurundi (India); buntut tikus (Indonesia); nha nguong xang (Laos); rumpus ekor kunching, seri bumi, jinkin kala (Malaysia); sin-hna-maung (Myanmar); dimibiya (Sri Lanka); oont chara (Pakistan); cabra cabra, elepanta, kambra kambra (the Philippines); ya nguang chang, yaa nguang chaang noi (Thailand); cây vòi voi (Vietnam)

Habitat: Roadsides, wastelands, seashores, drains, and villages

Distribution: Tropical Asia and Pacific

Botanical observation: This poisonous and strange-looking herb grows to a height of 50 cm and has an aura of dryness. The leaves are simple, spiral, and exstipulate. The petiole is up to 1.2 cm long. The blade is ovate or ovate-lanceolate, 3–5 cm × 1.5–5 cm, marked with about 4–5 pairs of secondary nerves sunken above, tapering at base, and acute at apex. The inflorescences are about 5 cm long scorpioid cymes characteristically recurved and shaped like some sort of tentacles or elephant trump. The calyx is tubular and develops 5 lobes. The corolla is pale purple to white with a yellow throat, about 5 mm long. The ovary is 4-lobed and the style is minute and develops a conical stigma. The anthers are attached to the corolla tube. The nutlets are conical, pear-shaped, beaked, and about 3 mm long.

Medicinal uses: Boils, fever, pimple, ulcers, wounds (India); diarrhea (Thailand); gonorrhea (Myanmar), ringworm (Cambodia); ophthalmia, typhoid (Bangladesh); boils, ulcers, wounds (Myanmar); viral infection (Thailand)

Broad-spectrum antibacterial halo developed by polar extract: Ethanol extract of leaves (100 µg/disc) inhibited the growth of *Bacillus cereus*, *Staphylococcus aureus*, *Bacillus subtilis*, *Escherichia coli*, *Salmonella typhi*, *Pseudomonas aeruginosa*, *Salmonella typhi*, *Vibrio mimicus*, and *Shigella dysenteriae* with the inhibition zone diameters of 18, 11, 13, 9, 8, 12, 9, 8, and 9 mm, respectively (Samira et al., 2016).

Broad-spectrum antifungal halo developed by polar extract: Ethanol extract of leaves inhibited the growth of *Candida albicans*, *Aspergillus flavus*, and *Aspergillus oryzae* (Wanu et al., 2018).

Commentaries

i. The plant produces pyrrolizidine alkaloids such as lycopsamine (Souza et al., 2005). The antiviral properties of this herb are apparently unknown. *Heliotropium subulatum* Hochst. ex DC. produces antiviral pyrrolizidine alkaloids: heliotrine and subulacine-*N*-oxide at the concentration of 10 µg/mL reduced the titer of the Poliovirus (type 1 strain 1A/S) by 1000 in Hela cells (Singh et al., 2002). 7-Angeloyl heliotrine at the concentration of 10 µg/mL reduced the titer of the Coxsackie virus by 1000 in in Hela cells (Singh et al., 2002). Subulacine at the concentration of 10 µg/mL reduced the titer of the Semliki forest virus by 1000 in Hela cells (Singh et al., 2002). Heliotrine and 7-angeloyl heliotrine at the concentration of 10 µg/mL reduced the titer of the vesicular stomatitis virus by 1000 in Vero cells (Singh et al., 2002). 7-Angeloyl heliotrine at the concentration of 10 µg/mL reduced the titer of the Measle virus (Edmonston A) by 1000 in Vero cells (Singh et al., 2002).
ii. Pyrrolizidine alkaloids are hepatotoxic (Neuman et al., 2015).
iii. In India, *Heliotropium curassavicum* L. (local name: *Banhatisur*) is used to treat jaundice and conjunctivitis, and in Pakistan it is used for boils. *Heliotropium strigosum* is used to treat urinary tract infections in Pakistan (local name: *Gorakh pan*).

References

Neuman, M.G., Cohen, L., Opris, M., Nanau, R.M. and Jeong, H., 2015. Hepatotoxicity of pyrrolizidine alkaloids. *Journal of Pharmacy & Pharmaceutical Sciences*, *18*(4), pp. 825–843.

Samira, K., Laboni, F.R., Julie, A.S., Jalal, U. and Labu, Z.K., 2016. Biological investigations of medicinal plants of Heliotropium indicum indigenous to Bangladesh. *Journal of Coastal Life Medicine*, *4*(11), pp. 874–878.

Singh, B., Sahu, P.M., Jain, S.C. and Singh, S., 2002. Antineoplastic and antiviral screening of pyrrolizidine alkaloids from Heliotropium subulatum. *Pharmaceutical Biology*, *40*(8), pp. 581–586.

Souza, J.S.N., Machado, L.L., Pessoa, O.D., Braz-Filho, R., Overk, C.R., Yao, P., Cordell, G.A. and Lemos, T.L., 2005. Pyrrolizidine alkaloids from Heliotropium indicum. *Journal of the Brazilian Chemical Society*, *16*, pp. 1410–1414.

Wani, P.A., Tolu, A.M. and Wahid, S., 2018. Antioxidant, antimicrobial and antibiotic resistance modifying effect of Heliotropium indicum. *Biocatalysis and Agricultural Biotechnology*, *15*, pp. 113–118.

10.1.1.8 *Lithospermum erythrorhizon* Siebold & Zucc.

Synonyms: *Lithospermum officinale* var. *erythrorhizon* (Siebold & Zucc.) Maxim.

Common names: Zi cao (China)

Habitat: Meadows

Distribution: China, Korea, Japan

Botanical observation: This beautiful herb grows up to about 1 m tall from reddish roots. The stems are hairy. The leaves are simple, spiral, sessile, and exstipulate. The blade is ovate-lanceolate, 3–8 cm × 0.7–1.7 cm, attenuate at base, marked with about 3 pairs of secondary nerves sunken above and somewhat parallel to the margin, and acuminate at apex. The cymes are terminal and about 5 cm long. The calyx is tubular and develops 5 linear and about 4 mm lobes. The corolla is membranous, pure white, 7–9 mm long, and develops 5 broadly ovate lobes. The 5 stamens are attached to the corolla tube. The style is slender. The nutlets are white to brownish, ovoid, about 4 mm long, smooth, and glossy.

Medicinal use: Smallpox (Korea)

Broad-spectrum antibacterial halo developed by naphthoquinone: Shikonin (50 µg/disc) inhibited the growth of *Bacillus subtilis* (613R), *Bacillus thuringiensis*, *Clavibacter michiganensis* subsp. *nebraskensis* (CN74-1), *Agrobacterium radiobacter* (K84), *Agrobacterium tumefaciens* (C58), *Burkholderia cepacia*, *Ralstonia solanacearum*, and *Serratia marcescens* (Brigham et al., 1999).

Weak antibacterial (Gram-positive) naphthoquinones: Deoxyshikonin inhibited the growth of *Staphylococcus aureus* (209P IFO 12732), *Micrococcus roseus* (IFO 3764), *Micrococcus luteus* (IFO 12708), and *Bacillus subtilis* (PCI 219 IFO 3513) with the MIC values of 200, 100, 50, and 200 µg/mL, respectively. Shikonin inhibited the growth of *Micrococcus roseus* (IFO 3764) and *Micrococcus luteus* (IFO 12708) with the MIC values of 200 and 100 µg/mL, respectively (Brigham et al., 1999).

Strong antifungal (filamentous) amphiphilic naphthoquinones: Deoxyshikonin (LogD 3.9 at pH 7.4; molecular mass = 272.2 g/mol) inhibited the growth of *Saccharomyces sake* (IFO 0305), *Trichophyton rubrum* (IFO 5808), *Trichophyton mentagrophytes* (IFO 5809), *Trichophyton tonsurans* (IFO 5945), *Microsporum gypseum* (IFO 8307), and *Epidermophyton floccosum* (IFO 9045) with the MIC values of 6.2, 25, 12.5, 12.5, 6.2, and 6.2 µg/mL, respectively. Shikonin (LogD 3.2 at pH 7.4; molecular mass = 288.2 g/mol) inhibited the growth of *Saccharomyces sake* (IFO 0305), *Trichophyton rubrum* (IFO 5808), *Trichophyton mentagrophytes* (IFO 5809), *Trichophyton tonsurans* (IFO 5945), *Microsporum gypseum* (IFO 8307), and *Epidermophyton floccosum* (IFO 9045) with the MIC values of 3.1, 25, 25, 25, 12.5, and 6.2 µg/mL, respectively (Honda et al., 1988).

Strong broad-spectrum antifungal amphiphilic naphthoquinones: Acetylshikonin (LogD 3.7 at pH 7.4; molecular mass = 330.3 g/mol) inhibited the growth of *Candida albicans* (YFC 497), *Candida albicans* (YFC 803-azole resistant), *Candida glabrata* (YFC 501), *Candida krusei* (YFC 827), *Candida tropicalis* (YFC 052), *Candida parapsilosis* (YFC 826), *Saccharomyces cerevisiae* (YFC 250), *Cryptococcus neoformans* (YFC 513), *Trichosporon cutaneum* (YFC 517), and *Aspergillus fumigatus* (YFC 526) with the MIC values ranging from 16 to 32 µg/mL, respectively (Sasaki et al., 2002). β-Hydroxyisovaleryl shikonin inhibited the growth of *Candida albicans* (YFC 497), *Candida albicans* (YFC 803-azole resistant), *Candida glabrata* (YFC 501), *Candida krusei* (YFC 827), *Candida tropicalis* (YFC 052), *Candida parapsilosis* (YFC 826), *Saccharomyces cerevisiae* (YFC 250), *Cryptococcus neoformans* (YFC 513), *Trichosporon cutaneum* (YFC 517), and *Aspergillus fumigatus* (YFC 526) with the MIC values of 32, 32, 32, 16, 32, 32, 16, 32, 32, and 64 µg/mL, respectively. Shikonin (LogD 3.2 at pH 7.4; molecular mass = 288.2 g/mol) inhibited the growth of *Candida albicans* (YFC 497), *Candida glabrata* (YFC 501), *Candida krusei* (YFC 827), *Candida tropicalis* (YFC 052), *Candida parapsilosis* (YFC 826), *Saccharomyces cerevisiae* (YFC 250), *Cryptococcus neoformans* (YFC 513), and *Trichosporon cutaneum* (YFC 517) with the MIC values of 8, 8, 4, 8, 16, 4, 8 and 8 µg/mL, respectively (Sasaki et al., 2002).

Antifungal (filamentous) naphthoquinone: Shikonin (50 µg/mL) abrogated the mycelial growth of *Monosporascus cannonballus*, *Fusarium oxysporum*, *Nectria haematococca*, and *Phytophthora parasitica* (Brigham et al., 1999).

Strong antiviral (enveloped monopartite linear single-stranded (+) RNA) amphiphilic naphthoquinone: Shikonin inhibited the replication of Japanese encephalitis virus in TE-671 cells with the IC_{50} value of 2.1 µg/mL and selectivity index value of 835.2 (Lin et al., 2008).

Weak antiviral (enveloped monopartite linear single-stranded (+) RNA) polar extract: Aqueous extract of roots inhibited the replication of the Dengue virus serotype 2 in Vero cells with the IC_{50} value of 375.8 µg/mL and a selectivity index of 1.2 (Maryam et al., 2020).

Viral enzyme inhibition by polar extract: Aqueous extract of roots at the concentration of 250 µg/mL inhibited human immunodeficiency virus type-1 protease by 55% (Xu et al., 1996).

Viral enzyme inhibition by naphthoquinones: Shikometabolin E and shikometabolin F inhibited neuraminidase (Yang et al., 2015)

Commentaries

i. The plant produces a series of naphthoquinones including deoxyshikonin, acetylshikonin, β-hydroxyisovaleryl shikonin, and shikonin (Sasaki et al., 2002) which could be examined for their activity against coronaviruses which like the Japanese encephalitis virus are enveloped monopartite linear single-stranded (+) RNA viruses.
ii. In India, *Lithospermum officinale* L. (local name: *Lubisfirmum*) is used for smallpox and measles.

References

Brigham, L.A., Michaels, P.J. and Flores, H.E., 1999. Cell-specific production and antimicrobial activity of naphthoquinones in roots of Lithospermum erythrorhizon. *Plant Physiology*, *119*(2), pp. 417–428.

Honda, G., Sakakibara, F., Yazaki, K. and Tabata, M., 1988. Isolation of deoxyshikonin, an antidermatophytic principle from Lithospermum erythrorhizon cell cultures. *Journal of Natural Products*, *51*(1), pp. 152–154.

Lin, C.W., Wu, C.F., Hsiao, N.W., Chang, C.Y., Li, S.W., Wan, L., Lin, Y.J. and Lin, W.Y., 2008. Aloe-emodin is an interferon-inducing agent with antiviral activity against Japanese encephalitis virus and enterovirus 71. *International Journal of Antimicrobial Agents*, *32*(4), pp. 355–359.

Maryam, M., Te, K.K., Wong, F.C., Chai, T.T., Gary, K.K. and Gan, S.C., 2020. Antiviral activity of traditional Chinese medicinal plants Dryopteris crassirhizoma and Morus alba against dengue virus. *Journal of Integrative Agriculture*, *19*(4), pp. 1085–1096.

Sasaki, K., Abe, H. and Yoshizaki, F., 2002. In vitro antifungal activity of naphthoquinone derivatives. *Biological and Pharmaceutical Bulletin*, *25*(5), pp. 669–670.

Xu, H.X., Wan, M., Loh, B.N., Kon, O.L., Chow, P.W. and Sim, K.Y., 1996. Screening of traditional medicines for their inhibitory activity against HIV-1 protease. *Phytotherapy Research*, *10*(3), pp. 207–210.

Yang, Y., Zhao, D., Yuan, K., Zhou, G., Wang, Y., Xiao, Y., Wang, C., Xu, J. and Yang, W., 2015. Two new dimeric naphthoquinones with neuraminidase inhibitory activity from Lithospermum erythrorhizon. *Natural Product Research*, *29*(10), pp. 908–913.

10.1.1.9 *Onosma echioides* (L.) L.

Synonym: *Cerinthe echioides* L.

Common name: Goldendrop, Indian alkanet root, vipers' bugloss; kapotacharana, ratanjot (India); havach i chubah (Iran)

Habitat: Meadows

Distribution: Turkey, Iran, Afghanistan, Pakistan, India, Siberia

Botanical observation: This beautiful herb grows up to about 30 cm tall and has a positive aura of life and at very first glance is comfrey-like. The taproots are woody, reddish, and elongated. The stems are hairy. The leaves are simple, sessile, and exstipulate. The blade is narrowly elliptic, hairy, and up to about 0.3–1 cm × 1–7 cm. The cymes scorpioid, terminal, with a few pendulous flowers, and up to about 3 cm long. The calyx is tubular and 5-lobed, hairy, the lobes linear and up to about 1 cm long. The corolla is tubular, somewhat inflated at apex, up to about 2 cm long, yellowish, and develops 5 minute and recurved lobes. The 5 stamens are attached to the corolla tube. The style is slender and develops a capitate stigma. The nutlets are white, glossy, and about 5 mm long.

Medicinal uses: Bronchitis, burns, fever, tuberculosis ulcers, wounds (India); cough (Iran)

Antibacterial (Gram-positive) halo developed by naphthoquinone: Alkannin (25 μg/disc) inhibited the growth of *Staphylococcus aureus* (ATCC 6538), *Staphylococcus aureus* (NGS), *Staphylococcus aureus* (34289), *Staphylococcus aureus* (38418), *Staphylococcus epidermidis* (473), *Staphylococcus epidermidis* (847), and *Staphylococcus haemolyticus* (148) with the inhibition zone diameters of 12, 14, 22, 16, 12, 10, and 10 mm, respectively (Rajbhandar et al., 2007).

Commentaries

i. The plant yields the antimicrobial naphthoquinones alkannin and shikonin (see earlier) which are stereoisomers (Sagratini et al., 2008) and it may contain pyrrolizidine alkaloids (Roeder & Wiedenfeld, 2013).
ii. A plant in the genus *Onosma* L. has been described in the magnificent *Kräuterbuch* of our peer the German physician Adam Lonicer (1528–1586) published in 1557. Note that Adam Lonicer uses binomial denominations.
iii. In Pakistan, *Onosma hispida* Wall.ex G. Don (local names: *Gao zaban*, *Tul charung*) affords a treatment for diseases of the lungs, pneumonia, and typhoid fever. Naphthoquinones in Boraginaceae could be of value as source of leads for the treatment of multidrug-resistant *Mycobacterium tuberculosis*. In Turkey, *Onosma microcarpum* Steven ex DC. (local name: *Havaciva*) is used to heal burns, and both *Onosma armenum* DC. and *Onosma sericeum* Willd. (local name: *Yalanci havaciva*) are used for wounds. In India, *Onosma bracteatum* Wall. (local name: *Shankhahul*) is used for bronchitis, cold, jaundice, sore throat, lung diseases, gonorrhea, syphilis, and leprosy.

References

Rajbhandari, M., Schoepke, T.H., Mentel, R. and Lindequist, U., 2007. Antibacterial and antiviral naphthazarins from Maharanga bicolor. *Die Pharmazie-An International Journal of Pharmaceutical Sciences*, *62*(8), pp. 633–635.

Roeder, E. and Wiedenfeld, H., 2013. Plants containing pyrrolizidine alkaloids used in the Traditional Indian Medicine–including Ayurveda. *Die Pharmazie-An International Journal of Pharmaceutical Sciences*, *68*(2), pp. 83–92.

Sagratini, G., Cristalli, G., Giardinà, D., Gioventù, G., Maggi, F., Ricciutelli, M. and Vittori, S., 2008. Alkannin/shikonin mixture from roots of Onosma echioides (L.) L.: Extraction method study and quantification. *Journal of Separation Science*, *31*(6–7), pp. 945–952.

10.1.1.10 *Tournefortia sarmentosa* Lam.

Synonym: *Heliotropium sarmentosum* (Lam.) Craven

Common names: Tai wan zi dan (China); salakapo kapu (Papua New Guinea); calagung ug (the Philippines)

Habitat: Seashores, open forests

Distribution: Vietnam, Taiwan, Indonesia, the Philippines, Papua New Guinea, Australia

Botanical observation: This shrubby climber grows up to a length of 5 m tall. The stems are hairy. The leaves are simple, spiral, and exstipulate. The petiole is up to about 1 cm long. The blade is oblong-lanceolate to elliptic, somewhat wavy, marked with about 4–6 pairs of secondary nerves, 4–15 cm × 2–7 cm, somewhat hairy below, tapering or rounded or somewhat cordate at base, dull green, and acute or acuminate at apex. The cymes are terminal, scorpioid (somewhat *Heliotropium*-like), and about 3 cm long. The calyx is about 2 mm long, hairy, minute, and 5-lobed. The corolla is tubular, whitish, about 4 mm long, and develops 5 lobes. The 5 stamens are attached to the corolla tube. A disc is present. The drupes are 5 mm across, pure white, and somewhat squarish.

Medicinal uses: Fever (Papua New Guinea); post-partum protective remedy, ulcers (the Philippines)

Weak broad-spectrum antibacterial hydroxycinnamic acid: Rosmarinic acid inhibited the growth of *Staphylococcus epidermidis* (5001), *Stenotrophomonas maltophila*, *Enterococcus faecalis* (C159-6), *Staphylococcus lugdunensis* (T26A3), *Pseudomonas aeruginosa* (ATCC 27583), *Corynebacterium* sp. (T25-17), and *Staphylococcus warneri* (T12A12) with the MIC values of

300/300, 300/300, 300/600, 600/1200, 2500/>2500, 2500/>2500, and 1200/2500 μg/mL, respectively (Abedini et al., 2003).

Very weak antimycobacterial hydroxycinnamic acid: Rosmarinic acid inhibited the growth of *Mycobacterium smegmatis* (5003) with the MIC/MBC values of 1200/2500 μg/mL (Abedini et al., 2003).

Strong antiviral (enveloped linear monopartite dimeric single-stranded (+) RNA) hydrophilic hydroxycinnamic acid: Lithospermic acid (LogD –2.9 at pH 7.4; molecular mass = 538.4 g/mol) inhibited the replication of the human immunodeficiency virus in H9 cells with the IC_{50} value of 2 μM (Bailly & Cotelle, 2005).

Strong antiviral (enveloped monopartite linear double-stranded DNA) hydroxycinnamic acid :

Rosmarinic acid (LogD –2.1 at pH 7.4; molecular mass = 360.3 g/mol) inhibited the growth of Herpes simplex virus type-1 (KOS), Herpes simplex virus type-1 (Acyclovir-resistant strain-1), and Herpes simplex virus type-1 (Acyclovir-resistant strain-2) in RC-37 cells with the IC_{50} values of 10, 10, and 8 μg/mL and the selectivity indexes of 20, 20, and 25, respectively (Astani et al., 2014).

In vivo antiviral (enveloped monopartite linear single-stranded (+) RNA) hydroxycinnamic acid: Rosmarinic acid given intraperitoneally at the dose of 25 mg/Kg twice a day protected mice against the Japanese encephalitis virus (Swarup et al., 2007).

Viral enzyme inhibition by hydroxycinnamic acids: Rosmarinic acid and lithospermic acid inhibited the human immunodeficiency virus integrase with the IC_{50} value of 4 and 1.4 μM, respectively (Bailly & Cotelle, 2005). Lithospermic acid inhibited the human immunodeficiency virus type-1 nucleocapsid protein with the IC_{50} value of 42 μM (Mori et al., 2020)

Commentaries

i. The plant produces phenolic compounds including the hydroxycinnamic acid derivatives lithospermic acid and rosmarinic acid and also yields the imidazole alkaloid allantoin (see earlier) (Lin et al., 1999, 2002) and the pyrrolizidine alkaloid supinine (Crowley and Culvenor, 1955).
ii. Examination of antimicrobial activity of natural products by oral route should be selected over the use of parenteral route for toxicity reason and for the practical aim to develop oral antimicrobial agents. The intravenous LD_{50} of rosmarinic acid in mice of 561 mg/Kg (Bailly & Cotelle, 2005).
iii. The Japanese encephalitis virus is an enveloped monopartite linear single-stranded (+) RNA virus just like the coronavirus and rosmarinic acid could be of value against COVID-19 and its mutants. Consider that rosmarinic acid is available in rosemary. Whence, rosemary (*Rosmarinus officinalis* L.) teas could be beneficial against COVID-19 and mutants. Likewise, *Melissa officinalis* produces rosmarinic acid and as such could be of value against zoonotic coronaviruses.
iv. In Cambodia, Laos, Vietnam *Tournefortia montana* Lour. is used for the treatment of fever and tuberculosis. In Taiwan, *Tournefortia tomentosa* Mill. is used for sores. In Myanmar, *Tournefortia roxburghii* C.B. Clarke is used to treat urinary tract infections. In Papua New Guinea, *Tournefortia argentea* L. f. (local names: *Parah*, *Guinewa*) is used to treat eye infection and mouth wash.
v. Rosmarinic is neuroactive and as such probably able to inhibit bacterial and/or fungal efflux pumps.
vi. Hydroxycinnamic acid derivatives and oligomers are common phytoanticipins in the Clade Lamiid.

References

Abedini, A., Roumy, V., Mahieux, S., Biabiany, M., Standaert-Vitse, A., Rivière, C., Sahpaz, S., Bailleul, F., Neut, C. and Hennebelle, T., 2013. Rosmarinic acid and its methyl ester as antimicrobial components of the hydromethanolic extract of Hyptis atrorubens Poit.(Lamiaceae). *Evidence-Based Complementary and Alternative Medicine, 2013*, p. 604536.

Astani, A., Heidary Navid, M. and Schnitzler, P., 2014. Attachment and penetration of acyclovir-resistant herpes simplex virus are inhibited by melissa officinalis extract. *Phytotherapy Research*, *28*(10), pp. 1547–1552.
Bailly, F. and Cotelle, P., 2005. Anti-HIV activities of natural antioxidant caffeic acid derivatives: Toward an antiviral supplementation diet. *Current Medicinal Chemistry*, *12*(15), pp. 1811–1818.
Crowley, C. and Culvenor, C.C.J., 1955. Occurrence of supinine in Tournefortia sarmentosa Lam. *Australian Journal of Chemistry*, *8*(3), pp. 464–465.
Lin, Y.L., Tsai, Y.L., Kuo, Y.H., Liu, Y.H. and Shiao, M.S., 1999. Phenolic compounds from Tournefortia sarmentosa. *Journal of Natural Products*, *62*(11), pp. 1500–1503.
Lin, Y.L., Chang, Y.Y., Kuo, Y.H. and Shiao, M.S., 2002. Anti-lipid-peroxidative principles from Tournefortia s armentosa. *Journal of Natural Products*, *65*(5), pp. 745–747.
Mori, M., Ciaco, S., Mély, Y. and Karioti, A., 2020. Inhibitory effect of lithospermic acid on the HIV-1 Nucleocapsid Protein. *Molecules*, *25*(22), p. 5434.
Pereira, P., Tysca, D., Oliveira, P., da Silva Brum, L.F., Picada, J.N. and Ardenghi, P., 2005. Neurobehavioral and genotoxic aspects of rosmarinic acid. *Pharmacological Research*, *52*(3), pp. 199–203.
Swarup, V., Ghosh, J., Ghosh, S., Saxena, A. and Basu, A., 2007. Antiviral and anti-inflammatory effects of rosmarinic acid in an experimental murine model of Japanese encephalitis. *Antimicrobial Agents and Chemotherapy*, *51*(9), pp. 3367–3370.

10.1.1.11 *Trichodesma indicum* (L.) Lehm.

Synonyms: *Borago indica* L.; *Trichodesma amplexicaule* Roth; *Trichodesma hirsutum* Edgew.

Common names: Indian borage; andhapuspi (India); chinnathadagam, kawodia, tajrai, hattmutya (India); gaaozeban, juri (Pakistan); phak phaeo (Thailand)

Habitat: Wastelands, roadsides, meadows, fields

Distribution: Afghanistan, Pakistan, India, Myanmar, Thailand, the Philippines

Botanical observation: This graceful herb grows up to about 50cm tall. The stems are terete, reddish, and hairy. The leaves are simple, sessile, amplexicaul, spiral, and exstipulate. The blade is narrowly lanceolate to narrowly elliptic, somewhat coriaceous, 4–8.5 cm × 0.3–2.2cm, hairy, the midrib conspicuous and sunken above, marked with 2–3 pairs of secondary nerves, and dark dull green. The inflorescence is a terminal fascicle with peculiarly pointed flower buds and reddish flower pedicels. The calyx is tubular, up to about 1.2cm, hairy, and 5-lobed. The corolla is infundibuliform, delightfully light blue to pure white, brown at throat, and develops 5 lobes which are about 1.5cm long and apiculate at apex. The androecium includes 5 stamens joined at apex in a Solanaceous manner. The style is slender. The nutlets are up to 5mm long.

Medicinal uses: Cold, leucorrhea, urinary tract infection (India); cough (India; Pakistan; Thailand); dysentery, fever, sore (India; Pakistan); eye infection, fever, influenza (Pakistan)

Broad-spectrum antibacterial halo developed by non-polar extract: Benzene extract of aerial parts (8 mg/6mm disc) inhibited the growth of *Staphylococcus aureus* (ATCC 25923), *Klebsiella pneumoniae* (ATCC 10031), and *Bacillus thuringiensis* (Kurastaki) with the inhibition zone diameters of 10.3, 8.2, and 9.1mm, respectively (Singh & Singh, S., 2003).

Strong broad-spectrum antibacterial polar extract: Ethanol extract of roots inhibited the growth of *Staphylococcus aureus*, *Staphylococcus epidermidis*, *Bacillus subtilis*, *Klebsiella pneumoniae*, *Escherichia coli*, and *Pseudomonas aeruginosa* with the MIC values of 19.2/38.4. 38.4/76.8, 19.2/19.2, 76.8/153.6, 38.4/76.8, and 76.8/153.6µg/mL, respectively (Perianayagam et al., 2012).

Strong broad-spectrum antifungal polar extract: Ethanol extract of roots inhibited the growth of *Candida albicans*, *Aspergillus niger*, and *Aspergillus flavus* with the MIC/MBC values of 19.2/38.4, 38.4/76.8, and 38.4/76.8µg/mL, respectively (Perianayagam et al., 2012).

Antibacterial (filamentous) halo developed by non-polar extract: Benzene extract of aerial parts (8 mg/6mm disc) inhibited the growth of *Rhizoctonia phaseoli* and *Penicillium chrysogenum* with the inhibition zone diameters of 10.6 and 10.5mm, respectively (Singh & Singh, S., 2003).

Weak antiviral (non-enveloped monopartite linear single-stranded (+) RNA) extract: Methanol extract of aerial parts was virucidal for the Feline calicivirus (Vimalanathan et al., 2009).

Commentaries

i. The plant contains poisonous pyrrolizidine alkaloids (Roeder & Wiedenfeld, 2013) such as supinine, lycopsamine, and heliotrine (Ahmad et al., 2018). Pyrrolizidine alkaloids are phytoanticipins with broad-spectrum antibacterial and antifungal effects (Joosten & van Veen, 2011) as well as animal repellents (Fu et al., 2001). These alkaloids or metabolites in mammalian cells (and most probably in bacteria and fungi) target DNA including DNA binding, DNA cross-linking, DNA-protein cross-linking leading to mutations (Fu et al., 2004).
ii. In the Philippines, *Trichodesma zeylanicum* (Burm.f.) R. Br. (local names: *Dilang usa*, *Mabulo*) is used to induce sweating (break fever), and in India (local name: *Adah-pushpi*) the plant is used for wounds. In Iran, *Trichoderma africanum* (local name: *Chaarmaahang*) is used to treat cold, mouth ulcers, measles, scarlet fever, and chickenpox.
iii. *Borago officinalis* L. is used in Unani medicine for COVID-19 (Ansari et al., 2020).
iv. In the closely related family Hydrophyllaceae R. Brown (1817), *Hydrolea zeylanica* (L.) Vahl is used in Bangladesh (local names: *Kasschra*, *Isha-langula*) to treat leprosy, wounds, and ulcers.
v. Benzene should be avoided at all cost for plant extraction as it is carcinogenic.

References

Ahmad, L., He, Y., Semotiuk, A.J. and Liu, Q.R., 2018. Qualitative determination of toxic pyrrolizidine alkaloids in Trichodesma indicum: A prevalent ethnomedicine of Northern Pakistan. *Journal of Complementary Medicine Research*, *9*(1), pp. 11–23.

Ansari, A.P., Ahmed, N.Z., Ahmed, K.K. and Khan, A.A., 2020. An insight on Wabāi Amrād (epidemic diseases) and COVID-19 like conditions–unani perspective. *International Journal of Current Research and Review*, *12*(17), p. 109.

Fu, P.P., Xia, Q., Lin, G. and Chou, M.W., 2004. Pyrrolizidine alkaloids—genotoxicity, metabolism enzymes, metabolic activation, and mechanisms. *Drug Metabolism Reviews*, *36*(1), pp. 1–55.

Joosten, L. and van Veen, J.A., 2011. Defensive properties of pyrrolizidine alkaloids against microorganisms. *Phytochemistry Reviews*, *10*(1), pp. 127–136.

Perianayagam, J.B., Sharma, S.K., Pillai, K.K., Pandurangan, A. and Kesavan, D., 2012. Evaluation of antimicrobial activity of ethanol extract and compounds isolated from Trichodesma indicum (Linn.) R. Br. root. *Journal of Ethnopharmacology*, *142*(1), pp. 283–286.

Roeder, E. and Wiedenfeld, H., 2013. Plants containing pyrrolizidine alkaloids used in the Traditional Indian Medicine–including Ayurveda. *Die Pharmazie-An International Journal of Pharmaceutical Sciences*, *68*(2), pp. 83–92.

Singh, B. and Singh, S., 2003. Antimicrobial activity of terpenoids from Trichodesma amplexicaule Roth. *Phytotherapy Research*, *17*(7), pp. 814–816.

Vimalanathan, S., Ignacimuthu, S. and Hudson, J.B., 2009. Medicinal plants of Tamil Nadu (Southern India) are a rich source of antiviral activities. *Pharmaceutical Biology*, *47*(5), pp. 422–429.

10.1.2 Family Icacinaceae Miers (1851)

The family Icacinaceae consists of about 50 genera and 400 species of trees or climbers. The leaves are simple, alternate, and exstipulate. The inflorescences are spikes, racemes, or cymes. The calyx and corolla are tubular and 4–5 lobed. The androecium comprises 4–5 stamens. The gynoecium consists of 2 carpels fused into a unilocular ovary, each locule sheltering a pair of ovules growing on apical placentas. The fruits are drupes.

10.1.2.1 *Gomphandra quadrifida* (Bl.) Sleumer

Synonym: *Stemonurus quadrifidus* Bl.

Common names: Kayu barik-barik (Indonesia); kayu gerang jantan (Malaysia); bor beso cher tuw (Vietnam)

Habitat: Forest

Distribution: Myanmar, Cambodia, Laos, Vietnam, Thailand, Malaysia

Botanical observation: This tree grows up to 20m tall. The leaves are simple, alternate, and exstipulate. The petiole is about 1cm long. The blade is elliptic, of a peculiar green, smooth, coriaceous, assymetrical and attenuate at base, 10–15 cm × 4–6cm, marked with 3–8 pairs of secondary nerves, and acuminate at apex. The cymes are axillary, up to about 7cm long, and few flowered, the whole with a Boragineous look. The calyx is tubular and 5-lobed. The corolla is tubular, whitish, about 5mm long, and 5-lobed. The androecium includes 5 stamens which are hairy. The style develops a dilated stigma. The drupes are pure white, ellipsoid, smooth, marked at apex with a disc, and about 1cm long.

Medicinal uses: Post-partum protective remedy (Malaysia)

Commentaries

i. In Malaysia, *Gomphandra salicifolia* Ridl. is used as post-partum remedy.

ii. Members of the genus *Gomphrena* L. produce the pyrrolo quinoline alkaloid camptothecin (Ramesha et al., 2013) which is a topoisomerase inhibitor (Pommier et al., 2010) and as such antibacterial, antifungal, and antiviral. Camptothecin (LogD 1.8 at pH 7.4; molecular mass = 348.3 g/mol) inhibited the growth of *Alternaria alternata*, *Epicoccum nigrum*, *Pestalotia guepinii*, *Drechslera* sp., and *Fusarium avenaceum* with the EC_{50} values of <25, <10, 10, <10, and 30µg/mL, respectively (Li et al., 2005). Camptothecin stabilizes topoisomerase I-DNA complex and inhibit the religation of DNA whence bacterial and fungal death (Del Poeta et al., 1999). Consider that camptothecin inhibited the synthesis of Vaccinia virus DNA in HeLa cells at the concentration of 40µM and adenoviruses (Horwitz et al., 1972). At the concentration of 10µM, camptothecin inhibited the replication of the (+) RNA virus Enterovirus 71 (Wu & Chu, 2017), and at 10µg/mL, it inhibited the synthesis of RNA in BHK cells infected by Influenza fowl plaque virus (A/FRV/Rostock/34 HavNi) (Kelly et al., 1974). It also inhibits the replication of the simian virus (Rubinstein & Rein, 1974) and the human immunodeficiency virus (Pantazis, 1996). Therefore, one could endeavor to examine the potentials of camptothecin against coronaviruses and its possible therapeutic uses against COVID-19 and variants. One can observe that broad-spectrum antiviral natural products acting against enveloped and non-enveloped virus target the replication of viral genetic material in host cells.

References

Del Poeta, M., Toffaletti, D.L., Rude, T.H., Dykstra, C.C., Heitman, J. and Perfect, J.R., 1999. Topoisomerase I is essential in Cryptococcus neoformans: Role in pathobiology and as an antifungal target. *Genetics*, *152*(1), pp. 167–178.

Horwitz, S.B., Chang, C.K. and Grollman, A.P., 1972. Antiviral action of camptothecin. *Antimicrobial Agents and Chemotherapy*, *2*(5), pp. 395–401.

Kelly, D.C., Avery, R.J. and Dimmock, N.J., 1974. Camptothecin: An inhibitor of influenza virus replication. *Journal of General Virology*, *25*(3), pp. 427–432.

Li, S., Zhang, Z., Cain, A., Wang, B., Long, M. and Taylor, J., 2005. Antifungal activity of camptothecin, trifolin, and hyperoside isolated from Camptotheca acuminata. *Journal of Agricultural and Food Chemistry*, *53*(1), pp.32–37.

Pantazis, P., 1996. Camptothecin: A promising antiretroviral drug. *Journal of Biomedical Science*, *3*(1), pp. 14–19.

Pommier, Y., Leo, E., Zhang, H. and Marchand, C., 2010. DNA topoisomerases and their poisoning by anticancer and antibacterial drugs. *Chemistry & Biology*, *17*(5), pp. 421–433.

Ramesha, B.T., Suma, H.K., Senthilkumar, U., Priti, V., Ravikanth, G., Vasudeva, R., Kumar, T.S., Ganeshaiah, K.N. and Shaanker, R.U., 2013. New plant sources of the anti-cancer alkaloid, camptothecine from the Icacinaceae taxa, India. *Phytomedicine*, *20*(6), pp. 521–527.

Rubinstein, L. and Rein, A., 1974. Effect of camptothecin on simian virus 40 DNA. *Nature*, *248*(5445), pp. 226–228.

Wu, K.X. and Chu, J.J.H., 2017. Antiviral screen identifies EV71 inhibitors and reveals camptothecin-target, DNA topoisomerase 1 as a novel EV71 host factor. *Antiviral Research*, *143*, pp. 122–133.

10.2 ORDER GARRYALES LINDLEY (1833)

10.2.1 Family Eucommiaceae Engler (1907)

The Family Eucommiaceae consists of the single genus *Eucommia* Oliv.

10.2.1.1 *Eucommia ulmoides* Oliv.

Common name: Du zhong (China); tochucha (Japan); duchung (Korea); do trong (Vietnam)

Habitat: Forest, cultivated

Distribution: Tibet, Vietnam, China, Korea, Kapan

Botanical observation: This tree can reach 20 m tall and has an unpleasant aura. The bark and stems yield a sticky latex upon incision. The leaves are simple, spiral, and exstipulate. The petiole is 1–2.5 cm long. The blade is elliptic, glossy, tapering at base to acute, acuminate at apex, 5–15 cm × 2.5–7 cm, marked with 6–9 pairs of secondary nerves sunken above, hairy below, serrate, and somewhat glossy and soft. The flower fascicles are cauliflorous. Perianth none. The androecium includes up to 12 stamens with elongated anthers. The ovary is about 1 cm long and develops a pair of stigmas. The samaras are about 2 cm long, smooth, coriaceous, and shelter a single seed.

Medicinal use: Fever (Tibet)

Strong broad-spectrum antibacterial mid-polar extract: Ethyl acetate of flowers inhibited the growth of *Staphylococcus aureus*, *Bacillus anthracis*, *Bacillus subtilis*, and *Escherichia coli* with the MIC values of was 15.6, 15.6, 31.3, and 31.3 µg/mL, respectively (Ji & Xu, 2008).

Very weak antibacterial (Gram-negative) hydroxycinnamic acid derivative: Chlorogenic acid inhibited *Staphylococcus aureus*, *Bacillus cereus*, *Staphylococcus epidermidis*, *Escherichia coli*, *Salmonella typhimurium*, *Shigella flexneri*, and *Pseudomonas aeruginosa* with MIC values ranging from 2500 to 5000 µg/mL (Fu et al., 2016).

Strong antifungal (filamentous) mid-polar extract: Ethyl acetate of flowers inhibited the growth of *Aspergillus niger* with the MIC values of 12.5 µg/mL (Ji & Xu, 2008).

Antibacterial iridoid glycoside: Aucubin (Wanxi et al., 2014).

Strong antiviral (enveloped circular double-stranded DNA) hydroxycinnamic acid derivative: Chlorogenic acid (LogD –3.9 at pH 7.4; molecular mass = 354.3 g/mol) inhibited the replication of the Hepatitis B virus with the IC_{50} value of 1.2 µM and a selectivity index above 1000 (Wang et al., 2009).

Moderate antiviral (enveloped linear segmented single-stranded (–) RNA) hydroxycinnamic acid derivative: Chlorogenic acid inhibited the replication of the Influenza A/PuertoRico/8/1934(H1N1) and Influenza A/Beijing/32/92(H3N2) with the EC_{50} values of 44.8 and 62.3 µM, respectively, by acting at the late stage of infectious cycle including neuraminidase activity (Ding et al., 2017).

Moderate antiviral (enveloped monopartite linear single-stranded (+) RNA) hydroxycinnamic acid derivative: Chlorogenic acid inhibited the replication of the human coronavirus NL63 with the IC_{50} of 43.4 µM (Weng et al., 2019).

Very weak antiviral (enveloped linear segmented single-stranded (−) RNA) furofuran-type lignan glycoside: (+) Pinoresinol *O* β D glucopyranoside inhibited the replication of the Influenza A/PR/8/34 (H1N1) and Influenza A/Guangzhou/GIRD07/09 (H1N1) strains (Li et al., 2019).

In vivo antiviral fruit: Fruits mixed in the food of chicken increased their resistance to viral infection (Hu et al., 2006).

In vivo antiviral (enveloped linear segmented single-stranded (−) RNA) hydroxycinnamic acid derivative: Intravenous injection of chlorogenic acid at the dose of 100 mg/kg/day protected mice against the Influenza viruses A H1N1 and H3N2 (Ding et al., 2017).

In vivo antiviral (enveloped circular double-stranded DNA) hydroxycinnamic acid derivative: Chlorogenic acid given orally the dose of 100 mg/kg twice a day for 5 days protected ducklings against the Hepatitis B virus as evidenced by a decrease in viral DNA in the sera (Wang et al., 2009).

Viral enzyme inhibition by polar extract: Methanol extract of stems at the concentration of 100 μg/mL inhibited RNA-dependent DNA polymerase activities of the human immunodeficiency virus type-1 reverse transcriptase and human immunodeficiency virus type-1 protease by 20.4% and 3.8%, respectively (Min et al., 2001).

Commentaries

i. The plant abounds with flavonoids such as quercetin, catechin derivatives, kaempferol-3-*O*-β-D-glucoside, and the pyran eucommiol (Watanabe et al., 1997) as well as lignans such as (+)-pinoresinol, (+)-syringaresinol, and iridoids such as genipin, geniposidic acid, and aucubin (Deyama et al., 1987). As such it is a typical example of phenolics-producing plants. Polar extracts from plants belonging to plants phenolics-producing plants are always antimicrobial *in vitro.*
ii. The plant is used in China for the treatment of COVID-19 in combination with oseltamivir phosphate (Luo, 2020).

References

Deyama, T., Ikawa, T., Kitagawa, S. and Nishibe, S., 1987. The constituents of *Eucommia ulmoides* Oliv. V. isolation of dihydroxydehydrodiconiferyl alcohol isomers and phenolic compounds. *Chemical and Pharmaceutical Bulletin*, *35*(5), pp. 1785–1789.

Ding, Y., Cao, Z., Cao, L., Ding, G., Wang, Z. and Xiao, W., 2017. Antiviral activity of chlorogenic acid against influenza A (H1N1/H3N2) virus and its inhibition of neuraminidase. *Scientific Reports*, *7*(1), pp. 1–11.

Fu, L., Lu, W. and Zhou, X., 2016. Phenolic compounds and in vitro antibacterial and antioxidant activities of three tropic fruits: Persimmon, guava, and sweetsop. *BioMed Research International*, *2016.*

Hu, J.Y., Zhang, Y.K., Ouyang, H. and Hu, Z.M., 2006. Application of Eucommia fruit meal in feed of broilers. *Journal of Jishou University (Natural Sciences Edition)*, *27*(1), p. 90.

Ji, Z.P. and Su, Y.Q., 2008. Study on antimicrobial activities of extracts from Eucommia ulmoides Oliv. leaves. *Chemistry and Industry of Forest Products*, 28(2), pp. 63–66.

Li, J., Liang, X., Zhou, B., Chen, X., Xie, P., Jiang, H., Jiang, Z., Yang, Z. and Pan, X., 2019. (+) - x5 pinoresinol O β D glucopyranoside from *Eucommia ulmoides* oliver and its antiinflammatory and antiviral effects against influenza A (H1N1) virus infection. *Molecular Medicine Reports*, *19*(1), pp. 563–572.

Luo, A., 2020. Positive SARS-Cov-2 test in a woman with COVID-19 at 22 days after hospital discharge: A case report. *Journal of Traditional Chinese Medical Sciences*, 7(4), pp. 413–417.

Wang, G.F., Shi, L.P., Ren, Y.D., Liu, Q.F., Liu, H.F., Zhang, R.J., Li, Z., Zhu, F.H., He, P.L., Tang, W. and Tao, P.Z., 2009. Anti-hepatitis B virus activity of chlorogenic acid, quinic acid and caffeic acid in vivo and in vitro. *Antiviral Research*, *83*(2), pp. 186–190.

Wanxi, P.E.N.G., Shengbo, G.E., Dongli, L.I., Bo, M.O., Daochun, Q. and Ohkoshi, M., 2014. Molecular basis of antibacterial activities in extracts of *Eucommia ulmoides* wood. *Pakistan Journal of Pharmaceutical Sciences*, *27.*

Watanabe, J., Kawabata, J., Kurihara, H. and Niki, R., 1997. Isolation and identification of α-glucosidase inhibitors from tochu-cha (*Eucommia ulmoides*). *Bioscience, Biotechnology, and Biochemistry*, *61*(1), pp. 177–178.

Weng, J.R., Lin, C.S., Lai, H.C., Lin, Y.P., Wang, C.Y., Tsai, Y.C., Wu, K.C., Huang, S.H. and Lin, C.W., 2019. Antiviral activity of Sambucus FormosanaNakai ethanol extract and related phenolic acid constituents against human coronavirus NL63. *Virus Research*, *273*, p. 197767.

Min, B.S., Kim, Y.H., Tomiyama, M., Nakamura, N., Miyashiro, H., Otake, T. and Hattori, M., 2001. Inhibitory effects of Korean plants on HIV-1 activities. *Phytotherapy Research*, *15*(6), pp. 481–486.

10.3 ORDER GENTIANALES JUSS. EX BERCHT. & J. PRESL (1820)

10.3.1 Family Apocynaceae A.L. de Jussieu (1789)

The family Apocynaceae consists of trees, treelets, shrubs, climbers, and herbs which produce a white latex. Plants in this family are often of breathtaking grace. The leaves are simple, exstipulate, and opposite or whorled. The blade is often glossy, somewhat thick, and often presents conspicuous secondary nerves which are somewhat straight. The flowers are arranged in cymes. The corolla is tubular and 5-lobed, the lobes contorted. The androecium includes 5 stamens attached to the corolla tube. The gynoecium consists of 2 carpels with parietal placentation. The fruits are characteristic pairs of follicles which are often fusiform and contain minute seeds which are not uncommonly comose. The family Asclepiadaceae has been included in the family Apocynaceae within the subfamily Asclepiadoideae and the subfamily Periplocoideae despite having obvious flower structures and chemical compositions. Members in the family Apocynaceae produce a bewildering array of indole alkaloids and steroid-type alkaloids.

10.3.1.1 *Aganosma dichotoma* (Roth) K. Schum.

Synonym: *Aganosma heynei* (Spreng.) I.M. Turner; *Echites dichotomus* Roth

Common names: Malati (Bangladesh); maloti, nallateega (India)

Habitat: Forests, cultivated

Distribution: India, Bangladesh, Myanmar, Thailand, Malaysia, Indonesia

Botanical observation: This stout laticiferous climber grows up to 10 m long. The leaves are simple, opposite, and exstipulate. The petiole is 3–5 mm long. The blade is elliptic, 5–12 cm × 2–4 cm, coriaceous, wedge-shaped at base, and acuminate or round at apex and presents 2–8 pairs of secondary nerves. The inflorescences are terminal and axillary lax cymes which are 5–20 cm long. The calyx produces 5 lanceolate lobes longer than the petals. The corolla is tubular, pure white, and 1.2–2 cm long and produces 5 contorted lobes which are sharply triangular. Five stamens are attached to the corolla tube. The nectary disc is 5-lobed. The gynoecium includes 2 carpels that form a columnar stigma. The follicles are of 15–40 cm × 2–5 mm and stuffed with linear oblong and plumed seeds.

Medicinal uses: Antiseptic, eye infection (India); leprosy, bronchitis, otitis (Bangladesh)

Commentaries

i. The plant is used in India for the treatment of insomnia and anxiety and thus could produce substances inhibiting bacterial and/or fungal efflux pumps. This plant is on its way to extinction owing to palm oil and logging business, and there is an urgent need for global legislation to preserve primary rainforest before disappearance.
ii. In Malaysia, *Aganosma marginata* (Roxb.) G. Don (local name: *Sekali lima*) is used for fever and likewise in Thailand (local name: *Maduea din*). Methanol extract of *Aganosma marginata* (Roxb.) G. Don inhibited the replication of the Herpes simplex virus type-1 with the IC_{50} value of 39 and the selectivity index of 1.1 (Akanitapichat et al., 2005).

Reference

Akanitapichat, P., Wangmaneerat, A. and Teerawatanasuk, N., 2005. Anti-herpes Simplex Virus Type 1 activity of local Northeastern plants. *Thai Journal of Pharmaceutical Sciences*, *29*(3–4), pp. 137–145.

10.3.1.2 *Allamanda cathartica* L.

Synonyms: *Allamanda aubletii* Pohl; *Allamanda cathartica* var. *grandiflora* (Aubl.) L. H. Bailey & Raffill; *Allamanda cathartica* var. *hendersonii* (W. Bull ex Dombrain) L.H. Bailey & Raffill; *Allamanda cathartica* var. *williamsii* (hort.) L.H. Bailey; *Allamanda grandiflora* (Aubl.) Lam.; *Allamanda hendersonii* W. Bull ex Dombrain; *Allamanda latifolia* C. Presl; *Allamanda linnei* Pohl;

Allamanda salicifolia Hort.; *Allamanda schottii* Hook.; *Allamanda wardleyana* Lebas; *Allamanda williamsii* Hort.; *Echites verticillatus* Sessé & Moc.; *Orelia grandiflora* Aubl.

Common names: Yellow allamanda, golden trumpet vine, malatilata (Bangladesh); shwe pan new (Myanmar)

Habitat: Cultivated

Distribution: Tropical Asia and Pacific

Botanical observation: This laticiferous and climbing shrub grows to 5 m tall. The stem is terete and yields a poisonous and white latex. The leaves are simples, verticillate, and exstipulate. The petiole is up to 5 cm long. The blade is obovate, 5.5–15 cm × 4–5 cm, glossy, marked with about 10–15 pairs of secondary nerves, coriaceous, acute at base, and acuminate at apex. The cymes are terminal and consist of a few showy flowers. The calyx presents 5 lobes. The perianth is tubular, plain dull yellow, 4–8 cm long, and produces 5 broad and orbicular lobes. Five stamens attached on the corolla are present. A disc vaguely 5-lobed is present. The ovary is ovoid and develops a bifid stigma. The capsules are spiny, 3–7 cm × 4–5 cm, very rarely observed, and contain winged seeds.

Medicinal use: Ringworm (India)

Antibacterial (Gram-positive) halo developed by polar extract: Aqueous extract inhibited the growth of *Staphylococcus aureus* (Osborn, 1943).

Broad-spectrum antibacterial halo developed by mid-polar extract: Chloroform extract of leaves (400 µg/disc) inhibited the growth of *Bacillus cereus*, *Bacillus megaterium*, *Bacillus subtilis*, *Staphylococcus aureus*, *Sarcina lutea*, *Escherichia coli*, *Pseudomonas aeruginosa*, *Salmonella paratyphi*, *Salmonella typhi*, *Shigella boydii*, *Shigella dysenteriae*, *Vibrio mimicus*, and *Vibrio parahaemolyticus* with the inhibition zone diameters of 10, 10, 12, 10, 10, 10, 12, 10, 10, 9, 13, 10, and 10 mm, respectively (Islam et al., 2010).

Broad-spectrum antifungal halo developed by mid-polar extract: Chloroform extract of leaves (400 µg/disc) inhibited the growth of *Candida albicans*, *Aspergillus niger*, and *Saccharomyces cerevisiae* with the inhibition zone diameters of 10, 12, and 9 mm, respectively (Islam et al., 2010).

Antifungal (filamentous) halo developed by iridoid glycoside: Plumieride isolated from the leaves inhibited the growth of *Epidermophyton floccosum* and *Microsporum gypseum* (Tiwari et al., 2002).

Strong antifungal (yeast) hydrophilic iridoids: Plumericin (LogD 0.03 at pH 7.4; molecular mass = 290.2 g/mol) isolated from the stem bark inhibited the growth of *Candida albicans* (ATCC 10231), *Candida krusei* (ATCC 14243), *Candida glabrata* (ATCC 15126), *Candida tropicalis* (ATCC 20336), and *Cryptococcus neoformans* (ATCC 90112) with the MIC/MFC values of 1.6/2.6., 0.3/0.4, 2/2.7, 2.6/3.2, and 0.3/0.6 µg/mL, respectively (Singh et al., 2011). Isoplumericin (LogD 0.03 at pH 7.4; molecular mass = 290.2 g/mol) isolated from the stem bark inhibited the growth of *Candida albicans* (ATCC 10231), *Candida krusei* (ATCC 14243), *Candida glabrata* (ATCC 15126), *Candida tropicalis* (ATCC 20336), and *Cryptococcus neoformans* (ATCC 90112) with the MIC/MFC values of 3.8/5.3, 0.4/0.8, 3.9/5.7, 3.7/5.6, and 0.5/0.9 µg/mL, respectively (Singh et al., 2011).

Commentaries

i. Ethanol extract of *Allamanda violacea* Gardner at the concentration of 200 µg/mL inhibited the Human immunodeficiency virus type-1 reverse transcriptase by 22% (Tan et al., 1991). The iridoid fulvoplumierin isolated from *Plumeria rubra* L. inhibited the Human immunodeficiency virus type-1 reverse transcriptase with the IC_{50} value of 45 µg/mL (Tan et al., 1991).

   Natural products with antileishmanial and cytotoxic activities *in vitro* are often antifungal *in vitro*, and this is the case for plumericin and isoplumericin (Sharma et al., 2011). The antifungal agent Amphotericin targets ergosterol of the membrane of *Leishmania donovani* promastigote (Saha et al., 1986). Plumericin induces apoptosis in mammalian cells (Sun et al., 2016), and one could suggest the same mechanism in yeasts (Madeo et al., 2004). Untreatable yeasts are emerging (Kainz et al., 2020), and there is a need for new anticandidal agents.

References

Islam, M.R., Ahamed, R., Rahman, M.O., Akbar, M.A., Al-Amin, M., Alam, K.D. and Lyzu, F., 2010. In vitro antimicrobial activities of four medicinally important plants in Bangladesh. *European Journal of Scientific Research*, *39*(Suppl 2), pp. 199–206.

Kainz, K., Bauer, M.A., Madeo, F. and Carmona-Gutierrez, D., 2020. Fungal infections in humans: The silent crisis. *Microbial Cell*, *7*(6), p. 143.

Madeo, F., Herker, E., Wissing, S., Jungwirth, H., Eisenberg, T. and Fröhlich, K.U., 2004. Apoptosis in yeast. *Current Opinion in Microbiology*, *7*(6), pp. 655–660.

Osborn, E.M., 1943. On the occurrence of antibacterial substances in green plants. *British Journal of Experimental Pathology*, *24*(6), p. 227.

Saha, A.K., Mukherjee, T. and Bhaduri, A., 1986. Mechanism of action of amphotericin B on Leishmania donovani promastigotes. *Molecular and Biochemical Parasitology*, *19*(3), pp. 195–200.

Sharma, U., Singh, D., Kumar, P., Dobhal, M.P. and Singh, S., 2011. Antiparasitic activity of plumericin & isoplumericin isolated from Plumeria bicolor against Leishmania donovani. *The Indian Journal of Medical Research*, *134*(5), p. 709.

Singh, D., Sharma, U., Kumar, P., Gupta, Y.K., Dobhal, M.P. and Singh, S., 2011. Antifungal activity of plumericin and isoplumericin. *Natural Product Communications*, *6*(11), p. 1567.

Sun, J., Wu, J., Zhang, L., Wan, Q.Y., Wang, L.C., Liu, Y.Y., Wang, R.M. and Sun, D.Q., 2016. Plumericin inhibits the viability of cervical cancer cells by induction of apoptosis through degradation of DNA structure. *Bangladesh Journal of Pharmacology*, *11*(2), pp. 337–341.

Tan, G.T., Pezzuto, J.M., Kinghorn, A.D. and Hughes, S.H., 1991. Evaluation of natural products as inhibitors of human immunodeficiency virus type 1 (HIV-1) reverse transcriptase. *Journal of Natural Products*, *54*(1), pp. 143–154.

Tiwari, T.N., Pandey, V.B. and Dubey, N.K., 2002. Plumieride from Allamanda cathartica as an antidermatophytic agent. *Phytotherapy Research*, *16*(4), pp. 393–394.

10.3.1.3 *Alstonia scholaris* (L.) R.Br.

Synonyms: *Echites scholaris* L.

Common names: Devil's tree, dita bark tree; chaitan, chatim, chatinidare, chattin (Bangladesh); lazarongpang, pala, satiana (India); kayu batu, lame (Indonesia); ditin pet, tin pet (Laos); pulai (Indonesia; Malaysia); leptan ba, let ban kha (Myanmar); chatiwan (Bangladesh; Nepal); aiting, budo, chimb, rambaka, watsil (Papua New Guinea); dita, karayum (the Philippines); sua, mo cua (Vietnam)

Habitat: Cultivated

Distribution: India, Bangladesh, Myanmar, Cambodia, Laos, Vietnam, China, Thailand, Malaysia, Indonesia, the Philippines, Papua New Guinea

Botanical observation: This tree grows to 25 m tall and emits somewhat a dark and lugubre aura (the branches) when fully grown. The bole is not straight and often massive. The bark is dark brown to light grey and yields a copious white latex upon incision. The leaves are simple, exstipulate, and whorled. The petiole is 1–2.3 cm long. The blade is elliptic to obovate, acute at base, acute to round at apex, dark green above and glossy, glaucous below, 5–15 cm × 2–5 cm, and marked with 25–50 pairs of parallel and straight secondary nerves. The cymes are terminal, somewhat round, up to about 15 cm across, and consist of innumerable, minute, and somewhat yellowish-white flowers with a subtle fragrance. The flowers are ephemeral and fall and cover the area around which the tree grows. The calyx is minute, tubular, and develops 5 triangular lobes. The corolla is whitish to cream, tubular, the tube 0.5–1 cm, and produces 5 contorted and somewhat obtriangularlobes. The 5 stamens are attached to the corolla. The ovary is globose, somewhat hairy, and develops a slender style. The follicles are slender, pendulous, slightly curling, light green, up to 25 cm long, and contain minute hairy seeds.

Medicinal uses: Abscesses, cuts, wounds (Myanmar); ulcers (Bangladesh, India, Myanmar); bronchitis, cholera, jaundice, leprosy, pneumonia, skin infections (India); diarrhea (India, Myanmar, Papua New Guinea); sores (Laos, Myanmar, Vietnam); leucorrhea (Bangladesh); dysentery (India, Indonesia, Papua New Guinea); syphilis (Indonesia, India); fever (Bangladesh, India, Indonesia, Myanmar, Papua New Guinea); cough (Papua New Guinea, the Philippines, China), cold (Bangladesh, Myanmar, the Philippines)

Broad-spectrum antibacterial polar extract: Butanol fraction of stem bark inhibited the growth of *Bacillus cereus*, *Staphylococcus aureus*, *Staphylococcus epidermidis*, *Escherichia coli*, *Klebsiella pneumoniae*, *Neisseria gonorrhoeae*, *Pseudomonas aeruginosa*, and *Salmonella typhimurium* (Khan et al., 2003).

Strong broad-spectrum antibacterial indole alkaloids: Alstoniascholarine A isolated from the leaves inhibited the growth of *Staphylococcus aureus*, *Pseudomonas aeruginosa*, *Enterococcus faecalis*, *Klebsiella pneumoniae*, *Providencia smaitii*, and *Escherichia coli* with the MIC values of 100, 25, 25, 12.5, 25, and 50 µg/mL, respectively (Qin et al., 2015). Alstoniascholarine E isolated from the leaves inhibited the growth of *Staphylococcus aureus*, *Pseudomonas aeruginosa*, *Enterococcus faecalis*, *Klebsiella pneumoniae*, *Providencia smaitii*, and *Escherichia coli* with the MIC values of 100, 50, 50, 25, 25, and 25 µg/mL, respectively (Qin et al., 2015). Alstoniascholarine J isolated from the leaves inhibited the growth of *Staphylococcus aureus*, *Pseudomonas aeruginosa*, *Enterococcus faecalis*, *Klebsiella pneumoniae*, *Providencia smaitii*, and *Escherichia coli* with the MIC values of 50, 3.1, 12.5, 12.5, 1.5, and 25 µg/mL, respectively (Qin et al., 2015).

Vallesamine isolated from the leaves inhibited the growth of *Staphylococcus aureus* (ATCC 25922), *Klebsiella pneumoniae* (AT13883), *Enterococcus faecalis* (ATCC 10541), *Escherichia coli* (ATCC 11775), and *Pseudomonas aeruginosa* (ATCC 27853) with the MIC values of 100, 50, 1.5, 50, and 0.7 µg/mL, respectively (Liu et al., 2015). 5-Hydroxy-19,20-*Z*-alschomine isolated from the leaves inhibited the growth of *Klebsiella pneumoniae* (AT13883), *Enterococcus faecalis* (ATCC 10541), *Escherichia coli* (ATCC 11775), and *Pseudomonas aeruginosa* (ATCC 27853) with the MIC values of 6.2, 1.5, 100, and 0.7 µg/mL, respectively (Liu et al., 2015).

Weak antimycobacterial indole alkaloid: Tubotaiwine isolated from the leaves inhibited the growth of *Mycobacterium tuberculosis* H37Rv with the MIC value of 100 µg/mL (Macabeo et al., 2008)

Strong antifungal (filamentous) indole alkaloids: Alstoniascholarine A isolated from the leaves inhibited the growth of *Microsporum gypseum*, *Epidermophyton floccosum*, and *Trichophyton mentagrophytes* with the MIC values of 125, 62.5, and 125 µg/mL, respectively (Qin et al., 2015). Alstoniascholarine E isolated from the leaves inhibited the growth of *Microsporum gypseum*, *Epidermophyton floccosum*, and *Trichophyton mentagrophytes* with the MIC values of 125, 62.5, 62.5 µg/mL, respectively (Qin et al., 2015). Alstoniascholarine J isolated from the leaves inhibited the growth of *Microsporum gypseum*, *Epidermophyton floccosum*, and *Trichophyton mentagrophytes* with the MIC values of 250, 31.2, and 62.5 µg/mL, respectively (Qin et al., 2015).

Strong antiviral (enveloped linear monopartite dimeric single-stranded (+) RNA) polar extract: Methanol extract of leaves inhibited the replication of the Human immunodeficiency virus type-1 (NL4.3 strain) in Human CD4+ T cell line CEM-GFP by 40.4% at the concentration of 10 µg/mL (Sabde et al., 2011).

Strong antiviral (enveloped monopartite linear double-stranded DNA) indole alkaloids: Strictamine isolated from the aerial parts inhibited the replication of the Herpes simplex virus with the IC_{50} value of 0.3 µg/mL and selectivity index of 13.9 (Zhang et al., 2014). 17-nor-excelsinidine isolated from the aerial parts inhibited the replication of the Herpes simplex virus with the IC_{50} value of 1 µg/mL and selectivity index of 6.4 (Zhang et al., 2014). Strictamine inhibited the replication of the Adenovirus with the IC_{50} value of 0.2 µg/mL and selectivity index of 11.8 (Zhang et al., 2014). 17-nor-excelsinidine inhibited the replication of the Adenovirus virus with the IC_{50} of 0.9 µg/mL and selectivity index of 3.5 (Zhang et al., 2014).

Moderate antiviral (enveloped linear segmented single-stranded (-) RNA) alkaloid fraction: Alkaloid fraction isolated from the leaves at the concentration of 100 µg/mL decreased the virus titer of the Influenza virus A/PR/8/34 (H1N1) in Madin-Darby canine kidney cells, A549 cells, and U937-derived macrophages (Zhou et al., 2020).

In vivo antiviral (enveloped linear segmented single-stranded (-) RNA) alkaloid fraction: Alkaloid fraction of leaves given orally to mice at the dose of 50 mg/kg/day for 5 days evoked a survival rate of 70% (oseltamivir and 50 mg/kg/day: 100%), mitigated body weight loss, and decreased pulmonary viral load during infection with the Influenza virus A/PR/8/34 (H1N1) (Zhou et al., 2020).

Commentaries

i. This tree is an example of plant used medicinally in various tropical Asian countries, and such plants tend to generate strong antimicrobial principles. The tree (and other members of the genus *Alstonia* R. Br.) was known of our peer Sushruta as *Sapta chaga* or *Sapta parna* which could mean 7 leaves and listed in the colonialist Pharmacopoeia of India of 1868 for malarial fever. We argue that the use of this tree in tropical Asia could be a vestige of Ayurvedic medicine from Indian kingdoms in the region including the Philippines (Houtart, 1976; Orlina, 2012). Ayurvedic practitioners use the plant for the management of COVID-19 (Zende and Amarprakash, 2020). As for bacterial and fungal resistance to antibiotic and antifungal agents indole alkaloids in this plant are neuroactive (Dutta et al., 1976) and as such could inhibit bacterial and/or fungal efflux pumps.
ii. *Alstonia angustiloba* Miq. is used in Malaysia to heal sores. *Alstonia spectabile* R. Br. is used to treat cold, cough, fever, and ulcers in Papua New Guinea (local name: *Tutu*)

References

Dutta, S.C., Bhattacharya, S.K. and Ray, A.B., 1976. Flower alkaloids of Alstonia scholaris. *Planta Medica*, *30*(05), pp. 86–89.

Houtart, F., 1976. Buddhism and politics in South-East Asia: Part one. *Social Scientist*, *5*, pp.3–23.

Khan, M.R., Omoloso, A.D. and Kihara, M., 2003. Antibacterial activity of Alstonia scholaris and Leea tetramera. *Fitoterapia*, *74*(7–8), pp. 736–740.

Liu, L., Chen, Y.Y., Qin, X.J., Wang, B., Jin, Q., Liu, Y.P. and Luo, X.D., 2015. Antibacterial monoterpenoid indole alkaloids from Alstonia scholaris cultivated in temperate zone. *Fitoterapia*, *105*, pp. 160–164.

Macabeo, A.P.G., Krohn, K., Gehle, D., Read, R.W., Brophy, J.J., Franzblau, S.G. and Aguinaldo, M.A.M., 2008. Activity of the extracts and indole alkaloids from Alstonia scholaris against Mycobacterium tuberculosis H37Rv. *The Philippine Agricultural Scientist*, *91*(3), pp. 348–351.

Qin, X.J., Zhao, Y.L., Lunga, P.K., Yang, X.W., Song, C.W., Cheng, G.G., Liu, L., Chen, Y.Y., Liu, Y.P. and Luo, X.D., 2015. Indole alkaloids with antibacterial activity from aqueous fraction of Alstonia scholaris. *Tetrahedron*, *71*(25), pp. 4372–4378.

Sabde, S., Bodiwala, H.S., Karmase, A., Deshpande, P.J., Kaur, A., Ahmed, N., Chauthe, S.K., Brahmbhatt, K.G., Phadke, R.U., Mitra, D. and Bhutani, K.K., 2011. Anti-HIV activity of Indian medicinal plants. *Journal of Natural Medicines*, *65*(3–4), pp. 662–669.

Zende, A.K. and Amarprakash, D., 2020. Role of Ayurveda in prevention of Covid-19 pandemic-An integrative overview. *Journal of Ayurveda and Integrated Medical Sciences*, *5*(05), pp. 350–364.

Zhang, L., Zhang, C.J., Zhang, D.B., Wen, J., Zhao, X.W., Li, Y. and Gao, K., 2014. An unusual indole alkaloid with anti-adenovirus and anti-HSV activities from Alstonia scholaris. *Tetrahedron Letters*, *55*(10), pp. 1815–1817.

Zhou, H.X., Li, R.F., Wang, Y.F., Shen, L.H., Cai, L.H., Weng, Y.C., Zhang, H.R., Chen, X.X., Wu, X., Chen, R.F. and Jiang, H.M., 2020. Total alkaloids from Alstonia scholaris inhibit influenza a virus replication and lung immunopathology by regulating the innate immune response. *Phytomedicine*, *77*, p. 153272.

10.3.1.4 *Calotropis gigantea* (L.) W.T. Aiton

Synonyms: *Asclepias gigantea* L.; *Madorius giganteus* (L.) Kuntze; *Periploca cochinchinensis* Lour.; *Streptocaulon cochinchinense* (Lour.) G. Don

Common name: Crown flower, gigantic swallow-wort, giant milkweed; angar pata, akanda, akund, akond mul, akondo (Bangladesh); hutu (Hawaii); aditya, anggot, mudar (India); khark (Iran); waduri, widuri (Indonesia); mayo (Myanmar); auk (Nepal); sasus (Papua New Guinea); manak kovi (Sri Lanka); rak khao (Thailand)

Habitat: Sun-exposed wastelands, sandy soils, roadsides

Distribution: From Pakistan to Hawaii

Botanical observation: This handsome Asclepiadaceous shrub grows to 3 m tall. The stem is terete, hairy, fibrous, and yields a poisonous latex upon incision. The leaves are simple, decussate, and exstipulate. The petiole is up to 2 cm long and stout. The blade is broadly elliptic, 6–20 cm × 3–10 cm, coriaceous, cordate at base, acute at apex, hairy, and marked with 4–8 pairs of conspicuous secondary nerves. The cymes are umbel-like, about 15 cm long, and present magnificent and characteristic flowers. The calyx produces 5 lobes and is about 5 mm long. The corolla is 3–4.5 cm × 1.5–2.2 cm, purplish, pinkish, or pure white, 5-lobed; and fleshy. A corona is present. The follicles are ovoid, green, rough, 5–10 cm × 2.5–4 cm, recurved, and contain numerous comose seeds.

Medicinal uses: Abscesses (Myanmar); boils, carries, dysentery, leukorrhea, otitis, tuberculosis, typhus (India); skin diseases (Myanmar); syphilis (China, India); sinusitis (Thailand); cholera (Pakistan); fever (India, Papua New Guinea); leprosy (Myanmar, India)

Broad-spectrum antibacterial halo developed by polar extract: Aqueous extract of leaves (100 µL of a 1 mg/mL solution/7 mm well) inhibited the growth of *Staphylococcus aureus*, *Klebsiella pneumoniae*, *Bacillus cereus*, *Pseudomonas aeruginosa*, *Micrococcus luteus*, and *Escherichia coli* with the inhibition zone diameters of 13.3, 12.6, 17.3, 16, 16, and 17.6 mm, respectively (Kumar et al., 2010).

Strong broad-spectrum antibacterial mid-polar extract: Ethyl acetate extract of flowers inhibited the growth of *Staphylococcus aureus*, *Bacillus subtilis*, *Bacillus megaterium*, *Sarcina lutea*, *Escherichia coli*, *Shigella sonnei*, *Shigella shiga*, and *Shigella dysenteriae* with the MIC value of 32, 16, 128, 16, 32, 32, 64, and 128 µg/mL, respectively (Habib & Karim, 2009).

Broad-spectrum antifungal halo developed by polar extract: Aqueous extract of latex inhibited the growth of *Candida albicans*, *Trichophyton mentagrophytes*, *Trichophyton rubrum*, *Aspergillus flavus*, *Aspergillus niger*, and *Penicillium chrysogenum* (Saratha and Subramanian, 2010).

Strong antiviral (enveloped linear monopartite dimeric single-stranded (+) RNA) polar extract: Acetonitrile extract of flowers inhibited the replication the Human immunodeficiency virus type-1 (strain IIIB) with the EC_{50} value of 0.6 µg/mL and the selectivity index of 1 (Locher et al., 1996).

Strong antiviral (enveloped monopartite linear double-stranded DNA) polar extract: Ethanol extract inhibited the replication of the Herpes simplex virus type-1 in Vero cells with the MIC value of 10 µg/mL (Hamidi et al., 1996).

Strong antiviral (enveloped monopartite linear single-stranded (-) RNA) polar extract: Ethanol extract inhibited the replication of the Vesicular stomatitis virus in Vero cells with the MIC value of 10 µg/mL (Hamidi et al., 1996).

Strong antiviral (enveloped linear segmented single-stranded (-) RNA) lignan-type glycoside(+): pinoresinol 4-*O*-[6″-*O*-vanilloyl]-*β*-D-glucopyranoside isolated from the latex inhibited the replication of the Influenza virus A/PR/8/34 (H1N1), A/FM/1/47 (H1N1), A/Aichi/2/68 (H3N2), and Influenza B/Lee/1940 with the IC_{50} of 24.5, 39.8, 16.3, and 13.4 µM, respectively (Parhira et al., 2014).

Strong antimycobacterial extract: Ethanol extract inhibited the growth of *Mycobacterium tuberculosis* (H37 Rv) at a dilution of 1:80 (Grange & Davey, 1990).

Commentaries

i. The plant was known of our peer the Indian physician Sushruta.
ii. The family Asclepiadaceae has been merged with the Apocynaceae and placed in the Subfamily Asclepiadoideae, although these families have clear different chemically and morphologically. Fusions and inclusions of families and genera create confusion for phytochemists and pharmacologists and even botanists.
iii. *Asclepias curassavica* L. is used to heal sores in China. *Calotropis procera* (Aiton) W.T. Aiton (local name: *Nika aak*) is used in India (local name: *Akkanda*) for abscesses, ringworm, and wounds; in Myanmar (local name: *Mayo*) for carries, cholera, leprosy, and otitis; and in Pakistan (local name: *Akh*, *Spalmakka*) for boils, cuts, chest infection, jaundice, and wounds. Consider that extract of latex *Calotropis procera* (Aiton) W.T. inhibited the growth of a broad-spectrum of bacteria and fungi (which is a sign of cytotoxicity for mammalian cells) and displayed antibiotic and antifungal potentiating effects (Nenaah & Ahmed, 2011). *Calotropis gigantea* (L.) W.T. Aiton yields cardiac glycosides such as calotropin (Kiuchi et al., 1998) which are dreadfully poisonous (Mishra et al., 2015). Cardiac glycosides targeting Na/K ATPase may very well inhibit bacterial and/or fungal efflux pumps. Further, cardiac glycosides inhibit the growth of a broad spectrum of viruses

(Amarelle & Lecuona, 2018). Cardiac glycosides inhibit the replication of the coronaviruses via the inhibition of Na^+/K^+-ATPase and inhibition of virus entry (Burkard et al., 2015). Thus, calotropin might have been able to inhibit coronaviruses replication and to inhibit bacterial and/or fungal efflux plumps.

References

Amarelle, L. and Lecuona, E., 2018. The antiviral effects of Na, K-ATPase inhibition: a minireview. *International Journal of Molecular Sciences*, *19*(8), p. 2154.

Burkard, C., Verheije, M.H., Haagmans, B.L., van Kuppeveld, F.J., Rottier, P.J., Bosch, B.J., de Haan, C.A. 2015. ATP1A1-mediated Src signaling inhibits coronavirus entry into host cells. *Journal of Virology*, *89*, pp. 4434–4448.

Grange, J.M. and Davey, R.W., 1990. Detection of antituberculous activity in plant extracts. *Journal of Applied Bacteriology*, *68*(6), pp. 587–591.

Habib, M.R. and Karim, M.R., 2009. Antimicrobial and cytotoxic activity of di-(2-ethylhexyl) phthalate and anhydrosophoradiol-3-acetate isolated from Calotropis gigantea (Linn.) flower. *Mycobiology*, *37*(1), pp. 31–36.

Hamidi, J.A., Ismaili, N.H., Ahmadi, F.B. and Lajisi, N.H., 1996. Antiviral and cytotoxic activities of some plants used in Malaysian indigenous medicine. *Pertanika Journal of Tropical Agricultural Science*, *19*(2/3), pp. 129–136.

Kumar, G., Karthik, L. and Rao, K.V.B., 2010. Antibacterial activity of aqueous extract of Calotropis gigantea leaves–an in vitro study. *International Journal of Pharmaceutical Sciences Review and Research*, *4*(2), pp. 141–144.

Kiuchi, F., Fukao, Y., Maruyama, T., Obata, T., Tanaka, M., Sasaki, T., Mikage, M., Haque, M.E. and Tsuda, Y., 1998. Cytotoxic principles of a Bangladeshi crude drug, akond mul (roots of Calotropis gigantea L.). *Chemical and Pharmaceutical Bulletin*, *46*(3), pp. 528–530.

Locher, C.P., Witvrouw, M., De Béthune, M.P., Burch, M.T., Mower, H.F., Davis, H., Lasure, A., Pauwels, R., De Clercq, E. and Vlietinck, A.J., 1996. Antiviral activity of Hawaiian medicinal plants against human immunodeficiency virus type-1 (HIV-1). *Phytomedicine*, *2*(3), pp. 259–264.

Mishra, A.K., George, A., Devakiruba, N.S. and Sathyendra, S., 2015. A rare case of calotropis poisoning. *Indian Journal of Forensic Medicine & Toxicology*, *9*, pp. 62–64.

Nenaah, E.G. and Ahmed, M.E., 2011. Antimicrobial activity of extracts and latex of Calotropis procera (Ait.) and synergistic effect with reference antimicrobials. *Research Journal of Medicinal Plant*, *5*(6), pp. 706–716.

Parhira, S., Yang, Z.F., Zhu, G.Y., Chen, Q.L., Zhou, B.X., Wang, Y.T., Liu, L., Bai, L.P. and Jiang, Z.H., 2014. In vitro anti-influenza virus activities of a new lignan glycoside from the latex of Calotropis gigantea. *PloS one*, *9*(8), p. e104544.

Saratha, V. and Subramanian, S.P., 2010. Evaluation of antifungal activity of Calotropis gigantea latex extract: An in vitro study. *International Journal of Pharmaceutical Sciences and Research*, *1*(9), pp. 88–96.

10.3.1.5 *Carissa carandas* L.

Synonyms: *Arduina carandas* (L.) K. Schum., *Carissa congesta* Wight, *Carissa edulis* (Forssk.) Vahl; *Damnacanthus esquirolii* H. Lév.; *Jasminonerium carandas* (L.) Kuntze

Common names: Bengal currant, Christ's thorn, jasmine flowered Carissa, wild black Carandas; karamcha, kormocha, kurumchi (Bangladesh); kalakai, karamurda, karinkara, kurunda, peria claka pullum (India)

Habitat: Cultivated

Distribution: Southeast Asia, Pacific Islands

Botanical observation: This handsome shrub grows to 3 m tall. The stems develop woody, needle-like, and up to 5 cm long spines. A latex exudes upon incision of the stems. The leaves are simple, sessile, opposite, and exstipulate. The blade is coriaceous, dark green above, 3–7.5 cm × 1.5–4.5 cm, round at base, minutely apiculate at apex, broadly lanceolate, and marked with about 8 pairs of secondary nerves. The cymes are terminal and present a few flowered flowers. The calyx comprises 5 lobes and reaches 7 mm long. The corolla is tubular, the tube 2 cm long, pinkish, and produces

5 pure white contorted lobes which are 1 cm long and lanceolate. The 5 stamens are attached to the corolla throat. The ovary is ovoid and produces a slender style. The berries are light-pink, olive-shaped, edible (India), 1.5–2.5 cm × 1–2 cm, and with a positive aura of life force.

Medicinal uses: Septicemia (Palau Island); ulcers, wounds (India)

Moderate broad-spectrum antibacterial polar extracts: Phenolic fraction of fruits inhibited the growth of *Staphylococcus aureus*, *Bacillus subtilis*, *Enterococcus faecalis*, *Listeria monocytogenes*, *Escherichia coli*, *Pseudomonas aeruginosa*, *Salmonella typhi*, and *Shigella dysenteriae* with MIC of 62.5, 62.5, 125, 125, 500, 1000, 500, and 1000 µg/mL, respectively (Aman et al., 2014). Polyphenolic fraction of fruits inhibited the growth of *Bacillus subtilis* (clinical isolate), *Staphylococcus aureus* (clinical isolate), *Staphylococcus epidermidis* (clinical isolate), and *Klebsiella pneumoniae* (clinical isolate) with the MIC values of 500, 250, 250, and 500 µg/mL, respectively (Siddiqi et al., 2011).

Broad spectrum antifungal polar extract: Polyphenolic fractions (100 µL of a 1 mg/mL solution/5 mm well) inhibited the growth of *Cryptococcus neoformans*, *Penicillium notatum*, *Aspergillus flavus*, *Aspergillus niger*, and *Microsporum gypseum* with the inhibition zone diameters of 25, 35, 40, 35, and 37 mm, respectively (Siddiqi et al., 2011).

Strong antiviral (enveloped monopartite linear double-stranded DNA) polar extracts: Aqueous extract of root bark and stems inhibited the replication of ganciclovir-sensitive clinical isolate of Human cytomegalovirus in human embryonic lung fibroblasts with the EC_{50} value of 74 µg/mL (Tolo et al., 2007). Aqueous extract of roots inhibited the replication of the Herpes simplex virus type-1, wild-type Herpes simplex virus type-2, acyclovir-resistant Herpes simplex virus type-1, and thymidine kinase–deficient Herpes simplex virus type-1 (B2006) with the EC_{50} values of 15.1, 6.9, 8.1, and 11.1 µg/mL, respectively (Tolo et al., 2020).

In vivo antiviral (enveloped monopartite linear double-stranded DNA) polar extract: Aqueous extract of roots given orally at the dose of 250 mg/day decreased the mortality rate of mice infected with the Herpes simplex virus type-1 (7401H) from 100% to 30% (Tolo et al., 2006).

Commentaries

i. In Bangladesh, *Carissa carandas* var. *congesta* (Wight) Bedd. is used to heal ulcers. In India, *Carissa opaca* Stapf. ex Haines (local name: *Garna*) is used to treat pneumonia, herpes, abscess, cough, and fever, and in Pakistan (local name: *Garanda*), it is used for jaundice and sore throat. In Myanmar, *Carissa spinarum* L. (local name: *Khan*) is used as antiseptic, and in India (local name: *Cherumully*), it is used for the treatment of gonorrhea, syphilis, rabies, fever, and ulcers.

ii. Aqueous extract of root of *Carissa carandas* L. has anticonvulsant effects (Ya'u et al., 2015) suggesting the presence of neuroactive and thus possible inhibitors of bacterial and/or fungal efflux pumps.

iii. One could contemplate the possibility of the fruits having some protective effects against viral infections. The plant yields 3-*O*-acetyl chlorogenic acid as well as flavonol glycosides, including kaempferol 3-*O*-*β*-D glucopyranoside, quercetin-3-*O*-*β*-D glucopyranoside, rhamnetin-3-*O*-*β*-D glucopyranoside, and isorhamnetin-3-*O*-*β*-D-glucopyranoside (Al-Youssef & Hassan, 2017). Polyphenolics by virtue of their mild astringency could precipitate viral particles and afford some sort of protection. These fruits like those of astringent plant species could be added to a list of functional antiviral foods.

iv. In the Subfamily Asclepiadoideae and Tribe Asclepiadeae, *Cynanchum atratum* Bunge is used in Korea to heal wounds. *Cynanchum auriculatum* Buch.-Ham. ex Wight is used in China to treat tuberculosis. In India, *Cynanchum arnottianum* Wight (local name: *Kuri*) is used to heal putrefied wounds. *Cynanchum extensum Jacq.* is used in India (local name: *Vayli parti*) to treat lung diseases. In Nepal, *Cynanchum callialata* Buch.-Ham. ex Wight (local name: *Kauwaloli*) is used to treat infection of nipples. Methanol extract of aerial parts of *Cynanchum wildfordii* Franch. & A. Sav. at the concentration of 100 µg/mL inhibited the Human immunodeficiency virus type-1 protease by 17.7%, respectively (Min et al., 2001). Plants in the genus *Cynanchum* L. are prone to yield phenanthroindolizidine alkaloids (Min et al., 2005) such as 7-demethoxytylophorine (also known as antofine; LogD 4.3 at pH 7.4; molecular mass = 363.4 g/mol) which inhibited the replication of the Tobacco mosaic virus by 65% at the concentration of 1 g/mL (An et al., 2001). The Tobacco mosaic virus being positive-stranded virus, 7-demethoxytylophorine could be of value against coronaviruses.

In the Subfamily Asclepiadoideae and Tribe Asclepiadeae *Metaplexis japonica* (Thunb.) Mak. is used to heal abscesses in Japan. In China, *Oxystelma esculentum* (L.f.) R. Br. is used to heal ulcers in China. In Korea, *Pycnostelma paniculatum* (Bge) K. Schim. is used to heal wounds in Korea.

v. In the Subfamily Asclepiadoideae and Tribe Marsdenieae *Dischidia acuminata* Cost. is used to treat blennorrhea in Cambodia, Laos, and Vietnam. In China, *Dischidia chinensis* Champ. ex Benth. (local name: *Gua nin*) is used to treat ulcers. In Thailand, *Dischidia bengalensis* Colebr. (local name: *O lop*) is used to treat ringworm. In Indonesia, *Dischidia imbricata* Blume (Steud.) (local name: *Gogotifo*) is used to treat gonorrhea, burns, yaws, and wounds, and it is used for fungal infections in Thailand (local name: *Kipma lom*). In Malaysia, *Dischidia major* (Vahl) Merr. (local name: *Akar kul*) is used to treat cough. In Indonesia, *Dischidia nummularia* R. Br. (local name: *Daun pitis kecil*) is used to treat mouth ulcers, wounds, and gonorrhea. In the Philippines, *Dischidia purpurea* Merr. (local name: *Talokop*) is used to treat Herpes. Plants in the genus *Dischidia* R. Br. have apparently not been much studied for antimicrobial constituents.

vi. In the Subfamily Asclepiadoideae and Tribe Marsdenieae *Gymnema acuminatum* Wall. is used to heal sores in Malaysia. Plants in the genus *Gymnema* R. Br. yield triterpene saponins such as gymnemic acid A that, at a dose of 75 mg/kg/day for 4 days intraperitoneally, protected mice against lung lesions induced by Influenza virus A by about 35% (Sinsheimer et al., 1968).

vii. In the Subfamily Asclepiadoideae and Tribe Periplocoideae, *Hemidesmus indicus* R. Br. is used to treat syphilis in Cambodia, Laos, and Vietnam, and in India, it is part of a formulation used in Siddha medicine for COVID-19. Hydroalcoholic extract of this plant at the concentration of 100 µg/mL abrogated the replication of the Herpes simplex virus type-2 in Vero cells (Bonvicini et al., 2018).

viii. In the Subfamily Asclepiadoideae and Tribe Ceropegieae, *Heterostemma cuspidatum* Decne. disinfects ulcers in the Philippines.

References

Al-Youssef, H.M. and Hassan, W.H., 2017. Chemical constituents of Carissa edulis Vahl. *Arabian Journal of Chemistry*, *10*(1), pp. 109–113.

Aman, S., Naim, A., Siddiqi, R. and Naz, S., 2014. Antimicrobial polyphenols from small tropical fruits, tea and spice oilseeds. *Food Science and Technology International*, *20*(4), pp. 241–251.

An, T.Y., Huang, R.Q., Yang, Z., Zhang, D.K., Li, G.R., Yao, Y.C. and Gao, J., 2001. Alkaloids from Cynanchum komarovii with inhibitory activity against the tobacco mosaic virus. *Phytochemistry*, *58*(8), pp. 1267–1269.

Bonvicini, F., Lianza, M., Mandrone, M., Poli, F., Gentilomi, G.A. and Antognoni, F., 2018. Hemidesmus indicus (L.) R. Br. extract inhibits the early step of herpes simplex type 1 and type 2 replication. *New Microbiologica*, *41*(3), pp. 187–194.

Min, B.S., Kim, Y.H., Tomiyama, M., Nakamura, N., Miyashiro, H., Otake, T. and Hattori, M., 2001. Inhibitory effects of Korean plants on HIV-1 activities. *Phytotherapy Research*, *15*(6), pp. 481–486.

Min, H.Y., Chung, H.J., Kim, S., Park, H.J., Park, E.J., Hong, J.Y., Kang, Y.J., Bae, I.K. and Lee, S.K., 2005. Suppressive effects of antofine, a phenanthroindolizidine alkaloid isolated from Cynanchum paniculatum, on the in vitro cancer cell proliferation and in vivo tumor growth.

Sinsheimer, J.E., Subba Rao, G., McIlhenny, H.M., Smith, R.V., Maassab, H.F. and Cochran, K.W., 1968. Isolation and antiviral activity of the gymnemic acids. *Experientia* 24(3): 302–303.

Siddiqi, R., Naz, S., Ahmad, S. and Sayeed, S.A., 2011. Antimicrobial activity of the polyphenolic fractions derived from Grewia asiatica, Eugenia jambolana and Carissa carandas. *International Journal of Food Science & Technology*, *46*(2), pp. 250–256.

Tolo, F.M., Rukunga, G.M., Muli, F.W., Njagi, E.N., Njue, W., Kumon, K., Mungai, G.M., Muthaura, C.N., Muli, J.M., Keter, L.K. and Oishi, E., 2006. Anti-viral activity of the extracts of a Kenyan medicinal plant Carissa edulis against herpes simplex virus. *Journal of Ethnopharmacology*, *104*(1–2), pp. 92–99.

Tolo, F.M., Rukunga, G.M., Muli, F.W., Ochora, J., Eizuru, Y., Muthaura, C.N., Kimani, C.W., Mungai, G.M. and Kofi-Tsekpo, M.W., 2007. In vitro anti-viral activity of aqueous extracts of Kenyan Carissa edulis Prunus africana and Melia azedarach against human cytomegalovirus. *African Journal of Health Sciences*, *14*(3), pp. 143–148.

Ya'u, J., Yaro, A.H., Malami, S., Musa, M.A., Abubakar, A., Yahaya, S.M., Chindo, B.A., Anuka, J.A. and Hussaini, I.M., 2015. Anticonvulsant activity of aqueous fraction of Carissa edulis root bark. *Pharmaceutical Biology*, *53*(9), pp. 1329–1338.

10.3.1.6 *Ervatamia divaricata* (L.) Burkill

Synonyms: *Ervatamia coronaria* (Jacq.) Stapf; *Ervatamia cumingiana* (A. DC.) Markgr.; *Ervatamia flabelliformis* Tsiang; *Nerium divaricatum* L.; *Taberna discolor* (Sw.) Miers; *Tabernaemontana coronaria* (Jacq.) Willd.; *Tabernaemontana cumingiana* A. DC.; *Tabernaemontana discolor* Sw.; *Tabernaemontana divaricata* (L.) R. Br. ex Roem. & Schult.; *Tabernaemontana flabelliformis* (Tsiang) P.T. Li

Common names: East Indian rosebay; kath-maloti, tagar (Bangladesh); gou ya hua (China)

Habitat: Cultivated

Distribution: India, Nepal, Bangladesh, Bhutan, Myanmar, Thailand, China

Botanical observation: This handsome shrub grows to 5 m tall. The stems exude a pure white and poisonous latex when incised. The leaves are simple, opposite, and exstipulate. The petiole is up to 1 cm long. The blade is elliptic, dark green and glossy, coriaceous, 3–17.5 cm × 1–5.5 cm, acuminate at apex, and marked with 5–17 pairs of secondary nerves. The cymes are terminal and few-flowered. The calyx presents 5 lobes which are hairy at margin. The corolla is tubular, ephemerous, pure white, the tube about 2.5 cm long, yellow at throat, and develops 5 contorted and oblong lobes which are 1.5–2.5 cm long. Five stamens are attached to the basal inner part of of the corolla tube. The follicles are ellipsoid, curved, seldom seen, and 2–7 cm × 0.6–1.5 cm.

Medicinal uses: Fever, wounds (India); eye infection (India, Bangladesh)

Moderate broad-spectrum antibacterial indole alkaloids: 5-Oxocoronaridine inhibited the growth of *Escherichia coli*, *Klebsiella pneumoniae*, and *Bacillus subtilis* with the MIC values of 80, 50, and 95 μg/mL, respectively (Singh et al., 2011). 3-Oxocoronaridine inhibited the growth of *Klebsiella pneumoniae* with the MIC value of 90 μg/mL (Singh et al., 2011). Coronaridine inhibited the growth of *Escherichia coli*, *Enterobacter cloacae*, *Bacillus subtilis*, and *Streptococcus pneumoniae* with the MIC values of 80, 60, 70, and 90 μg/mL, respectively (Singh et al., 2011). Ibogamine inhibited the growth of *Escherichia coli*, *Klebsiella pneumoniae*, *Staphylococcus aureus*, and *Streptococcus pneumoniae* with the MIC value of 120, 80, 100, and 90 μg/mL, respectively (Singh et al., 2011). Tabernamontamine inhibited the growth of *Escherichia coli*, *Klebsiella pneumoniae*, *Staphylococcus aureus*, *Bacillus subtilis*, and *Streptococcus pneumoniae* with the MIC values of 170, 100, 120, 70, and 90 μg/mL, respectively (Singh et al., 2011). Vobasine inhibited the growth of *Escherichia coli*, *Klebsiella pneumoniae*, and *Streptococcus pneumoniae* with the MIC values of 90, 70, and 100 μg/mL, respectively (Singh et al., 2011).

Moderate broad-spectrum antifungal indole alkaloids: The growth of *Aspergillus niger* was inhibited by 3-oxocoronaridine, coronaridine, and vobasine with the MIC values of 80, 100, and 130 μg/mL, respectively (Singh et al., 2011). The growth of *Aspergillus flavus* was inhibited by 3-oxocoronaridine, coronaridine, and vobasine with the MIC values of 80, 100, and 130 μg/mL, respectively (Singh et al., 2011). The growth of *Aspergillus flavus* was inhibited by 5-oxocoronaridine, 3-oxocoronaridine, ibogamine, vobasine, and voacamine with the MIC values of 100, 85, 130, 80, and 80 μg/mL, respectively (Singh et al., 2011). The growth of *Rhizoctonia phaseoli* was inhibited by 5-oxocoronaridine, coronaridine, ibogamine, vobasine, and voacamine with the MIC values of 100, 100, 80, 80, and 100 μg/mL, respectively (Singh et al., 2011). The growth of *Candida albicans* was inhibited by ibogamine, tabernamontamine, and voacamine with the MIC values of 80, 110, and 140 µg/mL, respectively (Singh et al., 2011). The growth of *Penicillium chrysogenum* was inhibited by 5-oxocoronaridine, 3-oxocoronaridine, tabernamontamine, and vobasine with the MIC values of 100, 100, 80, and 100 µg/mL, respectively (Singh et al., 2011).

Moderate antimycobacterial indole alkaloids: Voacangine and ibogaine inhibited the growth of *Mycobacterium tuberculosis* with the MIC values of 50 µg/mL (Rastogi et al., 1998). Voacangine and ibogaine inhibited the growth of *Mycobacterium avium* with the MIC values of 50 and 100 µg/mL, respectively (Rastogi et al., 1998). Voacangine and ibogaine inhibited the growth of *Mycobacterium kansasi* with the MIC values of 100 and 50 µg/mL, respectively (Rastogi et al., 1998).

Strong antiviral (enveloped monopartite linear single-stranded (+) RNA) indole alkaloid: Voacangine (LogD 3.2 at pH 7.4; molecular mass = 368.4 g/mol) inhibited the replication of the Dengue virus serotype-2 (NG strain) with the IC_{50} value of 10.1 µg/mL and a selectivity index of 113 in Vero cells (Gómez-Calderón et al., 2017). Voacangine inhibited the replication of the Chikungunya (ACol strain) virus with the IC_{50} value of 304.3 µg/mL and a selectivity index of 3.7 in Vero cells (Gómez-Calderón et al., 2017)

Commentaries

i. The plant produces a bewildering array of cytotoxic indole alkaloids (Bao et al., 2013; Singh et al., 2011; Arambewela & Ranatunge, 1991). Cytotoxic natural products are often active against Gram-positive bacteria and fungi. Of note, indoles in this plant inhibit the growth of Gram-negative bacteria. Hydrophilic and amphiphilic compounds can pass the outer membrane through water-filled transmembrane channels (porins or aquaporins), if they have a molecular size below or equal to 600 g/mol, and this is the case for indole alkaloids in this plant.
ii. Indole alkaloids in this plant are vasorelaxant (Hirasawa et al., 2019), and they might have some levels of bacterial and/or fungal efflux-pumps inhibition.
iii. Dengue and Chikungunya viruses are both enveloped monopartite linear and single-stranded (+)-RNA viruses making it probable for voacamine (and congeners) to have activity against coronaviruses.
iv. Being active against bacteria, mycobacteria, fungi, and viruses, indole alkaloids in this plant most probably target microbial genetic material.
v. The latex of *Ervatamia pandacaqui* (Poir.) Pichon (synonym: *Tabernaemontana pandacaqui* Poir.) is used to heal wounds by the Cebuanos (local name: *Tigpod*).

References

Arambewela, L.S. and Ranatunge, T., 1991. Indole alkaloids from Tabernaemontana divaricata. *Phytochemistry*, *30*(5), pp. 1740–1741.

Bao, M.F., Yan, J.M., Cheng, G.G., Li, X.Y., Liu, Y.P., Li, Y., Cai, X.H. and Luo, X.D., 2013. Cytotoxic indole alkaloids from Tabernaemontana divaricata. *Journal of Natural Products*, *76*(8), pp. 1406–1412.

Gómez-Calderón, C., Mesa-Castro, C., Robledo, S., Gómez, S., Bolivar-Avila, S., Diaz-Castillo, F. and Martínez-Gutierrez, M., 2017. Antiviral effect of compounds derived from the seeds of Mammea Americana and Tabernaemontana cymosa on Dengue and Chikungunya virus infections. *BMC Complementary and Alternative Medicine*, *17*(1), pp. 1–12.

Hirasawa, Y., Dai, X., Deguchi, J., Hatano, S., Sasaki, T., Ohtsuka, R., Nugroho, A.E., Kaneda, T. and Morita, H., 2019. New vasorelaxant indole alkaloids, taberniacins A and B, from Tabernaemontana divaricata. *Journal of Natural Medicines*, *73*(3), pp. 627–632.

Rastogi, N., Abaul, J., Goh, K.S., Devallois, A., Philogène, E. and Bourgeois, P., 1998. Antimycobacterial activity of chemically defined natural substances from the Caribbean flora in Guadeloupe. *FEMS Immunology & Medical Microbiology*, *20*(4), pp. 267–273.

Singh, B., A Sharma, R. and K Vyas, G., 2011. Antimicrobial, antineoplastic and cytotoxic activities of indole alkaloids from tabernaemontana divaricata (L.) R. Br. *Current Pharmaceutical Analysis*, *7*(2), pp. 125–132.

10.3.1.7 *Holarrhena pubescens* Wall. ex G. Don

Synonyms: *Chonemorpha antidysenterica* (Roth) G. Don, *Echites antidysentericus* Roth, *Echites pubescens* Buch.-Ham., *Echites pubescens* Willd. ex Roem. & Schult., *Holarrhena antidysenterica* (L.) Wall. ex A. DC., *Holarrhena antidysenterica* Roth, *Holarrhena antidysenterica* Roth; *Holarrhena codaga* G. Don; *Holarrhena malaccensis* Wight; *Holarrhena villosa* Aiton ex Loudon

Common names: Conessi bark, Easter tree, rosebay, tellicherry bark; bal duhak ki bipang, hartdare, indro jb, kuruk gaas, kurchi, kurochi (Bangladesh); anley khim, indrajaou, indrajal, kogar, kolamokhi, kura, kutaja (India); mivah i zaban i gunjishk, tukhm zaban i gunjishk i talkh (Iran); dangkyam (Myanmar); cold, cough (Islanders of Indian ocean)

Habitat: Cultivated

Distribution: From Pakistan to Bangladesh

Botanical observation: This laticiferous tree grows to 8 m tall. The leaves are simple, opposite, and exstipulate. The petiole is minute and channeled. The blade is elliptic, 10–20 cm × 4–11.5 cm, membranous, pubescent, round at base, acute at apex, and marked with 10–15 pairs of secondary nerves The inflorescences are terminal cymes. The 5 sepals are elliptic to linear and up to 1 cm long. The corolla is tubular, pure white, 0.9–1.9 cm long, and develops 5 contorted lobes which are elliptic-oblong and 1–3 cm long. The 5 stamens are attached to the corolla tube. The follicles are linear, dehiscent, up to 30 cm long, and stuffed with 0.9–1.6 cm long comose seeds.

Medicinal uses: Boils (Bangladesh, Myanmar); diarrhea, dysentery, fever, jaundice (Bangladesh); bites, bronchitis, cholera, cuts, diarrhea, dysentery, fever, leprosy, tuberculosis, ulcers, wounds (India); ear ache, cholera, dysentery, small pox, sore throat (Myanmar)

Moderate antimycobacterial extract: extract of seeds inhibited the growth of *Mycobacterium tuberculosis* H37Rv with the MIC value of 250 µg/mL and multidrug-resistant *Mycobacterium tuberculosis* (clinical isolate) with the MIC value of 62.5 µg/mL (Gupta et al., 2018). Methanol extract inhibited the growth of *Mycobacterium smegmatis* (MC155) with the MIC value of 500 µg/mL (Newton et al., 2002).

Moderate broad-spectrum antibacterial alkaloid fraction: Alkaloid fraction of bark inhibited the growth of *Staphylococcus aureus*, *Staphylococcus epidermidis*, *Streptococcus faecalis*, *Bacillus subtilis*, *Escherichia coli*, and *Pseudomonas aeruginosa* with the MIC/MBC values of 95/180, 100/200, 170/330, 600/1000, 55/900, and 420/900 µg/mL, respectively (Chakraborty & Brantner, 1999).

Strong antibacterial steroidal alkaloids: N-formylconessimine and conimine inhibited the growth of methicillin-sensitive *Staphylococcus aureus* and methicillin-resistant *Staphylococcus aureus* with the MIC values of 32 and 128 µg/mL, respectively (Li-Na et al., 2017).

Antibiotic potentiator steroidal alkaloid: Conimine increased the sensitivity of methicillin-resistant *Staphylococcus aureus* to vancomycin (Li-Na et al., 2017).

Antifungal (yeast) halo developed by polar extract: Methanol extract of bark (125 µg/7 mm disc) inhibited the growth of *Candida tropicalis* (ATCC 4563) and *Trichosporon beigelli* (NCIM 3404) with the inhibition zones of 10 and 11 mm, respectively (Li-Na et al., 2017).

Strong antiviral (enveloped monopartite linear single-stranded (+)-RNA) amphiphilic steroidal alkaloid: Conessine (LogD 1.6 at pH 7.4; molecular mass = 356.5 g/mol) inhibited the replication of the Human coronavirus (OC43), Human coronavirus (NL-63), Middle-East respiratory syndrome-associated coronavirus, and mouse hepatitis virus A59 with the EC_{50} values of 2.3, 10.7, 4.9, and 11.4 µM, respectively, and CC_{50} values above 20 (Shen et al., 2019). Conessine inhibited the replication of the Zika virus (strain PAN2016) with the IC_{50} value of 9.7 µM in Vero cells (Lima et al., 2021).

Strong antiviral (enveloped monopartite linear double-stranded DNA) amphiphilic steroidal alkaloid: Conessine at the concentration of 8 µM inhibited the replication of the Kaposi's sarcoma-associated herpesvirus (Chen et al., 2019).

Commentaries

i. The BCG vaccine was developed from the observation and dedication of our peers the French bacteriologists Albert Calmette (1863–1933) and Camille Guérin (1872–1961) at the Pasteur Institute in Lille at the beginning of the 19th century using *Mycobacterium bovis* and used to immunize against tuberculosis (Luca & Mihaescu, 2013), and one could enquire if the use of non-pathogenic betacoronaviruses to immunize against COVID-19 and variants could be a mean to eradicate this pandemic. Escobar et al. (2020) suggest that individuals that received the BCG vaccine are less likely to succumb to COVID-19. Nuovo et al. (2020) presented evidence for a strong homology between the severe acute respiratory syndrome-associated coronavirus 2 envelope protein and a *Mycobacterium* sp. antigen. Thus, there is an urgent need to examine the usefulness of BCG vaccine to prevent COVID-19 and its variant and other zoonotic coronavirus pandemics.

ii. Vaccination is a mean to protect populations against COVID-19, and we have to be grateful to the Pharmaceutical industries and government for that. However, one could wonder if vaccination alone would not give rise to the emergence of variants of increasing contamination powers (Chandramathi Samudi Raju personal communication). Also, one could wonder if long term side effect of RNA vaccines will not emerge after a few years. Thus, the development of effective oral medicines to prevent or mitigate COVID-19 (and other forthcoming coronaviruses) would seem to be inevitable, and in this context phytomedications or natural products from medicinal plants could be of value (when given synergistically with vaccines). Here, conessine (other candidates have been reviewed in this work) or even the seeds of *Holarrhena pubescens* Wall. ex G. Don with broad-spectrum anti-coronavirus effect could be candidates if proven to be not toxic. The Pharmaceutical Industry is not much excited about herbal remedies for patent and financial reasons, and it is perfectly understandable, and there is a need to have public research institutions (financed by public taxes) with the responsibility to develop and manufacture safe and effective herbal products. In this context, the training of pharmacy students in pharmacognosy is of extreme importance.

iii. *Chonemorpha fragrans* (Moon) Alston (local name: *Gothala phul*) is used in Nepal for fever, jaundice, ulcers, and wounds.

References

Chakraborty, A. and Brantner, A.H., 1999. Antibacterial steroid alkaloids from the stem bark of Holarrhena pubescens. *Journal of Ethnopharmacology*, *68*(1–3), pp. 339–344.

Chen, J., Dai, L., Goldstein, A., Zhang, H., Tang, W., Forrest, J.C., Post, S.R., Chen, X. and Qin, Z., 2019. Identification of new antiviral agents against Kaposi's sarcoma-associated herpesvirus (KSHV) by high-throughput drug screening reveals the role of histamine-related signaling in promoting viral lytic reactivation. *PLoS Pathogens*, *15*(12), p. e1008156.

Escobar, L.E., Molina-Cruz, A. and Barillas-Mury, C., 2020. BCG vaccine protection from severe coronavirus disease 2019 (COVID-19). *Proceedings of the National Academy of Sciences*, *117*(30), pp. 17720–17726.

Gupta, V.K., Kaushik, A., Chauhan, D.S., Ahirwar, R.K., Sharma, S. and Bisht, D., 2018. Anti-mycobacterial activity of some medicinal plants used traditionally by tribes from Madhya Pradesh, India for treating tuberculosis related symptoms. *Journal of Ethnopharmacology*, *227*, pp. 113–120.

Lima, N.M., de Jesus A.S.Andrade, T. and de Oliveira, M.A.L., 2021. Natural product bioactive as antiviral agents against Zika Virus. In: La Porta, F.A. and Taft, C.A. (eds.) *Functional Properties of Advanced Engineering Materials and Biomolecules* (pp. 765–780). Springer, Cham.

Li-Na, Z.H.O.U., Xiao-Lei, G.E., Ting-Ting, D.O.N.G., Hui-Yuan, G.A.O. and Bo-Hang, S.U.N., 2017. Antibacterial steroidal alkaloids from Holarrhena antidysenteriaca. *Chinese Journal of Natural Medicines*, *15*(7), pp. 540–545.

Luca, S. and Mihaescu, T., 2013. History of BCG vaccine. *Maedica*, *8*(1), pp. 53–58.

Newton, S.M., Lau, C., Gurcha, S.S., Besra, G.S. and Wright, C.W., 2002. The evaluation of forty-three plant species for in vitro antimycobacterial activities; isolation of active constituents from Psoralea corylifolia and Sanguinaria canadensis. *Journal of Ethnopharmacology*, *79*(1), pp. 57–67.

Nuovo, G., Tili, E., Suster, D., Matys, E., Hupp, L. and Magro, C., 2020. Strong homology between SARS-CoV-2 envelope protein and a Mycobacterium sp. antigen allows rapid diagnosis of Mycobacterial infections and may provide specific anti-SARS-CoV-2 immunity via the BCG vaccine. *Annals of Diagnostic Pathology*, *48*, p. 151600.

Shen, L., Niu, J., Wang, C., Huang, B., Wang, W., Zhu, N., Deng, Y., Wang, H., Ye, F., Cen, S. and Tan, W., 2019. High-throughput screening and identification of potent broad-spectrum inhibitors of coronaviruses. *Journal of Virology*, *93*(12), p. e00023.

10.3.1.8 *Hoya parasitica* Wall. ex Wight

Synonym: *Hoya parasitica* Wall. ex J. Traille; *Hoya verticillata* (Vahl) G. Don

Common names: Wax plant; cherapata, pargacha, faissa gash, fessya gach (Bangladesh)

Habitat: Forests, cultivated

Distribution: India, Bangladesh, Myanmar, Cambodia, Laos, Vietnam, Thailand, Malaysia, Indonesia

Botanical observation: This laticiferous climber grows to 8 m long. The stem is stout and terete. The leaves are simple, opposite, and exstipulate. The petiole is stout and up to 2.5 cm long. The blade is elliptic, coriaceous, fleshy, somewhat orchidaceous-like, about 5 cm × 12.5 cm, cuneate at base, and acute at apex. The cymes are magnificent, cauliflorous, umbel-like, globose, about 15 cm in diameter, and present numerous somewhat vaguely porcelain-like starry flowers. The calyx includes 5 sepals which are elongated. The corolla is fleshy, somewhat white and purple, coriaceous, develops 5 triangular lobes, and 1.5 cm in diameter. The androecium includes 5 stamens on reddish to pinkish stigmatal head. The follicles are fusiform and about 12 cm long.

Medicinal uses: Bronchitis (India); jaundice (Bangladesh); fever (Bangladesh; India)

Broad-spectrum antibacterial halo developed by polar extract: Ethanol extract of leaves (500 μg/well) inhibited the growth of *Staphylococcus aureus*, *Proteus* sp., *Escherichia coli*, *Shigella sonnei*, and *Shigella dysenteriae* with the inhibition zone diameters of 23, 19, 10, 8, and 20 mm, respectively (Ahmed et al., 2008)

Commentaries

i. The plant yields the androstane steroid hoyasterone, the sesquiterpene 15-bulnesolic acid, the phenolic 1-(4-hydroxy-3-methoxyphenyl)-1-methoxypropan-2-ol, and the triterpene dihydrocanaric acid (Sadhu et al., 2008).
ii. Plants in the genus *Hoya* R.Br. are used in the traditional medicinal systems of various Asia-Pacific countries pointing to the presence of antimicrobial principles apparently yet to be isolated. In India, *Hoya viridiflora* R. Br. (synonym: *Asclepias viridiflora* Roxb.) is used for abscesses, boils, and wounds, and *Hoya pendula* Wight. (local name: *Thigapappui*) is used to treat eye infections. *Hoya carnosa* (L.f.) R. Br. is used to maturate furuncles in Cambodia, Laos, and Vietnam. In Malaysia, *Hoya coriacea* Bl. is used for cough. In Indonesia, *Hoya rumphii* Bl. (local name: *Daun pitis*) is used to treat gonorrhea. In the Philippines, *Hoya imbricata* Callery ex Decne. (local name: *Pawi-pawikan*) is used for boils, wounds, and ulcers. In China, *Hoya pottsii* J. Traill (local name: *Ya ban*) is used for sores. In Papua New Guinea, *Hoya australis* R. Br. ex J. Traill (local name: *Kalavagala*) is used for burns.

References

Ahmed, F., Reza, M.S.H., Shahid, I.Z., Khatun, A., Islam, K.K. and Ali, M.R., 2008. Antibacterial and antinociceptive activity of hoya parasitica. *Hamdard Medicus*, *51*(3), pp. 22–26.

Sadhu, S.K., Khatun, A., Ohtsuki, T. and Ishibashi, M., 2008. Constituents from Hoya parasitica and their cell growth inhibitory activity. *Planta Medica*, *74*(07), pp. 760–763.

10.3.1.9 *Ichnocarpus frutescens* (L.) W.T. Aiton

Synonyms: *Apocynum frutescens* L.; *Echites frutescens* (L.) Roxb.; *Echites frutescens* Wall. ex Roxb.; *Ichnocarpus moluccanus* Miq.; *Ichnocarpus ovatifolius* A. DC.; *Ichnocarpus volubilis* (Lour.) Merr.; *Micrechites sinensis* Markgr.

Common names: Black creeper; red sarsaparilla; dutta ludi, shyama luta (Bangladesh); dudhi, gopa valli, moron ara, onol sing, palativa, pal valli (India); taw sabe (Myanmar); copralin (the Philippines); khruea pla (Thailand)

Habitat: Hedges

Distribution: India to Pacific Islands

Botanical observation: This climber grows to 8 m long. The stems are slender, fibrous, terete, smooth, glabrous at apex, and somewhat swollen at nodes. The leaves are simple, opposite, and exstipulate. The petiole is up to 1.5 cm long and curved at base. The blade is coriaceous, 5–11 cm × 2.5–5 cm, marked with 5–7 pairs of secondary nerves, and acute at base and apex. The inflorescences are axillary and terminal cymes. The calyx is 5-lobed, hairy, and minute. The corolla is tubular, minute, and produces 5 strongly contorted lobes which are hairy and about 5 mm long. A disc is present. The 5 stamens are attached to the corolla tube. The ovary is pubescent. The follicles are linear, 8–15 cm × 4–5 mm, pubescent, and contain numerous comose seeds.

Medicinal uses: Cholera, cough, dysentery, leprosy, measles, syphilis, ulcers of the tongue (India); fever (India; Myanmar)

Broad-spectrum antibacterial halo developed by mid-polar extract: Chloroform extract of roots (6 mg/mL solution/5 mm disc) inhibited the growth of *Bacillus pumilus*, *Staphylococcus aureus*, *Escherichia coli*, *Pseudomonas aeruginosa*, *Pseudomonas putida*, and *Proteus* sp. (Malathy & Sini, 2009).

Broad-spectrum antifungal halo developed by mid-polar extract: Chloroform extract of roots (6 mg/mL solution by 5 mm disc) inhibited the growth of *Aspergillus flavus*, *Aspergillus fumigatus*, and *Aspergillus niger* with the inhibition zone diameters of 14, 12, and 13 mm, respectively (Malathy & Sini, 2009).

Commentaries

i. The plant yields triterpene saponins (Minocha & Tandon, 1980).
ii. Plants in the family Apocynaceae are often used to treat syphilis in Asia Pacific which suggests either the presence of anti-*Treponema* principles or strong anti-inflammatory principles. Note that saponins are often anti-inflammatory.

References

Malathy, N.S. and Sini, S., 2009. Antimicrobial activities of Ichnocarpus frutescens (L.) R. Br. and Hemidesmus indicus R. Br. Roots. *Ancient Science of Life*, *28*(4), p. 13.

Minocha, P.K. and Tandon, R.N., 1980. A new triterpene glycoside from the stem of Ichnocarpus frutescens. *Phytochemistry*, *19*(9), pp. 2053–2055.

10.3.1.10 *Nerium oleander* L.

Synonym: *Nerium indicum* Mill.; *Nerium odoratum* Lam.; *Nerium odorum* Sol.; *Nerium verecundum* Salisb.

Common names: Indian oleander; white oleander, rose of Sharon; lal kharubi, rakta korobi (Bangladesh); lal gandila, lal kaneer (India); gish (Iran); nwei thargi (Myanmar); gandaraye, kaner, kuner (Pakistan); zakkum (Turkey)

Habitat: Cultivated

Distribution: Asia and the Pacific

Botanical observation: This handsome yet dreadfully poisonous shrub grows to 6 m tall. The stems are sappy. The leaves are simple, verticillate, and exstipulate. The petiole can reach 7 mm long. The blade is narrowly elliptic, coriaceous, acute at base and sharply acute at apex, marked with a conspicuous midrib, and 10–15 cm × 1–2 cm. The cymes are terminal. The calyx is 5-lobed and about 7 mm long. The corolla is tubular, of a heavenly light pink or pure white, the tube about 2 cm long, hairy within, narrow at throat, and produces 5 contorted lobes with scales at base. The 5 stamens are attached to the corolla tube and touch the stigma. The ovary develops a linear style and bifid stigma. The capsules are seldom produced, fusiform, 12–20 cm × 7.5 mm, and open to release numerous comose seeds.

Medicinal uses: AIDS, tuberculosis (India); eye infection (Pakistan); ringworm (India; Myanmar); leprosy (Pakistan; India; Bangladesh; Myanmar; China)

Broad-spectrum antibacterial polar extract: Methanol extract of leaves inhibited the growth of *Bacillus subtilis* (MTCC 441), *Escherichia coli* (MTCC 1687), *Pseudomonas aeruginosa* (MTCC 1688), and *Staphylococcus aureus* (MTCC 737) (Vyas et al., 2015).

Antibiotic potentiator polar extract: Methanol extract of leaves decreased the resistance of *Escherichia coli* (MTCC 1687) and *Staphylococcus aureus* (MTCC 737) towards erythromycin (Vyas et al., 2015).

Strong broad-spectrum antibacterial cardenolide-type steroid: 12β-Hydroxy-5β-carda-8,14,16,20(22)-tetraenolide isolated from the roots inhibited the growth of *Bacillus subtilis, Bacillus cereus, Escherichia coli,* and *Pseudomonas aeruginosa* (Huq et al., 1999).

Antimycobacterial polar extract: Aqueous extract of leaves (dilution below 1:20) inhibited the growth of *Mycobacterium tuberculosis* (Lucas et al., 1951).

Antiviral (non-enveloped monopartite linear single-stranded (+) RNA) polar extract: Aqueous extract inhibited the replication of the Poliovirus type-1 (Sb-1) (Sanna et al., 2021).

Strong antiviral (enveloped monopartite linear single-stranded (−) RNA) cardenolide: Oleandrin inhibited the replication of the Ebola virus (EBOV/Kik) with a IC_{50} value of 0.02 μM in Vero cells (Digoxin 0.09 μM) (James et al., 2017). Oleandrin inhibited the replication of the Marburg virus MARV/Ci67 in Vero cells with an EC_{50} value of 1.4 μM in Vero cells (Digoxin 0.5 μM) (James et al., 2017).

Strong antiviral (enveloped linear monopartite dimeric single-stranded (+) RNA) cardenolide: Oleandrin inhibited the concentrations ranging from 50 to 200 ng/mL reduced the reduced infectivity of the progeny Human immunodeficiency virus type-1 by 66%–77% in human PBMC cells, via at least in part, reduction in the concentration of viral envelope protein gp120 (Singh et al., 2013).

Strong antiviral (enveloped monopartite linear single-stranded (+)-RNA) amphiphilic cardenolide: Oleandrin (LogD 2.4 at pH 7.4; molecular mass = 576.7 g/mol) inhibited the replication of the severe acute respiratory syndrome-associated coronavirus 2 (USA-WA1/2020 strain) with the EC_{50} value as low as 11.9 ng/mL in Vero cells when added to culture 30 mins before infection (Plante et al., 2021).

Antiviral (enveloped monopartite linear single-stranded (+)-RNA) amphiphilic cardenolide: Oleandrin given to Hamsters sublingually at the dose of 25 μL of a 130 μg/mL once daily for 5 days evoked some levels of protection against SARS-CoV-2 P6 (Plante et al., 2021)

Commentaries

i. This poisonous plant is used for leprosy externally in 4 different Asian countries suggesting the presence of principle(s) active against *Mycobacterium leprae*. It is also used against tuberculosis due to *Mycobacterium tuberculosis*. Antimycobacterial principles most probably exist in this plant, and these are probably cardenolides.
ii. There is a trend to instruct undergraduate students to avoid using publications older than 10 years, and for whatever reason it is, one need however to recall to that ancient texts, and old papers are actually the best mean to approach a subject, as it approaches by its roots thus allowing a deep understanding of the subject. Here, a publication dating from 1949 reports activity on *Mycobacterium tuberculosis* which might assist in finding antimycobacterial principle from *Nerium oleander* L.
iii. Cardiac glycosides such as ouabain and digitoxin (which have a shallow therapeutic index) are broad-spectrum antiviral molecules acting on replication of the Chikungunya virus, Middle East Respiratory Syndrome, and Severe acute respiratory syndrome-associated coronavirus via inhibition of Na+/K+-ATPase (Amarelle & Lecuona, 2018; Yang et al., 2018). Inhibition of Na+/K+ ATPase blocks Zika virus infection in mice (Guo et al., 2020). Oleandrin is a dreadful poison (Wasfi et al., 2008). Digitalin from *Digitalis purpurea* L.

(family Scrofulariaceae) used in the therapeutic treatment of bradycardia could be safer as anti-COVID-19 agent or preventive broad-antiviral agent in time of viral pandemics, and this could be examined in clinical settings. Digitalin was used for the treatment of the Spanish flu of 1918 (Radusin, 2012). There is the need to develop oleandrin derivative with broader therapeutic index as an agent to combat pandemic viral zoonosis.

iv. Consider that membranes of bacteria, mycobacteria, and fungi have ATPases which could be targeted by cardenolides.

v. According to our peer the Armenian botanist Armen Takhtajan (1910–2009), the genus *Nerium* L. belongs to the Subfamily Apocynoideae, Tribe Wrightieae. Members of the Apocynoideae are often used for the treatment of microbial infectious diseases in Asia. In the Subfamily Apocynoideae, Tribe Apocyneae, *Anodendron tenuiflorum* Miq. is used in Malaysia to heal sores. In Papua New Guinea, *Anodendron aambe* (Warb.) Warb. (local name: *Kap*) is used to heal sores. In India, *Anodendron paniculatum* A. DC (local name: *Chedukura*) is used to treat leprosy. Members of the genus *Anodendron* A. DC. tend to produce the flavonol wogonin (Guo et al., 2007). Wogonin (also known as 5,7-dihydroxy-8-methoxyflavone) at the concentration of 50 µg/mL inhibited the secretion of Hepatitis B virus surface antigen and e-antigen by about 80% and 30%, respectively (Guo et al., 2007). Wogonin inhibited duck hepatitis B virus DNA polymerase with the IC_{50} value of 0.5 µg/mL (Guo et al., 2007). In ducks infected with Hepatitis B, wogonin given intravenously at the dose of 5 mg/kg/day for 10 days decreased Hepatitis B viremia and Hepatitis B virus serum DNA and improved hepatic cytoarchitecture (Guo et al., 2007). In Human Hepatitis B virus-transgenic mice from lineage 1.3.32, wogonin given intravenously at the dose of 28 mg/kg/day decreased plasmatic Hepatitis B surface-antigen and afforded hepatic protection (Guo et al., 2007). Wogonin inhibited the replication of the Influenza A and B virus in Madin-Darby canine kidney cells and human lung epithelial (A549) cells (Seong et al., 2018). Wogonin at 50 µg/mL inhibited the replication of the Herpes simplex virus type-1 and type-2 at post-entry stage (Chu et al., 2020). Wogonin inhibited the replication of the Respiratory syncytial virus with the IC_{50} value of 7.4 µg/mL and a selectivity index of 16 (Ma et al., 2002). Wogonin is an example of natural product with a broad array of pharmacological activities, and one could suggest that natural products acting mainly on viral replication are better therapeutic candidates. Being cytotoxic it probably hampers the replication of viruses by disturbing host cell transcription and/or translation processes. Understanding their mode of action allows to bring light to mechanisms of virus replication. It can be anticipated that, if enough efforts are produced, some strong antiviral molecules will be developed from flowering plants.

vi. Cardenolides are probably inhibitors of efflux pumps.

References

Amarelle, L. and Lecuona, E., 2018. The antiviral effects of Na, K-ATPase inhibition: A minireview. *International journal of Molecular Sciences*, 19(8), p. 2154.

Bai, Y., Zhu, W., Xu, Y., Xie, Z., Akihisa, T., Manosroi, J., Sun, H.

Guo, J., Jia, X., Liu, Y., Wang, S., Cao, J., Zhang, B., Xiao, G. and Wang, W., 2020. Inhibition of Na+/K+ ATPase blocks Zika virus infection in mice. *Communications Biology*, *3*(1), pp. 1–8.

Huq, M.M., Jabbar, A., Rashid, M.A. and Hasan, C.M., 1999. A novel antibacterial and cardiac steroid from the roots of Nerium oleander. *Fitoterapia*, *70*(1), pp. 5–9.

James, R., Dorosky, D., Stonier, S., Newman, R. and Dye, J., 2017. Antiviral potency of an extract from Nerium oleander. In *9th International Filovirus Symposium: Filo2017.*

Lucas, E.H., Lickfeldt, A., Gottshall, R.Y. and Jennings, J.C., 1951. The occurrence of antibacterial substances in seed plants with special reference to Mycobacterium tuberculosis. *Bulletin of the Torrey Botanical Club*, 78, pp.310–321.

Plante, K.S., Dwivedi, V., Plante, J.A., Fernandez, D., Mirchandani, D., Bopp, N., Aguilar, P.V., Park, J.G., Tamayo, P.P., Delgado, J. and Shivanna, V., 2021. Antiviral activity of oleandrin and a defined extract of Nerium oleander against SARS-CoV-2. *Biomedicine & Pharmacotherapy*, *138*, p. 111457.
Radusin, M., 2012. The Spanish flu, part II: The second and third wave. *Vojnosanitetski Pregled*, *69*(10), pp. 917–927.
Sanna, G., Madeddu, S., Serra, A., Collu, D., Efferth, T., Hakkim, F.L. and Rashan, L., 2021. Anti-poliovirus activity of Nerium oleander aqueous extract. *Natural Product Research*, *35*(4), pp. 633–636.
Singh, S., Shenoy, S., Nehete, P.N., Yang, P., Nehete, B., Fontenot, D., Yang, G., Newman, R.A. and Sastry, K.J., 2013. Nerium oleander derived cardiac glycoside oleandrin is a novel inhibitor of HIV infectivity. *Fitoterapia*, *84*, pp. 32–39.
Vyas, P., Suthar, A. and Joshi, D., 2015. Antibacterial activity of extracts of leaves of Nerium indicum in combination with antibiotics. *International Letters of Natural Sciences*, *46*, pp. 41–45.
Wasfi, I.A., Zorob, O. and Al Awadhi, A.M., 2008. A fatal case of oleandrin poisoning. *Forensic Science International*, *179*(2–3), pp. e31–e36.
Yang, C.W., Chang, H.Y., Lee, Y.Z., Hsu, H.Y. and Lee, S.J., 2018. The cardenolide ouabain suppresses coronaviral replication via augmenting a Na+/K+-ATPase-dependent PI3K_PDK1 axis signaling. *Toxicology and Applied Pharmacology*, *356*, pp. 90–97.

10.3.1.11 *Rauvolfia serpentina* (L.) Benth. ex Kurz

Synonyms: *Ophioxylon majus* Hassk; *Ophioxylon serpentinum* L.

Common names: Indian snakeroot; chandra, mahi hojai, sarpogonda (Bangladesh); sarpagandha (India; Nepal); jhaberbarua, garuda patala (India); bomma yaza (Myanmar); chandmaruwa (Nepal)

Habitat: Forests

Distribution: India, Sri Lanka, Nepal, Myanmar, Bangladesh, Thailand, Malaysia, Indonesia

Botanical observation: This laticiferous shrub grows up to 2 m tall from roots with somewhat vaguely a snake-like shape. The leaves are simple, exstipulate, and whorled. The petiole can reach 1.5 cm long. The blade is elliptic, 7–16 cm × 2.5–5 cm, acute to tapering at base, wavy, glossy, acute at apex, membranous, and marked with 8–12 pairs of secondary nerves. The inflorescences are terminal or axillary cymes with a unique type of bright red peduncles (could one see here a sign from Nature that the plant could be used for blood pressure?: Theory of Signatures). The calyx is pink, minute, and 5-lobed. The corolla is tubular, the tube pinking, 8–12 mm long, inflated in the middle, and develops 5 round lobes which are 5 mm long and pure white. A conspicuous disc is present. The 5 stamens are inserted in the middle of the corolla. The ovary is ovoid and develops a linear style and a capitate stigma. The drupes are greyish to black, glossy, olive-shaped, and about 1 cm long.

Medicinal uses: Fever, dysentery (India; Nepal); diarrhea, herpes (India); gonorrhea, fever, bronchitis (Myanmar)

Broad-spectrum halo developed by polar extract: Methanol extract of roots inhibited the growth of *Salmonella typhimurium* (MTCC 3224), *Escherichia coli* (MTCC 443), *Citrobacter freundii* (MTCC 4221), *Proteus vulgaris* (MTCC 1771), *Enterobacter faecalis* (MTTC 439), and *Staphylococcus aureus* (MTCC 3101) (Negi et al., 2014).

Moderate broad-spectrum antibacterial polar extract: Methanol extract of roots inhibited the growth of *Bacillus subtilis* (MTCC B2274), *Enterococcus faecalis* (MTCC B3159), *Micrococcus luteus* (MTCC B1538), *Staphylococcus aureus* (MTCC B3160), *Streptococcus pneumoniae* (MTCC B2672), *Escherichia coli* (MTCC B1560), *Pseudomonas aeruginosa* (MTCC B2297), and *Proteus vulgaris* (MTCC B7299), with the MIC values of 62.5, 500, 500, 31.2, 1000, 1000, 500, and 500 μg/mL, respectively (Owk & Lagudu, 2016).

Moderate broad-spectrum antifungal polar extract: Methanol extract of roots inhibited the growth of *Aspergillus niger* (MTCC F4325), *Candida albicans* (MTCC F7315), and *Saccharomyces cerevisiae* (MTCC F2567) with the MIC values of 125, 500, and 1000 μg/mL, respectively (Owk & Lagudu, 2016).

Antibiotic potentiator indole alkaloid: Reserpine decreased the resistance of *Staphylococcus aureus* (1199B, NorA hyperproducer) to ciprofloxacin and norfloxacin (Aeschlimann et al., 1999)

In vivo antiviral indole alkaloid: Reserpine in doses of 0.01–0.05 mg/kg evoked some levels of protection in mice and rabbits against Rabies (Zubovich et al., 1993).

Commentaries

i. The plant is close to extinction due to over collection. Consider that if abuses on the environment and life forms, such as logging and burning of the tropical rainforests, are not stopped, then global temperature will reach deadly levels and the seas will rise covering lowlands. Rise in global temperature will result in the disappearance of medicinal plants.

ii. The roots yield indole alkaloids including reserpine, ajmaline, and yohimbine (Itoh et al., 2005). Reserpine has been used for the treatment of high blood pressure and is able to inhibit bacterial efflux pumps (NorA competitive pump blocker) (Aeschlimann et al., 1999) as well as efflux pumps in Gram-negative bacteria and Mycobateria (Dwivedi et al., 2017). Ajmaline and yohimbine are neuroactive and efflux pump inhibitors in Gram-negative bacteria (Dwivedi et al., 2017). Efflux pump inhibitors of clinical value are needed to curb bacterial and fungal resistance.

iii. *Rauvolfia amsoniaefolia* A. DC. is used to treat yaws in the Philippines. *Rauvolfia tetraphylla* L. is used in India (local name: *Badaa chand*) for skin diseases and in Bangladesh for fever (local name: *Bara chadar*). In the Subfamily Rauvolfioideae, Tribe Willughbeieae, *Willughbeia cochinchinensis* Pierre is used to treat yaws in Malaysia. *Alyxia stellata* (J.R. Forst. & G. Forst.) Roem. & Schult. in the Subfamily Rauvolfioideae and the Tribe Alyxieae is used in Malaysia (local name: *Pulasari*) to treat thrush. In Indonesia, *Alyxia reinwardtii* Bl. (*Alyxia stellata* var. *latifolia* Bl.) is used to prevent flu. In the Tribe Alyxieae, *Chilocarpus alyxifolius* Pierre is used for wounds in Cambodia, Laos, and Vietnam. In India, *Chilocarpus malabaricus* Bedd. (local name: *Vallippala*) is used for skin diseases. In the Subfamily Rauvolfioideae, Tribe Hunterieae, *Hunteria zeylanica* (Retz.) Gardner ex Thwaites is used to treat yaws in Malaysia. In the Tribe Vinceae, *Kopsia fruticosa* (Ker) DC is used to heal syphilitic ulcerations of the nose, and *Leuconotis eugeniifolius* A. DC. in the Tribe Willughbeieae is used to treat yaws in Malaysia. The treatment of yaws and syphilis suggests the presence of anti-*Treponema* and/or strong anti-inflammatory principles in the Rauvolfioideae.

References

Aeschlimann, J.R., Dresser, L.D., Kaatz, G.W. and Rybak, M.J., 1999. Effects of NorA inhibitors on in vitro antibacterial activities and postantibiotic effects of levofloxacin, ciprofloxacin, and norfloxacin in genetically related strains of staphylococcus aureus. *Antimicrobial Agents and Chemotherapy*, *43*(2), pp. 335–340.

Dwivedi, G.R., Singh, D.P., Sharma, S.A. and Darokar, M.P., 2017. Efflux pumps: warheads of gram-negative bacteria and efflux pump inhibitors. In: Sinha, R.P. and Richa (eds.) *New Approaches in Biological Research. New York: Nova Science Publishers*, pp. 35–72.

Itoh, A., Kumashiro, T., Yamaguchi, M., Nagakura, N., Mizushina, Y., Nishi, T. and Tanahashi, T., 2005. Indole alkaloids and other constituents of Rauwolfia serpentina. *Journal of Natural Products*, *68*(6), pp. 848–852.

Negi, J.S., Bisht, V.K., Bhandari, A.K., Bisht, D.S., Singh, P. and Singh, N., 2014. Quantification of reserpine content and antibacterial activity of Rauvolfia serpentina (L.) Benth. ex Kurz. *African Journal of Microbiology Research*, *8*(2), pp. 162–166.

Owk, A.K. and Lagudu, M.N., 2016. In-vitro antimicrobial activity of roots of rauwolfia serpentina L. Benth Kurz. *Notulae Scientia Biologicae*, *8*(3), pp. 312–316.

Zubovich, I.K., Mishaeva, N.P., Votiakov, V.I., Kovalev, N.A., Shashen'ko, A.S. and Golovneva, G.P., 1993. The action of reserpine in experimental rabies infection. *Voprosy Virusologii*, *38*(2), pp. 81–83.

10.3.1.12 *Tylophora indica* (Burm.f.) Merr.

Synonyms: *Apocynum reticulatum* Lour.; *Asclepias asthmatica* L. f.; *Cynanchum indicum* Burm. f.; *Tylophora asthmatica* (L. f.) Wight & Arn.

Common names: Country ipecacuanha, emetic swallow wort, Indian ipecacuanha, untomul (Bangladesh); antamul, arkapatri, codegam (India); binnuga (Sri Lanka)

Habitat: Wastelands, forest edges

Distribution: India, Sri Lanka, Bangladesh

Botanical observation: This stout climber grows up to 10 m long. The stems are fibrous. The leaves are simple, opposite, and exstipulate. The petiole is about 1 cm long. The blade is broadly lanceolate, 3–10 cm × 2–6 cm, cordate at base, apiculate at apex, marked with about 4–5 pairs of secondary nerves, glossy, and fleshy. The cymes are terminal and lax. The calyx is hairy. The corolla is greenish purple, hairy, tubular, about 1 cm in diameter, and pink at throat. The androecium includes 5 stamens appressed to a columnar stigma. The follicles are up to 10 cm × 1–2 cm and contain numerous comose seeds.

Medicinal uses: Bronchitis, cough, diarrhea, dysentery (Bangladesh; India)

Broad-spectrum antibacterial polar extract: Aqueous extract (100 µL of a 5 mg/mL solution/8 mm well) inhibited the growth of *Staphylococcus aureus* with the inhibition zone of 9.6 mm (Khanna et al., 2020).

Strong broad-spectrum antifungal amphiphilic phenanthroindolizidine alkaloids: Tylophorinine (LogD 3.5 at pH 7.4; molecular mass = 379.4 g/mol) inhibited the growth of *Candida albicans* (14503), *Candida krusei*, *Candida glabrata*, and *Aspergillus fumigatus* with the MIC values of 0.6, 0.6, 2.5, and 5 µg/mL, respectively (Dhiman et al., 2012). Tylophorinidine (LogD 2.7 at pH 7.4; molecular mass = 365.4 g/mol) inhibited the growth of *Candida albicans* (14503), *Candida krusei*, *Candida glabrata*, and *Aspergillus fumigatus* with the MIC values of 2, 4, 8, and 8 µg/mL, respectively (Dhiman et al., 2012).

Commentaries

i. The plant yields phenanthroindolizidine alkaloids such as tylophorine, tylophorinidine, tylophorinine, tyloindicine A, tylocrebrine, and *O*-methyl tylophorinidine (Dhiman et al., 2013). Phenanthroindolizidine alkaloids are cytotoxic and being also antifungal and planar they probably target DNA and as such are broad-spectrum antimicrobial agents and this has been discussed earlier. Although being emetic, the plant could (but clinical trials are needed) have some value as phytomedication against zoonotic pandemic viruses, and added to a list with that could be called "first-line defense natural products" while waiting for vaccines or oral antiviral agents to be developed. Pandemic viruses will always develop variants forcing the industry to make new vaccines costing millions of USD to governments. The possibility to develop phytomedications to prevent or manage viral pandemic is a reality. There is the gut feeling that from now on, we will be forced to live under continuous pandemic threat and one of the ways out of this is to look into our medicinal plants and produce phytomedications with the same manufacturing rigor as Western medicines. There is a need to make clinical trials for medicinal plants affordable for developing countries. China and India are successfully utilizing their medicinal plant knowledge for the well-being of their communities. Note that in Sabah, some plants are used by locals claiming cure of COVID-19 (Julius Kulip, personal communication). In Bangladesh, the use of yogurts and *Capsicum annum* L. by remote tribes is claimed to protect community against COVID-19 (Mohammed Rahmatullah, personal communication).

ii. Our peer the Scottish botanist and physician William Roxburgh (1751–1815) who leaved in Calcuta most of his life observed the efficacy of this plant in treating cases of dysentery while serving at the Madras hospital from 1776 to 1778. Credit must be given to William Roxburgh for his monumental work on the botany of the plants of India including the "Flora indica, or, Descriptions of Indian plants". "Roxb." in botanical names indicates that the plant was described by William Roxburgh. The plant was listed in the Pharmacopoeia of India of 1868 as a substitute for Ipecacuanha for the treatment of dysentery.

iii. In India, *Tylophora fasciculata* Buch.-Ham. ex Wight is used for fever, ulcers, and wounds (local name: *Bhuidari*) and *Tylophora tenuis* Bl. is used for small pox.

iv. In the Subfamily Apocynoideae, Tribe Apocyneae *Trachelospermum jasminoides* (Lindl.) Lem. is used to treat gonorrhea in China (local name: *Luo shi*). From this plant the dibenzylbutyrolactone lignan trachelogenin inhibited the replication of the Hepatitis C virus with the IC_{50} value of 0.3 µg/mL via a mechanism involving, at least in part, blocking of virus entry in Huh7 cells through glycoprotein E2 binding to CD81 (Qian et al., 2016). In the Subfamily Apocynoideae, Tribe Apocyneae, *Parameria laevigata* (A. L. juss.) Moldenke is used to heal wounds in Indonesia. Aqueous extracts of bark at the concentration of 250 µg/mL inhibited the Human immunodeficiency type-1 protease by 19.1% (Xu et al., 1996). This plant produces cinnamtannin B1 (Kamiya et al., 2001) which inhibited the replication of the Severe acute respiratory syndrome-associated coronavirus with the IC_{50} value of 32.9 µM (Zhuang et al., 2009). In the Subfamily Apocynoideae, Tribe Apocyneae, the latex of *Epigynum perakense* King & Gamble is applied to wounds in Malaysia. In the Subfamily Apocynoideae, Tribe Echiteae, *Pottsia laxiflora* (Blume) Kuntze is used for boils in China (local name: *Lian zi teng*)

References

Chu, Y., Lv, X., Zhang, L., Fu, X., Song, S., Su, A., Chen, D., Xu, L., Wang, Y., Wu, Z. and Yun, Z., 2020. Wogonin inhibits in vitro herpes simplex virus type 1 and 2 infection by modulating cellular NF-κB and MAPK pathways. *BMC Microbiology*, *20*(1), pp. 1–11.

Dhiman, M., Khanna, A. and Manju, S., 2013. A new phenanthroindolizidine alkaloid from Tylophora indica. *Chemical Papers*, *67*(2), pp. 245–248.

Dhiman, M., Parab, R.R., Manju, S.L., Desai, D.C. and Mahajan, G.B., 2012. Antifungal activity of hydrochloride salts of tylophorinidine and tylophorinine. *Natural Product Communications*, *7*(9), pp. 1171–1172.

Guo, Q., Zhao, L., You, Q., Yang, Y., Gu, H., Song, G., Lu, N. and Xin, J., 2007. Anti-hepatitis B virus activity of wogonin in vitro and in vivo. *Antiviral Research*, *74*(1), pp. 16–24.

Kamiya, K., Watanabe, C., Endang, H., Umar, M. and Satake, T., 2001. Studies on the Constituents of Bark of Parameria laevigata MOLDENKE1. *Chemical and Pharmaceutical Bulletin*, *49*(5), pp. 551–557.

Khanna, C., Singh, S., Vyas, M. and Das, S., 2020. Evaluation of Antimicrobial behaviour of Tylophora indica leaf extracts against selected food microbial pathogens. *Research Journal of Pharmacy and Technology*, *13*(11), pp. 5224–5228.

Ma, S.C., Du, J., But, P.P.H., Deng, X.L., Zhang, Y.W., Ooi, V.E.C., Xu, H.X., Lee, S.H.S. and Lee, S.F., 2002. Antiviral Chinese medicinal herbs against respiratory syncytial virus. *Journal of Ethnopharmacology*, *79*(2), pp. 205–211.

Qian, X.J., Jin, Y.S., Chen, H.S., Xu, Q.Q., Ren, H., Zhu, S.Y., Tang, H.L., Wang, Y., Zhao, P., Qi, Z.T. and Zhu, Y.Z., 2016. Trachelogenin, a novel inhibitor of hepatitis C virus entry through CD81. *Journal of General Virology*, *97*(5), pp. 1134–1144.

Seong, R.K., Kim, J.A. and Shin, O.S., 2018. Wogonin, a flavonoid isolated from Scutellaria baicalensis, has anti-viral activities against influenza infection via modulation of AMPK pathways. *Acta Virologica*, *62*(1), pp. 78–85.

Xu, H.X., Wan, M., Loh, B.N., Kon, O.L., Chow, P.W. and Sim, K.Y., 1996. Screening of traditional medicines for their inhibitory activity against HIV-1 protease. *Phytotherapy Research*, *10*(3), pp. 207–210.

Zhuang, M., Jiang, H., Suzuki, Y., Li, X., Xiao, P., Tanaka, T., Ling, H., Yang, B., Saitoh, H., Zhang, L. and Qin, C., 2009. Procyanidins and butanol extract of Cinnamomi Cortex inhibit SARS-CoV infection. *Antiviral Research*, *82*(1), pp. 73–81.

10.3.2 Family Gentianaceae A.L. de Jussieu (1789)

The family Gentianaceae consists of 90 genera and 1000 species of cosmopolitan, trees, shrubs, and herbs. The leaves are simple, opposite connected at base, and exstipulate. The blade is often marked with few longitudinal secondary nerves. The inflorescences are racemes, cymes, verticillate clusters, or solitary. The calyx comprises 4–8 sepals free or united in a tube. The corolla is tubular, and 4–8 lobed. The androecium includes 4–5 stamens. The ovary consists of 2 carpels merged in a unilocular ovary sheltering ovules growing on parietal placentas. The fruits are dehiscent capsules. Plants in the family produce bitter iridoid glycosides (especially gentiopicroside).

10.3.2.1 *Enicostema axillare* (Lam.) A. Raynal

Synonyms: *Adenema hyssopifolium* (Willd.) G. Don; *Cicendia hyssopifolia* (Willd.) Wight & Arn.; *Gentiana axillaris* Lam.; *Gentiana hyssopifolia* L.; *Enicostema hyssopifolium* (Willd.) I. Verd.; *Enicostema littorale* Bl.; *Exacum hyssopifolium* Willd.

Common names: Indian Gentian; kirota (Bangladesh); chota-chirayata, naagjhvaa, vellerugu (India); ehba bisah (Indonesia)

Habitat: Sun-exposed coastal grasslands and open forests

Distribution: Tropical Asia

Botanical description: This herb grows up to about 1 m tall. It has at first glance a Lamiaceous look and emits positive aura. The stems are quadrangular. The leaves are simple, sessile, decussate, and exstipulate. The blade is narrowly elliptic to lanceolate, 4–16 cm × 0.5–4.5 cm, of a peculiar somewhat look of leaves of sage, hairy, clearer at margin, and acute at apex. The flowers are sessile, axillary, and verticillate (about 5 flowers) in a Lamiaceous manner (*Hyptis*-like). The calyx is tubular, 5-lobed, and about 5 mm long. The corolla is white (note that flower buds of Gentianaceae have a characteristic shape), 5-lobed, the tube 7 mm long, and the 5 lobes triangular and 3 mm long. The 5 stamens are attached to the corolla tube. The ovary develops a capitate stigma. The capsules are 7 mm long and open to release numerous seeds.

Medicinal uses: Blood purifier (India); fever (India; Indonesia)

Broad-spectrum antibacterial iridoid glycoside: Swertiamarin (1500 μg/5 mm disc) inhibited the growth of *Escherichia coli*, *Pseudomonas aeruginosa*, and *Bacillus subtilis* with the inhibition zone diameters of 7, 7, and 6.5 mm, respectively (Samaddar et al., 2013).

Strong broad-spectrum antibacterial iridoid glycoside: Swertiamarin (LogD –2.4 at pH 7.4, molecular mass = 374.3 g/mol) inhibited the growth of *Escherichia coli*, *Pseudomonas aeruginosa*, *Salmonella typhimurium*, *Enterobacter cloacae*, *Listeria monocytogenes*, *Bacillus cereus*, *Micrococcus flavus*, and *Staphylococcus aureus* with the MIC/MBC values of 20/40, 10/20, 20/40, 40/40, 40/40, 20/20, 20/20, and 20/20 μg/mL, respectively (Šiler et al., 2010).

Strong antiviral (enveloped linear monopartite dimeric single-stranded (+) RNA) polar extract: Methanol extract inhibited the replication of the Human immunodeficiency virus (IIIB) with the IC_{50} value of 85 μg/mL and the selectivity index of 4.8 (Praveena et al., 2014).

Strong antifungal (filamentous) iridoid glycoside: Swertiamarin (LogD –2.4 at pH 7.4, molecular mass = 374.3 g/mol) inhibited the growth of *Aspergillus versicolor*, *Aspergillus niger*, *Aspergillus fumigatus*, *Penicillium funiculosum*, and *Trichoderma viride* with the MIC/MBC values of 50/100, 50/100, 50/100, 50/100, and 1/25 μg/mL, respectively (Šiler et al., 2010).

Strong antiviral (enveloped circular double-stranded DNA) C-glycoside flavone: Swertisin (also known as 4',5-Dihydroxy 7-methoxyflavone 6-C-glucoside; LogD –0.4 at pH 7.4; molecular mass = 446.4 g/mol) inhibited the secretion of Hepatitis B surface antigen, e-antigen, and DNA with the IC_{50} values of about 2 μM in HepG2.2.15 cells (Xu et al., 2020).

In vivo antiviral (enveloped circular double-stranded DNA) C-glycoside flavone: Swertisin given intraperitoneally to mice at the dose of 5 mg/Kg every other day for 3 weeks decreased serum surface antigen, e-antigen, and viral DNA (Xu et al., 2020).

Commentaries

i. The plant yields iridoid glycosides and the flavone C-glycoside swertisin (also known as 4′,5-dihydroxy 7-methoxyflavone 6-C-glucoside; (Patel & Mishra, 2011).
ii. Consider that traditionally bitter plants are often used for fever in Asia. Bitter plants are often choleretic and swertiamarin stimulates the excretion of fecal bile acid (Vaidya et al., 2009). Is there a relationship between bile excretion and decrease of fever? Note that the

mystical book of Zohar also known as the book of radiance written by Rabbi Shimon bar Yochai during the 2nd century BC states that bile plays a fundamental role in human health.

iii. Swertiamarin has some effects on the angiotensin-converting enzyme (Li et al., 2016) and as such could be examined for its activity against coroviruses.

iv. Being depressant on the central nervous system (Bhattacharya et al., 1976) swertiamarin could have bacteria and/or fungal efflux pump inhibitory properties.

References

Bhattacharya, S.K., Reddy, P.K.S.P., Ghosal, S., Singh, A.K. and Sharma, P.V., 1976. Chemical constituents of gentianaceae XIX: CNS-depressant effects of swertiamarin. *Journal of Pharmaceutical Sciences*, *65*(10), pp. 1547–1549.

Li, S., Wang, Q., Tao, Y. and Liu, C., 2016. Swertiamarin attenuates experimental rat hepatic fibrosis by suppressing angiotensin II–angiotensin type 1 receptor–extracellular signal-regulated kinase signaling. *Journal of Pharmacology and Experimental Therapeutics*, *359*(2), pp. 247–255.

Patel, M.B. and Mishra, S.H., 2011. Hypoglycemic activity of C-glycosyl flavonoid from Enicostemma hyssopifolium. *Pharmaceutical Biology*, *49*(4), pp. 383–391.

Praveena, P., Godse, A., Pawar, M., Kulkarni, S. and Sudarsanam, D., 2014. In-Vitro Anti-HIV activity studies on Enicostemma littorale (Lam) Raynal. Whole plants. *World Journal of Pharmaceutical*, *3*(6), pp. 1343–1352.

Samaddar, T., Chaubey, B., Jha, S. and Jha, T.B., 2013. Determination of swertiamarin and amarogentin content and evaluation of antibacterial activity in Eastern Himalayan species of Swertia L. *Pharmacognosy Communications*, *3*(4), pp. 64–70.

Šiler, B., Mišić, D., Nestorović, J., Banjanac, T., Glamočlija, J., Soković, M. and Ćirić, A., 2010. Antibacterial and antifungal screening of Centaurium pulchellum crude extracts and main secoiridoid compounds. *Natural Product Communications*, *5*(10), pp. 1525–1530.

Vaidya, H., Rajani, M., Sudarsanam, V., Padh, H. and Goyal, R., 2009. Swertiamarin: A lead from Enicostemma littorale Blume. for anti-hyperlipidaemic effect. *European Journal of Pharmacology*, *617*(1–3), pp. 108–112.

Xu, H.Y., Ren, J.H., Su, Y., Ren, F., Zhou, Y.J., Jiang, H., Cheng, S.T., Zhang, C.R. and Chen, J., 2020. Anti-hepatitis B virus activity of swertisin isolated from Iris tectorum Maxim. *Journal of Ethnopharmacology*, *257*, p. 112787.

10.3.2.2 *Exacum tetragonum* Roxb.

Common names: Bicolor Persian violet; zao bai nian (China); kuchuri (Bangladesh); avachiretta, ava chirayata (India); pa deing gno (Myanmar); canto canto (the Philippines)

Habitat: Grass lands, roadsides

Distribution: Pakistan, India, Bhutan, Nepal, Bangladesh, Myanmar, Cambodia, Laos, Vietnam, China, Malaysia, the Philippines, Papua New Guinea, Australia

Botanical description: This magnificent herb grows up to 1 m tall. It has a positive and joyful aura of life force. The stems are quadrangular. The leaves are simple, opposite, and exstipulate. The blade is ovate to lanceolate, 3.5–12.5 cm × 0.5–3.5 cm, round at base, and marked with 2–3 pair of secondary nerves. The cymes are terminal or axillary and few flowered. The calyx is tubular, 5 mm long, and 4-lobed. The corolla develops 4 elliptic and acuminate lobes (the whole flower with somewhat vaguely with shapes of birds or butterflies) which are about 2 cm long, of a unique heavenly light blue to white and blue at apex. The 4 stamens are inserted at the throat of the corolla, bright yellow, pointed, and about 6 mm long. The ovary develops a slender style that reaches 1.2 cm long. The capsules are dehiscent and contain numerous warty seeds.

Medicinal uses: Fever (India; Myanmar)

Commentaries

i. In India, *Exacum bicolor* Roxb. (synonym: *Exacum tetragonum* var. *bicolor* (Roxb.) Hook. (local name: *Baracharayatah*) is used for fever and *Exacum lawii* C.B. Clarke is used for eye infections.

ii. Apparently the plant has not been examined for its antimicrobial properties. Consider that methanol extract of *Exacum affine* Balf. f. ex Regel inhibited the replication of the Influenza virus A A/WSN/33 (H1N1) in Madin-Darby canine kidney cells and Herpes simplex virus type-1 strain KOS) in Vero cells with the IC_{50} values of 0.7 and 12.5 µg/mL, respectively (Mothana et al., 2006). *Exacum affine* Balf. f. ex Regel produces the phenolic paeonol (also known as 2′-hydroxy-4′-methoxyacetophenone) (Matsumoto, 1994). Paeonol inhibited the replication of the Human rhinovirus type-2 and Human rhinovirus type-4 in MRC5 with the IC_{50} values of 83.6 and 81.7 µg/mL, respectively (CC_{50}: 522.5 µg/mL (Luong, 2013). Paeonol inhibited the replication of the Tobacco mosaic virus (Guokun et al., 2006). Both the Human rhinovirus and Tobacco mosaic viruses are non-enveloped monopartite linear single-stranded (+) RNA, and as such paenol could have activity against coronaviruses and other pandemic zoonotic (+) RNA viruses. Paeonol is produced by members of the genus *Paeonia* L. which are often used to treat COVID-19 in China (see earlier).

References

Guokun, L., Qijian, C. and Zujian, W., 2006. Inhibition of the infectivity of tobacco mosaic virus by paeonol. *Journal of Fujian Agriculture and Forestry University*, *35*, pp. 17–20.

Luong, T.M.N., 2013. *Antiviral and antimicrobial activities of Paeonia lactiflora root constituents and structurally related compounds against human rhinovirus and gastrointestinal bacteria* (Doctoral dissertation, 서울대학교 대학원).

Matsumoto, M., 1994. 2′-Hydroxy-4′-methoxyacetophenone (Paeonol) in Exacum affine cv. *Bioscience, Biotechnology, and Biochemistry*, *58*(10), pp. 1892–1893.

Mothana, R.A.A., Mentel, R., Reiss, C. and Lindequist, U., 2006. Phytochemical screening and antiviral activity of some medicinal plants from the island Soqotra. *Phytotherapy Research: An International Journal Devoted to Pharmacological and Toxicological Evaluation of Natural Product Derivatives*, *20*(4), pp. 298–302.

10.3.2.3 *Fragraea racemosa* Jack

Common names: False coffee tree; han tuk (Cambodia); engkudug biang, kay si markopi kopi (Indonesia); kl'u payam, mumpuli, setebal (Malaysia); balatbuaya (the Philippines); thumb ok (Thailand)

Habitat: Swampy forests, riverbanks, seashores, cultivated

Distribution: India, Sri Lanka, Bangladesh, Myanmar, Cambodia, Laos, Vietnam, Thailand, Andaman Islands, Malaysia, Indonesia, the Philippines, Papua New Guinea, Solomon Islands

Botanical description: This handsome tree grows up to 20 m tall. The bark is fissured and greyish-brown, and the inner bark is yellowish. The stems are smooth, terete, articulate, and glossy. The leaves are simple, decussate, and exstipulate. The petiole is up to 5 cm long and stout, and the interpetiolar region forms some sort of cup. The blade is fleshy, coriaceous, elliptic to obovate to broadly lanceolate or oblong, 5–50 cm × 1–20 cm, round to acute to tapering at base, acuminate at apex, and marked with about 5–9 pairs of secondary nerves. The racemes are pendulous and up to 60 cm long. The calyx is 5-lobed and 4 mm long. The corolla is bell-shaped, about 4 cm long, whitish, the tube 2 cm long, and develops 5 lobes. The 5 stamens are attached to the corolla tube and the anthers are white. The style is slender, 2 cm long, and protrudes out of the corolla. The berries are bluish to red, up to 2.5 cm long, yielding a sticking exudate, and glossy. The infructescence and even the inflorescence have somewhat vaguely an Alpiniaceous appearance.

Medicinal uses: Boils (Solomon Islands); fever (Malaysia)

Commentaries

i. The roots yield lignans such as pinoresinol and simple phenols like syringaldehyde (Okuyama et al., 1995). As seen earlier, pinoresinol is a broad-spectrum antibacterial and antifungal amphiphilic furofuran lignan. Syringaldehyde is an antibacterial (Gram-positive) phenolic.

ii. The bark contains the iridoid glycoside secologanoside and the alkaloid fagreoside (Lambert et al., 2011), which apparently have not been explored for antimicrobial effects. Note that most flavonoids, simple phenolics, lignans, and alkaloids produced by plants have a molecular size allowing entry into the active site of enzymes. Thus, such compounds will always have some levels of activities against enzymes. These compounds coming from the fixation of oxygen, carbon, and nitrogen via photosynthesis are themselves substrates for plant enzymes.

iii. *Fagraea fragrans* Roxb. ex Carey & Wall. is used to wash scabies in Cambodia, Laos, and Vietnam, and *Fagraea peregrina* (Reinw.) Bl. (synonym: *Cyrtophyllum peregrinum* Reinw.) is used for fever in Myanmar (local name: *Anan*). In Laos, *Fagraea fragrans* Roxb. ex Carey & Wall. (local name: *Man pa*) and in Vietnam (local name: *Trai*) it is used for fever and diarrhea. Ethylacetate extract of leaves of *Fragraea fragrans* Roxb. ex Carey & Wall. inhibited the growth of *Streptococcus mutans* (DMST 26095), *Streptococcus mitis* (ATCC 49456T), and *Streptococcus pyogenes* (DNAT 17020) (Meerungrueang and Panichayupakaranant, 2014). In Indonesia, *Fagraea* cf. *auriculata* Jack. (local name: Bako bako) is used for fever and mouth ulcers.

iv. There is the gut feeling that antiviral principles await discovery in members of the genus *Fagraea* Thunb.

References

Lambert, L.K., Ross, B.P., Deseo, M.A. and Garson, M.J., 2011. Phytochemical study of fagraea spp. Uncovers a new terpene alkaloid with anti-inflammatory properties1. *Australian Journal of Chemistry*, *64*(4), pp. 489–494.

Meerungrueang, W. and Panichayupakaranant, P., 2014. Antimicrobial activities of some Thai traditional medical longevity formulations from plants and antibacterial compounds from Ficus foveolata. *Pharmaceutical Biology*, *52*(9), pp. 1104–1109.

Okuyama, E., Suzumura, K. and Yamazaki, M., 1995. Pharmacologically active components of Todopon Puok (Fagraea racemosa), a medicinal plant from Borneo. *Chemical and Pharmaceutical Bulletin*, *43*(12), pp. 2200–2204.

10.3.2.4 *Gentiana macrophylla* Pallas

Synonym: *Dasystephana macrophylla* (Pall.) Zuev

Common name: Qin jiao (China)

Habitat: Meadows

Distribution: Kazakhstan, China, Siberia

Botanical observation: This magnificent herb grows up to about 60cm tall. The roots are large and elongated. The stems are stout, terete, and dark. The leaves are simple, in rosette or opposite, and exstipulate. The petiole is 3–5cm long. The blade is elliptic-lanceolate, 4–28 cm × 1.2–6cm, narrow to obtuse at base, acuminate at apex, and marked with 3–5 pairs of secondary nerves. The inflorescences are axillary and terminal whorls. The calyx develops 5 lobes. The corolla is tubular, blue-purple (typical of the family), tubular, 5-lobed, the lobes ovate, and 5mm long. The 5 stamens are attached on the corolla tube. The ovary is oblong and develops 2 stigmas. The capsules are ovoid-ellipsoid, about 2cm long, and shelter numerous tiny seeds.

Medicinal use: Tuberculosis (China).

Strong broad-spectrum polar extract: Methanol of roots inhibited the growth of *Micrococcus luteus*, *Enterococcus faecalis*, *Staphylococcus aureus*, *Staphylococcus epidermidis*, *Streptococcus uberis*, *Pseudomonas aeruginosa*, *Escherichia coli*, and *Klebsiella pneumoniae* with the MIC values of 76, 87, 85, 60, 120, 120, 140, and 100µg/mL, respectively (Yin et al., 2018).

Strong antibacterial (Gram-positive) lipophilic prenylated flavanone: Kurarinone (LogD 5.7 at pH 7.4; molecular mass = 438.5 g/mol) inhibited the growth of methicillin-resistant *Staphylococcus aureus*, Vancomycin-resistant *Enterococci*, and *Streptococcus mutans* (ATCC 25175) with the MIC values of 2µg/mL (Chen et al., 2005).

Antifungal (filamentous) coumarin: 2-Methoxyanofinic acid inhibited the growth of *Cladosporium cucumerinum* (Tan et al., 1996).

Antifungal flavanone: Kurarinone and kushenol I inhibited the growth of *Cladosporium cucumerinum*. Kurarinone inhibited also the growth of *Candida albicans* (Tan et al., 1996).

Strong antiviral (enveloped monopartite linear single-stranded (+) RNA) lipophilic prenylated flavanone: Kuranone inhibited the replication of the Human coronavirus (OC43) with the IC_{50} value of 3.4 µM in MRC-5 cells by acting at an early stage of virus infection and inhibiting the impairment of the virus-induced autophagic flux (Min et al., 2020)

Viral enzyme inhibition by polar extract: Aqueous extract of roots at the concentration of 250 µg/mL inhibited Human immunodeficiency virus type-1 protease by 60.7% (Xu et al., 1996).

Commentaries

i. Kuranone is being tested in China for the treatment of Hepatitis B (Pan et al., 2005) and could be of value against COVID-19 and other pandemic coronaviruses.

ii. Members of the genus *Gentiana* L. are often used for medicinal purpose in Temperate and Northern Asia or mountainous Tropical Asia.

In Pakistan, *Gentiana dahurica* Fisch. is used as sudorific (induce sweating to break fever) (local name: *Agherpanrai*) and *Gentiana kurroo* Royle is used to treat liver diseases (might include jaundice) (local name: *Neel kanth*). In India, *Gentiana kurroo* Royle is used for diarrhea, fever, jaundice, phtysis (tuberculosis?), syphilis, and urinary tract infections (local names: *Anujaa*, *Nilkant*), *Gentiana tenella* Rottb. (local name: *Tita*) is used to treat fever, *Gentiana argentea* (D.Don) Griseb. is used to treat sore throat, *Gentiana nubigena* Edgew. is used for cough, and *Gentiana olivieri* Griseb. is used for fever.

In Bhutan, *Gentiana urnula* Harry Sm. (local name: *Gang ga chung*) is used to treat cough, diarrhea, fever, septicemia, tuberculosis, and wounds.

In Nepal, *Gentiana ornata* (G. Don) Wall. ex Griseb. is used to treat cold and cough (local name: *Karu*).

Gentiana depressa Bourg. is used in Myanmar (local name: *Sao kho*) for cough.

In China, *Gentiana rhodantha* Franch. is used to treat gonorrhea and *Gentiana triflora* Pall. is used to heal abscesses. In China (local name: *Lóng dǎn*) and Korea, *Gentiana scabra* Bunge is used to treat viral infections, and in Vietnam, it is used for fever, jaundice, sore throat, and eye infections. Hexane extract of roots of *Gentiana scabra* Bunge inhibited the replication of the Severe acute respiratory syndrome-associated coronavirus in Vero cells with the IC_{50} value of 8.7 µg/mL and a selectivity index above 57.5 (Wen et al., 2011). In Vietnam, *Gentiana loureiroi* (G. Don) Griseb. is used to treat pimples, dysentery, and liver diseases (jaundice?).

In Indonesia, *Gentiana quadrifaria* Bl. is used to treat dysentery (local name: *Sirawan bumi*).

iii. In "*De historia stirpium commentarii insignes*" of our peer the German physician and botanist Leonhardt Fusch (1501–1566) published in 1542, is found a description of *Gentiana lutea* L. as hot and dry (*radicem eius esse calidam et siccam*) and prescribed to heal wounds and to treat ophthalmia (as per Dioscorides). In *A Dialogue bothe pleasaunte and pietifull, wherein is a goodly regimente against the fever pestilence, with a consolacion and comfort against death* of our peer the English physician William Bullein (1515–1576) published in 1564, *Gentiana lutea* L. is recommended for for plague.

References

Chen, L., Cheng, X., Shi, W., Lu, Q., Go, V.L., Heber, D. and Ma, L., 2005. Inhibition of growth of Streptococcus mutans, methicillin-resistant Staphylococcus aureus, and vancomycin-resistant enterococci by kurarinone, a bioactive flavonoid isolated from Sophora flavescens. *Journal of Clinical Microbiology*, *43*(7), pp. 3574–3575.

Min, J.S., Kim, D.E., Jin, Y.H. and Kwon, S., 2020. Kurarinone inhibits HCoV-OC43 infection by impairing the virus-induced autophagic flux in MRC-5 human lung cells. *Journal of Clinical Medicine*, *9*(7), p. 2230.
Pan, Z.S., Yu, Q.H., Yan, H. and Zhang, Y., 2005. Clinical study on treatment of chronic hepatitis B by kurarinone combined with interferon alpha-1b. *Zhongguo Zhong xi yi jie he za zhi Zhongguo Zhongxiyi jiehe zazhi= Chinese Journal of Integrated Traditional and Western Medicine*, *25*(8), pp. 700–703.
Tan, R.X., Wolfender, J.L., Zhang, L.X., Ma, W.G., Fuzzati, N., Marston, A. and Hostettmann, K., 1996. Acyl secoiridoids and antifungal constituents from Gentiana macrophylla. *Phytochemistry*, *42*(5), pp. 1305–1313.
Wen, C.C., Shyur, L.F., Jan, J.T., Liang, P.H., Kuo, C.J., Arulselvan, P., Wu, J.B., Kuo, S.C. and Yang, N.S., 2011. Traditional Chinese medicine herbal extracts of Cibotium barometz, Gentiana scabra, Dioscorea batatas, Cassia tora, and Taxillus chinensis inhibit SARS-CoV replication. *Journal of Traditional and Complementary Medicine*, *1*(1), pp. 41–50.
Yin, C., Xie, L. and Guo, Y., 2018. Phytochemical analysis and antibacterial activity of Gentiana macrophylla extract against bacteria isolated from burn wound infections. *Microbial Pathogenesis*, *114*, pp. 25–28.
Xu, H.X., Wan, M., Loh, B.N., Kon, O.L., Chow, P.W. and Sim, K.Y., 1996. Screening of traditional medicines for their inhibitory activity against HIV-1 protease. *Phytotherapy Research*, *10*(3), pp. 207–210.

10.3.2.5 *Swertia chirayita* H. Karst.

Synonym: *Swertia chirata* Buch.-Ham. ex C.B. Clarke

Common names: Bitter stick, cleaning nut ree, Indian gentian; chirota (Bangladesh); chiratika, siretta (India); nenilawandi (Iran); sekhagi (Myanmar); chiraito (Nepal)

Habitat: Meadows, grassy lands, mountains

Distribution: Pakistan, India, Nepal, Bangladesh, Myanmar

Botanical description: This is an erect herb which grows up to about 1 m tall. The stems are terete. The leaves are simple, sessile, opposite, and exstipulate. The blade is broadly lanceolate, 1.5–10 cm × 0.5–3.8 cm, amplexicaul, acute at apex. The inflorescences are axillary or terminal panicles of cymes. The calyx is tubular, minute, and 4-lobed. The corolla is tubular, 4-lobed, and yellowish-tinged with purple. The androecium includes 4 stamens which are about 5 mm long, the anthers 1 mm long. The ovary is ovoid and minute, and the stigma lobe capitate. The capsules are about 5 mm long.

Medicinal uses: Abscesses, cold, cough, diarrhea, tuberculosis, ulcers (India); septicemia (Bangladesh); typhoid (Vietnam); pimple (Nepal); fever (Bangladesh; India; Nepal); bronchitis, gonorrhea, leprosy (Bangladesh; India)

Broad-spectrum antibacterial polar extract: Methanol extract of leaves (50 µL of a 500 µg/mL solution/6 mm well) inhibited the growth of *Bacillus megaterium* (MTCC 8510), *Bacillus subtilis* (MTCC 441), *Bacillus flexus* (MTCC 7024), *Staphylococcus aureus* (MTCC 96), *Clostridium perfringens* (MTCC 450), *Lactobacillus rhamnosus* (MTCC 1423), *Pseudomonas oleovorans* (MTCC 617), *Klebsiella pneumoniae* (MTCC 7028), and *Salmonella enterica* (MTCC 1164) with the inhibition zone diameters of 13, 11, 12.5, 21.5, 17.5, 11.5, 12, 9.5, and 17.5 mm, respectively (Roy et al., 2015).

Antiviral (enveloped monopartite linear double-stranded DNA) polar extract: Aqueous extract at 1:64 dilution inhibited the replication of the Herpes simplex virus type-1 in Vero cells by more than 70% (Verma et al., 2008).

Strong antiviral (enveloped circular double-stranded DNA) mid-polar extract: Ethylacetate extract inhibited the secretions of Hepatitis B surface antigen and Hepatitis B e-antigen as well as Hepatitis B virus DNA replication (Zhou et al., 2015).

Very weak antiviral (enveloped circular double-stranded DNA) mid-polar xanthone glycoside: 1-*O*-*β*-D-glucopyranosyl-3, 5, 8-trihydroxyxanthone inhibited the secretions of Hepatitis B surface antigen and Hepatitis B e-antigen with IC_{50} values of 400 µM (Zhou et al., 2015).

Very weak antiviral (enveloped circular double-stranded DNA) mid-polar aryltetralin lignan glycoside: (+)-cycloolivil-4′-*O*-*β*-D-glucopyranoside inhibited the secretions of Hepatitis B surface antigen and Hepatitis B e-antigen with IC_{50} values of 310 µg/mL (selectivity index > 2) and 770 µg/mL (selectivity index > 19.6), respectively, as well as Hepatitis B virus DNA replication with an IC_{50} value of 290 µg/mL (Zhou et al., 2015)

Very weak antiviral (enveloped circular double-stranded DNA) mid-polar phenolic: 3, 4′-Dihydroxy-3′-methoxypropiophenone 3-*O*-*β*-D-glucopyranoside inhibited Hepatitis B DNA with IC_{50} value of 330 µg/mL (Zhou et al., 2015).

Commentaries

i. The plant produces xanthones (Ghosal et al., 1973) such as swerchirin (1,8-dihydroxy-3,5-dimethoxy xanthone), 7-*O*-methyl swertianin (1,8-dihydroxy-3,7-dimethoxy xanthone) (Asthana et al., 1991), as well as the xanthone *C*-glycoside mangiferin (Pandey et al., 2012). Mangiferin is a broad-spectrum antiviral agent *in vitro* (see earlier) as well as an inhibitor of lipopolysaccharide-induced inflammation and fever (Qu et al. 2017) and possibly active against the Severe acute respiratory syndrome-associated coronavirus 2. The plant is used for COVID-19 in India.

ii. Members of the genus *Swertia* L. are often used for fever in the Indian subcontinent. In India, *Swertia alata* C.B Clarkes and *Swertia thomsoni* C.B Clarkes are used for fever (local name: *Chiretta*), and *Swertia hookeri* (local name: *Rinku*) is used for diarrhea. In India and Nepal, *Swertia angustifolia* Buch.-Ham. ex D. Don (synonym: *Swertia affinis* C.B Clarkes local name in Nepal: *Chirayto*) is used for fever, and in Nepal it is also used for septicemia. In Nepal, *Swertia nervosa* (Wall. ex G. Don) C.B. Clarke is used for fever (local name: *Tite, Vale chirayto*). In Pakistan, *Swertia cordata* (Wall. ex G. Don) C.B. Clarke (local name: *Chairta, Karita*) is used to treat fever, jaundice, cough, cold, and typhoid; *Swertia paniculata* Wall. is used to treat fever (local name: *Malyjari*), and *Swertia petiolata* D. Don (local name: *Plamas*) is used for a cold, cough, fever, jaundice, and typhoid.

iii. Other members of the family Gentianaceae are used for microbial infections in Asia: In Iran, *Centaurium tenuifolium* (Hoffmanns. & Link) Fritsch ex Janch. (local name: *Ghontorion*) is used for jaundice and wounds. In Pakistan, *Comastoma boreale* (Bunge) T.N. Ho ex Omer (local name: *Shalay char*) is used for pneumonia and fever. *Gentianodes eumarginata* Omer (local name: *Shotay lileo*) is used for jaundice. *Gentianodes tianschanica* (Rupr.) Omer, Ali & Qaiser is used for bronchitis, cough, fever, jaundice, pneumonia, septicemia, and eye infections. *Gentianopsis vvedenskyi* (Grossh.) Pissjauk. is used for cough, fever, pneumonia, sore throat, and typhoid. In Nepal, *Halenia elliptica* D.Don (local name: *Baba tite*) is used to treat cold, cough, and fever.

iv. In the closely related family Gelsemiaceae Struwe and V.A. Albert (1995) the poisonous *Gelsemium elegans* (Gardner & Champ.) Benth. is used to heal boils and ringworms in China.

References

Asthana, R.K., Sharma, N.K., Kulshreshtha, D.K. and Chatterjee, S.K., 1991. A xanthone from Swertia chirayita. *Phytochemistry*, *30*(3), pp. 1037–1039.

Ghosal, S., Sharma, P.V., Chaudhuri, R.K. and Bhattacharya, S.K., 1973. Chemical constituents of the gentianaceae V: Tetraoxygenated xanthones of Swertia chirata buch.-ham. *Journal of Pharmaceutical Sciences*, *62*(6), pp. 926–930.

Pandey, D.K., Basu, S. and Jha, T.B., 2012. Screening of different East Himalayan species and populations of Swertia L. based on exomorphology and mangiferin content. *Asian Pacific Journal of Tropical Biomedicine*, *2*(3), pp. S1450–S1456.

Qu, S., Wang, W., Li, D., Li, S., Zhang, L., Fu, Y. and Zhang, N., 2017. Mangiferin inhibits mastitis induced by LPS via suppressing NF-κB and NLRP3 signaling pathways. *International Immunopharmacology*, *43*, pp. 85–90.

Roy, P., Abdulsalam, F.I., Pandey, D.K., Bhattacharjee, A., Eruvaram, N.R. and Malik, T., 2015. Evaluation of antioxidant, antibacterial, and antidiabetic potential of two traditional medicinal plants of India: Swertia cordata and Swertia Chirayita. Pharmacognosy *Research*, 7(Suppl 1), p. S57.

Verma, H., Patil, P.R., Kolhapure, R.M. and Gopalkrishna, V., 2008. Antiviral activity of the Indian medicinal plant extract, Swertia chirata against herpes simplex viruses: A study by in-vitro and molecular approach. *Indian Journal of Medical Microbiology*, *26*(4), p. 322.
Zhou, N.J., Geng, C.A., Huang, X.Y., Ma, Y.B., Zhang, X.M., Wang, J.L. and Chen, J.J., 2015. Anti-hepatitis B virus active constituents from Swertia chirayita. *Fitoterapia*, *100*, pp. 27–34.

10.3.3 Family Loganiaceae R.Br. ex Mart. (1827)

The family Loganiaceae consists of about 25 genera and 500 species of trees, shrubs, climbers, or herbs. The leaves are simple, opposite, and stipulate. The blade is often characteristically marked with 2–3 pairs of longitudinal secondary nerves. The inflorescences are axillary or terminal cymes. The calyx is tubular and 5-lobed. The corolla is tubular and 5-lobed. The 5 anthers are attached to the corolla tube. The gynoecium consists of 2 carpels fused into a bilocular ovary, each locule sheltering 1 to many ovules growing from axillary or parietal placentas. The style develops a capitate stigma. The fruits are capsules, berries, or drupes.

10.3.3.1 *Strychnos nux-vomica* L.

Synonym: *Strychnos spireana* Dop

Common names: Sleng (Cambodia); ma qian zi (China); etti, kanchiram, kanjiram, kanniram, kapilu, kuchila, kuchla (India); nirmali (Nepal); kuchulah (Iran); godakadura (Sri Lanka); cevzi mukayyi (Turkey)

Habitat: Forests

Distribution: India, Nepal, Sri Lanka, Bangladesh, Cambodia, Laos, Vietnam, Thailand, Malaysia

Botanical observation: This tree grows to about 30 m tall. The leaves are simple, opposite, and exstipulate. The petiole is 0.5–1.5 cm long. The blade is broadly elliptic, 5–15 cm × 4.5–7.5 cm, dark green and glossy, acute to somewhat round to almost cordate at base, acuminate at apex, and marked with 1–2 pairs of secondary nerves and a lax net of tertiary scalariform nerves. The cymes are axillary and up to about 10 cm long. The calyx is tubular, 4 mm long, 5-lobed, and hairy. The corolla is tubular, greenish, about 1.5 cm long, and develops 5 lobes which are 5 mm long, oblong, and acute. The 5 stamens are inserted in the corolla tube. The ovary is ovoid and develops a style which grows up to about 1 cm long and protrudes out of the corolla and a capitate stigma giving the flower a Rubiaceous look. The berries are globose, orange, glossy (vaguely from a distance has a look of *Citrus* tree), and 2–8 cm across The seeds are up to 5, flat, 2 cm across, brownish to somewhat silvery, somewhat glossy, minutely hairy, discoid, immersed in a white pulp, and horrifyingly poisonous (!).

Medicinal uses: Diarrhea, dysentery, fever, jaundice, ringworm, ulcers, wounds (India); leprosy (Cambodia); cholera (India; Laos); typhoid (China)

Broad-spectrum antibacterial halo developed by polar extract: Ethanol extract of stem bark inhibited the growth of *Micrococcus luteus* MTCC (106), *Bacillus cereus* MTCC (430), *Escherichia coli* MTCC (443), *Streptococcus pneumonia* MTCC (2672) with MIC value of 100, 100, 300, and 300 µg/mL, respectively (Singh et al 2010).

Strong antiviral (enveloped linear segmented single-stranded (−) RNA) polar extract: Methanol extract inhibited the replication of the Influenza A virus (AP/R/8 virus) with the IC_{50} value of 23.6 µg/mL and a selectivity index of 0.5 in Madin-Darby canine kidney cells (Enkhtaivan et al., 2015).

Commentaries

i. The plant yields the poisonous indole alkaloids brucine and strychnine which being neuroactive might be able to inhibit bacterial and/or fungal efflux pumps but too toxic to be used as therapeutic agents.
ii. Strychnobiflavone (molecular mass = 630.5 g/mL) inhibited the replication of the Herpes simplex virus type-1 (KOS) with the IC_{50} value of 2 µg/mL and selectivity index of 131.1

and Herpes simplex virus type-2 (333) with the IC_{50} value of 0.9 µg/mL and selectivity index of 270.1 (Boff et al., 2016). This biflavone is active against *Leishmania* (Lage et al., 2015) and as such is probably antifungal (see earlier).

iii. The seeds of *nux vomica* are said to be known of our peer the physician Garcia de Orta (1501–1568) in his "*Coloquios dos simples e drogas he cousas mediçinais da India*" published in 1563, and the seeds were sold in England in the 17th century to poison dogs.

iv. In India, *Strychnos columbrina* L. (local name: *Kuchilalata*) is used for diarrhea, fever, and small pox; *Strychnos potatorum* L.f (local name: *Amhuprasadanaphala*) is used for diarrhea, dysentery, eye infection, gonorrhea, and jaundice, and in Myanmar, it is used for dysentery and eye infection (local names: *Mango-taukpa-tit*, *Khabaung yay-kyi*). In India, *Strychnos rheedei* C.B Clarkes (synonym: *Strychnos wallichiana* Steud. ex A. DC; *Strychnos cinnamomifolia* Thwaites; *Strychnos bourdillonii* Brandis; local name: *Valli kanjiram*) is used for fever and ulcers as well as in Sri Lanka for ulcers (local name: *Etakirindiwe*). In India, *Strychnos minor* Dennst. (local name: *Cherukankiravally*) is used for dysentery. In Thailand, *Strychnos nux-blanda* A.W. Hill is used for fever. In Iraq and Iran, *Strychnos ignatia* P.J. Bergius (local name: *Paptiyal*) is used for plagues. In the family Loganiaceae, *Mitrasacme pygmaea* R. Br (local name: *Si ye lian*) is used to treat laryngitis.

References

Boff, L., Silva, I.T., Argenta, D.F., Farias, L.M., Alvarenga, L.F., Pádua, R.M., Braga, F.C., Leite, J.P.V., Kratz, J.M. and Simões, C.M.O., 2016. Strychnos pseudoquina A. St. Hil.: a Brazilian medicinal plant with promising in vitro antiherpes activity. *Journal of Applied Microbiology*, *121*(6), pp. 1519–1529.

Enkhtaivan, G., John, K.M., Ayyanar, M., Sekar, T., Jin, K.J. and Kim, D.H., 2015. Anti-influenza (H1N1) potential of leaf and stem bark extracts of selected medicinal plants of South India. *Saudi Journal of Biological Sciences*, *22*(5), pp. 532–538.

Lage, P.S., Chávez-Fumagalli, M.A., Mesquita, J.T., Mata, L.M., Fernandes, S.O., Cardoso, V.N., Soto, M., Tavares, C.A., Leite, J.P., Tempone, A.G. and Coelho, E.A., 2015. Antileishmanial activity and evaluation of the mechanism of action of strychnobiflavone flavonoid isolated from Strychnos pseudoquina against Leishmania infantum. *Parasitology Research*, *114*(12), pp. 4625–4635.

Singh, M., Khatoon, S., Singh, S., Kumar, V., Rawat, A.K.S. and Mehrotra, S., 2010. Antimicrobial screening of ethnobotanically important stem bark of medicinal plants. *Pharmacognosy Research*, *2*(4), p. 254.

10.3.4 Family Rubiaceae A.L. de Jussieu (1789)

The family Rubiaceae is a vast taxon consisting of about 650 species and 12,000 species of trees, shrubs, climbers, and herbs. The leaves are opposite, simple, and exstipulate. The stipules are characteristically interpetiolar. The inflorescences are cymes, spikes or heads, corymbs, panicules, or solitary. The calyx and corolla are 4–5 lobed and the corolla is often tubular. A disc is present. Four to five anthers are present. The gynoecium consists of 2 locules, each locule sheltering numerous ovules growing on axile or basal placenta. The style often protrudes out of the corolla and develops a linear, bifid, filiform or capitate or bilobed stigma. The fruits are capsules, berries, or drupes. Plants in this family produce a fascinating array antimicrobial iridoids, iridoid glycosides, indole alkaloids, anthraquinones, and proanthocyanidins.

10.3.4.1 *Adina cordifolia* (Roxb.) Hook. f. ex Brandis

Synonyms: *Haldina cordifolia* (Roxb.) Ridsdale, *Nauclea cordifolia* Roxb.

Common names: Yellow teak; kelikadum, pahaiphang; pang kha bang; pro om fang (Bangladesh); manjakadambu (India); yangmaw (Myanmar); kwow (Thailand)

Habitat: Forest

Distribution: India, Bangladesh, Myanmar, Cambodia, Laos, Vietnam, Malaysia

Botanical observation: These magnificent timber trees grow to 40 tall. The bole is straight and buttressed. The bark is reddish brown and the inner bark is reddish. The leaves are simple, opposite, and stipulate. The stipules are 10–12 mm × 5–12 mm, interpetiolar, and broadly lanceolate. The petiole is up to 12 cm long. The blade is broadly cordate, 5–25 cm in diameter, hairy below, marked with 6–10 pairs of conspicuous secondary nerves, wavy, and acuminate at apex. The flowers are arranged in axillary globose heads about 1 cm diameter on pedicels which are up to 10 cm long. The globose heads look very much like a coronaviruses. The calyx is minute, tubular, and develops 5 linear lobes. The corolla is tubular, yellowish-white, 5 mm long, and produces 5 triangular lobes. Five stamens attached to the corolla tube are present. The ovary develops a slender style that protrudes out of the corolla tube. The fruiting heads are about 1 cm across. The capsules are 5 mm long and containing minute seeds.

Medicinal uses: Diarrhea, dysentery (Cambodia; Laos; Vietnam); infections, fever (Bangladesh); sores (Myanmar)

Antibacterial (Gram-positive) halo developed by polar extract: Butanol extract of leaves (2 mg/mL/ 5 mm well) inhibited the growth of *Staphylococcus aureus* with the inhibition zone diameter of 7.3 mm (Kumari et al., 2017).

Moderate antibacterial (Gram-positive) indole alkaloid: Cadambine inhibited the growth of *Staphylococcus epidermidis*, *Staphylococcus aureus*, *Bacillus cereus*, and *Bacillus subtilis* with the MIC values of 120, 154.7, 164.5, and 88.3 µg/mL, respectively (Karaket et al., 2012).

Weak anticandidal indole alkaloid: Cadambine inhibited the growth of *Candida albicans* with the MIC value of 107.3 µg/mL (Karaket et al., 2012).

Commentaries

i. The plant yields the coumarin umbelliferone, the coumarin glycoside skimmin (Iqbal et al., 2009), as well as flavanones (Srivastava & Gupta, 1983), and indole alkaloids such as adifoline, cordifoline, and cadambine (Cannon et al., 1980). Alkaloids give to the wood a yellow color. Cadambine inhibited topoisomerase 1B in *Leishmania* with an IC_{50} value of 10 µg/mL (Kumar et al., 2015) which makes sense since it is planar. Planar heterocyclic alkaloids are often antiviral and broad-spectrum antibacterial. Topoisomerases are needed for the replication of (+)-RNA viruses (Prasanth et al., 2020). Is cadambine able to inhibit the replication of the Severe acute respiratory syndrome-associated coronavirus? This molecule could be added to the list of first-line natural product to test for forthcoming viral zoonotic pandemic waves.

ii. *Adina sessilifolia* (Roxb.) Hook. f. ex Brandis (synonym: *Nauclea sericea* Wall. ex G. Don; *Nauclea sessilifolia* Roxb.) is used in Bangladesh (local names: *Kam* gaas, *Kam gash*, *Kum*) to treat ringworm and skin infection, and in Cambodia, Laos, and Vietnam for tuberculosis.

iii. In Subfamily Cinchonoideae, Tribe Naucleeae, *Nauclea latifolia* Sm. (synonym: *Sarcocephalus latifolius* (Sm.) Bruce) is used in Bangladesh (local name: *Jongli-gulap jaam*) for the treatment of gonorrhea. Aqueous extract of this plant inhibited the growth of *Corynebacterium diphtheriae*, *Streptococcus* sp., *Neisseria* sp., *Pseudomonas aeruginosa*, *Salmonella* sp., *Streptobacillus* sp., and *Mucor* sp. with the MIC/MBC values of 200/500, 400/500, 400/500, 400/10000, 500/10000, 500/10000, and 500/500 µg/mL, respectively (Deeni & Hussain, 1991). A methanol/dichloromethane extract of root bark inhibited the replication of the acyclovir-resistant Herpes simplex virus type-2 strains with IC_{50} values of 7.1 µg/mL (Donalisio et al., 2013). Aqueous extract inhibited the replication of the Respiratory syncytial virus with the IC_{50} value of 58.1 µg/mL and the selectivity index of 3.2 (Odimegwu & Esimone, 2018).

iv. *Timonius timon* (Spreng.) Merr. in the Subfamily Cinchonoideae, Tribe Guettardeae, is used to heal wounds in New Britain.

v. In the Subfamily Cinchonoidea, Tribe Naucleeae, *Uncaria gambier* (W. Hunter) Roxb. is the source of an aqueous dry extract of leaves (known in Europe as pale catechu) used to treat diarrhea in Malaysia. In Indonesia (local name: *Getah gambir*) it is used for diarrhea, and in Iran (local name: *Kat-i-gulabi*) it is used for cough. It was called "*Terra Japonica*" by European phycisians of the 17th century and was used for the treatment of dysentery in Europe because of high amounts of (+)-catechin, (+)-epicatechin, procyanidin B1, procyanidin B3, gambiriin C, and gambiriin A1 (Taniguchi et al., 2007). Our peer the English physician John Pringle (1707–1782) recommended "*Terra Japonica*" as a "resister to putrefaction" (Dunker, 1938). Being strongly astringent, pale gambir is a broad-spectrum antimicrobial material. Ethanol extract of *Uncaria gambier* (W. Hunter) Roxb. inhibited the growth of 35 clinical isolates of methicillin-resistant *Staphylococcus aureus* and *Staphylococcus aureus* with MIC/MBC values of 0.4–0.8 /3.2 and 0.4 /3.2 µg/mL, respectively (Voravuthikunchai & Kitpipit, 2005).

In Thailand, *Uncaria macrophylla* Wall. (local name: *Smeng kr bey*) is used for postpartum. *Uncaria ferrea* (Blume) DC. is used to heal wounds in Malaysia. In Papua New Guinea, *Uncaria lanosa* Wall. (local name: *Marangi*) is used for cough, diarrhea, dysentery, and fever.

References

Cannon, J.R., Ghisalberti, E.L. and Lojanapiwatna, V., 1980. Isolation of cadambine from Adina cordifolia benth. and hook. F. *Journal of the Science Society of Thailand*, *6*(1), pp. 54–58.

Deeni, Y.Y. and Hussain, H.S.N., 1991. Screening for antimicrobial activity and for alkaloids of Nauclea latifolia. *Journal of Ethnopharmacology*, *35*(1), pp. 91–96.

Donalisio, M., Nana, H.M., Ngane, R.A.N., Gatsing, D., Tchinda, A.T., Rovito, R., Cagno, V., Cagliero, C., Boyom, F.F., Rubiolo, P. and Bicchi, C., 2013. In vitro anti-Herpes simplex virus activity of crude extract of the roots of Nauclea latifolia Smith (Rubiaceae). *BMC Complementary and Alternative Medicine*, *13*(1), pp. 1–8.

Dunker, M.F., 1938. A history of early antiseptics. *Journal of Chemical Education*, *15*(2), p. 58.

Iqbal, P.F., Bhat, A.R. and Azam, A., 2009. Antiamoebic coumarins from the root bark of Adina cordifolia and their new thiosemicarbazone derivatives. *European Journal of Medicinal Chemistry*, *44*(5), pp. 2252–2259.

Karaket, N., Supaibulwatana, K., Ounsuk, S., Bultel-Poncé, V., Pham, V.C. and Bodo, B., 2012. Chemical and bioactivity evaluation of the bark of Neonauclea purpurea. *Natural Product Communications*, *7*(2), pp. 169–170.

Kumar, A., Chowdhury, S.R., Jatte, K.K., Chakrabarti, T., Majumder, H.K., Jha, T. and Mukhopadhyay, S., 2015. Anthocephaline, a new indole alkaloid and cadambine, a potent inhibitor of DNA topoisomerase IB of Leishmania donovani (LdTOP1LS), isolated from Anthocephalus cadamba. *Natural Product Communications*, *10*(2), pp. 297–299.

Kumari, S., Verma, S.M., Kumar, H. and Kyal, C.K., 2017. Evaluation of antibacterial, antioxidant, wound healing properties of different solvent fractions of adina cordifolia leaves in experimental animals. *Advances in Research*, *12*, pp. 1–13.

Odimegwu, D.C. and Esimone, C.O., 2018. In vitro antiviral activity of nauclea latifolia root bark extract against the respiratory syncytial virus. *European Journal of Medicinal Plants*, *22*, pp. 1–7.

Prasanth, K.R., Hirano, M., Fagg, W.S., McAnarney, E.T., Shan, C., Xie, X., Hage, A., Pietzsch, C.A., Bukreyev, A., Rajsbaum, R. and Shi, P.Y., 2020. Topoisomerase III-β is required for efficient replication of positive-sense RNA viruses. *Antiviral Research*, *182*, p. 104874.

Srivastava, S.K. and Gupta, H.O., 1983. A new flavanone from Adina cordifolia. *Planta Medica*, *48*(05), pp. 58–59.

Taniguchi, S., Kuroda, K., Doi, K.I., Tanabe, M., Shibata, T., Yoshida, T. and Hatano, T., 2007. Revised structures of gambiriins A1, A2, B1, and B2, chalcane-flavan dimers from gambir (Uncaria gambir extract). *Chemical and Pharmaceutical Bulletin*, *55*(2), pp. 268–272.

Voravuthikunchai, S.P. and Kitpipit, L., 2005. Activity of medicinal plant extracts against hospital isolates of methicillin-resistant Staphylococcus aureus. *Clinical Microbiology and Infection*, *11*(6), pp. 510–512.

10.3.4.2 *Borreria articularis* (L. f.) F.N. Williams

Synonyms: *Spermacoce articularis* L. f.

Common names: Jointed Buttonweed, shaggy buttonweed; si fang kuan (China); mudmala, todarjil sak (Bangladesh); madana-ghanti, vasuka (India)

Habitat: Waste lands, roadside, lawn

Distribution: Tropical Asia

Botanical observation: This somewhat erect herb grows up to about 30 cm tall. The stem is fleshy, quadrangular, and hairy. The leaves are simple, decussate, and stipulate. The stipules are minute and hairy. The blade is somewhat elliptic to spathulate, fleshy, 2–3.5 cm × 0.5–1.5 cm, hairy, and marked with 3–5 pairs of secondary nerves. The inflorescences are axillary verticillasters of tiny flowers. The calyx develops 4 linear lobes. The corolla is tubular, light pink, about 1 cm long, and develops 4 lobes. The androecium includes 4 stamens attached to the corolla tube. The ovary produces a slender style with a bifid stigma. The capsules are 3 mm long, globose, and contain tiny and rough seeds.

Medicinal uses: Abscesses (China); wounds (Malaysia); eye infection (Bangladesh); diarrhea, dysentery, toothache, fever (India)

Broad-spectrum antibacterial halo developed by polar extract: Ethanol extract (1 mg/disc) inhibited the growth of *Shigella dysenteriae* (AE 14396), *Shigella sonnei* (ICDDR,B), *Salmonella typhi* (AE 14612), *Salmonella paratyphi* (AE 14613), *Bacillus subtilis* (BTCC 17). *Bacillus cereus* (BTCC 19) *Bacillus megaterium* (BTCC 18), *Staphylococcus aureus* (ATCC 6538), *Pseudomonas aeruginosa*, *Escherichia coli* (ATCC 25922), and *Vibrio cholerae* (AE 14748) (Sultana et al., 2008).

Broad-spectrum antifungal polar extract: Ethanol extracts of aerial parts at the concentration of 100 μg/mL inhibited the growth of *Aspergillus niger*, *Aspergillus ochraceus*, *Aspergillus ustus*, and *Candida albicans* by 34.5%, 49%, 53.8%, and 50%, respectively (Sultana et al., 2008).

Commentaries

i. The plant yields 3α-acetoxy-oleana-12-en-29-oic and β-amyrin (also known as (3β)-olean-13(18)-en-3-ol) (Mukherjee et al., 2004).
ii. In India, *Borreria ocymoides* DC (local name: *Kodi aman*) is used for wounds and *Borreria verticillata* (L.) G. Mey. is used (local name: *Nathaisoori*) for gonorrhea and leucorrhea. *Borreria verticillata* (L.) G. Mey. (as well as members of the genus *Borreria* G. Mey.) produces indole alkaloids such as verticillatine A and bis-indole alkaloids such as borreverine (Conserva et al., 2012). Alkaloidal extract of this plant protected CEM-C113 cells against the Human immunodeficiency virus and inhibited reverse transcriptase (Balde et al., 2015). Borreverine (LogD 5.9 at pH 7.4; molecular mass = 480.3 g/mol) at the concentration of 50 μg/mL inhibited the growth of *Staphylococcus aureus* and at 6 μg/mL inhibited the growth of *Vibrio cholerae* (Maynart et al., 1980). Alkaloids in the genus *Borreria* G. Mey. are probably antiviral.

References

Balde, A.M., Pieters, L.A., Traore, M.S., Camara, A., Balde, M.A., Oulare, K., Barry, M.S., Diallo, M.S.T., Balde, E.S., Diane, S. and Vlietinck, A.J., 2015. Chemotherapeutical evaluation of Borreria verticillata extracts. *Journal of Plant Sciences*, *3*(1–2), pp. 28–31.

Conserva, L.M. and Jesu Costa Ferreira, J., 2012. Borreria and Spermacoce species (Rubiaceae): A review of their ethnomedicinal properties, chemical constituents, and biological activities. *Pharmacognosy Reviews*, *6*(11), p. 46.

Maynart, G., Pousset, J.L., Mboup, S. and Denis, F., 1980. Antibacterial effect of borreverine, an alkaloid isolated from Borreria verticillata (Rubiaceae). *Comptes Rendus des Seances de la Societe de Biologie et de ses Filiales*, *174*(5), pp. 925–928.

Mukherjee, K.S., Mukhopadhyay, B., Mondal, S., Gorai, D. and Brahmachari, G., 2004. Triterpenoid constituents of Borreria articularis. *Journal of the Chinese Chemical Society*, *51*(1), pp. 229–231.

Sultana, R., Rahman, M.S., Bhuiyan, M.N.I., Begum, J. and Anwar, M.N., 2008. In vitro antibacterial and antifungal activity of Borreria articularis. *Bangladesh Journal of Microbiology*, *25*(2), pp. 95–98.

10.3.4.3 *Coptosapelta tomentosa* Valeton ex K. Heyne

Synonym: *Coptosapelta flavescens* Korth

Common names: Voa thraeng (Cambodia); akar metedong (Indonesia); dok khat khao (Laos); perual (Malaysia); yan khlong (Thailand); dây kha, day van hoan (Vietnam)

Habitat: Forests

Distribution: Myanmar, Cambodia, Laos, Vietnam, Thailand, Malaysia, Indonesia

Botanical observation: This climber grows up to 10 m long. The stems are hairy at apex. The leaves are simple, opposite, and stipulate. The stipules are interpetiolar, triangular, and about 5 mm long. The petiole is 0.5–1.5 cm long, hairy, and channeled. The blade is elliptic to oblong, membranous, 5–15 cm × 2.5–7.5 cm, rounded to acute at base, acuminate at apex, wavy, and marked with 3–5 pairs of secondary nerves. The cymes are terminal. The calyx is tubular, 5-lobed, and minute. The corolla is yellowish-white, narrowly tubular, the tube up to 8 mm long, 5-lobed, the lobes about 1 cm long and narrowly oblong. Fve anthers are adnate to the corolla tube and develop linear anthers growing up to 1 cm long. The ovary develops a slender style. The capsules are globose and contain numerous minute and winged seeds.

Medicinal use: Syphilis (Malaysia)

Strong antibacterial (Gram-positive) anthraquinones: 1-Hydroxy-2-hydroxymethyl-anthraquinone (also known as digiferruginol) and 1-hydroxy-2-methoxycarbonyl-anthraquinone inhibited the growth of methicillin-resistant *Staphylococcus aureus* with the MIC values of 32 and 16 μg/mL, respectively (Kongyen et al., 2014).

Moderate antibacterial (Gram-positive) amide alkaloid: 2-Amino-3-methoxycarbonyl-1,4-naphthoquinone isolated from the roots inhibited the growth of methicillin-resistant *Staphylococcus aureus* with the MIC value of 64 μg/mL (Kongyen et al., 2014).

Commentaries

i. The color of dry leaves in herbarium samples can give a hint at the type of secondary metabolites present in the sample. The leaves that turn black in herbarium sample often indicate the presence quinones, and it is the case for this plant which yields naphthoquinones such 1-hydroxy-2-hydroxymethylanthraquinone (Hounkong et al., 2014) as well as anthraquinones (Kongyen et al., 2014). Naphthoquinones and anthraquinones are broad-spectrum antimicrobial agents *in vitro* (see earlier). Also, in some members of the family accumulating anthraquinones, an orangish coloration of inner bark and wood is observed.
ii. Anthraquinones and naphthoquinones account probably for the use of this plant against syphilis. Consider increased incidence of resistance of *Treponema pallidum* to antibiotics (Katz, & Klausner, 2008).

References

Hounkong, K., Sawangjaroen, N., Kongyen, W., Rukachaisirikul, V., Voravuthikunchai, S.P. and Phongpaichit, S., 2014. Anti-intestinal protozoan activities of 1-hydroxy-2-hydroxymethylanthraquinone from Coptosapelta flavescens. *Asian Pacific Journal of Tropical Disease*, *4*(6), pp. 457–462.

Kongyen, W., Rukachaisirikul, V., Phongpaichit, S., Sawangjaroen, N., Songsing, P. and Madardam, H., 2014. Anthraquinone and naphthoquinone derivatives from the roots of Coptosapelta flavescens. *Natural Product Communications*, *9*(2), pp. 219–220.

Katz, K.A. and Klausner, J.D., 2008. Azithromycin resistance in Treponema pallidum. *Current Opinion in Infectious Diseases*, *21*(1), pp. 83–91.

10.3.4.4 *Gardenia jasminoides* J. Ellis

Common names: Cape jasmine, common gardenia; gandharaj (Bangladesh); kuchinashi (Japan), shan zhi i, zhi zi (China); gandharaaja (India); kacapiring (Indonesia); chijanamu (Korea); kuchinashi, san shi shi (Japan); kadenia (Palau)

Habitat: Mountains, forests edges, cultivated

Distribution: Tropical Asia and Pacific

Botanical observation: This magnificent shrub native to China grows up to 3 m tall. The stem is terete, articulated, striated, and hairy by apex, with few lenticels. The nodes show prominent leaf-scars. The internodes are 1–5 cm long. The leaves are simple, opposite, and stipulate. The stipules are lanceolate, deciduous, membranous, about 7 mm long with discrete longitudinal nervations, and rugose. The petiole is flattish, hairy, and about 1 cm long. The blade is 8.4–15 cm × 3.2–5.4 cm, elliptic to elliptic spathulate, dark green, glossy, membraneous, tapering at base, acuminate to caudate at apex, wavy, and marked with 11–13 pairs of secondary nerves. The flowers are fragrant, and solitary. The calyx is about 9 mm long and 6-lobed, and the lobes lanceolate and about 3 cm long. The corolla is tubular, heavenly pure white, the tube 5 cm long, 6-lobed, and the lobes obovate and about 6 cm long (the corolla looks very much a white rose flower in some varieties). Six stamens with linear anthers are present. The berries are crimson red to bright orange, about 3.5 cm × 1.5 cm, sharply ribbed, ovoid to elongated, with persistent calyx lobes which are about 2 cm long and linear. The seeds are numerous, about 3 m long, and immersed in an orangish pulp. Looking like some sort of lanterns, the fruits are common site in the markets of Northern Asia including in particular Jeju Island in South Korea.

Medicinal uses: Dysentery (India; Cambodia; Laos; Vietnam); fever, jaundice (India); gonorrhea (China)

Antibacterial polar extract: Aqueous extract of fruits (30 µg/mL/well) inhibited the growth of *Proteus* sp. and *Bacillus subtilis* with the inhibition zone diameters of 19.3 and 17.4 mm, respectively (Fang et al., 2017).

Antimycobacterial carotenoid glycoside: Crocin inhibited the growth of *Mycobacterium avium* (Choi et al., 1986).

Strong antiviral (enveloped monopartite linear single-stranded (-) RNA) polar extract: Aqueous extract of roots inhibited the replication of the Respiratory syncytial virus (Long Strain) in Hep-2 cells with the IC_{50} value of 21 µg/mL and a selectivity index of 5.3 (Ma et al., 2002).

Weak antiviral (enveloped linear segmented single-stranded (-) RNA) iridoid glycoside: Geniposide at the concentration of 260 µM abrogated the replication of the pandemic A/Jiangsu/1/2009 (H1N1) Influenza virus (NT0901) in Madin-Darby canine kidney cells (Zhang et al., 2017).

Strong antifungal hydrophilic iridoid: Cerbinal (LogD –0.2 at pH 7.4; molecular mass = 204.1 g/mol) isolated from the leaves at the concentration of 1 µg/mL inhibited the germination of *Bipolaris sorokiniana*, *Helminthosporium oryzae*, *Pyricularia oryzae*, and *Colletotrichum lagenarium* by 60%, 69%, 91%, and 44%, respectively (Ohashi et al., 1986).

In vivo antiviral (enveloped linear segmented single-stranded (-) RNA) iridoid glycoside: Geniposide given intraperitoneally at the dose of 20 mg/Kg/day increased the survival rate of mice experimentally infected with the pandemic A/Jiangsu/1/2009 (H1N1) Influenza virus (NT0901) (Zhang et al., 2017).

Viral enzyme inhibition by polar extract: Methanol extract of fruits at the concentration of 100 µg/mL inhibited ribonuclease H activities of Human immunodeficiency type-1 reverse transcriptase by 22.2%, respectively (Min et al., 2001).

Commentaries

i. The plant abounds with iridoid glycosides such as gardenoside, geniposide, genipin-1-β-D-gentobioside, and geniposidic acid, and the iridoids genipin (Endo & Taguchi, 1973; Tsm et al., 1994). It is used for COVID-19 in China (Huang et al., 2020; Wang et al., 2020) and could be added to the list of first-line phytomedication for pandemic coronaviral zoonosis.

ii. In India, *Gardenia coronaria* Ham (local name: *Kurdu*) is used for wounds; *Gardenia latifolia* Ait. (local name: *Papra*) is used for sores; *Gardenia gummifera* L. f. (local names: *Bikki*, *Venupatrikaa*) is used for cough, diarrhea, dysentery, fever, and ulcers; and *Gardenia turgida* (local name: *Mahaapindi*) is used for fever. The resin of *Gardenia lucida* Roxb.

(also known as *Dikamali* resin) has been used to heal putrefied wounds. In Bangladesh, *Gardenia latifolia* Ait. (local name: *Bor sudma*) is used for dental caries. In Myanmar, *Gardenia resinifera* (local name: *Shein kho)* is used to check diarrhea. In China, *Gardenia stenophylla* Merr. (local name: *Gŏng rŏng*) is used for fever. In the Philippines, *Gardenia pseudopsidium* (Blanco) Fern.-Vill. (local names: *Kalanigi*, *Malabayabas*) is used to treat smallpox.

iii. The plant belongs to the Subfamily Ixoroideae, Tribe Pavetteae. *Pavetta indica* L. Subfamily Ixoroideae, Tribe Pavetteae is used to heal boils in Malaysia. In Indonesia, *Pavetta multiflora* Bl. (local name: *Bunga susun* kelapa) is used to treat colds.

iv. *Randia esculenta* (Lor.) Merr. in the Subfamily Ixoroideae, Tribe Gardenieae is used to heal boils in Malaysia. *Randia dumetorum* (Retz.) Lam. is used in Bangladesh (local name: *Moyna-kata*) to treat abscesses and sore throat.

v. In the Subfamily Ixoroideae, Tribe Pavetteae, *Tarenna attenuata* Pitard is used for sores in Cambodia, Laos, and Vietnam

vi. In the Subfamily Ixoroideae, Tribe Octotropideaea, *Petunga roxburghii* DC is a remedy for small pox in Malaysia.

References

Choi, C.S., Shin, S.S., Chung, S.I. and Yang, Y.T., 1986. Antibacterial activity of soluble extracts of Gardenia jasminoides against mycobacteria other than Tubercle bacilli. *The Journal of the Korean Society for Microbiology*, *21*(1), pp. 53–62.

Endo, T. and Taguchi, H., 1973. The constituents of Gardenia jasminoides geniposide and genipin-gentiobioside. *Chemical and Pharmaceutical Bulletin*, *21*(12), pp. 2684–2688.

Fang, S.L., Guan, J., Zhou, J.B., Xie, Y.Q. and Meng, X., 2017. Studies on antibacterial activity of gardenia yellow pigment. In *Proceedings of the 7th International Conference on Bioscience, Biochemistry and Bioinformatics* (pp. 50–55).

Huang, J., Tao, G., Liu, J., Cai, J., Huang, Z. and Chen, J.X., 2020. Current prevention of COVID-19: Natural products and herbal medicine. *Frontiers in Pharmacology*, *11*, P. 588508.

Ma, S.C., Du, J., But, P.P.H., Deng, X.L., Zhang, Y.W., Ooi, V.E.C., Xu, H.X., Lee, S.H.S. and Lee, S.F., 2002. Antiviral Chinese medicinal herbs against respiratory syncytial virus. *Journal of Ethnopharmacology*, *79*(2), pp. 205–211.

Min, B.S., Kim, Y.H., Tomiyama, M., Nakamura, N., Miyashiro, H., Otake, T. and Hattori, M., 2001. Inhibitory effects of Korean plants on HIV-1 activities. *Phytotherapy Research*, *15*(6), pp. 481–486.

Ohashi, H., Tsurushima, T., Ueno, T. and Fukami, H., 1986. Cerbinal, a pseudoazulene iridoid, as a potent antifungal compound isolated from Gardenia jasminoides Ellis. *Agricultural and Biological Chemistry*, *50*(10), pp. 2655–2657.

Tsm, T.H., Westly, J., Lee, T.F. and Chen, C.F., 1994. Identification and determination of geniposide, genipin, gardenoside, and geniposidic acid from herbs by HPLC/photodiode-array detection. *Journal of Liquid Chromatography & Related Technologies*, *17*(10), pp. 2199–2205.

Wang, S.X., Wang, Y., Lu, Y.B., Li, J.Y., Song, Y.J., Nyamgerelt, M. and Wang, X.X., 2020. Diagnosis and treatment of novel coronavirus pneumonia based on the theory of traditional Chinese medicine. *Journal of Integrative Medicine*, *18*, pp. 275–283.

Zhang, Y., Yao, J., Qi, X., Liu, X., Lu, X. and Feng, G., 2017. Geniposide demonstrates anti-inflammatory and antiviral activity against pandemic A/Jiangsu/1/2009 (H1N1) influenza virus infection in vitro and in vivo. *Antiviral Therapy*, *22*(7), pp. 599–611.

10.3.4.5 *Guettarda speciosa* L.

Synonyms: *Gardenia speciosa* A. Rich., *Guettarda vermicularis* Blco.
Nyctanthes hirsuta L.

Common names: Beach gardenia; hai an tong (China); jati Pasir (Indonesia); hinma, nil pitcha, puneer marum, utirpannir (India), ketapang pasir (Malaysia); utilomar (Marshall Islands); bagaolan (the Philippines); puopua (Tonga), bwa kasan bordmer (Seychelles); belau, warao (Yap)

Habitat: Beaches, tidal forest, islands, atolls

Distribution: Tropical Asia and Pacific

Botanical observation: This handsome tree grows to about 6 m tall The bole is not straight. The bark is smooth gray brown and the wood soft. The stem is squarish, cracked, lenticelled, articulated, hairy at apex, and conspicuous leaf scars. The internodes are about 1 cm long. The leaves are simple, decussate, and stipulate. The stipules are interpetiolar, hairy at the middle, 8 × 4 mm, broadly lanceolate, and caducous. The petiole is.1.5–4 cm long, flattish, hairy at first, and clasps the stem. The blade is broadly obovate, 9–24 cm × 6–20 cm, thinly coriaceous, glossy, hairy below, of a peculiar green, marked with 6–11 pairs of stout yellowish secondary nerves, cordate at base, and acute to slightly acuminate at apex. The cymes are on a 4–9 cm long velvety pedicel. The flowers are fragrant and open at dawn and fall by the next morning. The calyx is cupular, 4 mm long, hairy, and without lobes. The corolla is tubular, the tube 1.5–2.5 cm long and white to light brown, and develops 4–9 lobes which are obovate, wavy, and up to about 1 cm long. The androecium consists of 4–9 stamens at the throat of the corolla tube. The stigma is capitate and protrudes out of the corolla. The drupes are globose, greenish, marked at apex with a tiny disc, 1.5–2.5 cm in diameter, lobed, somewhat fig-like or some sort of monstruous eye, and contain 4–6 pyrenes fused into a hard stone. The drupes are dispersed by the sea.

Medicinal uses: Abscess, wounds (Cambodia, Laos, Vietnam, India, Indonesia); dysentery (India; Indonesia)

Broad-spectrum antibacterial polar extract: Methanol extract (200 μg/6 mm disc) inhibited the growth of *Pseudomonas aeruginosa* (ATCC 37166) and *Salmonella choleraesuis* (ATCC 10708) (Muangrom & Vajrodaya, 2018).

Strong antimycobacterial indole alkaloid glycoside: 5α-Carboxystrictosidine (LogD –3.1 at pH 7.4; molecular mass = 574.5 g/mol) inhibited the growth of *Mycobacterium tuberculosis* (H37Rv) with the MIC value of 26.3 μg/mL (de Carvalho Junior et al., 2019).

Strong antiviral (enveloped monopartite linear single-stranded (+) RNA) iridoid glycoside: Loganic acid (LogD −4.7 at pH 7.4; molecular mass = 376.3 g/mol) at the concentration of 20 μg/mL inhibited the replication of the Hepatitis C virus by about 50% (Zhang et al., 2009).

Strong antiviral (enveloped linear monopartite dimeric single-stranded (+) RNA) hydroxycinnamic acid derivative: 4,5-di-*O*-caffeoylquinic acid inhibited the replication of the Human immunodeficiency virus type 1 (strain IIIB), Human immunodeficiency virus type-2 (HOD), SIM (strain MAC), and Herpes with the EC_{50} values of 0.6, 8, 2, and 0.1 μg/mL, respectively via a mechanism involving, at least in part, binding to gp120 (Mahmood et al., 1993).

Viral enzyme inhibition by hydroxycinnamic acid derivatives: 4,5-di-*O*-caffeoylquinic acid inhibited Human immunodeficiency reverse transcriptase with the IC_{50} value of 9 μg/mL (Mahmood et al., 1993).

Antiviral (enveloped circular double-stranded DNA) cinnamic acid derivative: 3-*O*-caffeoylquinic inhibited the replication of the Hepatitis B virus (Ge et al., 2018).

Commentaries

i. The plant yields indole glycoside alkaloids such as 5α-carboxystrictosidinic acid and 5α-carboxystrictosidine as well as the iridoid loganic acid and the hydroxycinnamic acid derivatives 3-*O*-caffeoylquinic acid and 4,5-di-*O*-caffeoylquinic acid (Muangrom et al., 2021).

ii. Ursane-type triterpenes isolated from a member of the genus *Guettarda* L. inhibited the replication of the Vesicular stomatitis virus with the MIC values of 33.1 μg/mL in CER cells (Aquino et al., 1989). Aqueous extract of a member of the genus *Guettarda* L. inhibited the replication of the Bovine herpesvirus 1, Suid herpesvirus 1, and Equine herpesvirus 1 with the IC_{50} values of 22.7, 91.3, and 19.9 μg/mL and the selectivity indexes of 40.3, 10, and 20 (Barros et al., 2012). There is the gut feeling that plants in the genus *Guettarda* L. produce strong antiviral agents yet to be characterized.

References

Aquino, R., De Simone, F., Pizza, C., Conti, C. and Stein, M.L., 1989. Plant metabolites. structure and in vitro antiviral activity of quinovic acid glycosides from uncaria tomentosa and Guettarda platyipoda. *Journal of Natural Products*, *52*(4), pp. 679–685.

Barros, A.V., Araújo, L.M., De Oliveira, F.F., Da Conceicao, A.O., Simoni, I.C., Fernandes, M.J.B. and Arns, C.W., 2012. In vitro evaluation of the antiviral potential of guettarda angelica against animal herpesviruses [avaliação In Vitro Do Potencial Antiviral De Guettarda Angelica Contra Herpesvírus Animais]. *Acta Scientiae Veterinariae*, 40(4), pp. 1–7.

de Carvalho Junior, A.R., Oliveira Ferreira, R., de Souza Passos, M., da Silva Boeno, S.I., Glória das Virgens, L.D.L., Ventura, T.L.B., Calixto, S.D., Lassounskaia, E., de Carvalho, M.G., Braz-Filho, R. and Curcino Vieira, I.J., 2019. Antimycobacterial and nitric oxide production inhibitory activities of triterpenes and alkaloids from psychotria nuda (Cham. & Schltdl.) Wawra. *Molecules*, *24*(6), p. 1026.

Ge, L., Wan, H., Tang, S., Chen, H., Li, J., Zhang, K., Zhou, B., Fei, J., Wu, S. and Zeng, X., 2018. Novel caffeoylquinic acid derivatives from Lonicera japonica Thunb. flower buds exert pronounced anti-HBV activities. *RSC Advances*, *8*(62), pp. 35374–35385.

Mahmood, N., Moore, P.S., De Tommasi, N., De Simone, F., Colman, S., Hay, A.J. and Pizza, C., 1993. Inhibition of HIV infection by caffeoylquinic acid derivatives. *Antiviral Chemistry and Chemotherapy*, *4*(4), pp. 235–240.

Muangrom, W. and Vajrodaya, S., 2018. Comparative Phytochemistry and Antibacterial Properties of Guettarda speciosa L. In *56. Kasetsart University Annual Conference, Bangkok (Thailand), 30 Jan–2 Feb 2018.*

Muangrom, W., Bacher, M., Berger, A., Valant-Vetschera, K., Vajrodaya, S. and Schinnerl, J., 2021. A novel tryptophan-derived alkaloid and other constituents from Guettarda speciosa (Rubiaceae: Cinchonoideae–Guettardeae). *Biochemical Systematics and Ecology*, *95*, p. 104239.

Zhang, H., Rothwangl, K., Mesecar, A.D., Sabahi, A., Rong, L. and Fong, H.H., 2009. Lamiridosins, hepatitis C virus entry inhibitors from Lamium album. *Journal of Natural Products*, *72*(12), pp. 2158–2162.

10.3.4.6 *Hedyotis scandens* Roxb.

Synonyms: *Dimetia scandens* (Roxb.) R.J. Wang; *Oldenlandia scandens* (Roxb.) Kuntze

Common names: Bishma; bashpuisak (Bangladesh); pan jing er cao, she she cao (China); n'rim, kalijar (India)

Habitat: Forest edges, roadsides, shady and wet soils

Distribution: India, Bhutan, Nepal, Bangladesh, Myanmar, Vietnam, China

Botanical observation: This herb grows to 2 m long. The stem is terete or angular and somewhat lignose and swollen at nodes. The leaves are simple, decussate, and stipulate. The stipules are up to 5 mm long. The petiole is minute. The blade is narrowly elliptic, dark green, glossy, somewhat coriaceous, 5–9 cm × 1.5–3 cm, acute at base, acuminate at apex, and marked with 3–5 pairs of secondary nerves with somewhat vaguely a Cornaceous look. The cymes are terminal, grow up to about 10 cm long, and bear small white flowers. The calyx is cupular and develops 4 triangular lobes. The corolla is tubular, hairy at throat, and whitish, with 4 oblong lobes which are about 5 mm long. The androecium includes 4 stamens which are inserted on and protrude out of the corolla tube. The ovary develops a slender style and a bifid stigma. The capsules are ovoid, about 5 mm long, and contain numerous minute black seeds.

Medicinal uses: Abscesses, aphthae (Bangladesh); fever (China); boils, eye infection, toothache (India)

Strong antiviral (enveloped monopartite linear single-stranded (-) RNA) phenolic glycosides: Maltol 6′-β-D-apiofuranosyl-β-D-glucopyranoside and grevilloside G inhibited the replication of the Respiratory syncytial virus with the IC_{50} values of 20 and 25 μg/mL, respectively (Wang et al., 2013).

Commentaries

i. The plant produces hydroxycinnamic acid derivatives and phenolic glycosides (Wang et al., 2013).
ii. Rubiaceae tend to accumulate glycosides of phenols, hydroxycinnamic acids, iridoids, and even alkaloids.

iii. Members of the genus *Hedyotis* L. are often used for fever in Asia. In India, *Hedyotis burmanniana* Roxb. (local name: *Papra*) is used for fever, and *Hedyotis umbellata* (L.) Lam. (local name: Inbooral) is used to treat bronchitis. *Hedyotis corymbosa* (L.) Lam. is used for fever and wounds in Thailand (local name: *Lin ngu*) and to heal wounds in Malaysia, and in China (local name: *Cè gōu jí*; *Shuixiancao*) it is used for skin diseases and jaundice. *Hedyotis uncinella* Hook. & Arn. is used in India (local name: *Sher rit along*) for fever, and *Hedyotis umbellata* (L.) Lam. (local name: *Emboorel cheddie*) is used to treat cough. In Myanmar, *Hedyotis capitellata* Wall. ex G. Don is used to treat dysentery, and in Vietnam (local name: *An điền mềm*) it is used for diarrhea and dysentery and yields a series of beta-carboline alkaloids (Phuong et al., 1999). *Hedyotis diffusa* Willd (synonym: *Oldenlandia diffusa* (Willd.) Roxb.) is used in China (local name: *Baihuasheshecao*) for cough, fever, hepatitis, and toothache. Aqueous extract of *Hedyotis diffusa* Willd at the concentration of 250 µg/mL inhibited Human immunodeficiency virus type-1 protease by 81.9% (Xu et al., 1996). *Hedyotis hedyotidea* (DC.) Merr. (local name: *Niu bai teng*) is used for cough, and *Hedyotis pinifolia* Wall. ex G. Don is used for hepatitis, and *Hedyotis verticillata* (L.) Lam. (local = *Jie jie hua*) is used for a cold, fever, and sore throat. In Indonesia, *Hedyotis leucocarpa* Elmer (local name: *Siguntur*) is used for fever.

iv. *Anotis urophylla* Wall. in the Subfamily Rubioideae, Tribe hedyotideae is used to heal sores in China. *Dentella repens* is used in Malaysia (local name: *Bunga karang*) to heal sores.

v. *Paederia foetida* L. in the Subclass Rubioideae, Tribe Paederieae affords a remedy for infected sores and diarrhea (in India local names: *Gandhali*, *Padbailudi*) and in Indonesia is used for herpes. Extract of this smelly plant inhibited the replication of the Herpes simplex virus type-1 with the IC_{50} value of 400 µg/mL (Thompson et al., 2005). *Serissa japonica* (Thunb.) Thunb. in the Subclass Rubioideae, is a remedy for carbuncles in China. Aqueous extract of stem and leaf of *Serissa japonica* inhibited Herpes simplex virus type-1, Herpes simplex virus type-2, and Adenovirus-3 and 8 with the EC_{50} of 67.5, 92.1, 72.9, and 24.6 µg/mL, and the selectivity indexes of 23.3, 1.1, 21.6, and 63.9, respectively (Chiang et al., 2003).

vi. In the Subfamily Rubioideae, Tribe Psychotrieae, *Psychotria curviflora* Wall. is used to heal sores in Malaysia, and *Psychotria luzoniensis* Fern.-Vill. is a remedy for wounds in the Philippines. In the Andamans, *Psychotria platyneura* Kurz is used for skin diseases (local name: *Nameta*). In Papua New Guinea, an unidentified member of the genus *Psychotria* L. (local name: *Sisikupa*) is used for boils. The indole alkaloid glycoside vincosamide (LogD 0.2 at pH 7.4; molecular mass = 498.5 g/mol) at the concentration of 25 µg/mL inhibited the replication of the Dengue virus serotype-1 (16681) in HepG2 cells (Costa et al., 2020). Methanol fraction of *Psychotria serpens* L. at the concentration of 100 µg/mL abrogated the replication of the Herpes simplex virus type-1 (KOS) in Vero cells (Kuo et al., 2001). In Hawaii, *Psychotria haitiensis* Urb. (local name: *Kopiko*) is used to treat cuts and wounds, and aqueous extract of bark inhibited the replication of the Human immunodeficiency virus (strain IIIB) with the EC_{50} value of 9.7 and selectivity index of 23 in MT-4 cells (Locher et al., 1996). Dengue, Herpes, and Human immunodeficiency viruses are (+) RNA viruses pointing to the presence of anti-coronavirus indole alkaloids awaiting discovery from the genus *Psychotria* L.

vii. Indole alkaloids (being often neuroactive) (Farias et al., 2012) in the Rubiaceae could be examined for their effects on bacterial and fungal efflux pumps.

References

Chiang, L.C., Cheng, H.Y., Liu, M.C., Chiang, W. and Lin, C.C., 2003. In vitro anti-herpes simplex viruses and anti-adenoviruses activity of twelve traditionally used medicinal plants in Taiwan. *Biological and Pharmaceutical Bulletin*, *26*(11), pp. 1600–1604.

Costa, J.O., Barboza, R.S., Valente, L.M., Wolff, T., Gomes, M., Gallo, B., Berrueta, L.A., Guimarães-Andrade, I.P., Gavino-Leopoldino, D. and Assunção-Miranda, I., 2020. One-Step isolation of monoterpene indole alkaloids from psychotria leiocarpa leaves and their antiviral activity on dengue virus type-2. *Journal of the Brazilian Chemical Society*, *31*, pp. 2104–2113.

Farias, F.M., Passos, C.S., Arbo, M.D., Barros, D.M., Gottfried, C., Steffen, V.M. and Henriques, A.T., 2012. Strictosidinic acid, isolated from Psychotria myriantha Mull. Arg.(Rubiaceae), decreases serotonin levels in rat hippocampus. *Fitoterapia*, *83*(6), pp. 1138–1143.

Kuo, Y.C., Chen, C.C., Tsai, W.J. and Ho, Y.H., 2001. Regulation of herpes simplex virus type 1 replication in Vero cells by Psychotria serpens: Relationship to gene expression, DNA replication, and protein synthesis. *Antiviral Research*, *51*(2), pp. 95–109.

Locher, C.P., Witvrouw, M., De Béthune, M.P., Burch, M.T., Mower, H.F., Davis, H., Lasure, A., Pauwels, R., De Clercq, E. and Vlietinck, A.J., 1996. Antiviral activity of Hawaiian medicinal plants against human immunodeficiency virus type-1 (HIV-1). *Phytomedicine*, *2*(3), pp. 259–264.

Phuong, N.M., Van Sung, T., Porzel, A., Schmidt, J., Merzweiler, K. and Adam, G., 1999. β-Carboline alkaloids from Hedyotis capitellata. *Phytochemistry*, *52*(8), pp. 1725–1729.

Thompson, K.D., Ather, A. and Khan, M.T.H., 2005. Antiviral activities of three Bangladeshi medicinal plant extracts against herpes simplex viruses. *Minerva Biotecnologica*, *17*(3), p. 193.

Wang, G.C., Li, T., Deng, F.Y., Li, Y.L. and Ye, W.C., 2013. Five new phenolic glycosides from Hedyotis scandens. *Bioorganic & Medicinal Chemistry Letters*, *23*(5), pp. 1379–1382.

Xu, H.X., Wan, M., Loh, B.N., Kon, O.L., Chow, P.W. and Sim, K.Y., 1996. Screening of traditional medicines for their inhibitory activity against HIV-1 protease. *Phytotherapy Research*, *10*(3), pp. 207–210.

10.3.4.7 *Hymenodictyon orixense* (Roxb.) Mabb.

Synonyms: *Cinchona excelsa* Roxb., *Cinchona orixensis* Roxb., *Hymenodictyon excelsum* (Roxb.) Wall.

Common names: Bhramar-chhalikaa; latikarum (Bangladesh); mao tu lian qiao (China); bhramar-chhalikaa, pundaru (India); dumsa gyaw (Myanmar); haligango (the Philippines); u lok, som kop (Thailand)

Habitat: Forests

Distribution: India, Nepal. Bangladesh, Myanmar, Cambodia, Laos, Vietnam, Thailand, Malaysia, Indonesia, the Philippines

Botanical observation: This timber tree grows to 30 m tall. The bark is fissured and thick. The stems are hairy. The leaves are simple, opposite, and stipulate. The stipules are triangular, about 1 cm long, hairy, and serrate. The petiole is 5–8 cm long, stout, and hairy. The blade is 15–30 cm × 10–17 cm, glossy, marked with 6–7 pairs of secondary nerves, hairy below, elliptic, tapering at base, and acute to acuminate at apex. The inflorescences are spikes. The calyx is 5-lobed, hairy, and minute. The corolla is tubular, hairy, about 5 mm long, and 5-lobed. Five stamens are inserted on the corolla tube. The ovary develops a slender style which extrudes out of the corolla (common feature of plants in the family Rubiaceae). The capsules are about 1.5 cm long, brownish, hairy, elliptic, somewhat like the capsules of *Cinchona*, and contain numerous minute winged seeds.

Medicinal uses: Fever (India, Myanmar, Thailand, the Philippines); skin diseases (Cambodia, Laos, Vietnam)

Strong broad-spectrum antifungal anthraquinone: Damnacanthal inhibited the growth of *Candida albicans*, *Cryptococcus neoformans*, *Sporothrix schenckii*, *Trichophyton mentagrophytes*, and *Aspergillus fumigatus* with the MIC values of 50, 12.5, 6.2, 1.5, and 50 μg/mL, respectively (Singh et al., 2006).

Strong antibacterial (Gram-positive) anthraquinone: Rubiadin inhibited the growth of *Staphylococcus aureus* (ATCC 29213) with the MIC value of 32 μg/mL (Comini et al., 2011).

Moderate broad-spectrum antibacterial coumarin: Scopoletin inhibited the growth of *Staphylococcus aureus* (ATCC 43300), *Enterococcus faecium* (UCLA 192), and *Stenotrophomonas maltophilia* (DMST 19079) with the MIC values of 128, 128, and 256 μg/mL, respectively (Buathong et al., 2019).

Strong antiviral (enveloped circular double-stranded DNA) anthraquinone: Rubiadin at the concentration of 6 μg/mL inhibited the secretion of Hepatitis B virus surface-antigen and e-antigen from HepG2.2.15 cells by about 15% and 30%, respectively (Peng et al., 2017).

Commentaries

i. The plant produces anthraquinones such as rubiadin, lucidin, nordamnacanthal, damnacanthal, 2-benzylxanthopurpurin, anthragallol (2,3,4-trihydroxyanthraquinone), 6-methylalizarin, soranjidiol, and morindone (Brew & Thomason, 1971) and coumarins such as scopoletin and its glycosides (Rao et al., 1988), and toxic alkaloids including hymenodictine (Rativanich & Dietrichs, 1971).

ii. Anthraquinones intercatale DNA and are broad-spectrum antimicrobial natural products (Bhakta & Siva, 2012). DNA intercalation is synonym with DNA mutations and carcinogenesis (Chakarov et al., 1968).

iii. In the Subfamily Rubioideae, Tribe Morindae, the beautiful *Prismatomeris tetrandra* (Roxb.) K. Sch. is a remedy for bronchitis in Cambodia, Laos, and Vietnam and produces rubiadin (Peng et al., 2017).

References

Bhakta, D. and Siva, R., 2012. Morindone, an anthraquinone, intercalates DNA sans toxicity: a spectroscopic and molecular modeling perspective. *Applied Biochemistry and Biotechnology*, *167*(4), pp. 885–896.

Brew, E.J.C. and Thomason, R.H., 1971. Naturally occurring quinones. Part XIX. Anthraquinones in hymenodictyon excelsum and damnacanthus major. *Journal of the Chemical Society C: Organic*, pp. 2001–2007.

Chakarov, E., Krachmarov, I., Kiljovska, M., Popov, D. and Skocheva, M., 1968. Specificity and possible mechanism of reaction between 9-anthraldehyde and unsaturated aliphatic compounds during ultraviolet irradiation. Scripta Scientifica Medica, 7, pp. 3–8.

Comini, L.R., Montoya, S.N., Páez, P.L., Argüello, G.A., Albesa, I. and Cabrera, J.L., 2011. Antibacterial activity of anthraquinone derivatives from Heterophyllaea pustulata (Rubiaceae). *Journal of Photochemistry and Photobiology B: Biology*, *102*(2), pp. 108–114.

Peng, Z., Fang, G., Peng, F., Pan, Z., Su, Z., Tian, W., Li, D. and Hou, H., 2017. Effects of Rubiadin isolated from Prismatomeris connata on anti-hepatitis B virus activity in vitro. *Phytotherapy Research*, *31*(12), pp. 1962–1970.

Rao, P.S., Asheervadam, Y., Khaleelullah, M.D., Rao, N.S. and Murray, R.D.H., 1988. Hymexelsin, an apiose-containing scopoletin glycoside from the stem bark of Hymenodictyon excelsum. *Journal of Natural Products*, *51*(5), pp. 959–961.

Rativanich, T. and Dietrichs, H.H., 1971. Alkaloids from Thai trees used in folk medicine. *The Natural History Bulletin of the Siam Society*, *24*(1 and 2), pp. 145–151.

Singh, D.N., Verma, N., Raghuwanshi, S., Shukla, P.K. and Kulshreshtha, D.K., 2006. Antifungal anthraquinones from Saprosma fragrans. *Bioorganic & Medicinal Chemistry Letters*, *16*(17), pp. 4512–4514.

10.3.4.8 *Ixora nigricans* R.Br.

Common names: Kalashona (Bangladesh); sawkan (Myanmar), bong trang trang (Vietnam); udappu (India); soka (Indonesia); khem tuut maa (Thailand)

Habitat: Forests

Description: India, Myanmar, Vietnam, Thailand, Malaysia, Indonesia

Botanical observation: This treelet grows to a height of about 5 m. The stem is terete and darkish. The internodes are about 6 cm long. The leaves are simple, decussate, and stipulate. The stipule is interpetiolar, cuspidate, and about 4 mm long. The petiole is none to about 3 mm long and terete. The blade is thinly coriaceous, dark green above and glossy, broadly lanceolate, 5.5–9 cm × 3–4.5 cm, acute at base, acuminate at apex, and marked with 7–9 pairs of secondary nerves. The cymes are terminal, showy, somewhat burgundy, and present numerous pure white, yellowish, or pinkish and slender flowers. The calyx is light green, to dark brown, 2 mm long, and 4-lobed, the lobes oblong-triangular. The corolla tube is 0.8–1.2 cm long, glabrous, with 4 lobes which are about 4 mm long, and lanceolate. Flowers in the genus *Ixora* L. have typical cross-shaped corollas and type of dull plain red or yellow colors. The 4 stamens are sessile, adnate at the corolla throat, and with filiform anthers. The ovary is minute and develops a slender and protruding style. The berries are globose or bilobed, red to black, smooth, and contain plano-convex, rugose and black seeds.

Medicinal uses: Diarrhea, ear infections (Bangladesh); dysentery (Cambodia, Laos, Vietnam, Indonesia)

Antibacterial (Gram-positive) halo developed by mid-polar extracts: Ether extract of leaves (100 µg/disc) inhibited the growth of *Arthrobacter citrus*, *Bacillus cereus*, *Bacillus licheniformis*, *Bacillus polymixa*, *Bacillus subtilis*, *Clostridium* sp., *Staphylococcus aureus*, *Streptococcus* sp., *Escherichia coli*, *Klebsiella aerogenes*, *Pseudomonas aeruginosa*, *Pseudomonas putida*, *Salmonella typhimurium*, and *Sarcina lutea* with the inhibition zone diameters of 18, 18, 18, 18, 20, 22, 22, 16, 20, 18, 22, 25, 23, and 25 mm, respectively (Annapurna et al., 2003).

Chloroform extract of root bark (400 µg/6 mm well) inhibited the growth of *Staphylococcus aureus* (ATCC 43300), *Staphylococcus aureus* (ATCC 25923), and *Stenotrophomonas maltophila* (DMST 19079) with the inhibition zone diameters of about 10, 15, and 12 mm, respectively (Buathong et al., 2019).

Broad-spectrum antifungal mid-polar extracts: Ether extract of leaves (100 µg/disc) inhibited the growth of *Nocardia* sp., *Candida albicans*, and *Saccharomyces cerevisiae* with the inhibition zones of 18, 12, 12 mm, respectively (Annapurna et al., 2003).

Strong antifungal cyclopeptide: Frangulanine inhibited the growth of *Candida albicans* and *Saccharomyces cerevisiae* with the minimum dose of 25 and 50 µg/TLC (Emile et al., 2007)

Commentaries

i. The use of the plant for diarrhea and dysentery suggests the plant to be astringent. Members of the genus *Ixora* L. produce cyclopeptide alkaloids such as frangulanine, flavone glycosides, catechin, and proanthocyanidins such as procyanidin A2 and cinnamtannin B1 (Lee et al., 2010). Cyclopeptides are often strongly antibacterial. Are cyclopeptides in members of the genus *Ixora* L. produced by symbiotic fungi?

ii. In India, *Ixora coccinea* L. (local names: *Bandhuka*, *Thetti chelthi*, *Veechi poo*) is sacred and used to treat bronchitis, diarrhea, dysentery, infected scabies, leucorrhea, sores, and ulcers, and in Sri Lanka it is used for bronchitis, dysentery, and leucorrhea. The common ornamental shrub *Ixora coccinea* L. yields cinnamtannin B1 which inhibited the growth of *Bacillus subtilis* (NCTC 8236) with the inhibition zone diameter of 9 mm (100 µL of a 20 mg/mL/8 mm well) (Idowu et al., 2010). Cinnamtannin B1 inhibited the growth of *Porphyromonas gingivalis* and *Prevotella intermedia* with the MIC values of 100 and 100 µg/mL, respectively (Ho et al., 2001). *Ixora pavetta* Andr. is used in India (local name: *Nevaari*) for cough and diarrhea. In Sri Lanka, *Ixora parviflora* Lam. (local name: *Maha ratambala*) is used for cough. *Ixora chinensis* Lam. is used in Myanmar (local name: *Pon na yeik*) for tuberculosis; in Bangladesh, it is a remedy for bronchitis, gonorrhea, and leucorrhea (local name: *Rongon*), and in China for abscesses and wounds. In Bangladesh, *Ixora pubirama* Bremek is used to treat insect bite infections (local name: *Jarbhua puishak*)

iii. Ethanol extract of leaves of *Ixora paludosa* (Blume) Kurz inhibited the growth of *Staphylococcus aureus* and *Candida albicans* with MIC value of 512 and 512 µg/mL, respectively (Romulo et al., 2018).

iv. *Canthium horridum* Bl. in the Subfamily Ixoroideae, Tribe Vanguerieiae is used to heal wounds in Malaysia and produces the lignan (+)- syringaresinol which inhibited the growth of *Bacillus subtilis*, *Escherichia coli*, and *Staphylococcus aureus* with the inhibition zone diameters of 32.7, 23.7, 32.7 mm, respectively (disc impregnated with 10 mg/mL solution) (Yang et al., 2010).

References

Annapurna, J., Amarnath, P.V.S., Kumar, D.A., Ramakrishna, S.V. and Raghavan, K.V., 2003. Antimicrobial activity of Ixora coccinea leaves. *Fitoterapia*, *74*(3), pp. 291–293.

Buathong, R., Chamchumroon, V., Schinnerl, J., Bacher, M., Santimaleeworagun, W., Kraichak, E. and Vajrodaya, S., 2019. Chemovariation and antibacterial activity of extracts and isolated compounds from species of Ixora and Greenea (Ixoroideae, Rubiaceae). *PeerJ*, *7*, p. e6893.

Emile, A., Waikedre, J., Herrenknecht, C., Fourneau, C., Gantier, J.C., Hnawia, E., Cabalion, P., Hocquemiller, R. and Fournet, A., 2007. Bioassay-guided isolation of antifungal alkaloids from Melochia odorata. *Phytotherapy Research: An International Journal Devoted to Pharmacological and Toxicological Evaluation of Natural Product Derivatives*, *21*(4), pp. 398–400.

Ho, K.Y., Tsai, C.C., Huang, J.S., Chen, C.P., Lin, T.C. and Lin, C.C., 2001. Antimicrobial activity of tannin components from Vaccinium vitis-idaea L. *Journal of Pharmacy and Pharmacology*, *53*(2), pp. 187–191.

Idowu, T.O., Ogundaini, A.O., Salau, A.O., Obuotor, E.M., Bezabih, M. and Abegaz, B.M., 2010. Doubly linked, A-type proanthocyanidin trimer and other constituents of Ixora coccinea leaves and their antioxidant and antibacterial properties. Phytochemistry, 71(17–18), pp. 2092–2098.

Lee, C.L., Liao, Y.C., Hwang, T.L., Wu, C.C., Chang, F.R. and Wu, Y.C., 2010. Ixorapeptide I and ixorapeptide II, bioactive peptides isolated from Ixora coccinea. *Bioorganic & Medicinal Chemistry Letters*, *20*(24), pp. 7354–7357.

Romulo, A., Zuhud, E.A., Rondevaldova, J. and Kokoska, L., 2018. Screening of in vitro antimicrobial activity of plants used in traditional Indonesian medicine. Pharmaceutical *Biology*, 56(1), pp. 287–293.

10.3.4.9 *Morinda citrifolia* L.

Synonyms: *Morinda aspera* Wight & Arn.; *Morinda asperula* Standl.; *Morinda bracteata* Roxb.; *Morinda chachuca* Buch.-Ham.; *Morinda ligulata* Blanco; *Morinda littoralis* Blanco; *Morinda macrophylla* Desf.; *Morinda mudia* Buch.-Ham.; *Morinda multiflora* Roxb.; *Morinda nodosa* Buch.-Ham.; *Morinda quadrangularis* G. Don; *Morinda stenophylla* Spreng.; *Morinda teysmanniana* Miq.; *Morinda tinctoria* Noronha; *Morinda zollingeriana* Miq

Common names: Indian mulberry; hai bin mu ba ji (China); rangkathal (Bangladesh); lada (Guam); noni (Hawai); mashyuka, manja-pavattay, noona marum; nunaa (India); cangkudu (Indonesia); nho (Laos); dino, tombong aso, tumbong-aso (the Philippines); ageru (Palau); knuel, leebhooga (Papua New Guinea); nonu (Samoa); dino nhau, ngao (Vietnam); nhor srok (Thailand); magarwek (Yap)

Habitats: Sea shores, cultivated

Distribution: India to Pacific Islands

Botanical observation: This tree grows up to about 5 m tall. The stems are glossy and angled at apex. The leaves are simple, opposite, and stipulate. The stipules are interpetiolar, up to about 1 cm long, and oblong. The petiole is 1–2 cm long, stout, and curved. The blade is elliptic, 10–20 cm × 5–9 cm, elliptic, dark green, glossy, somewhat fleshy and soft marked with 5–7 pairs secondary nerves, attenuate at base, and obtuse or acute-acuminate at apex. The flowering heads are somewhat ovoid, up to about 3 cm long, fleshy, develop one or a few flowers at a time, and have a repulsive aura. The calyx is cup-shaped. The corolla is pure white, the tube about 1 cm long, and develop 5–6 lanceolate lobes which are about 5 mm long and recurved. Five to six stamens adnate to the corolla and with dark brown anthers are present. The style is slender, develops a bifid stigma, and protrudes out of the corolla. The infructescences are heads which are somewhat ovoid to potato-shaped, glossy, about 5 × 3 cm, larval whitish or pale-yellowish green, edible, and sheltering numerous seeds

Medicinal uses: Boils, fungal infection (India); chancre, repel ghosts (Palau); sore eyes (Samoa); cough (Indonesia; Vietnam); diarrhea (India; Palau; Papua New Guinea); dysentery (India; Vietnam); fever (India; Papua New Guinea); furuncles (Vietnam); leprosy, pneumonia, toothaches, tuberculosis, warts (Papua New Guinea); post-partum (Malaysia); ulcers (India; the Philippines); sore throat (India; Palau); wounds (Indonesia; India; Yap)

Antibacterial (Gram-negative) halo developed by anthraquinone: Damnacanthal inhibited the growth of *Pseudomonas aeruginosa* with the minimum inhibitory mass of 10 µg/disc (Ali et al., 2000).

Moderate antibacterial (Gram-positive) anthraquinone: 1,3-Dihydroxy-5-methoxy-2,6-bismethoxymethyl-9,10-anthraquinone inhibited the growth of *Staphylococcus aureus* with the MIC value of 24 µM (Wang et al., 2016).

Strong antimycobacterial diterpene: (E)-phytol isolated from the leaves inhibited the growth of *Mycobacterium tuberculosis* (H37Rv) with the MIC value of 32 µg/mL (Saikia et al., 2010).

Strong antimycobacterial anthraquinone: Damnacanthal inhibited the growth of *Mycobacterium tuberculosis* (H37Rv) with the MIC value of 13 µg/mL (Pollo et al., 2021).

Strong antimycobacterial steroids: Sitosterol and stigmasterol inhibited the growth of *Mycobacterium tuberculosis* (H37Rv) with the MIC values of 128 and 32 μg/mL, respectively (Saludes et al., 2002).

Broad-spectrum antifungal halo developed by anthraquinones: Damnacanthal inhibited the growth of *Aspergillus ochraceus*, *Aspergillus niger*, and *Candida lipolytica* with the minimum inhibitory masses of 20, 20, and 20 μg/disc, respectively (Ali et al., 2000).

Strong antiviral (non-enveloped monopartite linear single-stranded (+) RNA) polar extract: Ethanol extract inhibited the replication of the Foot and mouth disease virus (Disease type O (FMDV), local strain KPS/005/2545) at the concentration of 0.1 μg/μL (Chungsamarnyart et al., 2007).

Strong antiviral (enveloped linear segmented single-stranded (-) RNA) anthraquinone: 1,3-Dihydroxy-5-methoxy-2,6-bismethoxymethyl-9,10-anthraquinone isolated from the bark inhibited the replication of the Influenza virus A H1N1 and H3N2 with the IC_{50} values of 10.5 and 11.5 μM, respectively (Wang et al., 2016).

Strong antiviral (enveloped monopartite linear double-stranded DNA) iridoid glycoside: Asperuloside (LogD –1.9 at pH 7.4 molecular mass = 414.3 g/mol) inhibited the replication of the Epstein-Barr virus (Kapadia et al., 1996)

Moderate antiviral (enveloped linear segmented single-stranded (-) RNA) anthraquinone: 1,3-Dihydroxy-5-methoxy-6-methoxymethyl-2-methyl-9,10-anthraquinone isolated from the bark inhibited the replication of the Influenza virus A H1N1 with the IC_{50} values of 66.1 μM, respectively (Wang et al., 2016).

Moderate antiviral (enveloped monopartite linear double-stranded DNA) iridoid glycoside: Asperuloside inhibited the replication of the Herpes simplex virus type-1 with the IC_{50} value of 111.3 μg/mL (Manzione et al., 2020).

Commentaries

i. Plants with dyes are often used in Asia and the Pacific for the treatment of infectious diseases, and this is the case for *Morinda citrifolia* L. which is used as a source of red dye for the *batiks* of Indonesia. Red pigments in this plant are anthraquinones such as damnacanthal (see earlier), alizarin 1-methyl ether, and rubiadin (Lv et al., 2011).

ii. The fruits contain the iridoid glycoside asperuloside (Levand & Larson, 1979).

iii. Plants in the genus *Morinda* L. are frequently used for the treatment of microbial infections in Asia and the Pacific as well as source of food and dyes. *Morinda pubescens* J. E. Smith (local Name: *Manjanathi*) is used for AIDS in India as well as diarrhea and dysentery. In Bangladesh, *Morinda persicifolia* Buch.-Ham. is used for diarrhea and jaundice. *Morinda angustifolia* Roxb. is used for skin infections and urinary tract infection in Bangladesh (local name: *Koba bena*), cough and diarrhea in Myanmar (local name: *Hla ponyork*), and diarrhea in India (local name: *Kleeba*). 1,8-Dihydroxy-2-methyl-3,7-dimethoxyanthraquinone from *Morinda angustifolia* Roxb. (13.3 μg/6 mm disc) inhibited the growth of *Bacillus subtilis*, *Escherichia coli*, *Micrococcus aureus*, and *Sarcina lutea* with the inhibition zone diameters of 14, 12.5, 13, and 6.8 mm, respectively (Xiang et al., 2008). 1,8-Dihydroxy-2-methyl-3,7-dimethoxyanthraquinone (13.3 μg/6 mm disc) inhibited the growth of *Candida albicans* and *Saccharomyces sake* with the inhibition zone diameters of 7.5 and 6.3 mm, respectively (Xiang et al., 2008). In Myanmar, *Morinda coreia* Buch.-Ham. (synonym= *Morinda tinctoria* Roxb.)(local name: *Nee par hsay pin*) is used for fever and sores. In Vietnam, *Morinda umbellata* L. is used for dysentery (local name: *Nhafu tasn*), and in India, it is used for diarrhea and dysentery. In Malaysia, *Morinda elliptica* (Hook.f.) Ridley is used for fever, cholera, diarrhea, post-partum, and wounds, and in Thailand, it is used for fever (local name: *Yo pa*). Ethanol extract of the immature fruits of *Morinda elliptica* L. inhibited the replication of the Foot and mouth disease virus (Disease type O (FMDV), local strain KPS/005/2545.) at the concentration of 0.3 μg/μL (Chungsamarnyart et al., 2007).

iv. The Malays of Perak (of Bugis descent) fold the fresh leaves around stones heated in fire and apply the whole on the abdomen at post-partum. The traditional Malay system of medicine is, compared to other Asian system, careful about post-partum protection.

References

Ali, A.M., Ismail, N.H., Mackeen, M.M., Yazan, L.S., Mohamed, S.M., Ho, A.S.H. and Lajis, N.H., 2000. Antiviral, cytotoxic and antimicrobial activities of anthraquinones isolated from the roots of Morinda elliptica. *Pharmaceutical Biology*, *38*(4), pp. 298–301.

Chungsamarnyart, N., Sirinarumitr, T., Chumsing, W. and Wajjawalku, W., 2007. In vitro study of antiviral activity of plant crude-extracts against the foot and mouth disease virus. *Agriculture and Natural Resources*, *41*(5), pp. 97–103.

Kapadia, G.J., Sharma, S.C., Tokuda, H., Nishino, H. and Ueda, S., 1996. Inhibitory effect of iridoids on Epstein-Barr virus activation by a short-term in vitro assay for anti-tumor promoters. *Cancer Letters*, *102*(1–2), pp. 223–226.

Levand, O. and Larson, H.O., 1979. Some chemical constituents of Morinda citrifolia. *Planta Medica*, *36*(06), pp. 186–187.

Lv, L., Chen, H., Ho, C.T. and Sang, S., 2011. Chemical components of the roots of Noni (Morinda citrifolia) and their cytotoxic effects. *Fitoterapia*, *82*(4), pp. 704–708.

Manzione, M.G., Martorell, M., Sharopov, F., Bhat, N.G., Kumar, N.V.A., Fokou, P.V.T. and Pezzani, R., 2020. Phytochemical and pharmacological properties of asperuloside, a systematic review. *European Journal of Pharmacology*, *883*, p. 173344.

Pollo, L.A., Martin, E.F., Machado, V.R., Cantillon, D., Wildner, L.M., Bazzo, M.L., Waddell, S.J., Biavatti, M.W. and Sandjo, L.P., 2021. Search for antimicrobial activity among fifty-two natural and synthetic compounds identifies anthraquinone and polyacetylene classes that inhibit mycobacterium tuberculosis. *Frontiers in Microbiology*, *11*, p. 3619.

Saikia, D., Parihar, S., Chanda, D., Ojha, S., Kumar, J.K., Chanotiya, C.S., Shanker, K. and Negi, A.S., 2010. Antitubercular potential of some semisynthetic analogues of phytol. Bioorganic & *Medicinal Chemistry Letters*, 20(2), pp. 508–512.

Saludes, J.P., Garson, M.J., Franzblau, S.G. and Aguinaldo, A.M., 2002. Antitubercular constituents from the hexane fraction of Morinda citrifolia Linn. (Rubiaceae). Phytotherapy Research: An International Journal Devoted to Pharmacological and Toxicological Evaluation of Natural Product Derivatives, 16(7), pp. 683–685.

Wang, J., Qin, X., Chen, Z., Ju, Z., He, W., Tan, Y., Zhou, X., Tu, Z., Lu, F. and Liu, Y., 2016. Two new anthraquinones with antiviral activities from the barks of Morinda citrifolia (Noni). *Phytochemistry Letters*, *15*, pp. 13–15.

Xiang, W., Song, Q.S., Zhang, H.J. and Guo, S.P., 2008. Antimicrobial anthraquinones from Morinda angustifolia. *Fitoterapia*, *79*(7–8), pp. 501–504.

10.3.4.10 *Mussaenda glabrata* Hutch.

Synonyms: *Mussaenda frondosa* var. *glabrata* Hook. f.; *Mussaenda glabrata* (Hook. f.) Hutch. ex Gamble

Common names: Dhobi tree, white lady; gach ranita (Bangladesh); bedina (India)

Habitat: Cultivated

Distribution: Tropical Asia

Botanical observation: This handsome climbing shrub grows to about 2 m tall. The leaves are simple, opposite, and stipulate. The petiole is about 1 cm. The stipule is about 1 cm long and bifid. The blade is elliptic, green, and glossy above, soft, 8–12 cm × 3–6.5 cm, marked with about 11 pairs conspicuous secondary nerves, tapering at base, and acuminate at apex. The inflorescences are cymose and terminal. The calyx includes 4 linear sepals which are about 8 mm long and a calycophyll which is pure white, lanceolate, and up to 10 cm long. The corolla is tubular, about 2–2.5 cm long, and develops 5 triangular lobes which are bright orange. The berries are oblong, green, fleshy, and about 1 cm long.

Medicinal use: Leprosy (India)

Broad-spectrum antibacterial halo developed by iridoid: Mussaenin A (Sahadevan and Menon, 2017) inhibited the growth of *Bacillus subtilis*, *Escherichia coli*, and *Staphylococcus aureus* (Dai et al., 2002).

Commentaries

i. *Mussaenda frondosa* L. is used in Nepal (local name: *Asari*) for cough, jaundice, diarrhea, and dysentery, and in India (local name: *Shrivati*.) for cough, dysentery, fever, jaundice, leprosy, and ulcers. In Yap, *Mussaenda frondosa* L. (local name: *Petch*) is given for gonorrhea. In India, *Mussaenda glabra* Vahl (local name: *Sonaaruupaa*) is used for cough and fever. Dichloromethane extract of leaves (1 mg/6 mm disc) inhibited the growth of *Bacillus cereus*, *Bacillus subtilis*, *Escherichia coli*, and *Saccharomyces cerevisiae* with the inhibition zone diameters of 8, 11, 9, and 15 mm, respectively (Jayasinghe et al., 2002).

ii. In Myanmar, *Mussaenda macrophylla* Wall. (local name: *Lelu*) is used for dysentery. In Bangladesh *Mussaenda roxburghii* Hook f. it is used for boils (local name: *Ranirtak*), and in India it is used jaundice.

iii. In Thailand, *Mussaenda roxburghii* Hook.f. (local name: *Ranirtak*) is used for boils. In Indonesia, *Mussaenda frondosa* L. (local name: *Balik adat*) is used for the treatment of jaundice.

iv. In China, *Mussaenda pubescens* Ait.f. (local names: *Bái miàn téng*, *Xiān gān téng*) is a remedy for colds and cough. Aqueous extract of *Mussaenda pubescens* Ait.f. inhibited the replication of the Respiratory syncytial virus with the IC_{50} value of 32 µg/mL and a selectivity index of 16 (Li et al., 2004) and the replication of the Herpes simplex virus type-1 with the IC_{50} value of 200 µg/mL and a selectivity index of 3.8 (Li et al., 2004).

v. In the Philippines, *Mussaenda philippica* A. Rich. (local name: *Hamiling*) is a remedy for jaundice and sore throat, and in Palau (local name: *Cherecheroi*) it is used for cuts, gonorrhea, and tuberculosis.

vi. In Papua New Guinea, *Mussaenda ferruginea* K. Schum. (local name: *Aganapa*) is used for cough and fever.

vii. In the Subfamily Ixoroideae, Tribe Mussaendeae, *Schizomussaenda dehiscens* (Craib) Li is used to treat small pox in Cambodia, Laos, and Vietnam.

viii. Iridoids are anti-inflammatory (del Carmen Recio et al., 1994), hepatoprotective (Vidyalakshmi et al., 2009), and antiviral whence the use of plants in the genus *Mussaenda* L. (and other plants accumulating iridoids) for fever and jaundice.

References

Dai, J.Q., Liu, Z.L. and Yang, L., 2002. Non-glycosidic iridoids from Cymbaria mongolica. *Phytochemistry*, *59*(5), pp. 537–542.

del Carmen Recio, M., Giner, R.M., Manez, S. and Ríos, J.L., 1994. Structural considerations on the iridoids as anti-inflammatory agents. *Planta Medica*, *60*(03), pp. 232–234.

Jayasinghe, U.L.B., Jayasooriya, C.P., Bandara, B.M.R., Ekanayake, S.P., Merlini, L. and Assante, G.M., 2002. Antimicrobial activity of some Sri Lankan Rubiaceae and Meliaceae.

Li, Y., Ooi, L.S., Wang, H., But, P.P. and Ooi, V.E., 2004. Antiviral activities of medicinal herbs traditionally used in southern mainland China. *Phytotherapy Research*, *18*(9), pp. 718–722.

Sahadevan, L.D.M. and Menon, D.B., 2017. Mussaenin a isolated from mussaenda glabrata induces apoptosis in the liver cancer cells via mitochondrial pathway. *International Journal of Pharmacognosy and Phytochemical Research*, 9(9), pp. 1266–1273.

Vidyalakshmi, K.S., Nagarajan, S., Vasanthi, H.R. and Rajamanickam, V., 2009. Hepatoprotective and antioxidant activity of two iridoids from Mussaenda 'dona aurora'. *Zeitschrift für Naturforschung C*, *64*(5–6), pp. 329–334.

10.3.4.11 *Neolamarckia cadamba* (Roxb.) Bosser

Synonyms: *Anthocephalus cadamba* (Roxb.) Miq.; *Anthocephalus chinensis* (Lam.) Walp.; *Anthocephalus indicus* A. Rich.; *Nauclea cadamba* Roxb.; *Sarcocephalus cadamba* (Roxb.) Kurz

Common names: Burflower-tree; kadam; karam (Bangladesh); kadamba (India); jabon, kelempayan (Indonesia); hkala-shwang (Myanmar); kadam (Nepal); bagarilao (the Philippines)

Habitat: Open forest, cultivated

Distribution: India to Papua New Guinea

Botanical observation: This graceful timber tree grows to about 30 m tall. The bole is straight and buttressed. The bark is somewhat smooth and brownish and the inner bark is yellow. The leaves are simple, opposite, and stipulate. The stipules are 2 cm long and triangular. The petiole is 1.5–4 cm long. The blade is elliptic-oblong, 10–30 cm × 5–16 cm, coriaceous, rounded or cordate at base, and marked with 8–17 pairs of secondary nerves. The flowering heads are globose, orangish, up to 5 cm in diameter, and with a positive aura of life force. The calyx develops 5 teeth. The corolla is tubular, about 9 mm long, and 5-lobed. The 5 stamens are adnate to the corolla tube. The ovary develops a slender style that protrudes out of the corolla. The infructescences are capsules packed into a fleshy globose head which reaches about 7 cm in diameter, yellow when ripe, and edible when ripe (India).

Medicinal uses: Toothache (Bangladesh); fever, diarrhea, dysentery, sore throat, wounds (Nepal); fever, gargle (Myanmar); fever (Malaysia); diarrhea, fever, dysentery, wounds (India)

Broad-spectrum antibacterial halo developed by polar extract: Ethanol extract of bark of inhibited the growth of *Staphylococcus epidermidis*, *Streptococcus pyogenes*, *Escherichia coli* and *Pseudomonas aeruginosa* with zones of inhibition of 15, 8, 8, and 9 mm, respectively (500 µg/disc) (Uddin et al, 2008).

Antifungal (filamentous) polar extract: Aqueous extract of seeds (solution from 15 g dry powder of seeds in 100 mL of water in 15% of culture medium) inhibited the growth of *Fusarium moniliforme* (IARI 4824 F) by 37.1% (Bhardwaj, 2011).

Moderate antiviral (enveloped monopartite linear single-stranded (+) RNA) hydroxycinnamic derivative: chlorogenic and inhibited the replication of the Human coronavirus NL63 with the IC_{50} value of 43.4 µM via a mechanism involving, at least in part, blockage of virus entry in host cells (Weng et al., 2019)

Commentaries

i. The leaves yield chlorogenic acid (see earlier) as well as indole alkaloids such as neolamarckines A and B (Qureshi et al., 2011), 3β-dihydrocadambine (Yuan et al., 2020), anthrocephaline and cadambine (Takayama et al., 2003). Consider that anthrocephaline and cadambine are topoisomerase inhibitors (Kumar et al., 2015) suggesting that alkaloids in this tree could be strong broad-spectrum antibacterial, antifungal, and/or antiviral agents.
ii. According to our peer the Indian physician Atreya of the 6th century BC, the fruits are used for fever and the leaves for sore throat.
iii. The flower heads look very much like coronaviruses and as per the theory of signature may have some usefulness against COVID-19.
iv. In Nepal and India, the plant is sacred for Hindus and used to make holy garlands to be worn by Guruswamies.
v. Excellent description of the plant has been published by Shantha et al. (2008).
vi. *Anthocephalus macrophyllus* (Roxb.) Havil is used to treat leucorrhea in Indonesia.
vii. *Sarcocephalus coadunatus* (Roxb. Ex Sm) Druce is used to heal ulcers in the Philippines.

References

Bhardwaj, S.K., 2011. Antifungal potential of plant extracts against Fusarium moniliforme Sheldon. *ORYZA-An International Journal on Rice*, *48*(3), pp. 278–279.

Kumar, A., Chowdhury, S.R., Jatte, K.K., Chakrabarti, T., Majumder, H.K., Jha, T. and Mukhopadhyay, S., 2015. Anthocephaline, a new indole alkaloid and cadambine, a potent inhibitor of DNA topoisomerase IB of Leishmania donovani (LdTOP1LS), isolated from Anthocephalus cadamba. *Natural Product Communications*, *10*(2), pp. 297–299.

Qureshi, A.K., Mukhtar, M.R., Hirasawa, Y., Hosoya, T., Nugroho, A.E., Morita, H., Shirota, O., Mohamad, K., Hadi, A.H.A., Litaudon, M. and Awang, K., 2011. Neolamarckines A and B, new indole alkaloids from Neolamarckia cadamba. *Chemical and Pharmaceutical Bulletin*, *59*(2), pp. 291–293.

Shantha, T.R., Vasanthakumar, K.G. and Gopakumar, K., 2008. Neolamarckia cadamba (roxb.) bosser. *Journal of Economic and Taxonomic Botany*, *32*(1), pp. 128–148.

Takayama, H., Tsutsumi, S.I., Kitajima, M., Santiarworn, D., Liawruangrath, B. and Aimi, N., 2003. Gluco-indole alkaloids from Nauclea cadamba in Thailand and transformation of 3α-dihydrocadambine into the indolopyridine alkaloid, 16-carbomethoxynaufoline. *Chemical and Pharmaceutical Bulletin*, *51*(2), pp. 232–233.

Uddin, S.J., Rouf, R., Shilpi, J.A., Alamgir, M., Nahar, L. and Sarker, S.D., 2008. Screening of some Bangladeshi medicinal plants for in vitro antibacterial activity. *Oriental Pharmacy and Experimental Medicine*, *8*(3), pp. 316–321.

Weng, J.R., Lin, C.S., Lai, H.C., Lin, Y.P., Wang, C.Y., Tsai, Y.C., Wu, K.C., Huang, S.H. and Lin, C.W., 2019. Antiviral activity of Sambucus FormosanaNakai ethanol extract and related phenolic acid constituents against human coronavirus NL63. *Virus Research*, *273*, p. 197767.

Yuan, H.L., Zhao, Y.L., Qin, X.J., Liu, Y.P., Yu, H.F., Zhu, P.F., Jin, Q., Yang, X.W. and Luo, X.D., 2020. Anti-inflammatory and analgesic activities of Neolamarckia cadamba and its bioactive monoterpenoid indole alkaloids. *Journal of Ethnopharmacology*, *260*, p. 113103.

10.3.4.12 *Rubia cordifolia* L.

Synonyms: *Galium cordifolium* (L.) Kuntze; *Rubia manjith* Roxb. ex Fleming; *Rubia mitis* Miq.; *Rubia munjista* Roxb.; *Rubia pratensis* (Maxim.) Nakai; *Rubia scandens* Zoll. & Moritzi

Common name: India madder, ti-hiueh (China); manjista (India); ruras (Iran)

Habitat: Open forest

Distribution: Afghanistan, Pakistan, India, Bangladesh, Cambodia, Laos, Vietnam, Thailand, Malaysia, Indonesia, China, Mongolia, Korea, Japan

Botanical observation: This slender climber grows up to about 3 m long. The roots are elongated. The stems are quadrangular, glabrous, or minutely hooked. The leaves are simple, in whorls of 4, and exstipulate. The petiole is slender and up to about 3 cm long. The blade is lanceolate, 1.5–4 cm × 0.5–1.5 cm, hairy below, cordate at base, acute at apex, and marked with 1–2 pairs of secondary nerves. The cymes are axillary and about 10 cm long. The calyx is tubular. The corolla is tubular, whitish to green, somewhat cucurbitaceous, minutely hairy within, and develops 5 triangular lobes which are about 4 mm long. The 5 stamens are adnate to the corolla and the stamens are blackish. The ovary is bilobed and develops 2 styles and capitate stigmas. The berries are black, glossy, and about 5 mm across.

Medicinal uses: Dysentery, ulcers (India); jaundice (Korea)

Strong anticandidal amphiphilic anthraquinone: Purpurin (LogD 3.2 at pH 7.4; molecular mass = 256.2 g/mol) inhibited the growth of *Candida albicans* (ATCC 18804), *Candida albicans* (ATCC 28367), *Candida glabrata* (ATCC 90030), *Candida krusei* (ATCC 6258), *Candida parapsilosis* (ATCC 22014), and *Candida tropicalis* (ATCC 13803) with the MIC values of 5.1, 2.5, 2.5, 1.2, 5.1, and 5.1 µg/mL, respectively (Kang et al., 2010).

Strong antiviral (enveloped linear monopartite dimeric single-stranded (+) RNA) mid-polar extract: Ethyl acetate extract of roots inhibited the replication of the Human immunodeficiency virus type-1 (NL4.3 strain) in CD4+ cell lines by 67.3% at the maximum non-cytotoxic concentration of 15 µg/mL (Sabde et al., 2011).

Strong antiviral (enveloped linear monopartite dimeric single-stranded (+) RNA) amphiphilic anthraquinone: Xanthopurpurin (LogD 2.7 at pH 7.4; molecular mass = 240.2 g/mol) inhibited the replication of Human immunodeficiency virus-1 (NL4.3 strain) in Human CD4+ T cell line CEM-GFP by 42% with a maximum non-cytotoxic concentration of 15 µg/mL (Sabde et al., 2011).

Antiviral (enveloped monopartite linear double-stranded DNA) anthraquinone: Alizarin inhibited the replication of the Human cytomegalovirus (Barnard et al., 1992).

Commentaries

i. The plant yields anthraquinones such as alizarin, purpurin, xanthopurpurin, rubiadin (see earlier), and 1,4-dihydroxy-2-methyl-5-methoxyanthraquinone (Sabde et al., 2011) whence it's use as a source of dye known as Indian madder.
ii. The plant was advocated for syphilis by our peer Girolamo Fracastoro (see earlier).

References

Barnard, D.L., Huffman, J.H., Morris, J.L., Wood, S.G., Hughes, B.G. and Sidwell, R.W., 1992. Evaluation of the antiviral activity of anthraquinones, anthrones and anthraquinone derivatives against human cytomegalovirus. *Antiviral Research*, *17*(1), pp. 63–77.

Kang, K., Fong, W.P. and Tsang, P.W.K., 2010. Novel antifungal activity of purpurin against Candida species in vitro. *Medical Mycology*, *48*(7), pp. 904–911.

Sabde, S., Bodiwala, H.S., Karmase, A., Deshpande, P.J., Kaur, A., Ahmed, N., Chauthe, S.K., Brahmbhatt, K.G., Phadke, R.U., Mitra, D. and Bhutani, K.K., 2011. Anti-HIV activity of Indian medicinal plants. *Journal of Natural Medicines*, *65*(3–4), pp. 662–669.

10.4 ORDER LAMIALES BROMHEAD (1838)

10.4.1 Family Acanthaceae A.L de Jussieu (1789)

The family Acanthaceae consists of about 250 and 2500 species of herbs or shrubs. The stems are often lignose, somewhat articulated, and swollen at nodes. The leaves are simple, opposite, and exstipulate. The inflorescences are cymes, and the flowers are zygomorphic. The calyx is tubular and 4- to 5-lobed. The corolla is tubular, membranous, ephemeral or not, and 2-lipped or not. The androecium includes 4 stamens. The gynoecium consists of 2 carpels united into a bilocular ovary, each locule sheltering numerous ovules on axile placentas and slender style. The capsules are characteristically somewhat fusiform, dehiscent, and sheltering minute and often discoid seeds. Plants in this family produce antimicrobial lignans.

10.4.1.1 *Acanthus ilicifolius* L.

Synonyms: *Acanthus ebracteatus* var. *xiamenensis* (R.T. Zhang) C.Y. Wu & C.C. Hu; *Acanthus ilicifolius* var. *xiamenensis* (R.T. Zhang) Y.F. Deng, N.H. Xia & H.B. Chen; *Acanthus xiamenensis* R.T. Zhang; *Dilivaria ilicifolia* (L.) Juss.

Common names: Sea holly; holly-leaved Acanthus; hargosa, harkuch kanta (Bangladesh); lao shu le (China); harakancha, paina schulli (India); jeruju (Indonesia); kaya-chon (Myanmar); ikili (Sri Lanka); dilivario (the Philippines); kollil (Palau); tror cheak kranh samout (Thailand)

Habitat: Mangroves, muds

Distribution: Tropical Asia and Australia

Botanical observation: This erect herb grows up to about 1.5 m tall and has a positive aura of life force. The stem is terete and stout. The leaves are decussate, simple, and exstipulate. The petiole is about 5 mm long and stout. The blade is coriaceous, oblong, 5–15×2–5 cm, glabrous, and holly-like. The spikes are terminal and about 15 cm long. The flower buds are somewhat sigmoid. The calyx is 4-lobed and about 1 cm long. The corolla is bluish, tubular, 2–4 cm long, and develops a conspicuous, recurved, and somewhat bifid limb. The 4 stamens are adnate to the corolla and 1.5 cm long. The capsules are dehiscent, smooth, glossy, and green at first, 2.5 cm long, marked with vestige of style at apex, containing a few white seeds and later turns brownish and open looking like somewhat the open mouth of a snake (the plant is used for snake-bites in India).

Medicinal uses: Hepatitis, leprosy, wounds (Indonesia); leucorrhea (Bangladesh; India); fever, skin infection (India)

Broad-spectrum antibacterial mid-polar extract: Chloroform extract of leaves (200 µL of a 10 mg/mL solution per well) inhibited the growth of *Bacillus cereus*, *Staphylococcus aureus*, *Pseudomonas aeruginosa*, and *Proteus vulgaris* with the inhibition zone diameters of 22, 21, 20, and 19 mm, respectively (Bose & Bose, 2008).

Antibacterial simple phenolic: 3,5-Dimethoxy-4-hydroxy methyl benzoic acid (Ravikumar et al., 2012)

Antibacterial hydroxycinnamic acid derivative: (Z)-4-Coumaric acid 4-*O*-β-D-glucopyranoside (Ravikumar et al., 2012)

Antibacterial benzoxazolinone: 6-Hydroxy-benzoxazolinone (Ravikumar et al., 2012)

Broad-spectrum antifungal mid-polar extract: Candida albicans, *Aspergillus fumigatus*, and *Aspergillus niger* (Bose & Bose, 2008).

In vivo antifungal (filamentous) extract: Chloroform extract of leaves given orally at the dose of 500 mg/g to experimentally immunocompromised mice evoked some levels of protection against *Aspergillus fumigatus* (Kalakar et al., 2012).

In vivo antiviral (enveloped circular double-stranded DNA) polar extract: Alcohol extract given to duckling orally at a dose of 2 g/Kg/day for 14 days evoked a decrease in serum Hepatitis B surface antigen, AST, ALT, and improved hepatic cytoarchitecture (Wei et al., 2015).

Commentaries

i. The plant yields lignans and lignan glycosides (Kanchanapoom et al., 2001b) as well as rare series of benzoxazinoid alkaloid glycosides (Kanchanapoom et al., 2001a). Benzoxazinoid alkaloids are phytoalexins produced by plants in the family Poaceae in response to phytopathogenic microorganisms from benzoxazinoid glycosides (Jimenez-Garcia et al., 2018). The plant produces 4-hydroxy-2-benzoxazolone (Wei et al., 2015).
ii. In Turkey, *Acanthus spinosus* L. (local name: *Karadiken*) is used for diarrhea. In India, *Acanthus volubilis* Wall. (local name: *Hargocha*) is used for boils, cuts, and wounds. In Palau, *Acanthus ebracteatus* Vahl (local name: *Korir*) is used for ear infection.

References

Jimenez-Garcia, S.N., Garcia-Mier, L., Garcia-Trejo, J.F., Ramirez-Gomez, X.S., Guevara-Gonzalez, R.G. and Feregrino-Perez, A.A., 2018. Fusarium mycotoxins and metabolites that modulate their production. In: Askun, T. (Ed.) Fusarium: Plant Diseases, Pathogen Diversity, Genetic Diversity, Resistance and Molecular Markers. Books on Demand: Norderstedt, Germany, 18 p.

Kanchanapoom, T., Kamel, M.S., Kasai, R., Picheansoonthon, C., Hiraga, Y. and Yamasaki, K., 2001a. Benzoxazinoid glucosides from Acanthus ilicifolius. *Phytochemistry*, *58*(4), pp. 637–640.

Kanchanapoom, T., Kamel, M.S., Kasai, R., Yamasaki, K., Picheansoonthon, C. and Hiraga, Y., 2001b. Lignan glucosides from *Acanthus ilicifolius*. *Phytochemistry*, *56*(4), pp. 369–372.

Ravikumar, S., Raja, M. and Gnanadesigan, M., 2012. Antibacterial potential of benzoate and phenylethanoid derivatives isolated from *Acanthus ilicifolius* L. leaf extracts. *Natural Product Research*, *26*(23), pp. 2270–2273.

Wei, P.H., Wu, S.Z., Mu, X.M., Xu, B., Su, Q.J., Wei, J.L., Yang, Y., Qin, B. and Xie, Z.C., 2015. Effect of alcohol extract of *Acanthus ilicifolius* L. on anti-duck hepatitis B virus and protection of liver. *Journal of Ethnopharmacology*, *160*, pp. 1–5.

10.4.1.2 *Andrographis paniculata* (Burm. f.) Wall. ex Nees

Synonyms: *Justicia paniculata* Burm.f.

Common names: Create; maha tita, kalamegh (Bangladesh); kirata, nela vembu, vubati (India); sambiroto (Indonesia); nainehavand (Iran); hempedu pahit (Malaysia); ninhinkohomb (Sri Lanka)

Habitat: Sandy and sun-exposed soils, roadsides, wastelands

Distribution: From India to Indonesia

Botanical observation: This herb grows to 40 cm tall and has a positive aura. The stems are finely quadrangular, slender, and somewhat lignose. The leaves are simple, decussate, and exstipulate. The petiole is minute. The blade is lanceolate, dark green, somewhat glossy, bitter, marked with 3 to 5 pairs of secondary nerves, 1.5–5 × 0.5–2 cm, and shortly acuminate. The inflorescence is a terminal, leafy, and candelabra-like panicles of erect flowers. The calyx is hairy and develops 5 linear lobes. The corolla is tubular, pure white, up to about 1 cm long, the upper lip is 3-lobed and beautifully penciled with purple marks and the lower lip entire. The androecium includes 2 stamens exerted from corolla tube, the filaments white and hairy, and the anthers purple. The capsules are upright, dehiscent, up to 1.5 cm long and contain numerous tiny seeds.

Medicinal uses: Boils, bronchitis, fever, cholera, diarrhea, dysentery, influenza, sore throat, syphilis (India), cough, dysentery, fever, skin infections (Bangladesh); cough, fever (Indonesia); diarrhea, dysentery, fever, leprosy (Myanmar)

Weak broad-spectrum antibacterial labdane-type diterpene: Andrographolide inhibited the growth of *Escherichia coli* (AG100A), *Staphylococcus aureus* (MTCC 96), *Streptococcus pneumoniae* (ATCC BAA), *Enterococcus faecalis* (ATCC 29212), and *Bacillus subtilis* (ATCC 27370) with the MIC values of 250, 100, 250, 500, and 250 µg/mL, respectively (Banerjee et al., 2017). In *Staphylococcus aureus* (MTCC 96), andrographolide inhibited DNA synthesis and was bacteriostatic (Banerjee et al., 2017).

Moderate broad-spectrum antifungal labdane-type diterpenes: Andrographolide and 14-deoxy-11, 12-didehydroandrographolide inhibited the growth of *Microsporium canis* with the MIC/MFC values of 50/50 and 150/150 µg/mL, respectively (Sule et al., 2012). Andrographolide and 14-deoxy-11, 12-didehydroandrographolide inhibited the growth of *Candida albicans* with the MIC/MFC values of 150/200 and 100/150 µg/mL, respectively (Sule et al., 2012).

Weak antiviral (enveloped segmented linear single-stranded (-) RNA) labdane-type diterpene: Andrographolide at the concentration of 250 Mg/ml inhibited the replication of the Influenza A virus H1N1 A/E/13 by 43.9% in Human respiratory epithelial cells and prevented cell death via inhibition of NF-*k*B and cytokine secretions (Yu et al., 2014).

Strong antiviral (enveloped monopartite linear single-stranded (+) RNA) polar extract: Methanol extract at the concentration of 50 µg/mL inhibited the cytopathic effect of the Dengue virus serotype-1 in Vero cells by about 75% (Tang et al., 2012).

Clinical antiviral (enveloped linear monopartite single-stranded (+) RNA) labdane-type diterpene: Andrographolide given orally to COVID-19 patients at the dose of 20 mg 3 times per day for 5 days prevented the development of pneumonia (Wanaratna et al., 2021).

Clinical antiviral (enveloped linear monopartite dimeric single-stranded (+) RNA) labdane-type diterpene: Andrographolide given orally to AIDS patients at the dose of 5 mg/Kg, 3 times per day for 3 weeks and 10 mg/kg, 3 times per day evoked a mean increase in CD4+ cell counts of 96.3 cells/mm^3 at week 6 (Calabrese et al., 2000).

Viral enzyme inhibition by polar extract: Aqueous extract of bark at the concentration of 250 µg/mL inhibited the Human immunodeficiency virus type-1 protease by 29.6% (Xu et al., 1996).

Commentaries

i. In Sabah, locals drink infusion of the leaves to treat COVID-19 within a few days. *Andrographis paniculata* (Burm. f.) Wall. ex Nees could be of value in preventing complication in COVID-19 such as cytokine blast, blood cloth inhibition, and respiratory distress. Cytokine blast is responsible for fatalities (Петров et al., 2021) and anti-inflammatory plants could play a role in saving lives. It is one of the numerous medicinal plants unknown to the West that relies upon Big Pharma medicines and that could benefit people. If the West continues to rely solely on industrial medicines, there is the grim possibility that it may not do so well in the forthcoming pandemics. It is not impossible to see China and other Asian countries using their medicinal plants to survive global pandemics. It is time for the West to give herbal medicine a place in clinical practice under the control of well-trained (in terms of Pharmacognosy) pharmacy graduates. The West and the East need to unite and join efforts to combat a well-united spectrum of microbial enemies. In Bangladesh, where a vast number of people rely on herbs for their health, this plant is well known as king of bitter and is also used for COVID-19 like *Vitex negundo* L. (Mohamed Rhamatullah, personal communication).

ii. The plant is used in India for the treatment of Chikungunya virus infection (Viswanathan et al., 2008).

iii. Andrographolide relaxes smooth muscles (Ozolua et al., 2011) and thus most probably able to inhibit bacterial and/or fungal efflux pumps.

iv. *Andrographis echioides* Ness (local name: *Pitumba*) is used for fever in India.

v. In the Subfamily Acanthoideae, Tribe Ruellieae, *Blechum pyramidatum* (Lam.) Urb. is used for the treatment of blenorrhea in the Philippines and *Blechum brownei* Juss. is used for yaws in Yap (local name: *Malai*). In the Tribe Ruellieae, *Hemigraphis alternata* (Burm.f.) T. Anders is used for venereal diseases (Indonesia).

vi. In the Subfamily Acanthoideae, Tribe Justicieae, *Graptophyllum pictum* (L.) Griffith is used for wounds in the Philippines and at the concentration of 100 µg/mL decreased the Hepatitis B virus surface antigen by about 80% in HepG2 (Indrasetiawan et al., 2019). In the Tribe Justicieae *Peristrophe acuminata* Nees is a remedy for wounds in Malaysia. In the Tribe Justicieae *Pseuderanthemum bicolor* (Schrank) Radlk. ex Lindau is used for wounds in the Philippines. In the Tribe Justicieae *Hypoestes polythysa* Miq. is used to remove maggots from wounds (Indonesia)

vii. In the Subfamily Acanthoideae, the Tribe Justicieae, *Rhinacanthus nasutus* (L.) Kurz. is a remedy for herpes in Thailand and yields the naphthoquinones rhinacanthin C (Log D 2.7 at pH 7.4; molecular mass = 410.5 g/mol) and rhinacanthin D which inhibited the replication of the Cytomegalovirus with the EC_{50} values of 0.02 and 0.2 µg/mL, respectively (Sendl et al., 1996). Rhinacanthins E, F, and G inhibited the replication of the Cytomegalovirus with the IC_{50} values of 0.3, 0.9, and 1.9 µM, respectively (Ngoc et al., 2019). Rhinacanthins E, F, and G inhibited the replication of Human rhinovirus 1B with the IC_{50} values of 0.2, 0.2, and 0.9 µM and selectivity indexes of 103.3, 124.3, and 51.4 µM, respectively (Ngoc et al., 2019). Rhinacanthins E, F, and G inhibited the replication of Coxsackie virus type 3B with the IC_{50} values of 0.03, 1.4, and 1.5 µM and selectivity indexes of above 1666.6, 34.7, and 33.3, respectively (Ngoc et al., 2019). Rhinacanthinic acid C inhibited the replication of the Herpes simplex virus with the IC_{50} values of 60.3 µg/mL, respectively (Thongchuai et al., 2015). The dibenzylbutane lignans rhinacanthin E and rhinacanthin F (LogD 3.5 at pH 7.4; molecular mass = 444.4 g/mol) inhibited the replication of the Influenza virus, with EC_{50} values of 7.4 and 3.1 µg/mL, respectively (Kernan et al., 1997).

viii. In the Subfamily Acanthoideae, Tribe Lepidagathidae: *Lepidagathis incurva* Buch.-Ham. ex D.Don (synonym: *Lepidagathis hyalina* Nees) is used for cough in Malaysia. *Lepidagathis formosensis* C.B. Clarke is a remedy for wounds in Taiwan.

References

Banerjee, M., Parai, D., Chattopadhyay, S. and Mukherjee, S.K., 2017. Andrographolide: Antibacterial activity against common bacteria of human health concern and possible mechanism of action. *Folia Microbiologica*, *62*(3), pp. 237–244.

Calabrese, C., Berman, S.H., Babish, J.G., Ma, X., Shinto, L., Dorr, M., Wells, K., Wenner, C.A. and Standish, L.J., 2000. A phase I trial of andrographolide in HIV positive patients and normal volunteers. *Phytotherapy Research*, *14*(5), pp. 333–338.

Indrasetiawan, P., Aoki-Utsubo, C., Hanafi, M., Hartati, S.R.I., Wahyuni, T.S., Kameoka, M., Yano, Y., Hotta, H. and Hayashi, Y., 2019. Antiviral activity of cananga odorata against hepatitis B virus. *Kobe Journal of Medical Sciences*, *65*(2), p. E71.

Kernan, M.R., Sendl, A., Chen, J.L., Jolad, S.D., Blanc, P., Murphy, J.T., Stoddart, C.A., Nanakorn, W., Balick, M.J. and Rozhon, E.J., 1997. Two new lignans with activity against influenza virus from the medicinal plant *Rhinacanthus nasutus*. *Journal of Natural Products*, *60*(6), pp. 635–637.

Ngoc, T.M., Phuong, N.T.T., Khoi, N.M., Park, S., Kwak, H.J., Nhiem, N.X., Trang, B.T.T., Tai, B.H., Song, J.H., Ko, H.J. and Kim, S.H., 2019. A new naphthoquinone analogue and antiviral constituents from the root of *Rhinacanthus nasutus*. *Natural Product Research*, *33*(3), pp. 360–366.

Ozolua, R.I., Adejayan, A., Aigbe, O.P., Uwaya, D.O. and Argawal, A., 2011. Some characteristic relaxant effects of aqueous leaf extract of *Andrographis paniculata* and andrographolide on guinea pig tracheal rings. *Nigerian Journal of Physiological Sciences*, *26*(2), pp. 119–24.

Sendl, A., Chen, J.L., Jolad, S.D., Stoddart, C., Rozhon, E., Kernan, M., Nanakorn, W. and Balick, M., 1996. Two new naphthoquinones with antiviral activity from Rhinacanthus nasutus. *Journal of Natural Products*, *59*(8), pp. 808–811.

Sule, A., Ahmed, Q.U., Latip, J., Samah, O.A., Omar, M.N., Umar, A. and Dogarai, B.B.S., 2012. Antifungal activity of *Andrographis paniculata* extracts and active principles against skin pathogenic fungal strains in vitro. *Pharmaceutical Biology*, *50*(7), pp. 850–856.

Tang, L.I., Ling, A.P., Koh, R.Y., Chye, S.M. and Voon, K.G., 2012. Screening of anti-dengue activity in methanolic extracts of medicinal plants. *BMC Complementary and Alternative Medicine*, *12*(1), p. 3.

Thongchuai, B., Tragoolpua, Y., Sangthong, P. and Trisuwan, K., 2015. Antiviral carboxylic acids and naphthoquinones from the stems of *Rhinacanthus nasutus*. *Tetrahedron Letters*, *56*(37), pp. 5161–5163.

Viswanathan, M.V., Raja, D.K. and Khanna, S.D., 2008. Siddha way to cure Chikungunya. *Indian Journal of Traditional Knowledge*, *7*(2), pp. 345–346.

Wanaratna, K., Leethong, P., Inchai, N., Chueawiang, W., Sriraksa, P., Tabmee, A. and Sirinavin, S., 2021. Efficacy and safety of *Andrographis paniculata* extract in patients with mild COVID-19: A randomized controlled trial. medRxiv.

Xu, H.X., Wan, M., Loh, B.N., Kon, O.L., Chow, P.W. and Sim, K.Y., 1996. Screening of traditional medicines for their inhibitory activity against HIV-1 protease. *Phytotherapy Research*, *10*(3), pp. 207–210.

Yu, B., Dai, C.Q., Jiang, Z.Y., Li, E.Q., Chen, C., Wu, X.L., Chen, J., Liu, Q., Zhao, C.L., He, J.X. and Ju, D.H., 2014. Andrographolide as an anti-H1N1 drug and the mechanism related to retinoic acid-inducible gene-I-like receptors signaling pathway. *Chinese Journal of Integrative Medicine*, *20*(7), pp. 540–545.

Петров, В.И., Амосов, А.А., Герасименко, А.С., Шаталова, О.В., Пономарева, А.В., Акинчиц, А.Н., Кулакова, И.С. and Горбатенко, В.С., 2021. МЕХАНИЗМЫ РАЗВИТИЯ ЦИТОКИНОВОГО ШТОРМА ПРИ COVID-19 И НОВЫЕ ПОТЕНЦИАЛЬНЫЕ МИШЕНИ ФАРМАКОТЕРАПИИ. *Фармация и фармакология*, *8*(6), pp. 380–391.

10.4.1.3 *Barleria longifolia* L.

Synonyms: *Asteracantha longifolia* (L.) Nees; *Hygrophila auriculata* Heine; *Ruellia longifolia* Roxb.; *Hygrophila spinosa* T. Anders.

Common names: Long-leaved Barleria; kanta kulika; adyanda, kuruntaka, itchura, neer-moollie (India); le-padu (Myanmar)

Habitat: Meadows, paddy fields

Distribution: Pakistan, India, Sri Lanka, Nepal, Bangladesh, Myanmar, Cambodia, Laos, Vietnam, Thailand

Botanical observation: This stout herb grows to about 1 m tall. The stem is quadrangular, purplish or not, hairy, lignose, and develops 6 axillary, very sharp, and about 3 cm long thorns. The leaves are simple, sessile, exstipulate, decussate, and whorled. The blade is glossy, cuneate at base, acute at apex, serrulate, 5–12 cm × 1.5–2.5 cm, hairy, somewhat boraginaceous, marked with 5–12 pairs of secondary nerves, and narrowly oblong, lanceolate, to spathulate. The flowers are axillary, sessile, and verticillate. The calyx is 4-lobed, hairy. The corolla is tubular, light blue, about 2.5 cm long, the upper lip 2-lobed, the lower 3-lobed. The androecium includes 4 stamens. The ovary is oblong and develops a slender style which grows up to 2.5 cm long. The capsules are oblong, dehiscent, up to about 2 cm long, and contain 4 to 8 tiny and orbicular seeds.

Medicinal uses: Leprosy (Bangladesh); abscesses, boils, cough, flu, jaundice, tuberculosis (Myanmar)

Broad-spectrum antibacterial halo developed by polar extract: Methanol extract (6 mm disc impregnated with 250 μg/mL solution) inhibited the growth of *Staphylococcus aureus*, *Streptococcus*

Barleria Longifolia L.

pneumoniae, *Bacillus cereus*, *Escherichia coli*, *Klebsiella pneumoniae*, *Salmonella typhi*, *Proteus vulgaris*, and *Shigella flexneri* with the inhibition zone diameters of 18, 12, 11, 12, 10, 9, 10, and 10 mm, respectively (Dos & An, 2013).

Broad-spectrum antibacterial halo developed by polar extract: Methanol extract (6 mm disc impregnated with 250 µg/mL solution) inhibited the growth of *Candida albicans* and *Aspergillus niger* with the inhibition zone diameters of 9 and 9 mm, respectively (Dos & An, 2013).

Commentaries

i. In India, *Barleria prionitis* L. (local name: *Ananta*) is used for boils, bronchitis, cough, fever, pimple, toothache, wounds and in Myamar (local name: *Leik-su-ywe*) for boils and ringworm. In Bangladesh (local name: *Kanta jati*) it is used for fever and boils and in Thailand for ringworm. 6-*O*-p-coumaroyl-8-*O*-acetyl shanzhiside methyl ester from this plant inhibited the Respiratory syncytial virus (A2 strain) with EC_{50}=2.4 µg/mL and a selectivity index of 17 (Chen et al., 1998).

ii. In India, *Barleria cristata* L. (local name: *Jhinti*) is used for bronchitis, cough, cuts, fever, toothache, and wounds. *Barleria strigosa* Willd. (local name: *Bala*) is used for bronchitis and cough. *Barleria courtallica* Nees is used for ear infection, eye infection, and pneumonia. *Barleria buxifolia* L. is used for bronchitis and cough.
iii. In Pakistan, *Barleria acanthoides* Vahl (local name: *That*) is used for fever.
iv. In Myanmar, *Hygrophila phlomoides* Nees (local name: *Hsay-dan*) is used to treat eye infection, dermatomycosis, and festering sores.
v. Methanol extract of *Barleria lupulina* Lindl. inhibited the replication of the Herpes simplex virus type-2 (#408) with EC_{50} value of 442.1 µg/mL (Yoosook et al., 1999).

References

Chen, J.L., Blanc, P., Stoddart, C.A., Bogan, M., Rozhon, E.J., Parkinson, N., Ye, Z., Cooper, R., Balick, M., Nanakorn, W. and Kernan, M.R., 1998. New iridoids from the medicinal plant *Barleria prionitis* with potent activity against respiratory syncytial virus. *Journal of Natural Products*, *61*(10), pp. 1295–1297.

Doss, A. and An, S.P., 2013. Antimicrobial activity of *Hygrophila auriculata* (Schumach.) Heine and *Pergularia daemia* Linn. *African Journal of Plant Science*, *7*(4), pp. 137–142.

Yoosook, C., Panpisutchai, Y., Chaichana, S., Santisuk, T. and Reutrakul, V., 1999. Evaluation of anti-HSV-2 activities of *Barleria lupulina* and *Clinacanthus nutans*. *Journal of Ethnopharmacology*, *67*(2), pp. 179–187.

10.4.1.4 *Ecbolium linneanum* Kurz.

Synonyms: *Ecbolium viride* (Forsk.) Alston, *Justica ecbolium* L.

Common name: Udujati (Bangladesh)

Habitat: Moist and shady spots

Distribution: India, Bangladesh

Botanical observation: This herb grows up to 50 cm tall. The stem is terete. The leaves are opposite, simple, and exstipulate. The petiole is 2 cm long. The blade is elliptic, dark green, glossy, wavy, oblong, acute, at base and apex marked with about 6–10 pairs of secondary nerves, and 5–15 cm × 2–5 cm. The spikes are terminal, bracteated, and up to 20 cm long. The calyx comprises 5 linear lobes. The corolla is tubular, to 4 cm long, membranous, of a unique type of bluish-jade color, with a mysterious aura, the upper lip linear and the lower with 3 broadly elliptic lobes. The capsules are elongated, up to 1.5 cm long, and contain minute tuberculate seeds.

Medicinal uses: Chicken pox, measles (Bangladesh); jaundice (Bangladesh)

Broad-spectrum antibacterial mid-polar extract: Dichloromethane extract of leaves (50 mg/mL/6 mm well) inhibited the growth of *Staphylococcus aureus*, *Pseudomonas aeruginosa*, *Escherichia coli*, *Enterococcus faecalis*, and *Klebsiella pneumoniae* with the inhibition zone diameter of 7.3. 10.3, 10.2, 9.7, and 6.5 mm, respectively (Chaudhuri et al., 2011).

Strong antibacterial (Gram-positive) furofuran lignan: Ecbolin A inhibited the growth of *Erwinia amylovora* (MTCC 2760), *Klebsiella pneumoniae* (ATCC 15380), *Proteus vulgaris* (MTCC 1771), *Proteus mirabilis* (ATCC 49565), *Vibrio cholerae* (ATCC 14035), *Bacillus subtilis* (ATCC 441), *Staphylococcus aureus* (ATCC 25923), *Staphylococcus epidermidis* (MTCC 3615), *Shigella flexneri* (MTCC 1457) with the MIC values of 62.5, 15.6, 250, 250, 31.2, 62.5, 7.8, 250, and 250 µg/mL, respectively (Cecilia et al., 2012).

Moderate broad-spectrum antifungal furofuran lignan: Ecbolin A inhibited the growth of *Candida albicans* (MTCC 227), *Malassezia pachydermatis*, and *Scopulariopsis* sp. with the MIC values of 125, 62.5, and 250 µg/mL, respectively (Cecilia et al., 2012)

Commentary: Medicinal uses suggest the presence of antiviral principles in this plant. It produces the furofuran ecbolin A (Venkataraman and Gopalakrishnan, 2002), and lignans are often antiviral.

References

Cecilia, K.F., Ravindhran, R. and Duraipandiyan, V., 2012. Ecbolin A: A bioactive compound from the roots of *Ecbolium viride* (Forssk.) Alston. *Asian Journal of Pharmaceutical and Clinical Research*, *5*, pp. 99–101.

Chaudhuri, D., Pathak, A.K. and Sevanan, M., 2011. Phytochemical screening and antimicrobial activity of extracts from leaves and stem of *Ecbolium linneanum*. *Bangladesh Journal of Pharmacology*, *6*(2), pp. 84–91.

Venkataraman, R. and Gopalakrishnan, S., 2002. A lignan from the root of *Ecbolium linneanum* Kurz. *Phytochemistry*, *61*(8), pp. 963–966.

10.4.1.5 *Justicia adhatoda* L.

Synonyms: *Adhatoda vasica* Nees; *Adhatoda zeylanica* Medik.

Common names: Malabar nut; arusa, bashok, kalo durba (Bangladesh); ya zui hua (China); adatodai, baahak, basak, basing, brenkar (India); my yar gyi (Myanmar); asuro (Nepal); bahekar, bahker, vadak (Pakistan)

Habitat: Cultivated, wastelands

Distribution: Pakistan, India, Nepal, Bangladesh, Myanmar

Botanical observation: This handsome shrub grows to 4 m tall. The stems are quadrangular. The leaves are simple, opposite, and exstipulate. The petiole is 0.5–3 cm long. The blade is broadly

Justicia Adhatoda L.

lanceolate, wavy at margin, dark green above, 5–20×2–10 cm, with 9–15 pairs of secondary nerves, attenuate at base, and acuminate at apex. The spikes are terminal or axillary, oblong, and up to 10 cm long. The calyx presents 5 lobes which are linear and about 1 cm long. The corolla is pure white, showy, somewhat fleshy, up to 3 cm long, upper lip oblong, 1.5 cm long, and 2-lobed and the lower lip 3-lobed and colored at throat with light purplish capillary-like lines. The 2 stamens are about 1.5 cm long. The ovary develops a 2.5 cm long style. The capsules are oblanceolate, dehiscent, and about 4 cm long.

Medicinal uses: Cholera (Bangladesh); cold, eye infection, gonorrhea, leprosy (India); cough (Bangladesh; India; Myanmar; Nepal; Pakistan); diarrhea (Bangladesh; India; Pakistan); tuberculosis (Bangladesh; India; Myanmar); bronchitis (India; Nepal; Pakistan); leucorrhea (Pakistan); phlegm, urinary tract infection (Myanmar); fever (India; Myanmar; Nepal; Pakistan); dysentery, toothache (Myanmar; Pakistan)

Broad-spectrum antibacterial halo developed by alkaloid fraction: Alkaloid fraction of leaves inhibited the growth of *Staphylococcus aureus* (ATCC 25923), *Staphylococcus aureus* (NTCC 3750), *Escherichia coli* (ATCC 25922), *Proteus mirabilis* (clinical isolate), *Pseudomonas aeruginosa* (ATCC 27853), and *Salmonella typhi* (NTCC 786) (Sawant, 2013).

Antifungal (yeast) halo developed by alkaloid fraction: Alkaloid fraction of leaves inhibited the growth of *Candida albicans* (MTCC 183) and *Cryptococcus neoformans* (NCIM 3542) (Sawant, 2013).

Commentaries

i. The plant has apparently not been studied for antimicrobial effects. It produces pyrroloquinazoline alkaloid such as vasicol, vascinine, and vasicinone which are bronchondilators and therefore probably able to inhibit bacterial and/or fungal efflux pumps. Pyrroloquinazoline alkaloid are planar and as such is probably antibacterial, antifungal, and antiviral. The plant is used in India for the treatment of Chikungunya virus infection (Viswanathan et al., 2008) and as such could be of value against COVID-19 and other pandemic coronaviruses.

ii. In Nepal, *Justicia procumbens* L. (local name: *Bisaune jhar*) is used for cold, cough, cuts, fever, and wounds.

iii. *Justicia gendarussa* L. (synonym: *Gendarussa vulgaris* Bojer) is used in Bangladesh (local name: *Jogut-mudun*) to heal wounds and in China, it is a remedy for carbuncles (local name: *Xiao bo gu*). The arylnaphthalene lignan glycoside patentiflorin A inhibited the replication of the Human immunodeficiency virus type-1 (Zhang et al., 2017b), and the arylnaphthalene lignan diphyllin (LogD 2.5 at pH 7.4; molecular mass = 380.3 g/mol) inhibited the replication of the Zika Virus with the IC_{50} value of 0.06 µM (Cui et al., 2020). The arylnaphthalene lignan glycoside justiprocumin B inhibited the replication of the Human immunodeficiency virus (Bal: M-Tropic), the Human immunodeficiency virus (89.6: DUal-Tropic), and the Human immunodeficiency virus (SF162: M-Tropic) with the IC_{50} values of 15, 15, and 14 nM, respectively, and more potently than Azidothymidine (Zhang et al., 2017a).

References

Amin, A.H. and Mehta, D.R., 1959. A bronchodilator alkaloid (vasicinone) from *Adhatoda vasica* Nees. *Nature*, *184*(4695), pp. 1317–1317.

Cui, Q., Du, R., Liu, M. and Rong, L., 2020. Lignans and their derivatives from plants as antivirals. *Molecules*, *25*(1), p. 183.

Sawant, C.S., 2013. Antimicrobial activity of alkaloids extracted from Adhatoda vasica. *International Journal of Pharma and Bio Sciences* 2013; 4(3), pp. 803–807.

Viswanathan, M.V., Raja, D.K. and Khanna, S.D., 2008. Siddha way to cure Chikungunya. *Indian Journal of Traditional Knowledge*, *7*(2), pp. 345–346.

Zhang, H.J., Rumschlag-Booms, E., Guan, Y.F., Liu, K.L., Wang, D.Y., Li, W.F., Cuong, N.M., Soejarto, D.D., Fong, H.H. and Rong, L., 2017a. Anti-HIV diphyllin glycosides from Justicia gendarussa. *Phytochemistry*, *136*, pp. 94–100.

Zhang, H.J., Rumschlag-Booms, E., Guan, Y.F., Wang, D.Y., Liu, K.L., Li, W.F., Nguyen, V.H., Cuong, N.M., Soejarto, D.D., Fong, H.H. and Rong, L., 2017b. Potent inhibitor of drug-resistant HIV-1 strains identified from the medicinal plant Justicia gendarussa. *Journal of Natural Products*, *80*(6), pp. 1798–1807.

10.4.1.6 *Ruellia tuberosa* L.

Common names: Meadow-weed; alughanti, potpoti (Bangladesh); toi ting (Thailand); *Chatakani phalli* (India); tanh tasch (Vietnam)

Habitat: Moist meadows, roadsides, vacan lots, cultivated

Distribution: Tropical Asia and Pacific

Botanical observation: This tuberous herb native to tropical America grows to 80cm tall. The stems are swollen at nodes and slightly quadrangular. The petiole is up to about 1.5cm long. The blade is obovate to elliptic 3–8×1.5–5cm, cuneate and tapering at base, somewhat laxly crenate, somewhat glossy, marked with 5 to 10 pairs of secondary nerves, and acute to rounded at apex. The cymes are axillary. The calyx comprises 5 lobes which are linear and up to 2cm long. The corolla is infundibuliform, 5-lobed, bluish to purplish, membranous, 2–5cm long, the lobes oblong and round at apex. Four stamens are present. The style is slender, 2cm long, and the stigma bifid. The capsules are dehiscent, linear, dark brown, surrounded by persistent calyx lobes, and contain numerous tiny and flat seeds.

Medicinal uses: Abscesses, ulcers (Thailand); bronchitis, gonorrhea, wounds (India)

Broad-spectrum antibacterial mid-polar extract: Ethyl acetate extract (5 mg/disc) inhibited the growth of *Staphylococcus aureus* (ATCC 25923), *Staphylococcus aureus* (ATCC 29213), *Pseudomonas aeruginosa* (ATCC 27853), *Klebsiella pneumoniae*, *Shigella sonnei*, and *Shigella flexneri* with the inhibition zone diameters of 20, 13, 14, 9, 9, and 7mm, respectively (Wiart et al., 2005).

Broad-spectrum halo developed by polar extract: Methanol extract of roots (500 µg/6mm disc) inhibited the growth of *Aspergillus niger*, *Aspergillus ochraceus*, *Aspergillus ustus*, *Candida albicans*, *Rhizopus orizae*, and *Trichophyton rubrum* with the inhibition zone diameters of 16, 15, 14, 18, 10, and 8mm, respectively (Kader et al., 2012).

Moderate antiviral (enveloped monopartite linear single-stranded (+) RNA) flavones: Sorbifolin and pedalitin inhibited the replication of the Hepatitis C virus with the EC_{50} values of 37.9 and 51.9µM, respectively, in Huh-7.5 cell line (Shimizu et al., 2017).

Viral enzyme inhibition by polar extract: Ethanol extract at the concentration of 200µg/mL inhibited Human immunodeficiency reverse transcriptase by 16% (Tan et al., 1991).

Viral enzyme inhibition by flavone Pedalitin inhibited Zika virus NS2B-NS3 protease with an IC_{50} value of 5µM (Lima et al., 2021).

Commentaries

i. The plants yield the flavones cirsimaritin, cirsimarin, cirsiliol 4′-glucoside, sorbifolin, and pedalitin (Lin et al., 2006).
ii. In India, *Ruellia suffruticosa* Roxb. (local name: *Chaulia*) is used for venereal diseases. In India, *Ruellia prostrata* Poir. (local names: *Kodi urinchi*, *Upudali*) is used for gonorrhea. In India and Bangladesh, *Ruellia suffruticosa* Roxb. (local name: *Chaulia* in Bangladesh and *Chaarapaatu* in India) is used for eye infections, gonorrhea, and syphilis.
iii. *Ruellia repens* L. is used in Malaysia (local name: *Akar keremak*) for sores, ulcers, and cuts as well as toothaches. In China, *Ruellia repens* L. (local name: *Nan cao*) is used for cough and wounds.
iv. In the Subfamily Acanthoideae, Tribe Ruelliaea, *Strobilanthes flaccidifolia* Nees is used for fever in China. Methanol extract of *Strobilanthes cusia* (Nees) Kuntze inhibited the

replication of the Human coronavirus NL63 with the IC_{50} value of 0.6 μg/mL (Tsai et al., 2020). From the plant the indoloquinazoline alkaloid tryptanthrin (LogD 2.4 at pH 7.4; molecular mass = 248.2 g/mol) and indigodole B inhibited of the replication of the Human Coronavirus NL63 with the IC_{50} values of 0.06 and 2 μg/mL and selectivity indexes of > 6600 and > 191, respectively (Tsai et al., 2020).

v. Close to the family Acanthaceae is the family Avicenniaceae Endlicher (1841) in which the mangrove tree *Avicennia officinalis* L. is given for ulcers in the Philippines. An extract of this plant inhibited reverse transcriptase, Hepatitis B virus surface antigen, and Hepatitis B virus DNA polymerase inhibition with the IC_{50} values of 403.9, 489.3, and 372 μg/mL, respectively (Beula et al., 2012). *Avicennia marina* Vicrh. inhibited the replication of the Encephalomyocarditis virus and Hepatitis B virus (Premnathan et al., 1992).

References

Beula, J.M., Gnanadesigan, M., Rajkumar, P.B., Ravikumar, S. and Anand, M., 2012. Antiviral, antioxidant and toxicological evaluation of mangrove plant from South East coast of India. *Asian Pacific Journal of Tropical Biomedicine*, *2*(1), pp. S352–S357.

Kader, M.A., Parvin, S., Chowduri, M.A.U. and Haque, M.E., 2012. Antibacterial, antifungal and insecticidal activities of *Ruellia tuberosa* (L.) root extract. *Journal of Bio-Science*, *20*, pp. 91–97.

Lima, C.S., Mottin, M., de Assis, L.R., Mesquita, N.C.D.M.R., de Paula Sousa, B.K., Coimbra, L.D., Bispo-dos-Santos, K., Zorn, K.M., Guido, R.V., Ekins, S. and Marques, R.E., 2021. Flavonoids from pterogyne nitens as Zika virus NS2B-NS3 protease inhibitors. *Bioorganic Chemistry*, *109*, p. 104719.

Lin, C.F., Huang, Y.L., Cheng, L.Y., Sheu, S. and Chen, C., 2006. Bioactive flavonoids from *Ruellia tuberosa*. *The Journal of Chinese Medicine*, *17*(3), pp. 103–109.

Premnathan, M., Chandra, K., Bajpai, S.K. and Kathiresan, K., 1992. A survey of some Indian marine plants for antiviral activity. *Botanica Marina 35*(4), pp. 321–324.

Shimizu, J.F., Lima, C.S., Pereira, C.M., Bittar, C., Batista, M.N., Nazaré, A.C., Polaquini, C.R., Zothner, C., Harris, M., Rahal, P. and Regasini, L.O., 2017. Flavonoids from *Pterogyne nitens* inhibit hepatitis C virus entry. *Scientific Reports*, *7*(1), pp. 1–9.

Tan, G.T., Pezzuto, J.M., Kinghorn, A.D. and Hughes, S.H., 1991. Evaluation of natural products as inhibitors of human immunodeficiency virus type 1 (HIV-1) reverse transcriptase. *Journal of Natural Products*, *54*(1), pp. 143–154.

Tsai, Y.C., Lee, C.L., Yen, H.R., Chang, Y.S., Lin, Y.P., Huang, S.H. and Lin, C.W., 2020. Antiviral action of tryptanthrin isolated from *Strobilanthes cusia* leaf against human coronavirus NL63. *Biomolecules*, *10*(3), p. 366.

Wiart, C., Akaho, E., Hannah, M., Yassim, M., Hamirnah, H., Au, T.S. and Sulaiman, M., 2005. Antimicrobial activity of *Ruellia tuberosa* L. *American Journal of Chinese Medicine*, *33*(4), pp. 683–685.

10.4.2 Family Bignoniaceae A.L de Jussieu (1789)

The family Bignoniaceae consists of about 120 genera and 750 species of trees, shrubs, or climbers. The leaves are simple or compound, opposite, alternate, or whorled, and exstipulate The inflorescences are cymes, panicles, or racemes. The flowers are zygomorphic and often short. The calyx is tubular and 2- to 5-lobed. The corolla is ephemeral, tubular, and 5-lobed. A disc is observable. Four stamens are present. The gynoecium is made of 2 carpels fused into a 1- or 4-locular ovary, each locule sheltering numerous ovules growing from axile or parietal placentas. The style is often filiform and the stigma bifid. The capsules are woody, dehiscent, and contain numerous winged or hairy seeds. Plants in this family produce antimicrobial naphthoquinones and flavonols.

10.4.2.1 *Oroxylum indicum* (L.) Vent.

Synonyms: *Bignonia indica* L., *Calosanthes indica* Blume

Common names: Indian trumpet, tree of Damocles; khonha, kanaidingi, sona; taita, thona (Bangladesh); mano, talelo (Nepal); bapana, nata, palakappayyani, shamba (India); pokok bekah (Malaysia), kyaung shar, yawng li (Myanmar); pikat (Thailand)

Habitat: Forest margins roadsides

Distribution: Tropical Asia

Botanical observation: This lugubrious tree grows up to about 10 m tall. The bole is slender, the bark light gray, and the wood yellow. The leaves are 2- or 3-pinnately compound, opposite, and exstipulate. The rachis is fleshy, up to 1.8 m long, and is clasping and swollen at base. The folioles are opposite, elliptic-ovate, 7–14×5–8 cm, asymmetrical or rounded at base, marked with about 5 pairs of secondary nerves, and acuminate at apex. The calyx is purple, tubular, fleshy, and subtruncate. The corolla is purplish to pink to reddish, tubular, vaguely bilabiate, fleshy, and ephemeral, the lobes are wavy. Five stamens are present. The style is slender. The capsules are woody, somewhat like hanging long-boats, dark brown to grayish green, somewhat glossy, about 1 m long, and contain numerous winged seeds. The wings are membranous and translucent.

Medicinal uses: Boils, cuts, ear infection, sores (Myanmar); bronchitis (Myanmar; India); burns, cough, leucorrhoea, pneumonia, tuberculosis (India); cholera (Bangladesh; Myanmar); diarrhea (India; Myanmar; Nepal); dysentery (Nepal; Myanmar); fever (India; Laos; Myanmar; Nepal); jaundice (Bangladesh; India; Nepal); tonsillitis (Bangladesh); wounds (India; Myanmar; Nepal)

Antibacterial halo (Gram-positive) developed by naphthoquinone: Lapachol (5 μg/5 mm disc) inhibited the growth of *Bacillus subtilis* and *Staphylococcus aureus* with inhibition zones similar to streptomycin (5 μg/5 mm disc) (Ali et al., 1998).

Antibacterial halo (Gram-negative) developed by flavonol: Chrysin (5 μg/5 mm disc) inhibited the growth of *Pseudomonas aeruginosa* with an inhibition zone similar to streptomycin (5 μg/5 mm disc) (Ali et al., 1998).

Moderate antibacterial (Gram-negative) flavonol: Baicalein inhibited the growth of *Salmonella typhimurium* with the MIC/MBC values of 64/>128 μg/mL (Wu et al., 2018).

Moderate antifungal (filamentous) mid-polar extract: Dichloromethane extract of root bark inhibited the growth of *Trichophyton mentagophytes*, *Microsporium gypsum*, and *Epidermophyton floccosum* with the MIC values of 400, 200, and 200 μg/mL, respectively (Ali et al., 1998).

Antifungal (filamentous) halo developed by naphthoquinone: Lapachol (5 μg/5 mm disc) inhibited the growth of *Trichophyton mentagophytes*, *Microsporium gypsum*, and *Epidermophyton floccosum* with the inhibition zone diameters of 10, 10, and 10 mm, respectively (Ali et al., 1998).

Anticandidal halo developed by flavonol: Chrysin (5 μg/5 mm disc) inhibited the growth of *Candida albicans* with an inhibition zone similar to streptomycin (5 μg/5 mm disc) (Ali et al., 1998).

Strong antiviral (enveloped linear segmented single-stranded (−) RNA) amphiphilic flavonols: Chrysin (also known as 5,7-dihydroxyflavone; LogD 1.8 at pH 7.4; molecular mass = 254.2 g/mol) at the concentration of 10 μM abrogated the replication of the Influenza virus A/PR/8 virus in A549 cells by a mechanism involving, at least in part, early stage of viral replication (Kim et al., 2021). Oroxylin (also known as 6-methoxybaicalein; LogD 1.7 at pH 7.4; molecular mass = 284.2 g/mol) inhibited the replication of the Influenza virus A/FM/1/47 (H1N1), A/Beijing/32/92 (H3N2), and oseltamivir-resistant A/FM/1/47-H275Y (H1N1-H275Y) in Madin-Darby canine kidney cells with EC_{50} values of 44.6, 36.1, and 109.4 μM, respectively (Jin et al., 2018).

Strong antiviral (enveloped linear segmented single-stranded (−) RNA) flavonol glycoside: Baicalin (also known as baicalein 7-*O*-*β*-D-glucuronide; LogD −4.5 at pH 7.4; molecular mass = 444.3 g/mol) inhibited the replication of the Influenza virus A/FM1/1/47 (H1N1) virus with the IC_{50} value of 0.7 μg/mL and selectivity index of 22.9 in Madin-Darby canine kidney cells (Chen et al., 2011).

Strong antiviral (enveloped monopartite linear single stranded (+) RNA) amphiphilic flavonol Baicalein (also known as 5,6,7-trihydroxyflavone; LogD 1.6 at pH 7.4; molecular mass = 270.2 g/mol) inhibited the replication of the Dengue virus serotype-2 (New Guinea C strain) with the IC_{50} value of 5.3 μg/mL and a selectivity index of 21.3 at post virion entry in Vero cells and was virucidal with IC_{50} = 1.5 μg/mL (Zandi et al., 2012). Baicalein inhibited the replication of the Japanese encephalitis virus in Vero cells with the IC_{50} value of 14.2 µg/mL and was virucidal with the IC_{50} value of 3.4 µg/mL (Johari et al., 2012). Baicalein inhibited the replication of the Severe acute respiratory syndrome-associated coronavirus 2 with the IC_{50} value of 10 µM (Huang et al., 2020). Baicalein inhibited the replication of the Chikungunya virus (East/Central/South African genotype) in BHK-2 cells with the IC_{50} value of 1.8 µg/mL and a selectivity index of 188.4 at the intracellular stage (Lani et al., 2016).

Strong antiviral (enveloped monopartite linear double-stranded DNA) amphiphilic flavonol: Baicalein inhibited the replication of the Human cytomegalovirus with the IC_{50} value of 1.2 µM via a mechanism involving, at least in part, by inhibiting virus entry in HEL 299 cells (Evers et al., 2005).

Strong antiviral (enveloped monopartite linear double-stranded DNA) flavonol glycoside: Baicalin inhibited the replication of the Human cytomegalovirus with the IC_{50} value of 15 µM in HEL 299 cells (Evers et al., 2005).

Strong antiviral (non-enveloped monopartite linear single-stranded (+) RNA) amphiphilic flavonols: Baicalein inhibited the replication of the Enterovirus 71 (strain 41) in Vero cells with the IC_{50} value of 2.2 µg/mL and a selectivity index of 20 (Lalani et al., 2020). Oroxylin inhibited the replication of the Enterovirus 71 with the IC_{50} value of 14.1 and a selectivity index above 3.5 in Vero cells (Choi et al., 2016). Oroxylin inhibited the replication of the Coxsackie virus B3 (ATCC VR-30) with the IC_{50} value of 3.1 µg/mL in Vero cells and a selectivity index of 16 (Kwin et al., 2016).

Strong antiviral (enveloped monopartite linear single-stranded (−) RNA) amphiphilic flavonol: Baicalein inhibited the replication of the Vesicular stomatitis virus with the IC_{50} value of 15 µM (Huang et al., 2020).

Moderate antiviral (enveloped monopartite linear single-stranded (+) RNA) polar extract: Aqueous extract of leaves at the concentration of 150 µg/mL reduced by 10% the viral titer of Chikungunya virus in Vero cells (Mohamat et al., 2018).

Antiviral (enveloped monopartite linear double-stranded DNA) flavonol: Chrysin at 20.3 µg/mL inhibited the replication of the Herpes simplex virus type-1 in Vero cells (Amoros et al., 1992).

In vivo antiviral (enveloped linear segmented single-stranded (−) RNA) amphiphilic flavonols: Baicalein given to 10 mice at the dose of 400 mg/Kg/day for 5 days orally and twice daily allowed the survival of 2 mice against the Influenza A/FM1/1/47 (H1N1) and increased the antiviral potencies of ribavirin at 50 mg/Kg/day (Chen et al., 2011). Oroxylin given to mice at the dose of 100 mg/Kg/d for 5 days once a day orally allowed the survival of mice against the Influenza A/FM/1/47 (H1N1) (Jin et al., 2011).

Viral entry inhibition by flavonol: Oroxylin at the concentration of 20 µM inhibited the entrance of the Severe acute respiratory syndrome-associated coronavirus into HEK293T cells (Gao et al., 2021).

Commentaries

i. The plants yield flavonols (Ali et al., 1998; Yuan et al., 2008) which in combination with lapachol synergistically account for the medicinal uses.
ii. Flavonols are always active against at least one type of virus *in vitro* and often antibacterial and/or antifungal and are probably phytoanticipins and/or phytoalexins. Their antimicrobial mode of action *in vitro* depends on their concentrations. Flavonols are metabolized by intestinal bacteria and have low oral bioavailability. The antiviral activities of flavonols observed *in vivo* given orally are exerted by various types of intestinal and hepatic flavonols catabolites and metabolites. Amoros et al. (1992) observed that flavonoids could act synergistically against the Herpes simplex type-1 and made the suggestion that flavonoids with different modes of actions against this virus were synergistic and this can be extended to all natural products against all microorganisms.

iii. Oroxylin binds to estrogen receptors (Wang et al., 2013) and as such is probably able to inhibit bacterial and or fungal efflux pumps. It inhibits efflux pumps in mammalian cells (Zhu et al., 2013).
iv. The plant is sacred in India and Nepal and grown in Nepal around houses for protection.
v. Plants in the family Bignoniaceae are mainly used for the treatment of microbial infections in India. For instance, the mangrove tree *Dolichandrone spathacea* (L. f.) Seem. (local name: *Dongi*) is used for bronchitis, thrush in Indonesia, and for eye infection in Papua New Guinea (local name: *Touee*). From the leaves of this plant, the hydroxycinnamic acid glycoside derivatives decaffeoyl acteoside and verbascoside inhibited the growth of *Enterococcus faecalis* (ATCC 1034) and *Shigella sonnei* with MIC value of 31.2 μg/mL (Nguyen et al., 2019). Verbascoside (also known as acteoside; LogD 0.7 at pH 7.4; molecular mass=624.5 g/mol) inhibited the Respiratory syncytial virus (A2 strain) with the EC_{50} value of 0.8 μg/mL and the selectivity index of 85.4 (Chen et al., 1998). Note that in the Lamiales, tannins are replaced by hydroxycinnamic glycoside derivatives and phenylethanoid glycosides as phytoanticipins. Each Clade has its own arsenal of chemical defenses. Plants do not have immune system and respond to infections by producing secondary metabolites working synergistically to inhibit the growth of a broad array of phytopathogenic microorganisms.
vi. In India, *Dolichandrone falcata* Seem. (local name: *Kattu varsana*) is used for bronchitis. *Pajanelia longifolia* (Willd.) Sch. is used for boils (local name: *Jhingam*) and is a remedy for fever in Indonesia. In India, *Stereospermum suaveolens* DC (local name: *Parul*) is used for bronchitis and gonorrhea, and the extract of roots of this plant inhibited the replication of the Newcastle diseases virus (Babbar et al., 1982). *Stereopermum neuranthum* Kurz is used for ulcers in Myanmar and India (local name: *Zihaw*). In Myanmar, *Stereospermum chelonoides* (L.f.) DC. is used for fever. In India, it is used for cholera and fevers, and to cure dysentery and cholera (local name: *Mesusung*) and for tuberculosis (local name: *Hamarang gaas*) in Bangladesh. *Stereospermum colais* (Buch.-Ham. ex Dillwyn) Mabb. is used in Myanmar (local name: *Hingut-pho*) for cough, and in India (local names: *Pathiri*, *Ser phang*) it is a remedy for cough, fever, and enters in a mixture of several plants for COVID-19. *Stereospermum fimbriatum* DC. is given as post-partum protective remedy in Malaysia.
vii. In India, *Spathodea campanulata* P. Beauv. (local name: *Kottakarthai*) is used for skin diseases. Methanol extract of flowers of *Spathodea campanulata* P. Beauv. inhibited the growth of *Enterococcus faecalis* (clinical), *Staphylococcus aureus* (clinical strain), *Staphylococcus saprophyticus* (clinical strain), and *Staphylococcus epidermidis* (clinical strain) with the MIC/MBC values of 39.1/78.1, 312.5/625, 312.5/312.5, and 625/1250 μg/mL, respectively (Mbosso Teinkela et al., 2016).
viii. In India, the magnificent *Amphicome emodi* Royle ex Lindl. (local name: *Kaur*) is used for fever; *Heterophragma roxburghii* (Spreng.) DC. (local name: *Waarasa*) is used for skin diseases, sores; *Jacaranda acutifolia* Bonpl. (local name: *Nili-gulmohar.*) is used for blenorrhagia, syphilis, ulcers; *Kigelia pinnata* (Jacq.) DC. (local name: *Jhaar Phaanuus.*) is used for dysentery and venereal diseases. *Kigelia pinnata* (Jacq.) DC. produces the furanonaphthoquinone 2-(1- hydroxyethyl)naphtho[2,3-b]furan-4,9-dione (Weiss et al., 2000) which at the concentration of 3 μg/mL inhibited the replication of the Japanese encephalitis virus (JaGAr-01 strain) by about 80% in Vero cells (Takegami et al., 1998).
ix. The beautiful tree *Millingtonia hortensis* L. f. (local name: *Aakaasha neem*) is a remedy for fever and sinusitis, and in Myanmar (local name: *Mai-long-ka-hkami*) it is used for eye infection. It yields lapachol (Epifano et al., 2014). Ethanol extract of leaves of *Millingtonia hortensis* L. f. inhibited the Human immunodeficiency type-1 integrase with the IC_{50} value of 77.7 μg/mL (Suedee et al., 2014). Ethanol extract at the concentration of 200 μg/mL inhibited the Human immunodeficiency type-1 reverse transcriptase by 21% (Tan et al.,

1991). It yields lapachol (Epifano et al., 2014). *Radermachera xylocarpa* K. Schum. (local name: *Pathiri.*) is used for skin infections on probable account of the appearance of the fruits which look pimply and produces lapachol (Epifano et al., 2014).

x. The ornamental shrub *Tecoma stans* ((L.) Juss. ex Kunth (local name: *Sein-takyu*) is used in Myanmar for syphilis and produces the hydroxycinnamic acid glycoside derivative crenatoside which inhibited the replication of the Zika virus with the EC_{50} value of 34.7 µM and a selectivity index of 4.2 (Reis et al., 2020). Note that hydroxycinnamic acid and phenylethanoid glycosides glycoside derivative per se are of no systemic therapeutic use because of unfavorable oral bioavailability. Dichloromethane extract of leaves inhibited the growth of *Aspergillus parasiticus*, *Aspergillus niger*, *Phytophtora nicotiana*, *Penicillium janthinellum*, *Trichoderma harzianum*, *Rhizoctonia solani*, *Fusarium oxysporium*, *Penicillium expansum*, and *Pythium ultimum* with the MIC values of 160, 120, 120, 470, 20, 20, 20, 20, and 60 µg/mL, respectively (Meela et al., 2017).

xi. In India, the magnificent desert tree *Tecomella undulata* (Sm.) Seem. (local names: *Daadimchhada, Rugtrora*) is a remedy for leucorrhoea and syphilis and for eye infection and it produces lapachol (Mohsenzadeh et al., 2010). Note that plants used for syphilis, tuberculosis, or leprosy often produce quinones (benzoquinones, naphthoquinones, anthraquinones).

xii. *Catalpa bungei* C.A. Mey. is used for boils in China.

xiii. The strange ornamental tree *Crescentia cujete* L. is used for wounds in Indonesia. From this plant, 5-hydroxy-2-(1hydroxyethyl)naphtho[2,3-β]furan-4,9-dione inhibited the growth of *Saccharomyces cerevisiae* (RS322 YK) with inhibition zone diameter of 12 mm (100 µL of a solution at 3 µg/mL/well) (Heltzel et al., 1993).

xiv. *Incarvillea sinensis* Lam. is given for ulcers in China. This beautiful tree yields the monoterpene indole alkaloid incarvillateine which can act on adenosine receptors (Wang et al., 2015) and may inhibit bacterial and/or fungal efflux pumps.

References

Ali, R.M., Houghton, P.J., Raman, A. and Hoult, J.R.S., 1998. Antimicrobial and antiinflammatory activities of extracts and constituents of *Oroxylum indicum* (L.) Vent. *Phytomedicine*, *5*(5), pp. 375–381.

Amoros, M., Simõs, C.M.O., Girre, L., Sauvager, F. and Cormier, M., 1992. Synergistic effect of flavones and flavonols against herpes simplex virus type 1 in cell culture. Comparison with the antiviral activity of propolis. *Journal of Natural Products*, *55*(12), pp. 1732–1740.

Babbar, O.P., Joshi, M.N. and Madan, A.R., 1982. Evaluation of plants for antiviral activity. *Indian Journal of Medical Research*, *76*(S), pp. 54–65.

Chen, J.L., Blanc, P., Stoddart, C.A., Bogan, M., Rozhon, E.J., Parkinson, N., Ye, Z., Cooper, R., Balick, M., Nanakorn, W. and Kernan, M.R., 1998. New iridoids from the medicinal plant *Barleria prionitis* with potent activity against respiratory syncytial virus. *Journal of Natural Products*, *61*(10), pp. 1295–1297.

Chen, L., Dou, J., Su, Z., Zhou, H., Wang, H., Zhou, W., Guo, Q. and Zhou, C., 2011. Synergistic activity of baicalein with ribavirin against influenza A (H1N1) virus infections in cell culture and in mice. *Antiviral Research*, *91*(3), pp. 314–320.

Choi, H.J., Song, H.H., Lee, J.S., Ko, H.J. and Song, J.H., 2016. Inhibitory effects of norwogonin, oroxylin A, and mosloflavone on enterovirus 71. *Biomolecules & Therapeutics*, *24*(5), p. 552.

Epifano, F., Genovese, S., Fiorito, S., Mathieu, V. and Kiss, R., 2014. Lapachol and its congeners as anticancer agents: A review. *Phytochemistry Reviews*, *13*(1), pp. 37–49.

Evers, D.L., Chao, C.F., Wang, X., Zhang, Z., Huong, S.M. and Huang, E.S., 2005. Human cytomegalovirus-inhibitory flavonoids: Studies on antiviral activity and mechanism of action. *Antiviral Research*, *68*(3), pp. 124–134.

Heltzel, C.E., Gunatilaka, A.L., Glass, T.E., Kingston, D.G., Hoffmann, G. and Johnson, R.K., 1993. Bioactive furanonaphthoquinones from *Crescentia cujete*. *Journal of Natural Products*, *56*(9), pp. 1500–1505.

Huang, S., Liu, Y.E., Zhang, Y., Zhang, R., Zhu, C., Fan, L., Pei, G., Zhang, B. and Shi, Y., 2020. Baicalein inhibits SARS-CoV-2/VSV replication with interfering mitochondrial oxidative phosphorylation in a mPTP dependent manner. *Signal Transduction and Targeted Therapy*, *5*(1), pp. 1–3.

Jin, J., Chen, S., Wang, D., Chen, Y., Wang, Y., Guo, M., Zhou, C. and Dou, J., 2018. Oroxylin A suppresses influenza A virus replication correlating with neuraminidase inhibition and induction of IFNs. *Biomedicine & Pharmacotherapy*, *97*, pp. 385–394.

Johari, J., Kianmehr, A., Mustafa, M.R., Abubakar, S. and Zandi, K., 2012. Antiviral activity of baicalein and quercetin against the Japanese encephalitis virus. *International Journal of Molecular Sciences*, *13*(12), pp. 16785–16795.

Kim, S.R., Jeong, M.S., Mun, S.H., Cho, J., Seo, M.D., Kim, H., Lee, J., Song, J.H. and Ko, H.J., 2021. Antiviral activity of chrysin against influenza virus replication via inhibition of autophagy. *Viruses*, *13*(7), p. 1350.

Kwon, B.E., Song, J.H., Song, H.H., Kang, J.W., Hwang, S.N., Rhee, K.J., Shim, A., Hong, E.H., Kim, Y.J., Jeon, S.M. and Chang, S.Y., 2016. Antiviral activity of oroxylin A against coxsackievirus B3 alleviates virus-induced acute pancreatic damage in mice. *PLoS One*, *11*(5), p. e0155784.

Lalani, S.S., Anasir, M.I. and Poh, C.L., 2020. Antiviral activity of silymarin in comparison with baicalein against EV-A71. *BMC Complementary Medicine and Therapies*, *20*(1), pp. 1–12.

Lani, R., Hassandarvish, P., Shu, M.H., Phoon, W.H., Chu, J.J.H., Higgs, S., Vanlandingham, D., Bakar, S.A. and Zandi, K., 2016. Antiviral activity of selected flavonoids against Chikungunya virus. *Antiviral Research*, *133*, pp. 50–61.

Mbosso Teinkela, J.E., Assob Nguedia, J.C., Meyer, F., Vouffo Donfack, E., Lenta Ndjakou, B., Ngouela, S., Tsamo, E., Adiogo, D., Guy Blaise Azebaze, A. and Wintjens, R., 2016. In vitro antimicrobial and antiproliferative activities of plant extracts from *Spathodea campanulata*, *Ficus bubu*, and *Carica papaya*. *Pharmaceutical Biology*, *54*(6), pp. 1086–1095.

Meelah, M.M., Mdee, L.K. and Eloff, J.N., 2017. Tecoma stans (Bignoniaceae), leaf extracts, fractions and isolated compound have promising activity against fungal phytopathogens. *Suid-Afrikaanse Tydskrif vir Natuurwetenskap en Tegnologie*, *36*(1), pp. 1–7.

Mohamat, S.A., Shueb, R.H. and Mat, N.F.C. 2018. Anti-viral activities of *Oroxylum indicum* extracts on chikungunya virus infection. *Indian Journal of Microbiology* 58(1), pp. 68–75.

Mohsenzadeh, S.A.S.A.N. and Amiri, A.A., 2010. Lapachol extraction from inner bark stem of *Tecomella undulata* (Roxb.) Seem. *Iranian Journal of Medicinal and Aromatic Plants*, *26*(1), pp. 114–120.

Nguyen, P.D., Abedini, A., Gangloff, S.C. and Lavaud, C., 2018. Antimicrobial constituents from leaves of *Dolichandrone spathacea* and their relevance to traditional use. *Planta Medica International Open*, *5*(01), pp. e14–e23.

Reis, A.C.C., Silva, B.M., de Moura, H.M.M., Pereira, G.R. and Brandão, G.C., 2020. Anti-Zika virus activity and chemical characterization by ultra-high performance liquid chromatography (UPLC-DAD-UV-MS) of ethanol extracts in *Tecoma species*. *BMC Complementary Medicine and Therapies*, *20*(1), pp. 1–15.

Sithisarn, P., Nantateerapong, P., Rojsanga, P. and Sithisarn, P., 2016. Screening for antibacterial and antioxidant activities and phytochemical analysis of *Oroxylum indicum* fruit extracts. *Molecules*, *21*(4), p. 446.

Suedee, A., Tewtrakul, S. and Panichayupakaranant, P., 2014. Anti-HIV-1 integrase activity of *Mimusops elengi* leaf extracts. *Pharmaceutical Biology*, *52*(1), pp. 58–61.

Takegami, T., Simamura, E., Hirai, K.I. and Koyama, J., 1998. Inhibitory effect of furanonaphthoquinone derivatives on the replication of Japanese encephalitis virus. *Antiviral Research*, *37*(1), pp. 37–45.

Tan, G.T., Pezzuto, J.M., Kinghorn, A.D. and Hughes, S.H., 1991. Evaluation of natural products as inhibitors of human immunodeficiency virus type 1 (HIV-1) reverse transcriptase. *Journal of Natural Products*, *54*(1), pp. 143–154.

Wang, H., Guo, Y., Zhao, X., Li, H., Fan, G., Mao, H., Miao, L. and Gao, X., 2013. An estrogen receptor dependent mechanism of Oroxylin A in the repression of inflammatory response. *PLoS One*, *8*(7), p. e69555.

Wang, M.L., Yu, G., Yi, S.P., Zhang, F.Y., Wang, Z.T., Huang, B., Su, R.B., Jia, Y.X. and Gong, Z.H., 2015. Antinociceptive effects of incarvillateine, a monoterpene alkaloid from Incarvillea sinensis and possible involvement of the adenosine system. *Scientific Reports*, *5*(1), pp. 1–11.

Wu, S.C., Chu, X.L., Su, J.Q., Cui, Z.Q., Zhang, L.Y., Yu, Z.J., Wu, Z.M., Cai, M.L., Li, H.X. and Zhang, Z.J., 2018. Baicalin protects mice against *Salmonella typhimurium* infection via the modulation of both bacterial virulence and host response. *Phytomedicine*, *48*, pp. 21–31.

Yuan, Y., Hou, W., Tang, M., Luo, H., Chen, L.J., Guan, Y.H. and Sutherland, I.A., 2008. Separation of flavonoids from the leaves of *Oroxylum indicum* by HSCCC. *Chromatographia*, *68*(11), pp. 885–892.

Zandi, K., Teoh, B.T., Sam, S.S., Wong, P.F., Mustafa, M.R. and AbuBakar, S., 2012. Novel antiviral activity of baicalein against dengue virus. *BMC Complementary and Alternative Medicine*, *12*(1), pp. 1–9.

Zhu, L., Zhao, L., Wang, H., Wang, Y., Pan, D., Yao, J., Li, Z., Wu, G. and Guo, Q., 2013. Oroxylin A reverses P-glycoprotein-mediated multidrug resistance of MCF7/ADR cells by G2/M arrest. *Toxicology Letters*, *219*(2), pp. 107–115.

10.4.3 Family Gesneriaceae Richard et A.L. de Jussieu ex A.P. de Candolle (1816)

The family Gesneriaceae consists of about 150 genera and 3500 species of tropical herbs, climbers, or shrubs or seldom trees which are often somewhat fleshy, soft, and water and shade loving. The leaves are simple, opposite, or whorled or basal in rosette and exstipulate. The blade is often serrate or incised or lobed. The inflorescences are axillary or terminal cymes or racemes. The calyx is tubular and 5-lobed. The corolla is tubular, membranous, ephemeral, and bilobed. A disc is present. Four stamens are present. A ring-like to cupular disc is present. The ovary is uni- or bilocular with ovules growing from parietal or axile placentas. The fruits are dehiscent capsules or berries.

10.4.3.1 *Aeschynanthus longicaulis* Wall. ex R. Br.

Synonym: *Aeschynanthus marmoratus* T. Moore

Common name: Chang jing mang mao ju tai (China)

Habitat: Forests, cultivated

Distribution: Vietnam, Thailand, Malaysia, China

Botanical observation: This epiphytic herb develops pendulous and terete stems which reach about 50 cm long. The plant has somewhat an aura of dryness. The leaves are simple, opposite, and exstipulate. The petiole is none to about 1.5 cm long and curved. The blade is elliptic to lanceolate, 1.5–12.5 × 0.5–3.3 cm, fleshy, somewhat minutely crenate at base, acuminate to caudate at apex, obscurely marked with 5 to 6 pairs of secondary nerves, and variegated green above and variegated pink below. The cymes are terminal and few flowered. The calyx is green to orange-brownish or purplish and develops 5 narrowly triangular lobes which reach about 2 cm in length. The corolla is yellow or green, up to 3 cm long, curved, and develops 5 round lobes. The 4 stamens protrude out of the corolla. The style is slender and protrudes out of the corolla. The capsules are linear, 15–30 cm long, and open to release an innumerable number of comose seeds.

Medicinal uses: Boils (Malaysia)

Commentaries

i. The antimicrobial properties of the plant have apparently not been examined nor its active principles. Members of the genus *Aeschynanthus* yield hydroxycinnamic acid derivatives and phenylethanoid glycosides (Tian et al., 2021).
ii. In India, *Aeschynanthus sikkimensis* Stapf. (synonym: *Aeschynanthus maculatus* var. *sikkimensis* C.B. Clarke; local name: *Bawltehlantai*) is used for fever, and in Nepal (local name: *Khagija*) it is used for fever and tonsillitis. In Myanmar, *Aeschynanthus parasiticus* (Roxb.) Wall. (synonym: *Incarvillea parasitica* Roxb.; local name: *Long shei*) is used for boils, cut, ear infection, leprosy, toothache, tonsillitis, tuberculosis, and wounds.

Reference

Tian, X.Y., Li, M.X., Lin, T., Qiu, Y., Zhu, Y.T., Li, X.L., Tao, W.D., Wang, P., Ren, X.X. and Chen, L.P., 2021. A review on the structure and pharmacological activity of phenylethanoid glycosides. *European Journal of Medicinal Chemistry*, *209*, p. 112563.

10.4.3.2 *Cyrtandra cupulata* Ridl.

Common names: Bebangun, meroyan kabut (Malaysia); ta po leh mai (Thailand)

Habitat: Forests along rocky streams

Distribution: Malaysia, Thailand

Botanical observation: This herb grows to about 1 m tall. The stem is woody and at apex hairy. The leaves are simple, opposite, and exstipulate. The petiole is about 1–4 cm long and stout or none. The blade is oblanceolate to elliptic, dark green, somewhat glossy, 5–13 × 18–35 cm, hairy below, oblanceolate, serrate, membranous, tapering at base, acuminate at apex, and marked with 10–15 pairs of secondary nerves. The cymes are axillary and surrounded in a greenish-white and cupular bracts reaching a length of 6 cm and somewhat hairy. The calyx is bilobed, white, and about 2 cm long. The corolla is whitish, pinkish to yellow at throat, membranous, ephemeral, up to about 4 cm long, develops 2 round lobes at apex, and 3 round lobes at base. Two stamens adnate to the corolla and with white and 1 cm long filaments are present. The ovary is oblong and develops a slender, 2.5 cm long, and hairy style. The capsules are bullet-shaped, fleshy, and about 1.5 cm long.

Medicinal uses: Fever, post-partum protective remedy (Malaysia)

Commentaries

i. The plant has apparently not been studied for antimicrobial properties. It will disappear if deforestation is not stopped in Malaysia.
ii. In Malaysia, *Cyrtandra wallichii* (C.B Clarke) B.L. Burtt (local name: *Lemak batu*) is used as post-partum remedy. Not that that traditional Malay system of medicine, compared to other Asian systems, uses several plants to prevent infection after childbirth.
iii. Plants in the genus *Cyrtandra* J.R. Forst. & G. Forst. are common in the last patched rainforest of Thailand and Malaysia and have apparently not been phytochemically studied.

10.4.3.3 *Cyrtandromoea grandis* Ridl.

Common names: Bayam beraleh (Malaysia); cha sam kaeo (Thailand)

Habitat: Forests

Distribution: Thailand, Malaysia, Indonesia

Botanical observation: This herb grows up to 3 m tall. The leaves are simple, opposite, and exstipulate. The blade is oblong, 5–12 × 20–30 cm. The cymes are many flowered. The corolla is pure white with a yellow throat, with 2-lobes above and 3 lobes below. The capsules are about 1 cm across.

Medicinal use: Fever (Malaysia)

Broad-spectrum antibacterial halo developed by polar extract: Ethanol extract of roots (6 mm disc impregnated with a solution at 100 mg/mL) inhibited the growth of *Bacillus megaterium*, *Escherichia coli*, and *Pseudomonas aeruginosa* with the inhibition zone diameters of 8, 8, and 7 mm, respectively (Jasril et al., 1999).

Antifungal (yeasts) halo developed by polar extract: Ethanol extract of roots (6 mm disc impregnated with a solution at 100 mg/mL) inhibited the growth of *Saccharomyces cerevisiae* with the inhibition zone diameter of 8 mm (Jasril et al., 1999).

Commentaries

i. The plant will disappear if deforestation continues in Thailand, Malaysia, and Indonesia. Plants in the genus *Cyrtandromoea* Zoll. have not been apparently studied for antimicrobial properties and constituents.
ii. *Cyrtandromoea subsessilis* (Miq.) B.L. is used in Malaysia (local name: *Pokok sampu menahun*) for fever.

Reference

Jasril, J., Mohamed, S.M., Mackeen, M.M., Lajis, N.H., Rahman, A.A. and Ali, A.M., 1999. Antimicrobial and cytotoxic activities of some Malaysian flowering plants. *Natural Product Sciences*, 5(4), pp. 172–176.

10.4.3.4 *Rhynchotechum ellipticum* (Wall. ex D. Dietr.) A. DC.

Synonyms: *Chiliandra obovata* Griff.; *Corysanthera elliptica* Wall. ex D. Dietr.; *Rhynchotechum latifolium* Hook. f. & Thomson ex C.B. Clarke; *Rhynchotechum obovatum* (Griff.) B.L. Burtt

Common names: Gankarek, manipuri, yenbum (India); xian zhu ju tai (China)

Habitat: Forests along streams

Distribution: India, Nepal, Bhutan, Bangladesh, Myanmar, Laos, Vietnam Thailand, Malaysia

Botanical observation: This herb grows to 2 m long. The stem is hairy. The leaves are simple, opposite, and exstipulate. The petiole is 0.8–5 cm long and hairy. The blade is oblanceolate, dark green, glossy, 9.5–32 × 3–10 cm, hairy below, edible (India), attenuate at base, serrate, acute to acuminate at apex, and marked with 10–20 pairs of conspicuous secondary nerves. The cymes are axillary, hairy, and many flowered. The calyx is tubular, 5-lobed, pink, hairy, and minute. The corolla is white or pink, membranous, tubular, 5–6 mm long, and develops 5 rounded lobes. Four stamens are present. The ovary is minute and develops a slender, white, up to about 1 cm long and fleshy style. The berries are supernaturally pure white and about 5 mm long.

Medicinal uses: Wounds (India)

Commentaries

i. The plant has apparently not been examined for antimicrobial properties.
ii. In the Subfamily Didymocarpoideae, a number of plants are used for the treatment of microbial infections. *Didymocarpus subalternans* Wall. ex R. Br. as well as *Didymocarpus primulifolius* D. Don (synonym: *Henckelia primulifolia* (D. Don) Spreng.; local name: *Paharo ko kan*) are used for flu in Nepal. Extract of Didymocarpus aromaticus Wall. ex D. Don the replication of the Sindbis virus (Tuladhar, et al., 2020). The lignans didymocarpuslignan A, didymocarpuslignan B, and didymocarpuslignan C from *Didymocarpus hedyotideus* DC inhibited the replication of the Hepatitis B virus (Tang et al., 2014).
iii. In Malaysia, *Didymocarpus frutescens* Jack (synonym: *Didissandra frutescens* (Jack) C.B. Clarke; local name: *Tarom hutan*) is used as post-partum protective remedy, *Didimocarpus platypus* C.B. Clarke (synonym: *Henckelia platypus* (C.B. Clarke) A. Weber; local name: *Julang rimba*) affords a remedy for cough, and *Didymocarpus crinita* (Jack) Sprengel (synonym: *Henckelia crinita* (Jack) Sprengel; local name: *Pokok ekor kuncing*) is used as a post-partum protective remedy and in Thailand it is used for wounds.
iv. *Didymocarpus griffithii* Wight (synonym: *Loxostigma griffithii* (Wight) C.B. Clarke) is a remedy for venereal diseases in China.
v. Medicinal Gesneriaceae are mainly vulnerable primary rainforest plants which will be soon extinct if deforestation for palm oil and logging are not stopped. Those plants have not been apparently much studied for their antimicrobial constituents. The COVID-19 pandemic and other forthcoming waves of pandemic viruses, bacteria, or fungi require a deep examination of forest plants for the development of antimicrobial drugs or phytomedications and the burning and logging of the rainforest in Southeast Asia (which is the largest and richest source of drugable molecules) must be stopped immediately. It is perhaps time to implement "the universal rights of trees."

References

Tang, W.J., Chen, X. and Yang, L.H., 2014. New lignans from the roots of didymocarpus hedyotideus and their antiviral effect on hepatitis B virus. *Latin American Journal of Pharmacy, 33*, pp. 948–53.

Tuladhar, A., Kunwar, R.M. and Bussmann, R.W., 2020. Didymocarpus aromaticus Wall. ex D. Don G esneriaceae. In: Kunwar, R.M., Sher, H., and Bussmann, R.W. (Eds). Ethnobotany of the Himalayas. Springer International Publishing: New York, pp. 1–7.

10.4.4 Family Lamiaceae Martynov (1820)

The family Lamiaceae consists of about 220 genera and 3500 species of herbs or shrubs which are often aromatic. The stems are 4-angled and not uncommonly hairy (the hairs often packed with essential oils). The leaves are simple, opposite, and exstipulate. The inflorescences are verticillasters. The flowers are zygomorphic. The calyx is 5-lobed. The corolla is bilabiate, the upper lip 2-lobed and the lower lip 3-lobed. The androecium comprises 2 or 4 stamens. The ovary is 2 loculed, quadrangular, each locule contains 2 ovules. The fruits are 4 nutlets packed in persistent calyx. Plants in this family synthetize a fascinating array of antimicrobial monoterpenes, diterpenes, and triterpenes.

10.4.4.1 *Ajuga bracteosa* Wall.

Common names: jingucao (China); neel kanthi (India); rati booti, soi beetaie (Pakistan)

Habitat: Mountain meadows

Distribution: Afghanistan, Pakistan, India, Nepal, Bangladesh, Myanmar, Cambodia, Laos, Vietnam, China, Indonesia, the Philippines, Japan

Botanical observation: This beautiful herb grows up to about 30 cm tall from a rhizome. The stems are hairy and quadrangular. The leaves are simple, opposite, and exstipulate. The petiole is none or up to 1.5 cm long and hairy. The blade is ovate, oblanceolate, or spathulate, 4–10 × 2–3 cm, hairy, cuneate-decurrent at base, somewhat shaped like the blade of an oak tree, wavy and lobed, obtuse to acute at apex, and with about 5 pairs of secondary nerves. The inflorescences are verticillaster. The calyx is tubular, 6 mm long, and produces 5 triangular lobes. The corolla is whitish blue to purple, tubular, 7–9 mm, and bilabiate. The ovary is 4-lobed and develops a slender style and bifid stigma. Four stamens protrude out of the corolla. The nutlets are minute, rugose, and enclosed in persistent calyx.

Medicinal uses: Cough, diarrhea (China); fever (India); ulcers (India); wounds (Pakistan; India; Taiwan); diarrhea (Pakistan)

Broad-spectrum antibacterial halo developed by diterpene: 14, 15-Dihydroajugapitin isolated from the leaves inhibited the growth of *Staphylococcus aureus*, *Pseudomonas aeruginosa*, *Klebsiella pneumoniae*, *Escherichia coli*, *Bacillus subtilis*, and *Salmonella typhi* (Ganaie et al., 2017).

Broad-spectrum antibacterial halo developed by diterpene: 8-*O*-Acetylharpagide isolated from the leaves inhibited the growth of *Staphylococcus aureus*, *Pseudomonas aeruginosa*, *Klebsiella pneumoniae*, *Escherichia coli*, *Bacillus subtilis*, and *Salmonella typhi* (Ganaie et al., 2017).

Weak broad-spectrum antibacterial polar extract: Methanol extract inhibited the growth of *Staphylococcus aureus*, *Pseudomonas aeruginosa*, *Klebsiella pneumoniae*, *Escherichia coli*, *Bacillus subtilis*, and *Salmonella typhi* with the MIC values of 3000, 3000, 1500, 3000, 1500, and 3000 µg/mL, respectively (Ganaie et al., 2017).

Strong antiviral (enveloped monopartite linear single-stranded (+) RNA) nonpolar extract: Hexane extract inhibited the replication of the Dengue virus serotype-2 (New Guinea C strain) in C6/36 cells with the IC_{50} value of 0.3 µg/mL and a selectivity index of 852 (Batool et al., 2018).

Moderate antiviral (enveloped monopartite linear single-stranded (+) RNA) polar extract: Ethanol extract at the concentration of 200 µg/mL halved the replication of the Hepatitis C virus in HepG2 cells (Yousaf et al., 2018).

Commentaries

i. *Ajuga chamaepitys* (L.) Schreb. is used in Turkey for fever (local name: *Mayasilotu*). Hexane fraction of *Ajuga chamaepitys* (L.) Schreb. (1.2 mg/well) inhibited the growth of *Staphylococcus aureus*, *Micrococcus luteus*, *Mycobacterium smegmatis*, *Bacillus subtilis*, *Bacillus subtilis* var. *niger*, *Aeromonas hydrophila*, and *Klebsiella pneumonae* with inhibition zone diameter of 14, 15, 16, 17, 16, 14, and 16 mm, respectiely (Ulukanli et al., 2005). Hexane extract of this plant at the concentration of 100 µg/mL inhibited the replication of the Herpes simplex virus type-1 and the Parainfluenza-3 virus in Vero cells (Ohran et al., 2009).

ii. A member of the genus *Ajuga* L. is used for cough and tuberculosis in Pakistan.
iii. In China, *Ajuga decumbens* Thunb. is a remedy for diarrhea and an aqueous extract of this plant inhibited the replication of the Respiratory syncytial virus (Long strain) with the IC_{50} value of 131.6 µg/mL and a selectivity index above 1000 in Hep-2 cells (Ma et al., 2002). From this herb, the iridoid glycoside 8-acetyharpagide at the concentration of 500 µg/mL yielded 32.1% cellular viability in the presence of the Vesicular stomatitis virus (Bermejo et al., 2002). *Ajuga lupulina* Maxim. is used for fever and anthrax in China (local name: *Jin gu cao*).

References

Bermejo, P., Abad, M.J., Díaz, A.M., Fernández, L., De Santos, J., Sanchez, S., Villaescusa, L., Carrasco, L. and Irurzun, A., 2002. Antiviral activity of seven iridoids, three saikosaponins and one phenylpropanoid glycoside extracted from Bupleurum rigidum and Scrophularia scorodonia. *Planta Medica*, *68*(02), pp. 106–110.

Ganaie, H.A., Ali, M.N., Ganai, B.A., Meraj, M. and Ahmad, M., 2017. Antibacterial activity of 14, 15-dihydroajugapitin and 8-o-acetylharpagide isolated from Ajuga bracteosa Wall ex. Benth against human pathogenic bacteria. *Microbial Pathogenesis*, *103*, pp. 114–118.

Ma, S.C., Du, J., But, P.P.H., Deng, X.L., Zhang, Y.W., Ooi, V.E.C., Xu, H.X., Lee, S.H.S. and Lee, S.F., 2002. Antiviral Chinese medicinal herbs against respiratory syncytial virus. *Journal of Ethnopharmacology*, *79*(2), pp. 205–211.

Orhan, I., Deliorman-Orhan, D. and Özçelik, B., 2009. Antiviral activity and cytotoxicity of the lipophilic extracts of various edible plants and their fatty acids. *Food Chemistry*, *115*(2), pp. 701–705.

Ulukanli, Z., Ulukanli, S., Ozbay, H., Ilcim, A. and Tuzcu, M., 2005. Antimicrobial activities of some plants from the Eastern Anatolia region of Turkey. *Pharmaceutical Biology*, *43*(4), pp. 334–339.

Yousaf, T., Rafique, S., Wahid, F., Rehman, S., Nazir, A., Rafique, J., Aslam, K., Shabir, G. and Shah, S.M., 2018. Phytochemical profiling and antiviral activity of *Ajuga bracteosa*, *Ajuga parviflora*, *Berberis lycium* and *Citrus lemon* against hepatitis C virus. *Microbial Pathogenesis*, *118*, pp. 154–158.

10.4.4.2 *Anisomeles indica* (L.) O.K

Synonyms: *Ajuga disticha* (L.) Roxb.; *Anisomeles ovata* R. Br.; *Ballota disticha* L.; *Epimeredi indicus* (L.) Rothm.; *Marrubium indicum* (L.) Burm. f.; *Nepeta indica* L.

Common names: Indian catmint; hial-muching, harinchi, kukur shuka (Bangladesh); chi' en tsao, guang fang feng (China); basinga, gopali (India); talingharap, kabling-parang (the Philippines)

Habitat: Forest edges, disturbed thicket, waste lands

Distribution: India, Bangladesh, Myanmar, Cambodia, Laos, Vietnam, China, Thailand, Malaysia, the Philippines

Botanical observation: This herb grows to a height of 1.5 m and has an aura of dryness. The stem is hairy, sharply quadrangular, branched, with 9.3–14 cm long internodes, and slightly swollen right above the internodes. The leaves are simple, decussate, and without stipule. The petiole is 1.3–2.2 cm long. The leaf blade is broadly lanceolate, crenate to serrate, marked with 3 to 4 pairs of secondary nerves, 1.2–3.8 × 4.5–6.8 cm, dull green, cuneate at base, and acute at apex. The verticillasters are 3.5–5.5 cm in diameter. The calyx is membranous, hairy, conspicuously nerved, about 6 mm long, and develops 5 purple-red, triangular-lanceolate teeth. The corolla is 1.6 cm long, tubular, bilabiate. The upper lip is short. Both sides of the upper lip and filaments are pale light greenish. The lower lip is whitish tinted violet. The 4 stamens and style protrude out of the corolla. The anthers are dark maroon, and style and stigma are light maroon. The nutlets are minute and packed in persistent calyx.

Medicinal use: Dysentery (Bangladesh); wounds (India); fever (Bangladesh; India; the Philippines)

Broad-spectrum antibacterial polar extract: Methanol extract of leaves (disc impregnated with 100 mg/mL) inhibited the growth of *Staphylococcus albus*, *Staphylococcus aureus*, *Escherichia coli*, and *Pseudomonas aeruginosa* with the inhibition zone diameters of 17.3, 17, 10, and 16 mm, respectively (Patel & Patel, 2013).

Moderate antibacterial (Gram-negative) cembrane-type diterpene: Ovatodiolide inhibited the growth of *Helicobacter pylori* (ATCC 700392), *Helicobacter pylori* (v633), *Helicobacter pylori* (v1254), and *Helicobacter pylori* (v1354) with the MIC values of 100, 50, 100, and 50 µM, respectively (Rao et al., 2012).

Strong antimycobacterial amphiphilic cembrane-type diterpene: Ovatodiolide (LogD 3.2 at pH 7.4; molecular mass = 328.4 g/mol) inhibited the growth of *Mycobacterium tuberculosis* with the IC_{90} value of 6.5 µg/mL (Kulkarni et al., 2012).

Moderate antifungal essential oil: Essential oil of leaves inhibited the growth of *Rhizoctonia solani* (ITCC 5251), *Rhizoctonia bataticola* (ITCC 4721), *Fusarium oxysporum* (ITCC 5656), *Fusarium udum* (ITCC 2169), *Alternaria solani* (ITCC 4342), *Sclerotium rolfsii* (ITCC 2330), *Aspergillus flavus* (ITCC 1742), and *Pithium aphanidermatum* (ITCC 3339) with the ED_{50} values of 57.9, 51.2, 99.3, 98, 70.5, 72.8, 108.1, and 43.2 µg/mL, respectively (Kindu et al., 2013).

Strong antiviral (enveloped monopartite linear single-stranded (+) RNA): Ethanol extract inhibited the replication of the Porcine epidemic diarrhea virus in Vero cells with the IC_{50} value of 0.6 µg/mL (Trinh et al., 2021).

Antiviral (enveloped linear monopartite dimeric single-stranded (+) RNA) cembrane-type diterpene: Ovatodiolide inhibited the replication of the Human immunodeficiency virus (Alam et al., 2000).

Commentaries

i. The plants produce series of cembrane-type diterpenes such as ovatodiolide (Chen et al., 2008) which could be examined for their anti-Severe acute respiratory syndrome-associated coronavirus and other coronaviruses.
ii. In the Family Lamiaceae; Subfamily Lamioideae; Tribe Pogostemeae, *Anisomeles heyneana* Benth. is used in Bangladesh (local name: *Gobhura*) to treat dysentery. It yields ovatodiolide (Kulkarni et al., 2012).
iii. In India, *Anisomeles malabaria* Roxb. (local name: *Oshthaphala*) is used for fever as well as in the Philippines.

References

Alam, M.S., Quader, M.A. and Rashid, M.A., 2000. HIV-inhibitory diterpenoid from *Anisomeles indica*. *Fitoterapia*, *71*(5), pp. 574–576.

Chen, Y.L., Lan, Y.H., Hsieh, P.W., Wu, C.C., Chen, S.L., Yen, C.T., Chang, F.R., Hung, W.C. and Wu, Y.C., 2008. Bioactive cembrane diterpenoids of *Anisomeles indica*. *Journal of Natural Products*, *71*(7), pp. 1207–1212.

Kulkarni, R.R., Shurpali, K., Gawde, R.L., Sarkar, D., Puranik, V.G. and Joshi, S.P., 2012. Phyllocladane diterpenes from *Anisomeles heyneana*. *Journal of Asian Natural Products Research*, *14*(12), pp. 1162–1168.

Kundu, A., Saha, S., Walia, S. and Kour, C., 2013. Antioxidant and antifungal properties of the essential oil of *Anisomeles indica* from India. *Journal of Medicinal Plants Research*, *7*(24), pp. 1774–1779.

Patel, N.B. and Patel, K.C., 2013. Antimicrobial activity of *Anisomeles indica* Linn. against UTI pathogens. *The Microbes*, *4*, pp. 1–6.

Rao, Y.K., Lien, H.M., Lin, Y.H., Hsu, Y.M., Yeh, C.T., Chen, C.C., Lai, C.H. and Tzeng, Y.M., 2012. Antibacterial activities of *Anisomeles indica* constituents and their inhibition effect on Helicobacter pylori-induced inflammation in human gastric epithelial cells. *Food Chemistry*, *132*(2), pp. 780–787.

Trinh, T.B.N., Le, D.H., Nguyen, T.T.K., Nguyen, M.H., Muller, M. and Nguyen, T.K.N., 2021. In vitro antiviral activities of ethanol and aqueous extracts of Vietnamese traditional medicinal plants against Porcine Epidemic Diarrhea virus: A coronavirus family member. *VirusDisease*, pp. 1–7.

10.4.4.3 *Hyptis capitata* Jacq.

Synonyms: *Clinopodium capitatum* Sw., *Mesosphaerum capitatum* (Jacq.) Kuntze, *Pycnanthemum decurrens* Blco, *Thymus virginicus* Blco

Common names: Knobweed; pukain-tong-phang (Bangladesh), botonesan (the Philippines); nh thoi (Vietnam)

Habitat: Roadsides, disturbed areas, wastelands, open forests

Distribution: Thailand, Malaysia, Indonesia, Vietnam, Papua New Guinea, Australia, Pacific Islands

Botanical observation: This herb native to tropical America grows to about 2 m tall. The stem is hollow, strongly quadrangular, and with 7–10 cm long internodes. The leaves are simple, decussate, and exstipulate. The petiole is about 2–3 cm long. The blade is 4–10.5×0.5–4 cm, broadly lanceolate, dull green, irregularly serrate, cuneate to tapering at base, acute at apex, and marked with about 5 pairs of secondary nerves sunken above. The flowering heads are globose, 1.5–2.5 cm across, and sustained by 3–5 cm long and slender pedicels. The calyx is tubular, membranous, hairy at base, with striations, 8 mm long, and develops 5 linear lobes. The corolla is tubular, bilobed, pure white, and about 5 mm long. The 4 stamens and stigma protrude out of the corolla. The nutlets are brownish-black and packet in the persistent calyx within globose infructescences.

Medicinal use: Dysentery (Bangladesh); influenza, tuberculosis (Taiwan); wounds (the Philippines)

Antibacterial halo (Gram-positive) developed by mid-polar extract: Ethyl acetate extract (50 µL/well) inhibited the growth of *Propionibacterium acnes* with the inhibition zone diameter of 16 mm (Kusima et al., 2014).

Antiviral (enveloped linear monopartite dimeric single-stranded (+) RNA) lipophilic oleanane-type triterpene: Oleanolic acid (LogD 5.9; molecular mass=456.7 g/mol) inhibited the replication of the Human immunodeficiency virus type-1 in H9 cells with an EC_{50} value of 1.7 µg/mL (Kashiwada et al., 1998).

Antiviral (enveloped linear monopartite dimeric single-stranded (+) RNA) amphiphilic ursane-type triterpene: Pomolic acid (also known as (3β)-3,19-dihydroxyurs-12-en-28-oic acid; LogD 4.3 at pH 7.4; molecular mass=472.7 g/mol) inhibited the replication of the Human immunodeficiency virus type-1 in H9 cells with an EC_{50} value of 1.4 µg/mL and selectivity index of 16.6 (Kashiwada et al., 1998).

Commentaries

i. In the Subfamily Nepetoideae; Tribe Ocimeae, *Hyptis brevipes* Poit. is used for wounds in Indonesia (local name: *Godong puser*) and the extract of this herb inhibited the growth of a broad spectrum of bacteria and fungi *in vitro* (Nugraha & Keller, 2011).
ii. *Hyptis suaveolens* (L.) Poit. (local names: *Jangulijungol*; *Changa dana*, *Tokma*) is used to treat boils, fever, and gonorrhea in Bangladesh, and in India (local names: *Ban tulsai, Vantulsi*) it is a remedy for cuts and wounds. In the Philippines (local name: *Loko loko*) is used for wounds and skin diseases; in Papua New Guinea (local name: *Iliplua*) it is used for cuts and wounds and fever; and in China (local name: *Chan xian*) it is used for a colds. Extract of this plant inhibited the replication of the Chikungunya virus (Kothandan et al., 2014), and the essential oil at the concentration of 500 µg/mL inhibited the growth of *Saccharomyces cerevisiae*, *Fusarium moniliforme*, and *Mucor* sp. (Malele et al., 2003).

References

Kashiwada, Y., Wang, H.K., Nagao, T., Kitanaka, S., Yasuda, I., Fujioka, T., Yamagishi, T., Cosentino, L.M., Kozuka, M., Okabe, H. and Ikeshiro, Y., 1998. Anti-AIDS agents. 30. Anti-HIV activity of oleanolic acid, pomolic acid, and structurally related triterpenoids. *Journal of Natural Products*, *61*(9), pp. 1090–1095.

Kothandan, S. and Swaminathan, R., 2014. Evaluation of in vitro antiviral activity of *Vitex Negundo* L., *Hyptis suaveolens* (L) poit., *Decalepis hamiltonii* Wight & Arn., to Chikungunya virus. *Asian Pacific Journal of Tropical Disease*, *4*, pp. S111–S115.
Kusuma, I.W., Arung, E.T. and Kim, Y.U., 2014. Antimicrobial and antioxidant properties of medicinal plants used by the Bentian tribe from Indonesia. *Food Science and Human Wellness*, *3*(3–4), pp. 191–196.
Malele, R.S., Mutayabarwa, C.K., Mwangi, J.W., Thoithi, G.N., Lopez, A.G., Lucini, E.I. and Zygadlo, J.A., 2003. Essential oil of *Hyptis suaveolens* (L.) Poit. from Tanzania: Composition and antifungal activity. *Journal of Essential Oil Research*, *15*(6), pp. 438–440.
Nugraha, A.S. and Keller, P.A., 2011. Revealing indigenous Indonesian traditional medicine: Anti-infective agents. *Natural Product Communications*, *6*(12), pp. 1953–1966.

10.4.4.4 *Leonurus sibiricus* L.

Synonyms: *Leonurus manshuricus* Yabe

Common names: Siberian motherwort; dondo-kolosh, dorpie (Bangladesh); ch' ung wai, xi ye yi mu cao (China); raktadron (India); seranting (Malaysia); camariangsongson (the Philippines)

Habitat: Meadows

Distribution: Tropical Asia

Botanical observation: This herb grows to about 1 m tall and has an unpleasant aura. The stems are quadrangular and somewhat lignose. The leaves are simple, decussate, and exstipulate. The petiole is up to 5 cm long. The blade is 3-palmatisect, and up to 10 cm long. The verticillasters are many flowered. The calyx is tubular, 8 mm long, hairy, obscurely 2-lipped; upper lip straight, 3-toothed and lower lip bifid. The corolla is tubular, bilabiate, purple-red, about 1.8 cm long. The upper lip is oblong, straight, concave, about 1 cm long, The lowerlip is about 7 mm long and emarginate. The androecium includes 4 stamens. The ovary develops a slender style and a bifid stigma. The nutlets are triquetrous and minute.

Medicinal use: Fever (India); wounds (China)

Broad-spectrum antibacterial halo developed by midpolar extract: Chloroform extract of aerial parts (500 µg/disc) inhibited the growth of *Staphylococcus aureus*, *Staphylococcus epidermidis*, *Streptococcus pyogenes*, *Escherichia coli*, *Vibrio cholerae*, *Shigella dysenteriae*, and *Shigella boydii* with the inhibition zone diameters of 19, 14, 28, 20, 15, 17, and 19 mm, respectively (Ahmed et al., 2006).

Broad-spectrum antifungal polar extract: Ethanol extract (10 mg/mL/4 mm well) inhibited the growth of *Candida albicans* (MTCC 183) and *Lasiodiplodia theobromae* (ITCC 5446.02) with the inhibition zone diameters of 10.6 and 14 mm, respectively (Saha et al., 2019).

Strong antiviral (enveloped circular double-stranded DNA) polar extract: Aqueous extract of leaves and stems inhibited the replication of the Hepatitis B virus with an EC_{50} value below 0.7 µg/mL in 2.2.15 cells and a selectivity index above 20 (Nam et al., 1996).

Commentaries

i. The plant generates a series of labdane-type diterpenes as well as the flavonol genkwanin (Boalino et al., 2004).
ii. *Leonurus japonicus Houtt* is used for fever in India and for diarrhea in Korea (local name: *Ikmocho*). In Turkey, *Leonurus glaucescens* Bunge (local name: *Öküzguyrugu*) is used for wounds.

References

Ahmed, F., Islam, M.A. and Rahman, M.M., 2006. Antibacterial activity of *Leonurus sibiricus* aerial parts. *Fitoterapia*, *77*(4), pp. 316–317.
Boalino, D.M., McLean, S., Reynolds, W.F. and Tinto, W.F., 2004. Labdane diterpenes of *Leonurus sibiricus*. *Journal of Natural Products*, *67*(4), pp. 714–717.
Nam, K.W., Chang, I.M., Choi, J.S., Hwang, K.J. and Mar, W.C., 1996. Antiviral effects of natural products on the inhibition of hepatitis B virus DNA replication in 2.2. 15 cell culture system. *Natural Product Sciences*, *2*(2), pp. 130–136.

10.4.4.5 *Leucas lavendulifolia* Sm.

Synonyms: *Leonurus indicus* L.; *Leucas indica* (L.) Sm.; *Leucas linifolia* (Roth) Spreng.; *Phlomis linifolia* Roth

Common names: Xian ye bai rong cao (China); dhandakolosh (Bangladesh); dronpushpi (India)

Habitat: Sun exposed soils, roadsides, riverbanks, forest margins

Distribution: Tropical Asia

Botanical observation: This herb grows up to 1 m tall. The stem is slender and quadrangular. The petiole is none to 1cm long. The blade is narrowly lanceolate or oblong-linear, 2.5–7×1.5–1.2cm, hairy, cuneate at base, somewhat serrate, obtuse at apex, and marked with 2 or 3 pairs of secondary nerves. The verticillasters are many flowered and 1.5–2.5cm across. The calyx is tubular, 5mm long, hairy, truncate, and 10-lobed. The corolla is pure white, 1.5cm long, bilabiate. The upper lip is hairy, and the lower lip is 3-lobed. A disc is present. There are 4 anthers. The style is slender and bilobed. The nutlets are brown and ovoid.

Medicinal use: Sores (India); wounds (India; the Philippines)

Weak broad-spectrum antibacterial friedoolean-type triterpene: Taraxerone (Jayaraman & Variyar, 2021) inhibited the growth of *Staphylococcus aureus* (ATCC 25923), *Bacillus subtilis* (ATCC 6633), *Lactobacillus plantarum* (ATCC 8014), *Escherichia coli* (ATCC 25922), *Salmonella typhimurium* (ATCC 14028), *Klebsiella pneumoniae* (ATCC 10031), and *Pseudomonas aeruginosa* (ATCC 27853) with the MIC values of 400, 1400, 1400, 200, 2800, 1400, and 2800µg/mL, respectively (Manríquez-Torres et al., 2007).

Broad-spectrum antifungal halo developed by friedoolean-type triterpene: Taraxerone (200 µg/disc) inhibited the growth of *Aspergillus flavus*, *Penicillium* sp., *Candida albicans*, and *Botryodiplodia theobromae* with the inhibition zone diameters of 12, 8, 13, and, 10mm, respectively (Abu-Sayeed et al., 2005).

Commentaries

i. Taxarenone being a triterpene with a ketone moiety is probably antiviral.
ii. Asian systems of medicine utilize plants in the genus *Leucas* R.Brown for the treatment of microbial infections pointing to the existence of strong antimicrobial principles in these plants. *Leucas aspera* (Willd.) Link is used in Bangladesh (local name: *Kan-sisha*) for cough, fever, and skin infection, in India (local name: *Dronpushpi*, Thumbai) for fever and dysentery, and in the Philippines (local name: Pansi-pansi) for diarrhea. Methanol extract of roots (20µL of a 100mg/mL solution/6mm disc) inhibited the growth of *Escherichia coli*, *Staphylococcus aureus*, *Salmonella choleraesuis*, *Salmonella typhimurium*, *Shigella flexneri*, and *Pseudomonas aeruginosa* with the inhibition zone diameters of 9, 10, 11, 11, 11, and 11mm, respectively (Chew et al., 2012). Dichloromethane extract of this plant (1 mg/disc) inhibited the growth of *Aspergillus flavus*, *Candida albicans*, *Epidermophyton floccosum*, and *Trichophyton viride* with the inhibition zone diameters of 10.4, 12.8, 7.5, and 16mm, respectively (Pavunraj et al., 2017).
iii. In India, *Leucas plukenetti* (Roth) Spreng. (local name: *Mayang lenbum*) is used for a cold and cough, fever and pneumonia; *Leucas linifolia* (Spreng). is used for a cold; *Leucas lanata* Benth. (local name: *Chinnapoola tummi*) is used for toothache, *Leucas cephalotes* (Roth.) Spreng. is used for cough, diarrhea, dysentery, fever, and jaundice; and *Leucas martinicensis* R. Br. (local name: Guumaa) is used for a cold.
iv. *Leucas urticefolia* (Vahl) R. Br. is used in Pakistan for cold and fever.
v. In Myanmar, *Leucas cephalotes* (Roth) Spreng. (local name: *Pin-gu-hteik-peik*.) is used for bronchitis, fever, and jaundice.
vi. In China, *Leucas mollissima* Wall. ex Benth. (local name: *Bái huā cháshi hóng*) is used for cough.
vii. In Thailand, *Leucas zeylanica* (L.) R. Be. (local name: *Brek*) is used for wounds, and in Bangladesh (local name: *Guma saag*) it is used for cough, cuts, and sores.

References

Abu-Sayeed, M., Alib, M.A., Bhattacharjeeb, P.K., Islamb, A. and Astaqb, G.R.M., 2005. Biological evaluation of extracts and triterpenoids of Euphorbia hirta. *Biological Sciences-PJSIR*, *48*(2), pp. 112–125.

Chew, A.L., Jessica, J.J.A. and Sasidharan, S., 2012. Antioxidant and antibacterial activity of different parts of *Leucas aspera*. *Asian Pacific Journal of Tropical Biomedicine*, *2*(3), pp. 176–180.

Jayaraman, S. and Variyar, E.J., 2021. Role of taraxerone isolated from *Leucas lavandulifolia*, as an immunomodulator. *Journal of Ethnopharmacology*, p. 114307.

Mangathayaru, K., Lakshmikant, J., Sundar, N.S., Swapna, R.G.X.F., Grace, X.F. and Vasantha, J., 2005. Antimicrobial activity of *Leucas aspera* flowers. *Fitoterapia*, *76*(7–8), pp. 752–754.

Manríquez-Torres, J.J., Zúñiga-Estrada, A., González-Ledesma, M. and Torres-Valencia, J.M., 2007. The antibacterial metabolites and proacacipetalin from *Acacia cochliacantha*. *Journal of the Mexican Chemical Society*, *51*(4), pp. 228–231.

Pavunraj, M., Ramasubbu, G. and Baskar, K., 2017. Leucas aspera (Willd.) L.: Antibacterial, antifungal and mosquitocidal activities. *Trends in Phytochemical Research*, *1*(3), pp. 135–142.

10.4.4.6 *Mentha canadensis* L.

Synonyms: *Mentha arvensis* L.; *Mentha haplocalyx* Briq.; *Mentha pedunculata* H.H. Hu & H.T. Tsai; *Mentha sachalinensis* (Briq.) Kudô

Common names: Japanese mint; bo he (China); nana, nungshi-hidak (India); payoke aye (Myanmar); pudina (Nepal); herba Buena (the Philippines)

Habitat: Watery and shady meadows

Distribution: Pakistan, India, Bangladesh, Myanmar, Cambodia, Laos, Vietnam, Thailand, Malaysia, China, Japan, Korea

Botanical observation: This beautiful herb grows up to 50cm tall and has a positive aura. The stems are quadrangular. The leaves are simple, decussate, and exstipulate. The petiole is 2–10mm long. The blade is blade is ovate, aromatic, lanceolate or oblong, 2.5–5×0.5–3.5cm, hairy, cuneate to rounded at base, fragrant when crushed, marked with about 4 to 5 pairs of secondary nerves, serrate, and acute at apex. The verticillasters are globose and about 2cm across. The calyx is tubular, hairy, about 3mm long, marked with longitudinal nervations, and develops 5 triangular lobes. The corolla is tubular, hairy within, white to purplish, bilabiate, and about 5mm long. The upper lip is minutely bilobed and the lower lip is 3-lobed. Four stamens are adnate to and protrude out of the corolla. The ovary is somewhat globose and develops a slender style and a bifid stigma. The capsules are minute.

Medicinal uses: Cold (Korea); cough (India; Pakistan); fever (Myanmar; the Philippines); diarrhea (Nepal)

Broad-spectrum antibacterial essential oil: Essential oil of leaves (15µL/disc) inhibited the growth of *Salmonella typhi*, *Streptococcus enteritis*, *Escherichia coli*, *Clostridium perfringens*, and *Campylobacter jejuni* (Wannissorn et al., 2005).

Strong antibacterial (Gram-negative) polar extract: Ethanol extract of aerial parts inhibited the growth of *Acinetobacter baumannii* (ATCC ATCC 10545) with the MIC/MBC value of 23.5/72.1µg/mL (Zhang et al., 2015).

Strong antiviral (enveloped monopartite linear double-stranded DNA) polar extract: Extract inhibited the replication of the Herpes simplex virus type-1 with the IC_{50} value of 2µg/mL in Vero cells (Hamidi et al., 1996).

Strong antiviral (enveloped monopartite linear single-stranded (−) RNA) polar extract: Ethanol extract inhibited the replication of the Vesicular stomatitis virus with the IC_{50} value of 100µg/mL in Vero cells (Hamidi et al., 1996).

Strong antiviral (enveloped monopartite linear double-stranded DNA) amphiphilic cyclic monoterpene epoxide: Piperitenone oxide (LogD 1.5 at pH 7.4; molecular mass=166.2 g/mol) inhibited the replication of the Herpes simplex virus type-1 with an IC_{50} value of 1.4µg/mL (Civitelli et al., 2014).

Commentaries

i. The plant is used for the treatment of COVID-19 in China (Wang et al., 2020). Anti-severe acute respiratory syndrome-associated coronavirus principles in this plant are unknown but could be essential oil constituents such as piperitenone oxide and piperitone oxide (Umemoto & Tsuneya, 1988).

ii. Plants in the genus *Mentha* L, are used for diarrhea and fever in Asia. In Turkey, *Mentha longifolia* L. is given for cholera and diarrhea (local name: *Tüylü nane*), in Pakistan (local names: *Safedpodina, Velanay*) it is used for diarrhea and dysentery, and in India (local names: *Pudina bari, Veina*) it is used as a remedy for fever. *Mentha pulegium* L. is classified as hot and dry in Iran (local name: *Fodenj*), used in Turkey for diarrhea (local name: *Nane, Punk*), and in Morocco for COVID-19 prevention (El Alami et al., 2020). *Mentha sylvestris* L. is used for dysentery in Iran (local name: *Gul-i-punah*) and in India. In India, *Mentha viridis* L. (local name: *Pudina*) is given for bronchitis and fever, *Mentha sylvestris* L. (local name: *Ajirnahara*) is used for bronchitis, diarrhea, and dysentery, *Mentha aquatica* L. is used in India for diarrhea, and *Mentha piperita* L. (local name: *Vilayati pudina*) is used for cough and fever. Essential oil of *Mentha piperita* L. inhibited *Staphylococcus aureus* (ATCC 25923), *Klebsiella pneumoniae* (NCTC 9633), and *Candida albicans* (ATCC 10231) with MIC values of 8, 8, and 6 µg/mL, respectively (Van Vuuren et al., 2009). Essential oil of *Mentha piperita* L. inhibited the growth of *Streptococcus mutans* with MIC of 420 µg/mL (Miller et al., 2015). Extract of *Mentha piperita* inhibited he replication of avian infectious bronchitis virus with the IC_{50} value of 0.04 µg/mL (Lelešius et al., 2019). Extract of this herb inhibited the replication of the Newcastle disease virus, Herpes simplex virus, Vaccinia virus, Semliki Forest, and West Nile viruses (Herrmann & Kucera, 1967).

In Nepal, *Mentha spicata* L. (local name: *Daunne*) is used for diarrhea and dysentery. It is used for cough and dysentery (local name: *Poodna*) in Pakistan and for diarrhea in Turkey (local name: *Kivircik nane*). Aqueous extract of *Mentha spicata* L. inhibited the replication of the Porcine parvovirus with the IC_{50} value of 0.03 µg/mL and a selectivity index of 79.1 (WeiLi et al., 2011).

iii. Extract of a member of the genus *Mentha* L. inhibited the replication of the Human immunodeficiency virus (de Paz-Silava et al., 2021).

iv. Essential oil of *Mentha suaveolens* Ehrh. (local name in China: *Yuan ye bo he*) inhibited the replication of the Herpes simplex virus type-1 with the IC_{50} value of 5.1 µg/mL, and this oil contains piperitenone oxide. In Morocco, it is used for COVID-19 prevention (El Alami et al., 2020).

v. Essential oils of plants in the genus *Mentha* L. have antibiotic potentiating properties (Van Vuuren et al., 2009).

References

Civitelli, L., Panella, S., Marcocci, M.E., De Petris, A., Garzoli, S., Pepi, F., Vavala, E., Ragno, R., Nencioni, L., Palamara, A.T. and Angiolella, L., 2014. In vitro inhibition of herpes simplex virus type 1 replication by *Mentha suaveolens* essential oil and its main component piperitenone oxide. *Phytomedicine*, *21*(6), pp. 857–865.

de Paz-Silava, S.L.M., Victoriano-Belvis, A.F.B., Gloriani, N.G., Hibi, Y., Asamitsu, K. and Okamoto, T., 2021. In vitro antiviral activity of *Mentha cordifolia* plant extract in HIV-1 latently infected cells using an established human cell line. *AIDS Research and Human Retroviruses*.

El Alami, A., Fattah, A. and Chait, A., 2020. Medicinal plants used for the prevention purposes during the COVID-19 pandemic in Morocco. *Journal of Analytical Sciences and Applied Biotechnology*, *2*(1), pp. 2–1.

Herrmann Jr, E.C. and Kucera, L.S., 1967. Antiviral substances in plants of the mint family (labiatae). III. peppermint (*Mentha piperita*) and other mint plants. *Proceedings of the Society for Experimental Biology and Medicine*, *124*(3), pp. 874–878.

Lelesius, R., Karpovaite, A., Mickiene, R., Drevinskas, T., Tiso, N., Ragazinskiene, O., Kubiliene, L., Maruska, A., Salomskas, A., 2019. In vitro antiviral activity of fifteen plant extracts against avian infectious bronchitis virus. *BMC Veterinary Research* 15(1), p. 178.

Umemoto, K. and Tsuneya, T., 1988. Mentha arvensis containing piperitenone oxide and piperitone oxide as major components (studies on chemical constituents of wild mints. Part XIV). *Journal of the Agricultural Chemical Society of Japan*, *62*, pp. 1073–1076.

Van Vuuren, S.F., Suliman, S. and Viljoen, A.M., 2009. The antimicrobial activity of four commercial essential oils in combination with conventional antimicrobials. *Letters in Applied Microbiology*, *48*(4), 440–446.

Wang, S.X., Wang, Y., Lu, Y.B., Li, J.Y., Song, Y.J., Nyamgerelt, M. and Wang, X.X., 2020. Diagnosis and treatment of novel coronavirus pneumonia based on the theory of traditional Chinese medicine. *Journal of Integrative Medicine*, *18*(4), pp. 275–283.

WeiLi, H., BaoAn, C., HongYing, Z., XueBing, W., DuanHong, X. and RuiLiang, C., 2011. Antiviral activity of *Mentha spicata* Linn. extracts against porcine parvovirus in vitro. *Jiangsu Journal of Agricultural Sciences*, *27*(3), pp. 556–560.

Zhang, L., Xu, S.G., Liang, W., Mei, J., Di, Y.Y., Lan, H.H., Yang, Y., Wang, W.W., Luo, Y.Y. and Wang, H.Z., 2015. Antibacterial activity and mode of action of *Mentha arvensis* ethanol extract against multidrug-resistant *Acinetobacter baumannii*. *Tropical Journal of Pharmaceutical Research*, *14*(11), pp. 2099–2106.

10.4.4.7 *Ocimum basilicum* L.

Synonym: *Ocimum thyrsiflorum* L.

Common names: Sweet basil; luo le (Bangladesh); arjaka, tirunitru, pachchai (India); selaseh, ruku-ruku (Malaysia); balanoi (the Philippines); hung que (Vietnam)

Habitat: Cultivated

Distribution: Tropical Asia

Botanical observation: This shrubby herb grows up to a height of about 1 m. The stems are purplish or green, squarish, succulent and glabrous or hairy and woody at base. The leaves are simple, decussate, and without stipules. The petiole is 1.5–2 cm long and hairy. The blade is lanceolate, aromatic, 2.5–9 cm × 1.2–5 cm, glossy, fragrant when crushed, serrate, somewhat fleshy, round to acute at base, acute to acuminate at apex, glossy, and marked with 4–6 pairs of secondary nerves sunken above. The inflorescences are terminal racemes of verticillasters, 7–20 cm long, purplish or not, and cylindrical. The flowers and the plant in general give off somewhat an aura of immaculate sanctity. The calyx is hairy, tubular, purplish or green, 4 mm long, bilobed, the upper lobe broad and round, the lower lobe with 4 tiny teeth. The corolla is pure white or purplish, bilobed, about 8 mm long, the lower lobe projecting forward and nearly flat, the upper one with 4 blunt teeth. The 4 stamens are in 2 pairs or different length and adnate to the corolla tube. The style is pinkish. The nutlets are included in persistent calyx.

Medicinal uses: Dysentery (India; Vietnam); fever (Vietnam); ulcers (China)

Broad-spectrum antibacterial essential oil: Essential oil of leaves (15 µL/disc) inhibited the growth of *Salmonella typhi*, *Streptococcus enteritis*, *Escherichia coli*, *Clostridium perfringens* and *Campylobacter jejuni* (Wannissorn et al., 2005).

Strong broad-spectrum antibacterial essential oil: Essential oil of leaves inhibited the growth of *Bacillus cereus*, *Bacillus subtilis*, *Bacillus megaterium*, *Staphylococcus aureus*, *Listeria monocytogenes*, *Escherichia coli*, *Shigella boydii*, *Shigella dysenteriae*, *Vibrio parahaemolyticus*, *Vibrio mimicus*, *Salmonella typhi*, and *Salmonella paratyphi* (Hossain et al., 2010).

Strong broad-spectrum antibacterial amphiphilic linear monoterpene: Linalool inhibited the growth of *Escherichia coli* (ATCC 35218), *Pseudomonas aeruginosa* (ATCC 10145), *Klebsiella pneumoniae*, *Acinetobacter baumanii*, *Staphylococcus aureus* (ATCC 25923), *Enterococcus faecalis* (ATCC 29212), and *Bacillus subtilis* (ATCC 6633) (Ohran et al., 2012).

Strong anticandidal amphiphilic linear monoterpene: Linalool inhibited the growth of *Candida albicans* (ATCC 10231) and *Candida parapsilosis* (ATC 22019) with the MIC values of 8 and 16 µg/mL, respectively (Orhan et al., 2012).

Strong antifungal (filamentous) essential oil: Essential oil of aerial parts inhibited the growth of *Fulvia fulva*, *Glomerella cingulate*, *Alternaria alternata*, and *Fusarium solani* with the EC_{50} values of 15.5, 363.9, 376.7, and 14.2 ppm, respectively (Zhang et al., 2009).

Moderate antiviral (enveloped monopartite linear double-stranded DNA) polar extract: Aqueous extract inhibited the replication of the Herpes simplex virus type-1 (strain KOS) and type-2 (strain 196) with the EC_{50} values of 90.9 and 51.4 µg/mL, respectively, and the selectivity indexes of 16.2 and 28.6 in BCC-1/KMC cells (Chiang et al., 2005).

Strong antiviral (enveloped monopartite linear double-stranded DNA) amphiphilic flavonol: Apigenin (LogD 1.2 at pH 7.4; molecular mass=270.2 g/mol) inhibited the replication of the Herpes simplex virus type-1 (strain KOS) and type-2 (strain 196) with the EC_{50} values of 6.7 and 9.7 µg/mL, respectively, and the selectivity indexes of 9 and 6.2, respectively BCC-1/KMC cells (Chiang et al., 2005).

Strong antiviral (enveloped monopartite linear double-stranded DNA) polar amphiphilic ursane-type triterpene: Ursolic acid (LogD 5.8 at pH 7.4; molecular mass=456.7 g/mol) inhibited the replication of the Herpes simplex virus type-1 (strain KOS) with the EC_{50} value of 6.6 and the selectivity index of 15.2 (Chiang et al., 2005).

Moderate antiviral (non-enveloped monopartite linear double-stranded DNA) polar extract: Aqueous extract inhibited the replication of the Adenovirus type-3, type-8, and type-11 with the EC_{50} values 174.1, 129.6, and 129.1 µg/mL, respectively, and the selectivity indexes of 8.4, 11.3, and 11.4, respectively, BCC-1/KMC cells (Chiang et al., 2005).

Strong antiviral (non-enveloped monopartite linear double-stranded DNA) linear monoterpene: Linalool (LogD 3.2 at pH 7.4; molecular mass=154.2 g/mol) inhibited the replication of the Adenovirus type-3, type-8, and type-11 with the EC_{50} values 24.4, 26.4, and 16.9 µg/mL, respectively, and the selectivity indexes of 7.3, 6.7, and 10.5, respectively, BCC-1/KMC cells (Chiang et al., 2005).

Strong antiviral (non-enveloped monopartite linear double-stranded DNA) amphiphilic flavonol: Apigenin inhibited the replication of the Adenovirus type-3, type-8, and type-11 with the EC_{50} values 11.1, 8, and 20.9 µg/mL, respectively, and the selectivity indexes of 5.4, 7.5, and 2.9, respectively, BCC-1/KMC cells (Chiang et al., 2005).

Strong antiviral (non-enveloped monopartite linear double-stranded DNA) amphiphilic ursane-type triterpene: Ursolic acid inhibited the replication of the Adenovirus type-8, and type-11 with the EC_{50} values of 4.2, and 24.5 µg/mL, respectively, and the selectivity indexes of 23.8 and 4.1, respectively, in BCC-1/KMC cells (Chiang et al., 2005).

Moderate antiviral (enveloped circular double-stranded DNA) polar extract: Aqueous extract inhibited the secretion of the Hepatitis B virus surface antigen and e-antigen with the EC_{50} values 85.4 and 60.1 µg/mL, respectively, and the selectivity indexes of 8.9 and 12.7, respectively (Chiang et al., 2005).

Strong antiviral (enveloped circular double-stranded DNA) amphiphilic flavonol: Apigenin inhibited the secretion of the Hepatitis B virus surface antigen and e-antigen with the EC_{50} values 7.1 and 12.8 µg/mL, respectively, and the selectivity indexes of 2.3 and 1.3, respectively (Chiang et al., 2005).

Strong antiviral (enveloped circular double-stranded DNA) amphiphilic ursane-type triterpene: Ursolic acid inhibited the secretion of the Hepatitis B virus surface antigen and e-antigen with the EC_{50} values 21 and 30 µg/mL, respectively, and the selectivity indexes of 1.8 and 1.2, respectively (Chiang et al., 2005).

Moderate antiviral (enveloped circular double-stranded DNA) amphiphilic linear monoterpene: Linalool inhibited the secretion of the Hepatitis B virus surface antigen and e-antigen with the EC_{50} values 145.5 and 219.1 µg/mL, respectively, and the selectivity indexes of 0.9 and 0,6, respectively (Chiang et al., 2005).

Moderate antiviral (non-enveloped monopartite linear single-stranded (+) RNA) polar extract: Aqueous extract inhibited the replication of the Coxsackie virus B1 and the Enterovirus 71 with the EC_{50} values 105.7 and 200.2 µg/mL, respectively, and the selectivity indexes of 13.9 and 7.3, respectively (Chiang et al., 2005).

Strong antiviral (non-enveloped monopartite linear single-stranded (+) RNA) amphiphilic flavonol: Apigenin inhibited the Replication of the Coxsackie virus B1 and the Enterovirus 71 with the EC_{50} values 34.9 and 42.2 µg/mL, respectively, and the selectivity indexes of 7.5 and 8.7, respectively, in BCC-1/KMC cells (Chiang et al., 2005).

Strong antiviral (non-enveloped monopartite linear single-stranded (+) RNA) amphiphilic ursane-type triterpene: Ursolic acid inhibited the replication of the Coxsackie virus B1 and the Enterovirus 71 with the EC_{50} values 0.4 and 0.5 µg/mL, respectively, and the selectivity indexes of 251.3 and 201, respectively, in BCC-1/KMC cells (Chiang et al., 2005).

Strong antiviral (non-enveloped monopartite linear single-stranded (+) RNA) amphiphilic linear monoterpene: Linalool inhibited the replication of the Coxsackie virus B1 and the Enterovirus 71 with the EC_{50} values 34.9 and 42.2 µg/mL, respectively, and the selectivity indexes of 5.1 and 4.2, respectively, in BCC-1/KMC cells (Chiang et al., 2005).

Commentaries

i. Essential oil contains 1,8-cineole (see earlier) and linalool (Vieira et al., 2014), and note that numerous varieties exist and for each, various proportion of constituents exist, and also vary with time of collection, exposure to the sun, and soils. Note that essential oil containing linalool inhibited the expression of angiotensin-converting enzyme *in vitro* (Senthil Kumar et al., 2020) and as such could be of value against COVID-19. There is a possibility that sweet basil teas could assist in managing COVID-19 and this needs to be verified.

ii. The plant was recommended, with many others including *Melissa officinalis* L., against the Black Plague in the "*Compendium de epidimia*" published by the Faculty of Medicine of Paris in the year 1348 (Arveiller, 1973).

iii. Aqueous extracts of Lamiaceae are often antiviral *in vitro.*

iv. *Ocimum gratissimum* L. is used for gonorrhea in Malaysia. Essential oil of leaves of *Ocimum gratissimum* L. (15 µL/disc) inhibited the growth of *Salmonella typhi*, *Streptococcus enteritis*, *Escherichia coli*, *Clostridium perfringens*, and *Campylobacter jejuni* with inhibition zone diameter of 13.5, 20.3, 11, 11 and 20 mm, respectively (Wannissorn et al., 2005). Aqueous extract of leaves of *Ocimum gratissimum* L. inhibited the replication of the Human immunodeficiency virus type-1 (strain IIIB) in Molt-4 clone 8 with EC_{50} of 10 µg/mL and selectivity index of 110 (Ayisi & Nyadedzor, 2003). This extract inhibited the replication of the Human immunodeficiency virus type-1 (strain GH1) with the EC_{50} value of 7 µg/mL and selectivity index of 14.7 (Ayisi and Nyadedzor, 2003). This extract blocked *in vitro* Human immunodeficiency virus type-1 at a concentration of 10 µg/mL (Ayisi & Nyadedzor, 2003).

v. *Ocimum sanctum* L. (synonym: *Ocimum tenuiflorum* L.) is used in Bangladesh (local names: *Tulshi; Torshi; Tulsi*) for a cold, ringworm, wounds, fever, bronchitis, cough, skin diseases, influenza, tuberculosis, respiratory tract infections, and measles. The plant is used in India for the treatment of Chikungunya virus infection (Viswanathan et al., 2008). The phenylpropanoid eugenol and essential oil from this plant inhibited angiotensin-converting enzyme (Chaudhary et al., 2014) and as such could be of value against COVID-19.

vi. In the Philippines, *Ocimum sanctum* L. is used for gonorrhea Ethanol extract of leaves inhibited *Candida albicans* (clinical strain) (Vaijayanthimala et al., 2000).

vii. Essential oil of *Ocimum micrathum* Willd. inhibited the growth of *Streptococcus mutans* with MIC of 420 µg/mL (Miller et al., 2015).

References

Arveiller, R., 1973. Textes médicaux français d'environ 1350. *Romania*, *94*(374 (2), pp. 157–177.

Ayisi, N.K. and Nyadedzor, C., 2003. Comparative in vitro effects of AZT and extracts of *Ocimum gratissimum*, *Ficus polita*, *Clausena anisata*, *Alchornea cordifolia*, and *Elaeophorbia drupifera* against HIV-1 and HIV-2 infections. *Antiviral Research*, *58*(1), pp. 25–33.

Chaudhary, S.K., Mukherjee, P.K., Maity, N., Nema, N.K., Bhadra, S. and Saha, B.P., 2014. *Ocimum sanctum* L. a potential angiotensin converting enzyme (ACE) inhibitor useful in hypertension. *Indian Journal of Natural Products and Resources*, 5(1), pp. 83–87.
Chiang, L.C., Ng, L.T., Cheng, P.W., Chiang, W. and Lin, C.C., 2005. Antiviral activities of extracts and selected pure constituents of *Ocimum basilicum*. *Clinical and Experimental Pharmacology and Physiology*, *32*(10), pp. 811–816.
Hossain, M.A., Kabir, M.J., Salehuddin, S.M., Rahman, S.M., Das, A.K., Singha, S.K., Alam, M.K. and Rahman, A., 2010. Antibacterial properties of essential oils and methanol extracts of sweet basil Ocimum basilicum occurring in Bangladesh. *Pharmaceutical Biology*, *48*(5), pp. 504–511.
Miller, A.B., Cates, R.G., Lawrence, M., Soria, J.A.F., Espinoza, L.V., Martinez, J.V. and Arbizú, D.A., 2015. The antibacterial and antifungal activity of essential oils extracted from Guatemalan medicinal plants. *Pharmaceutical Biology*, *53*(4), pp. 548–554.
Orhan, İ.E., Özçelik, B., Kartal, M. and Kan, Y., 2012. Antimicrobial and antiviral effects of essential oils from selected Umbelliferae and Labiatae plants and individual essential oil components. *Turkish Journal of Biology*, *36*(3), pp. 239–246.
Senthil Kumar, K.J., Gokila Vani, M., Wang, C.S., Chen, C.C., Chen, Y.C., Lu, L.P., Huang, C.H., Lai, C.S. and Wang, S.Y., 2020. Geranium and lemon essential oils and their active compounds downregulate angiotensin-converting enzyme 2 (ACE2), a SARS-CoV-2 spike receptor-binding domain, in epithelial cells. *Plants*, *9*(6), p. 770.
Shynu, M., Saini, M., Sharma, B., Gupta, L.K. and Gupta, P.K., 2006. Ocimum tenuiflorum possesses antiviral activity against bovine herpes virus-1. *Indian Journal of Virology*, *17*, p. 28.
Vaijayanthimala, J., Anandi, C., Udhaya, V. and Pugalendi, K.V., 2000. Anticandidal activity of certain South Indian medicinal plants. *Phytotherapy Research*, *14*(3), pp. 207–209.
Vieira, P.R., de Morais, S.M., Bezerra, F.H., Ferreira, P.A.T., Oliveira, Í.R. and Silva, M.G.V., 2014. Chemical composition and antifungal activity of essential oils from *Ocimum species*. *Industrial Crops and Products*, *55*, pp. 267–271.
Viswanathan, M.V., Raja, D.K. and Khanna, S.D., 2008. Siddha way to cure Chikungunya. *Indian Journal of Traditional Knowledge*, *7*(2), pp. 345–346.
Yamani, H.A., Pang, E.C., Mantri, N. and Deighton, M.A., 2016. Antimicrobial activity of Tulsi (*Ocimum tenuiflorum*) essential oil and their major constituents against three species of bacteria. *Frontiers in Microbiology*, *7*, p. 681.
Zhang, J.W., Li, S.K. and Wu, W.J., 2009. The main chemical composition and in vitro antifungal activity of the essential oils of *Ocimum basilicum* Linn. var. pilosum (Willd.) Benth. *Molecules*, *14*(1), pp. 273–278.

10.4.4.8 *Salvia plebeia* R.Br

Synonyms: *Lumnitzera fastigiata* (Roth) Spreng.; *Ocimum fastigiatum* Roth; *Ocimum virgatum* Thunb.; *Salvia brachiata* Roxb.; *Salvia minutiflora* Bunge

Common names: Li zhi cao (China); buthulsi, hath mutha (Bangladesh); kammarka (India); banbawari, kalo pati (Nepal); giowsi daji (Vietnam)

Habitat: cultivated fields, meadows, roadsides

Distribution: Afghanistan, Pakistan, India, Bhutan, China, Japan, Malaysia

Botanical description: This discrete herb grows up to 30 tall. The stems are quadrangular and hairy. The leaves are simple, opposite, and exstipulate. The petiole is narrowly winged and 5–30 mm long. The blade is oblong to ovate, 2.5–7 × 1–3.5 mm, hairy below, attenuate at base, serrate at margin, and acute at apex, and presents 3 to 4 pairs of secondary nerves sunken above. The racemes of verticils are terminal and about 25 cm long, each verticil 3 to 8-flowered. The calyx is 2 mm long, hairy, bilabiate, the upper lip 3-toothed, and the lower lip 2-toothed. The corolla is tubular, bilabiate, light-blue, pale pink or mauve, about 5 mm long. The upper lip is straight and oblong and the lower lip 3-lobed. Two stamens are present. The nutlets are ovoid and minute.

Medicinal uses: Abscesses, boils (China); cholera, dysentery (Cambodia, Laos, Vietnam); diarrhea, gonorrhea (Bangladesh; India), toothache (India); wounds (Nepal)

Strong antiviral (enveloped linear segmented single-stranded (−) RNA) flavones: Hispidulin and nepetin (LogD 0.9 at pH 7.4; molecular mass = 316.2 g/mol) inhibited the replication of the Influenza A virus (H1N1 A/PR/8/34) with the EC_{50} values of 22.6 and 17.5 µM and the selectivity indexes of

8.9 and 11.4, respectively, in Madin-Darby canine kidney cells via a mechanism involving, at least in part, neuraminidase inhibition (Bang et al., 2016).

Moderate antiviral (enveloped linear segmented single-stranded (−) RNA) hydroxycinnamic acid derivative: Rosmarinic acid methyl ester inhibited the replication of the Influenza A virus (H1N1 A/PR/8/34) with the EC_{50} value of 22.6 μM and the selectivity index of 8.9 in Madin-Darby canine kidney cells via a mechanism involving, at least in part, neuraminidase inhibition (Bang et al., 2016).

Commentaries

i. The plant yields eudesmane-type sesquiterpenes (Jang et al., 2017).

ii. In Iran, *Salvia macrosiphon* Boiss. (local name: *Kenocheh*) is used for cough and jaundice.

iii. In Turkey, several plants in the genus *Salvia* L. are used to treat cold of which *Salvia candidissima* Vahl. subsp. *candidissima* (Local name: *Galabor*), *Salvia hydrangea* DC. ex Bentham (local name: *Koç out*), *Salvia multicaulis* Vahl. (local name: *Adaçayi*), *Salvia nemorosa* L. (local name: *Gemta*), *Salvia sclarea* L., *Dag çayi*, *Salvia verticillata* L. (Freyn. & Bornm.) Bornm. (local name: *Karabas otu*), and *Salvia tomentosa* Mill. (local name: *Calba*).

iv. In Pakistan, *Salvia aegyptiaca* L. (Local name: *Tukhum malanga*) is used for diarrhea and gonorrhea, *Salvia cabulica* Benth. is used for a cold and fever, and *Salvia spinosa* L is given toothaches (local name: *Ganach*). *Salvia moorcroftiana* wall ex Benth. (local name: *Gadakan*) is given for boil, cough, and dysentery in Pakistan and dysentery in India.

v. In India, *Salvia officinalis* L. (sage) is used for thrush, sore throat, and fever. Essential oil of *Salvia officinalis* L. inhibited the growth of *Bacillus subtilis* (ATCC 6633), *Bacillus cereus* (ATCC 11778), *Staphylococcus aureus* (ATCC 6538), *Staphylococcus aureus* (ATCC 29213), and *Escherichia coli* (ATCC 35218) (Rosato et al., 2007). Essential oil of *Salvia officinalis* L. inhibited the replication of the Severe acute respiratory-associated coronavirus and the Herpes simplex virus type-1 (Loizzo et al., 2008) as well as Influenza virus A (H1N1) (Jalali et al., 2016). Ethanol extract of *Salvia officinalis* L. inhibited the replication of the Avian infectious bronchitis virus with the IC_{50} value as low as 0.003 μg/mL (Lelešius et al., 2019). Aqueous extract of this herb inhibited the growth of *Mycobacterium tuberculosis* (Fitzpatrick, 1954). *Salvia officinalis* L. held great medicinal repute by European physicians of the Middle Ages and Renaissance and the 12th century "*Regimen Sanitatis Salnitanum*" states "*cur moriatur homo cui Salvia crescit in horto*?" meaning "why should he die, whose garden growth sage?"

vi. *Salvia aculata* R.Br is given for leucorrhoea in Bangladesh (local name: *Hath mutha*)

vii. In China, *Salvia miltiorrhiza* Bunge (local name: *Dan shen*) is used for wounds and for the treatment of COVID-19. Aqueous extract of root at the concentration of 250 μg/mL inhibited the Human immunodeficient type-1 protease by 34.9% (Xu et al., 1996). This herb yield phenanthrofuran diterpenes such as tanshinone IIB, methyl tanshinonate, tanshinone I, and rosmariquinone inhibiting the Severe acute respiratory syndrome-associated coronavirus 3-chymotrypsin-like protease with the IC_{50} values of 24.8, 21.1, 14.4, and 21. 1 μM, respectively (Park et al., 2012). Tanshinone IIB, cryptotanshinone, and dihydrotanshinone I inhibited Severe acute respiratory syndrome-associated coronavirus papain-like protease with the IC_{50} values of 10.7, 0.8, and 4.9 μM, respectively (Park et al., 2012). *Salvia japonica* Thunb. is used for abscesses in China.

viii. In the Subfamily Nepetoideae, Tribe Satureieae: Acetone extract of *Acinos rotundifolius* Pers. (1.2 mg/well in agar) inhibited the growth of *Staphylococcus aureus*, *Micrococcus luteus*, *Mycobacterium smegmatis*, *Bacillus subtilis*, *Bacillus subtilis* var. *niger*, *Klebsiella pneumonae*, and *Pseudomonas aeruginosa* with inhibition zone diameter of 24, 27, 21, 28, 40, 22 and 23 mm, respectively (Ulukanli et al., 2005).

ix. In the Tribe Satureieae, the extract of *Melissa officinalis* L. (Lemon balm) inhibited the replication of the Avian infectious bronchitis virus with the IC_{50} value of 0.01 µg/mL (Lelešius et al., 2019). Aqueous extract of this herb inhibited the replication of the Semliki Forest, Newcastle, vaccinia, and herpes simplex viruses (Cohen et al., 1964), and there is the gut feeling that it could be of value against COVID-19. Eradication of COVID-19 will be attained by vaccination (which offer some level of protection against COVID-19 and hopefully against its variants) and oral medicines including phytomedication.

x. In the Tribe Satureieae, *Origanum vulgare* L. (oregano) is used for tuberculosis in Cambodia, Laos, Vietnam. Essential oil of *Origanum vulgare* L. inhibited the growth of *Candida albicans* (ATCC 14053), *Candida glabrata* (NRRL 417), *Candida krusei* (NRRL 7179), *Candida parapsilosis* (NRRL 22019), and *Candida tropicalis* (ATCC 750) with the MIC values of 0.7, 0.3, 0.7, 0.1, and 0.7 µg/mL, respectively (Rosato et al., 2009). Methanol extract (50 mL per well/1 mg/5 mL) of *Origanum vulgare* L. inhibited the growth of *Escherichia coli*, *Klebsiella pneumoniae*, *Pseudomonas aeruginosa*, *Proteus mirabilis*, *Salmonella typhi*, *Salmonella paratyphi* A, *Salmonella paratyphi* B, *Salmonella typhimurium*, *Shigella boydii*, *Shigella flexneri*, *Shigella sonnei*, *Staphylococcus aureus*, and *Streptococcus faecalis* with inhibition zone diameter of 14.5, 16.4, 18.7, 19.5, 21.8, 20.5, 16.7, 18.4, 17.9, 19.6, 19.5, 22.6, and 20.5 mm, respectively (Satish et al., 2009). Methanol extract of leaves of *Origanum vulgare* L. (6 mg/ well) inhibited the growth of *Bacillus cereus*, *Listeria monocytogenes*, *Staphylococcus aureus*, *Escherichia coli*, and *Salmonella anatum* with inhibition zone diameter of 11, 13.7, 24.2, 7.3, and 10.5 mm, respectively (Shan et al., 2007). Essential oil inhibited the growth of *Bacillus subtilis* (ATCC 6633), *Bacillus cereus* (ATCC 11778), *Staphylococcus aureus* (ATCC 6538), *Staphylococcus aureus* (ATCC 29213), and *Escherichia coli* (ATCC 35218) with MIC values of 0.3, 0.7, 0.7, 0.7, and 0.7 mg/mL, respectively (Rosato et al., 2007). Essential oil inhibited the growth of *Escherichia coli*, *Staphylococcus aureus*, and *Streptococcus mutans* (Miller et al., 2015). Carvacrol inhibited the growth of *Streptococcus mutans* (ATCC 25175) with an MIC value of 125 µg/mL (Park et al., 2003), inhibited *Candida albicans* with an MIC value of 100 ppm (Tampieri et al., 2005), and inhibited the growth of *Bacillus subtilis* (ATCC 6633), *Bacillus cereus* (ATCC 11778), *Staphylococcus aureus* (ATCC 6538), *Staphylococcus aureus* (ATCC 29213), and *Escherichia coli* (ATCC 35218) with MIC values of 0.3, 2.8, 0.7, 0.7, and 1.4 mg/mL, respectively (Rosato et al., 2007). Carvacrol at 0.2% inhibited the replication of the Murine norovirus (MNV (strain S7-PP3) in RAW 264·7 cells (Gilling et al., 2014) and strongly inhibited the replication of the Herpes simplex virus type-1, Respiratory syncytial virus, Human rotavirus, and Bovine diarrhea virus (Pialu et al., 2011).

xi. In the Tribe Satureieae *Thymus linearis* Benth. is used in Nepal (local name: *Ghoda marcha*) for toothaches. Essential oil of *Thymus serphyllum* L. inhibited the growth of *Bacillus subtilis* (ATCC 6633), *Bacillus cereus* (ATCC 11778), *Staphylococcus aureus* (ATCC 6538), *Staphylococcus aureus* (ATCC 29213), and *Escherichia coli* (ATCC 35218) with MIC values of 0.7, 1.4, 0.2, 0.7, and 0.7 mg/mL, respectively (Rosato et al., 2007). Chloroform extract of aerial part of *Thymus capitatus* (L.) Hoffmanns. & Link weakly inhibited the growth of *Bacillus cereus*, *Bacillus subtilis*, *Enterococcus faecalis*, *Staphylococcus aureus*, *Staphylococcus epidermidis* (Abdelah Bogdadi et al., 2007). Essential oil of *Thymus vulgaris* L. (thyme) inhibited *Staphylococcus aureus* (ATCC 25923), *Klebsiella pneumoniae* (NCTC 9633), and *Candida albicans* (ATCC 10231) with MIC values of 4.7, 4.0, and 3.3 µg/mL, respectively (Van Vuuren et al., 2009). Essential oil of *Thymus vulgaris* L. inhibited the growth of *Legionella pneumophila* (strain GC-4, serogroup 5/8), *Legionella pneumophila* (strain GC-5, serogroup 4), *Legionella pneumophila* (strain GC-6 strain, serogroup 8), *Legionella pneumophila* (strain GC 6, serogroup 8), and *Legionella pneumophila* (strain Paris, serogroup 1) with MIC values inferior to 0.03 mg/

mL, respectively (Chaftar et al., 2015). Ethanol extract of *Thymus vulgarus* L. inhibited he replication of the Avian infectious bronchitis virus with the IC_{50} value of 0.01 µg/mL (Lelešius et al., 2019). Thymol inhibited the growth of *Streptococcus mutans* (ATCC 25175) with MIC value 500 µg/mL (Park et al., 2003), and inhibited the growth of *Bacillus subtilis* (ATCC 6633), *Bacillus cereus* (ATCC 11778), *Staphylococcus aureus* (ATCC 6538), *Staphylococcus aureus* (ATCC 29213), and *Escherichia coli* (ATCC 35218) with MIC values of 0.7, 0.7, 0.7, 0.7, and 1.4 mg/mL, respectively (Rosato et al., 2007). Thymol inhibited the growth of *Streptococcus pyogenes* with MIC of 25 µg/mL and MBC > 100 µg/mL (Macé et al., 2017).

xii. In the Tribe Satureieae, Methanol fraction of aerial part of *Satureja nervosa* Desf. inhibited the growth of *Staphylococcus aureus*, *Staphylococcus epidermidis*, *Staphylococcus saprophyticus* with MIC values of 1.2, 0.1, and 0.3 mg/mL, respectively (Sassi et al., 2007). Methanol extract (50 mL per well/ 1 mg/5 mL) of *Satureja hortensis* L. inhibited the growth of *Escherichia coli*, *Klebsiella pneumoniae*, *Pseudomonas aeruginosa*, *Proteus mirabilis*, *Salmonella typhi*, *Salmonella paratyphi* A, *Salmonella paratyphi* B, *Salmonella typhimurium*, *Shigella boydii*, *Shigella flexneri*, *Shigella sonnei*, *Staphylococcus aureus*, and *Streptococcus faecalis* with inhibition zone diameter of 11.4, 12.2, 11.9, 9.6, 10.1, 10.3, 11.7, 10.4, 10.5, 9.8, 10.4, 13.3, and 12.7 mm, respectively (Satish et al., 2009). Extract of *Satureja hortensis* L. inhibited he replication of the Avian infectious bronchitis virus with the IC_{50} value of 0.04 µg/mL (Lelešius et al., 2019).

xiii. Subfamily Nepetoideae, Tribe Nepeteae, *Agastache rugosa* (Fisch. & Mey.) O Ktze. is used for influenza, cholera, fever, and COVID-19 (Lem et al., 2020). In the Tribe Nepeteae, *Schizonepeta tenuifolia* (Bth.) Briq. is used for boils and tuberculosis in China.

xiv. In the Subfamily Nepetoideae, Tribe Prunelleae, *Prunella vulgaris* L. is used for abscesses in Korea and from this plant betulinic acid and 2α, 3α-dihydroxyurs-12-en-28-oic acid inhibited Herpes simplex virus with the EC_{50} values of 30 and 8 µg/mL, respectively (Ryu et al., 1992).

xv. *Nepeta elliptica* Royle ex Benth. in the Subfamily Nepetoideae; Tribe Nepeteae is used for dysentery in India, *Glechoma brevibuta* Kuprian is a remedy for a cold in China, and *Glechoma hederacea* L. is used for pneumonia in Japan. *Glechoma longituba* (Nakai) Kuprian. is used in China (local name: *Lianqiao*) for the treatment of COVID-19 (Luo, 2020; Wang et al., 2020).

xvi. Subfamily Nepetoideae Tribe Satureieae: *Calamintha clinopodium* Benth is used for gonorrhea in Taiwan. Ethanol extract of *Hyssopus officinalis* L. in the Tribe Satureieae inhibited the replication of the Avian infectious bronchitis virus with the IC_{50} value of 0.07 µg/mL (Lelešius et al., 2019).

xvii. In the Subfamily Nepetoideae, Tribe Ocimeae, Methanol extract of leaves of *Basilicum polystachyon* inhibited the growth of *Escherichia coli* (AG100 *Atet*: AG100 super-expressing *AcrAB* pumps, containing TET^R gene actif) with MIC/MBC of 512/1024 µg/mL (Touani et al., 2014). The extract inhibited the growth of *Enterobacter aerogenes* (ATCC 13048), *Klebsiella pneumoniae* (K2; MDR clinical strain), and *Providencia stuarti* (ATCC 29916), with MIC values of 256, 512, and 260 µg/mL (Touani et al., 2014). In the Tribe Ocimeae *Mesona chinensis* Benth. is used for gonorrhea in China. In the Tribe Ocimeae, *Plectranthus macranthus* Hook f. is a remedy for sores in China.

xviii. Subfamily Nepetoideae, Tribe Ocimeae, *Coleus amboinicus* Lour. is used for flu in Cambodia, Laos, and Vietnam. *Coleus tuberosus* (Bl.) Benth. is a remedy for cholera in Indonesia. *Coleus scutallerioides* (L.) Benth. is used to treat infections in the Philippines. Ethyl acetate fraction of aerial parts of *Coleus forskolii* Briq. inhibited the replication of the Human immunodeficiency virus type-1 (NL4.3 strain) virus in CD4+ cells by 66.5% with a maximum non-cytotoxic concentration of 5 µg/mL (Sabde et al., 2011).

xix. In the Tribe Ocimeae, ethanol extract of leaves of *Orthosiphon aristatus* (Blume) Miq. inhibited the growth of *Staphylococcus aureus, Enterococcus faecalis, Pseudomonas aeruginosa,* and *Candida albicans* with MIC values of 256, 256, 256, and 128 µg/mL, respectively (Romulo et al, 2018).

xx. Subfamily Nepetoideae, Tribe Elsholtzieae: *Elsholtzia haichowensis* Sun. is a remedy for typhoid in China. In the Tribe Elsholtzieae *Lycopus lucidus* Turcz. ex Benth. is a remedy for abscesses in Korea. Methanol extract of roots of a member of the genus *Lycopus* L. at the concentration of 100 µg/mL inhibited RNA-dependent DNA polymerase activity of Human immunodeficiency virus type-1 reverse transcriptase by 10.7%, respectively (Min et al., 2001). In the Tribe Elsholtzieae, *Perilla frutescens* (L.) Britt. is used for cold in China. The plant is used in China for the treatment of COVID-19 in combination with oseltamivir phosphate (Luo, 2020). Extract of leaves inhibited the Human immunodeficiency type-1 protease (Park, 2003).

xxi. In the Subfamily Nepetoideae, Tribe Rosmarineae, ethanol fraction of leaves of *Rosmarimus officinalis* L. (rosemary) weakly inhibited the growth of *Bacillus cereus, Bacillus subtilis, Enterococcus faecalis, Staphylococcus aureus, Staphylococcus epidermidis, Escherichia coli,* and *Pseudomonas aeruginosa* (Abdelah Bogdadi et al., 2007). Essential oil of *Rosmarinus officinalis* L. inhibited the growth of *Bacillus subtilis* (ATCC 6633), *Bacillus cereus* (ATCC 11778), *Staphylococcus aureus* (ATCC 6538), *Staphylococcus aureus* (ATCC 29213), and *Escherichia coli* (ATCC 35218) (Rosato et al., 2007). Essential oil of *Rosmarinus officinalis* L. inhibited *Staphylococcus aureus* (ATCC 25923), *Klebsiella pneumoniae* (NCTC 9633), and *Candida albicans* (ATCC 10231) with MIC values of 6, 8, and 6 µg/mL, respectively (Van Vuuren et al., 2009). Essential oil of inhibited *Legionella pneumophila* (strain GC-4, serogroup 5/8), *Legionella pneumophila* (strain GC-5, serogroup 4), *Legionella pneumophila* (strain GC-6 strain, serogroup 8), *Legionella pneumophila* (strain GC 6, serogroup 8), *Legionella pneumophila* (strain GC13, serogroup 6), and *Legionella pneumophila* (strain Paris, serogroup 1 (Chaftar et al., 2015). Essential oil of inhibited the growth of *Streptococcus mutans* with MIC of 0.5 µg/mL (Miller et al., 2015).

xxii. In the Subfamily Lamioideae, Tribe Pogostemeae, *Dysophylla auricularia* (L.) Bl. is a remedy for fever in Cambodia, Laos, and Vietnam and is used for wounds in China. In the Tribe Pogostemeae, *Pogostemon heyneanus* Bth. is a remedy for boils in Malaysia and wounds in the Philippines.

xxiii. In the Subfamily Lamioidease, Tribe Prasieae, *Gomphostemma phlomoides* Benth. is a remedy for wounds in Indonesia. In the Subfamily Lamioideae, Tribe Stachydeae, *Stachys sieboldii* Miq. is given for influenza in China. In the Subfamily Lamioideae, Tribe Lamieae, *Lamium chinense* Benth. is a remedy for influenza in China. In the Tribe Lamieae, methanol extract of roots of *Phlomis umbrosa* Turcz. at the concentration of 100 µg/mL inhibited RNA-dependent DNA polymerase activity of Human immunodeficiency virus type-1 reverse transcriptase and Human immunodeficiency virus type-1 protease by 13.2% and 11.6%, respectively (Min et al., 2001).

xxiv. In the Subfamily Scutellarioideae, *Scutellaria indica* L. in used in China for mycoses. *Scutellaria baicalensis* Georgi is used in China for the treatment of COVID-19 in combination with oseltamivir phosphate (Luo, 2020; Yang et al., 2020). *Scutellaria baicalensis* Georgi is used for pneumonia in Japan. Aqueous extract of root at the concentration of 250 µg/mL inhibited the Human immunodeficiency virus-1 protease by 96.8% (Xu et al., 1996). Aqueous extract of roots of *Scutellaria baicalensis* Georgi inhibited the replication of Respiratory Syncitial Virus (Long strain) in Hep-2 cells with IC_{50} value of 22.8 µg/mL and a selectivity index of 10.7 (Ma et al., 2002). From this extract, wogonin, oroxylin, baicalein, scutellarein, and baicalin inhibited the replication of the Respiratory syncytial virus with the IC_{50} values of 7.4, 14.5, 20.8, 20.8, and 20.8 µg/mL and selectivity indexes of 16.1, 4.0, 12.0, 16.0, and 12.0, respectively (Ma et al., 2002). Methanol extract of root of *Scutellaria baicalensis* Georgi (6 mg/well) inhibited the growth of *Bacillus cereus, Listeria monocytogenes, Staphylococcus aureus, Escherichia coli,* and *Salmonella anatum* with inhibition zone diameter of 10.9, 11, 15.5, 6.2, and 8.6 mm, respectively (Shan et al., 2007). Baicalein at the concentration of 300 µM inhibited Aminopeptidase N by 57% and angiotensin-converting hormone by 10% (Bauvois & Dauzonne, 2006).

xxv. In Nepal, *Caryopteris nepalensis* Moldenke (local name: *Hedera*) is given to treat ulcers in the mouth. Methanol extract of aerial parts of *Caryopteris divaricata* Maxim. in the Subfamily Ajugoideae, Tribe Caryopterideae, at the concentration of 100 µg/mL inhibited the Human immunodeficiency type-1 protease by 25.7%, respectively (Min et al., 2001).

References

Bang, S., Ha, T.K.Q., Lee, C., Li, W., Oh, W.K. and Shim, S.H., 2016. Antiviral activities of compounds from aerial parts of *Salvia plebeia* R. Br. *Journal of Ethnopharmacology, 192*, pp. 398–405.

Chaftar, N., Girardot, M., Quellard, N., Labanowski, J., Ghrairi, T., Hani, K., Frère, J. and Imbert, C., 2015. Activity of six essential oils extracted from Tunisian plants against *Legionella pneumophila. Chemistry & Biodiversity, 12*(10), pp. 1565–1574.

Fitzpatrick, F.K., 1954. Plant substances active against *Mycobacterium tuberculosis. Antibiotics and Chemotherapy* 4, 528–536.

Gilling, D.H., Kitajima, M., Torrey, J.R. and Bright, K.R., 2014. Antiviral efficacy and mechanisms of action of oregano essential oil and its primary component carvacrol against murine norovirus. *Journal of Applied Microbiology, 116*(5), pp. 1149–1163.

Jalali, P., Moattari, A., Mohammadi, A., Ghazanfari, N. and Pourghanbari, G., 2016. Melissa officinalis efficacy against human influenza virus (New H1N1) in comparison with oseltamivir. *Asian Pacific Journal of Tropical Disease, 6*(9), pp. 714–717.

Jang, H.J., Lee, S., Lee, S.J., Lim, H.J., Jung, K., Kim, Y.H., Lee, S.W. and Rho, M.C., 2017. Anti-inflammatory activity of eudesmane-type sesquiterpenoids from *Salvia plebeia. Journal of Natural Products, 80*(10), pp. 2666–2676.

Lelešius, R., Karpovaitė, A., Mickienė, R., Drevinskas, T., Tiso, N., Ragažinskienė, O., Kubilienė, L., Maruška, A. and Šalomskas, A., 2019. In vitro antiviral activity of fifteen plant extracts against avian infectious bronchitis virus. *BMC Veterinary Research, 15*(1), p. 178.

Loizzo, M.R., Saab, A.M., Tundis, R., Statti, G.A., Menichini, F., Lampronti, I., Gambari, R., Cinatl, J. and Doerr, H.W., 2008. Phytochemical analysis and in vitro antiviral activities of the essential oils of seven Lebanon species. *Chemistry & Biodiversity, 5*(3), pp. 461–470.

Miller, A.B., Cates, R.G., Lawrence, M., Soria, J.A.F., Espinoza, L.V., Martinez, J.V. and Arbizú, D.A., 2015. The antibacterial and antifungal activity of essential oils extracted from Guatemalan medicinal plants. *Pharmaceutical Biology, 53*(4), pp. 548–554.

Min, B.S., Kim, Y.H., Tomiyama, M., Nakamura, N., Miyashiro, H., Otake, T. and Hattori, M., 2001. Inhibitory effects of Korean plants on HIV-1 activities. *Phytotherapy Research, 15*(6), pp. 481–486.

Park, K.M., You, J.S., Lee, H.Y., Baek, N.I. and Hwang, J.K., 2003. Kuwanon G: an antibacterial agent from the root bark of Morus alba against oral pathogens. *Journal of Ethnopharmacology, 84*(2–3), pp. 181–185.

Park, J.Y., Kim, J.H., Kim, Y.M., Jeong, H.J., Kim, D.W., Park, K.H., Kwon, H.J., Park, S.J., Lee, W.S. and Ryu, Y.B., 2012. Tanshinones as selective and slow-binding inhibitors for SARS-CoV cysteine proteases. *Bioorganic & Medicinal Chemistry, 20*(19), pp. 5928–5935.

Pilau, M.R., Alves, S.H., Weiblen, R., Arenhart, S., Cueto, A.P. and Lovato, L.T., 2011. Antiviral activity of the *Lippia graveolens* (Mexican oregano) essential oil and its main compound carvacrol against human and animal viruses. *Brazilian Journal of Microbiology, 42*, pp. 1616–1624.

Qu, X.J., Xia, X., Wang, Y.S., Song, M.J., Liu, L.L., Xie, Y.Y., Cheng, Y.N., Liu, X.J., Qiu, L.L., Xiang, L. and Gao, J.J., 2009. Protective effects of *Salvia plebeia* compound homoplantaginin on hepatocyte injury. *Food and Chemical Toxicology, 47*(7), pp. 1710–1715.

Rosato, A., Vitali, C., De Laurentis, N., Armenise, D. and Milillo, M.A., 2007. Antibacterial effect of some essential oils administered alone or in combination with Norfloxacin. *Phytomedicine, 14*(11), pp. 727–732.

Rosato, A., Vitali, C., Gallo, D., Balenzano, L. and Mallamaci, R., 2008. The inhibition of Candida species by selected essential oils and their synergism with amphotericin B. *Phytomedicine*, *15*(8), pp. 635–638.

Rosato, A., Vitali, C., Piarulli, M., Mazzotta, M., Argentieri, M.P. and Mallamaci, R., 2009. In vitro synergic efficacy of the combination of Nystatin with the essential oils of *Origanum vulgare* and Pelargonium graveolens against some Candida species. *Phytomedicine*, *16*(10), pp. 972–975.

Ryu, S.Y., Lee, C.K., Lee, C.O., Kim, H.S. and Zee, O.P., 1992. Antiviral triterpenes from *Prunella vulgaris. Archives of Pharmacal Research*, *15*(3), pp. 242–245.

Satish, S., Mohana, D.C., Raghavendra, M.P., Babu, S., Raveesha, K.A. (2009) Antibacterial evaluation of some Iranian medicinal plants against some human pathogenic bacteria. *Asian Journal of Microbiology, Biotechnology and Environmental Sciences,* 11(4), pp. 735–738.

Shan, B., Cai, Y.Z., Brooks, J.D. and Corke, H., 2007. The in vitro antibacterial activity of dietary spice and medicinal herb extracts. *International Journal of Food Microbiology*, *117*(1), pp. 112–119.

Wang, S.X., Wang, Y., Lu, Y.B., Li, J.Y., Song, Y.J., Nyamgerelt, M. and Wang, X.X., 2020. Diagnosis and treatment of novel coronavirus pneumonia based on the theory of traditional Chinese medicine. *Journal of Integrative Medicine*, *18*(4), pp. 275–283

Xu, H.X., Wan, M., Loh, B.N., Kon, O.L., Chow, P.W. and Sim, K.Y., 1996. Screening of traditional medicines for their inhibitory activity against HIV-1 protease. *Phytotherapy Research*, *10*(3), pp. 207–21

Abdelah Bogdadi, H.A., Kokoska, L., Havlik, J., Kloucek, P., Rada, V. and Vorisek, K., 2007. In Vitro. antimicrobial activity of some Libyan medicinal plant extracts. *Pharmaceutical Biology*, *45*(5), pp. 386–391.

Bauvois, B. and Dauzonne, D., 2006. Aminopeptidase-N/CD13 (EC 3.4. 11.2) inhibitors: Chemistry, biological evaluations, and therapeutic prospects. *Medicinal Research Reviews*, *26*(1), pp. 88–130.

Chaftar, N., Girardot, M., Quellard, N., Labanowski, J., Ghrairi, T., Hani, K., Frère, J. and Imbert, C., 2015. Activity of six essential oils extracted from Tunisian plants against *Legionella pneumophila. Chemistry & Biodiversity*, *12*(10), pp. 1565–1574.

Cohen, R.A., Kucera, L.S. and Herrmann Jr, E.C., 1964. Antiviral Activity of Melissa officinalis (Lemon Balm) Extract. *Proceedings of the Society for Experimental Biology and Medicine*, *117*(2), pp. 431–434.

Lelešius, R., Karpovaitė, A., Mickienė, R., Drevinskas, T., Tiso, N., Ragažinskienė, O., Kubilienė, L., Maruška, A. and Šalomskas, A., 2019. In vitro antiviral activity of fifteen plant extracts against avian infectious bronchitis virus. *BMC Veterinary Research*, *15*(1), p. 178.

Lem, F.F., Opook, F., Herng, D.L.J., Na, C.S., Lawson, F.P. and Tyng, C.F., 2020. Molecular mechanism of action of repurposed drugs and traditional Chinese medicine used for the treatment of patients infected with COVID-19: A systematic review. *medRxiv*

Luo, A., 2020. Positive SARS-Cov-2 test in a woman with COVID-19 at 22 days after hospital discharge: A case report. *Journal of Traditional Chinese Medical Sciences.*

Ma, S.C., Du, J., But, P.P.H., Deng, X.L., Zhang, Y.W., Ooi, V.E.C., Xu, H.X., Lee, S.H.S. and Lee, S.F., 2002. Antiviral Chinese medicinal herbs against respiratory syncytial virus. *Journal of Ethnopharmacology*, *79*(2), pp. 205–211.

Macé, S., Truelstrup Hansen, L. and Rupasinghe, H.P., 2017. Anti-Bacterial Activity of Phenolic Compounds against Streptococcus pyogenes. *Medicines*, *4*(2), p. 25.

Miller, A.B., Cates, R.G., Lawrence, M., Soria, J.A.F., Espinoza, L.V., Martinez, J.V. and Arbizú, D.A., 2015. The antibacterial and antifungal activity of essential oils extracted from Guatemalan medicinal plants. *Pharmaceutical Biology*, *53*(4), pp. 548–554.

Min, B.S., Kim, Y.H., Tomiyama, M., Nakamura, N., Miyashiro, H., Otake, T. and Hattori, M., 2001. Inhibitory effects of Korean plants on HIV-1 activities. *Phytotherapy Research*, *15*(6), pp. 481–486.

Mohammed, M.J. and Al-Bayati, F.A., 2009. Isolation and identification of antibacterial compounds from Thymus kotschyanus aerial parts and Dianthus caryophyllus flower buds. *Phytomedicine*, *16*(6), pp. 632–637.

Park, K.M., You, J.S., Lee, H.Y., Baek, N.I. and Hwang, J.K., 2003. Kuwanon G: an antibacterial agent from the root bark of Morus alba against oral pathogens. *Journal of Ethnopharmacology*, *84*(2–3), pp. 181–185.

Rosato, A., Vitali, C., De Laurentis, N., Armenise, D. and Milillo, M.A., 2007. Antibacterial effect of some essential oils administered alone or in combination with Norfloxacin. *Phytomedicine*, *14*(11), pp. 727–732.

Sabde, S., Bodiwala, H.S., Karmase, A., Deshpande, P.J., Kaur, A., Ahmed, N., Chauthe, S.K., Brahmbhatt, K.G., Phadke, R.U., Mitra, D. and Bhutani, K.K., 2011. Anti-HIV activity of Indian medicinal plants. *Journal of Natural Medicines*, *65*(3–4), pp. 662–669.

Sassi, A.B., Harzallah-Skhiri, F. and Aouni, M., 2007. Investigation of some medicinal plants from Tunisia for antimicrobial activities. *Pharmaceutical Biology*, *45*(5), pp. 421–428.
Shan, B., Cai, Y.Z., Brooks, J.D. and Corke, H., 2007. The in vitro antibacterial activity of dietary spice and medicinal herb extracts. *International Journal of Food Microbiology*, *117*(1), pp. 112–119.
Tampieri, M.P., Galuppi, R., Macchioni, F., Carelle, M.S., Falcioni, L., Cioni, P.L. and Morelli, I., 2005. The inhibition of Candida albicans by selected essential oils and their major components. *Mycopathologia*, *159*(3), pp. 339–345.
Touani, F.K., Seukep, A.J., Djeussi, D.E., Fankam, A.G., Noumedem, J.A. and Kuete, V., 2014. Antibiotic-potentiation activities of four Cameroonian dietary plants against multidrug-resistant Gram-negative bacteria expressing efflux pumps. *BMC Complementary and Alternative Medicine*, *14*(1), p. 258.
Ulukanli, Z., Ulukanli, S., Ozbay, H., Ilcim, A. and Tuzcu, M., 2005. Antimicrobial activities of some plants from the Eastern Anatolia region of Turkey. *Pharmaceutical Biology*, *43*(4), pp. 334–339.
Van Vuuren, S.F., Suliman, S. and Viljoen, A.M., 2009. The antimicrobial activity of four commercial essential oils in combination with conventional antimicrobials. *Letters in Applied Microbiology*, *48*(4), pp. 440–446.
Wang, S.X., Wang, Y., Lu, Y.B., Li, J.Y., Song, Y.J., Nyamgerelt, M. and Wang, X.X., 2020. Diagnosis and treatment of novel coronavirus pneumonia based on the theory of traditional Chinese medicine. *Journal of Integrative Medicine*, *18*(4), pp. 275–283
Xu, H.X., Wan, M., Loh, B.N., Kon, O.L., Chow, P.W. and Sim, K.Y., 1996. Screening of Traditional Medicines for their Inhibitory Activity Against HIV-1 Protease. *Phytotherapy Research*, *10*(3), pp. 207–210.
Yang, Q.X., Zhao, T.H., Sun, C.Z., Wu, L.M., Dai, Q., Wang, S.D. and Tian, H., 2020. New thinking in the treatment of 2019 novel coronavirus pneumonia. *Complementary Therapies in Clinical Practice*, *39*, p. 101131.

10.4.5 Family Oleaceae Hoffmannsegg et Link (1809)

The family Oleaceae consists of about 28 genera and 400 species of trees or shrubs. The leaves are simple or pinnate, opposite, and exstipulate. The inflorescences are cymes, panicles, racemes, umbels, or fascicles. The flowers are actinomorphic and ephemeral. The calyx is tubular and 4 or more lobed. The corolla is none or tubular and develops 4 to 10 lobes. The androecium comprises 2 anthers. The gynoecium consists of 2 carpels united into a 2-locular ovary, each locule shelter 2 ovules. The fruit is a capsule, a drupe, or a berry, containing 1 to 4 seeds. Plants in this family produce glycosides of phenethyl and hydroxycinnamic acid derivatives glycosides and iridoid glycosides.

10.4.5.1 *Forsythia suspensa* (Thunb.) Vahl

Common names: Weeping Forsythia, weeping golden bells; lian qiao (China); yeon-gyo, dang gae na ri (Korea)

Synonyms: *Forsythia fortunei* Lindl., *Ligustrum suspensum* Thunb; *Rangium suspensum* (Thunb.) Ohwi, *Syringa suspensa* (Thunb.) Thunb. ex Murray

Habitat: Cultivated, thickets, meadows

Distribution: China, Korea

Botanical observation: This beautiful shrub grows up to about 5 m tall. The stems are pendulous. The leaves are simple, opposite, and exstipulate. The petiole is up to 1.5 cm long. The blade is lanceolate, 1.5–2 cm × 5–10 cm, papery, round at base, serrate, and acute at apex. The flowers are axillary. The calyx is deeply 4-lobed, the lobes oblong, 6 mm long and ciliate. The corolla is 4-lobed, and the lobes oblong, 2 cm long, and yellow. The androecium includes 2 stamens which are about 4 mm long. The gynoecium is 6 mm long with a bilobed stigma. The capsules are ovoid, 2.5 cm × 1.2 cm, and lenticelled.

Medicinal uses: Antiseptic, fever (China)

Strong broad-spectrum antibacterial hydroxycinnamic acid derivative glycosides: Forsythiaside isolated from the fruits inhibited the growth of *Escherichia coli*, *Pseudomonas aeruginosa*, and *Staphylococcus aureus* with the MIC values of 38.3, 38.3, and 76.6 µg/mL, respectively

(Qu et al., 2008). Lianqiaoxinoside B inhibited the growth of *Micrococcus* sp. and *Bacillus dysenteriae* the MIC values of 28.5 and 36.7 μg/mL (Kuang et al., 2011).

Strong antiviral (non-enveloped monopartite linear single-stranded DNA) dibenzylbutyrolactone lignan: Arctigenin (Kang et al., 2008) at the concentration of 62.5 μg/mL inhibited the replication of the Porcine circovirus type 2 (WH strain) in the PK-15 *cells* (Cheng et al., 2016).

Strong antiviral (enveloped monopartite linear single-stranded (−) RNA) labdane-type diterpenes: Forsypensins A-E isolated from the fruits inhibited the replication of the Respiratory syncytial virus with the EC_{50} values of 14.6, 15.4, 13.7, 11.8 and 10.5 μM, respectively, in Hep-2 cells (Xiang et al., 2020; Zhao et al., 2020).

Moderate antiviral (enveloped linear segmented single-stranded (−) RNA) hydroxycinnamic acid glycoside derivative: Forsythoside A at the concentration of at 160 μM abrogated the replication of the Influenza A H1N1 virus, oseltamivir-resistant Influenza A H1N1 virus pandemic Influenza A H1N1 Influenza A H9N2/G1 and Influenza A H3N2 via a mechanism involving inhibition of M1 protein and blockage of release of newly formed viruses from the host cells (Law et al., 2017).

Moderate antiviral (enveloped monopartite linear single-stranded (+) RNA) furofuran lignan glycoside: Phillyrin dose-dependently inhibited the replication of the Severe acute respiratory syndrome-associated coronavirus 2 and Human coronavirus (229E) in Vero cells with the IC_{50} values if 63.9 and 64.5 μg/mL and the selectivity indexes of 30.6 and 16.2, respectively (Ma et al., 2020).

Very weak (enveloped monopartite linear single-stranded (+) RNA) antiviral hydroxycinnamic acid derivative glycoside: Forsythoside at the concentration of 640 μM abrogated the replication of the Avian infectious bronchitis virus strain M41 in chicken embryo kidney cells (Li et al., 2011).

In vivo antiviral (non-enveloped monopartite linear single-stranded DNA) dibenzylbutyrolactone lignan: Arctigenin given to mice intraperitoneally at the dose of 200 μg/Kg evoked protection against the Porcine circovirus type 2 (Cheng et al., 2016).

In vivo (enveloped monopartite linear single-stranded (+) RNA) dibenzylbutyrolactone lignan: Arctigenin given intraperitoneally, twice daily at the dose of 10 mg/kg for 7 days protected mice against Japanese encephalitis virus infection (Swarup et al., 2008).

In vivo (enveloped linear segmented single-stranded (−) RNA) hydroxycinnamic acid glycoside derivative: Forsythoside A given orally at the dose of 10 μg/g to mice infected with influenza virus A/PR/8/34 evoked about 80% survival (Law et al., 2017).

Viral enzyme inhibition by labdane-type diterpenes: Forsypensins A-E inhibited the activity of Influenza A virus (A/PR/8/34/H1N1) neuraminidase with the IC_{50} values of 21.8, 23.2, 22.9, 27.4, and 24.6 μM, respectively (Xiang et al., 2020; Zhao et al., 2020).

Commentaries

i. Intraperitoneal administration is not a convenient way to administer drugs and oral route is preferable. When taking infusions or decoctions of plants, the absorption of active principles can be ameliorated by the presence of other principles and examples have been given earlier.
ii. The COVID-19 pandemic is not the last because the burning of the primary rainforests, pollution of the seas, intensive farming, industrial activities, global warming, and other means to continuously and increasingly exploit resources (until exhaustion) for financial benefits are not apparently likely to end. The next pandemics are bound to be even more lethal including Influenza virus pandemics and other viral pandemics but also bacterial or fungal pandemics touching not only humans but also crops, poultries, and other domesticated animals. Thus, more than ever, knowledge on medicinal plants from traditional healers in Asia and the Pacific is to be used to formulate oral phytomedication allowing some sort of protection against zoonotic organisms. The fruits of *Forsythia suspensa* (Thunb.) Vahl are used in China for the treatment of COVID-19 and other viral pneumonia, and this plant is one of the numerous plants in this series with possible usefulness against

COVID-19. Why are these plants and phytomedication not known of and used by the Western world as antiviral phytomedication? This is a subject for discussion and action by law makers in the West world to allow affordable clinical testing of herbal remedies for the well-being and protection of their citizens. Also, there is a need to re-implement the teaching of pharmacognosy in order to train pharmacy graduates with the skills required to develop and dispense herbal products safely to their patients.

iii. Phillyrin (also known as forsythin) is produced by *Colletotrichum gloeosporioides* growing on *Forsythia suspensa* (Thunb.) Vahl (Zhang et al., 2012) suggesting that fungi can metabolize natural products produced by their host perhaps in order to eliminate other competing microorganisms.

iv. Arctigenin being able to protect mice against the Japanese encephalitis virus could have some value against the Severe acute respiratory syndrome-associated coronavirus 2. This lignan abounds in *Arctium lappa* L. in the Family Asteraceae is used for COVID-19 in China.

References

Chen, J., Li, W., Jin, E., He, Q., Yan, W., Yang, H., Gong, S., Guo, Y., Fu, S., Chen, X. and Ye, S., 2016. The antiviral activity of arctigenin in traditional Chinese medicine on porcine circovirus type 2. *Research in Veterinary Science*, *106*, pp. 159–164.

Kang, H.S., Lee, J.Y. and Kim, C.J., 2008. Anti-inflammatory activity of arctigenin from Forsythiae Fructus. *Journal of Ethnopharmacology*, *116*(2), pp. 305–312.

Kuang, H.X., Xia, Y.G., Liang, J., Yang, B.Y. and Wang, Q.H., 2011. Lianqiaoxinoside B, a novel caffeoyl phenylethanoid glycoside from *Forsythia suspensa*. *Molecules*, *16*(7), pp. 5674–5681.

Law, A.H.Y., Yang, C.L.H., Lau, A.S.Y. and Chan, G.C.F., 2017. Antiviral effect of forsythoside A from *Forsythia suspensa* (Thunb.) Vahl fruit against influenza A virus through reduction of viral M1 protein. *Journal of Ethnopharmacology*, *209*, pp. 236–247.

Li, H., Wu, J., Zhang, Z., Ma, Y., Liao, F., Zhang, Y. and Wu, G., 2011. Forsythoside a inhibits the avian infectious bronchitis virus in cell culture. *Phytotherapy Research*, *25*(3), pp. 338–342.

Ma, Q., Li, R., Pan, W., Huang, W., Liu, B., Xie, Y., Wang, Z., Li, C., Jiang, H., Huang, J. and Shi, Y., 2020. Phillyrin (KD-1) exerts anti-viral and anti-inflammatory activities against novel coronavirus (SARS-CoV-2) and human coronavirus 229E (HCoV-229E) by suppressing the nuclear factor kappa B (NF-κB) signaling pathway. *Phytomedicine*, *78*, p. 153296.

Qu, H., Zhang, Y., Wang, Y., Li, B. and Sun, W., 2008. Antioxidant and antibacterial activity of two compounds (forsythiaside and forsythin) isolated from Forsythia suspensa. *Journal of Pharmacy and Pharmacology*, *60*(2), pp. 261–266.

Swarup, V., Ghosh, J., Mishra, M.K. and Basu, A., 2008. Novel strategy for treatment of Japanese encephalitis using arctigenin, a plant lignan. *Journal of Antimicrobial Chemotherapy*, *61*(3), pp. 679–688.

Zhang, Q., Wei, X. and Wang, J., 2012. Phillyrin produced by *Colletotrichum gloeosporioides*, an endophytic fungus isolated from *Forsythia suspensa*. *Fitoterapia*, *83*(8), pp. 1500–1505.

Zhao, L., Xiang, K.L., Liu, R.X., Xie, Z.P., Zhang, S.M. and Dai, S.J., 2020. Anti-inflammatory and anti-viral labdane diterpenoids from the fruits of *Forsythia suspensa*. *Bioorganic Chemistry*, *96*, p. 103651.

10.4.5.2 *Nyctanthes arbor-tristis* L.

Synonym: *Parilium arbor-tristis* (L.) Gaertn.

Common names: Coral jasmine, night jasmine; shefali, shinguri phul (Bangladesh); sri gading (Indonesia, Malaysia); harsingar parijata (India); salika (Laos); hseik hpalu, seik-palu (Myanmar); kuri (Pakistan); kanika (Thailand); lai tau (Vietnam)

Habitat: Cultivated

Distribution: Pakistan to Indonesia

Botanical observation: This handsome tree grows to about 8 m tall. The stems are angled at apex. The leaves are simple, opposite, and exstipulate. The petiole is about 1 cm long. The blade grows to 10 cm × 5 cm, coriaceous, and hairy below. The inflorescences are axillary and terminal cymes which are few flowered and with a positive aura. The calyx is minute, tubular and truncate. The corolla is tubular, ephemeral, fragrant, the tube orange, 1 cm long, with about 8 lobes which are about 7 mm long and pure white. The androecium consists of 2 anthers. The ovary develops a cylindrical style and a bifid stigma. The fruits are capsules.

Medicinal uses: Dysentery (Bangladesh, Myanmar); fever (Indonesia; Myanmar)

Broad-spectrum antibacterial halo developed by mid-polar extract: Chloroform extract inhibited the growth of a broad spectrum of bacteria (Khatune et al., 2001).

Very weak antifungal (yeast) polar extract: Ethanol extract of leaves inhibited the growth of *Malassezia furfur* (1878), *Malassezia restricta* (7877), *Malassezia globose* (7966), and *Malassezia sympodialis* (9974) with the MIC values of 1220, 1470, 1050, and 1350 µg/mL, respectively (Mishra et al., 2016).

Strong antiviral (non-enveloped monopartite linear single-stranded (+) RNA) polar extract: Ethanol extract of seeds inhibited the replication of the Encephalomyocarditis virus and the Semliki forest virus at the concentration of 62.5 µg/mL, respectively, and the selectivity indexes of 4, respectively, in Vero cells (Gupta et al., 2005).

Strong antiviral (non-enveloped monopartite linear single-stranded (+) RNA) iridoid glycosides: Abortristoside A and C isolated from the seeds inhibited the replication of the Encephalomyocarditis virus at the concentration of 3.9 and 7.8 µg/mL, respectively, and the selectivity indexes of 6 and 4, respectively, in Vero cells (Gupta et al., 2005). Abortristoside A and C isolated from the seeds inhibited the replication of the Semliki forest virus at the concentration of 1.9 and 3.9 µg/mL, respectively, and the selectivity indexes of 8 and 6, respectively, in Vero cells (Gupta et al., 2005).

Commentary: The plant yields iridoid glycosides (Rathore et al., 1990) and is recommended by Ayurvedic physicians for the treatment of COVID-19 (Das, 2021).

References

Das, K., 2021. Herbal plants as immunity modulators against COVID-19: A primary preventive measure during home quarantine. *Journal of Herbal Medicine*, p. 100501.

Gupta, P., Bajpai, S.K., Chandra, K., Singh, K.L. and Tandon, J.S., 2005. Antiviral profile of *Nyctanthes arbortristis* L. against encephalitis causing viruses. *Indian Journal of Experimental Biology (IJEB)*, 47, pp. 1156–1160.

Khatune, N.A., Mosaddik, M.A. and Haque, M.E., 2001. Antibacterial activity and cytotoxicity of *Nyctanthes arbor-tristis* flowers. *Fitoterapia*, *72*(4), pp. 412–414.

Mishra, R.K., Mishra, V., Pandey, A., Tiwari, A.K., Pandey, H., Sharma, S., Pandey, A.C. and Dikshit, A., 2016. Exploration of anti-Malassezia potential of *Nyctanthes arbor-tristis* L. and their application to combat the infection caused by Mala s1 a novel allergen. *BMC Complementary and Alternative Medicine*, *16*(1), p. 114.

Rathore, A., Rivastava, V., Srivastava, K.C. and Tandon, J.S., 1990. Iridoid glucosides from *Nyctanthes arbor-tristis. Phytochemistry*, *29*(6), pp. 1917–1920.

10.4.5.3 *Jasminum multiflorum* (Burm. f.) Andrews

Synonyms: *Jasminum pubescens* (Retz.) Willd.; *Nyctanthes multiflora* Burm. f.; *Nyctanthes pubescens* Retz

Common names: Downy jasmine, star jasmine, angel hair jasmine, mao mo li (China); saumanasyaayani (India); melati gambir (Indonesia); sa thaan (Thailand); chi-nhài (Vietnam)

Habitat: Cultivated

Distribution: Tropical Asia and Pacific

Botanical description: It is a climbing shrub native to India which is cultivated round the tropical and subtropical world as an ornamental plant. It grows up to about 4 m tall. The stems are terete and hairy. The leaves are opposite, simple, and exstipulate. The petiole is up to 1 cm long and hairy. The blade is cordate at base, acuminate at apex, 1.5–3 cm × 15–10 cm, papery, and shows 3–4 pairs of secondary nerves. The inflorescence is terminal and densely flowered. The calyx is hairy, tubular, and develops 6–9 lobes which are filiform and up to 7 mm long. The corolla is pure white, tubular, the tube up to 1.5 cm long, fragrant, and with up to 9 lobes which are oblong lanceolate and up to 1.5 cm long. The androecium comprises 2 stamens. The style is slender and the stigma bifid. The fruits are berries.

Medicinal uses: Fever (India; Myanmar)

Commentaries

i. The plant has apparently not been studied for antimicrobial properties.
ii. *Jasminum sambac* (L.) Aiton is used for eye infection in Bangladesh (local name: *Beli*), sores in Cambodia, venereal disease in Malaysia, and wounds in the Philippines. Oleuropein isolated from the flowers of *Jasminum officinale* L. inhibited the secretion of the Hepatitis B surface antigen by HepG2 2.2.15 cells with the IC_{50} of 23.2 µg/mL . Oleuropein given intraperitoneally to Ducks at the dose of 80 mg/Kg evoked some levels of protection against Hepatitis B virus (Zhao et al., 2009). Oleuropein is produced by *Olea europea* L. (Ranalli et al., 2006). Olive oil was used as ointment to prevent plague in the 1837 "A New Supplement to the Latest Pharmacopoeias" of London, Edinburgh, Dublin. In the Midrash on Banidbar Rabbah, olive oil is to be used against sores and wounds, chills, and sore throat (Goor, 1966). Olive oil has been found to relieve cough and sorethroat in India.
iii. *Jasminum bifarium* Wall. ex DC. is used in Malaysia for yaws.
iv. In the Tribe, Fraxineae, *Fraxinus rhynchophylla* Hance is used in Korea for leucorrhea and *Fraxinus pubinervis* Bl. is given for diarrhea in China.
v. In the Tribe Oleeae, *Osmanthus fragrans* Lour. is given in Taiwan for boils. Aqueous extract of leaves of *Syringa dilatat* Nakai inhibited the replication of the Human immunodefiency virus-1 protease (Park, 2003).

References

Goor, A., 1966. The place of the olive in the holy land and its history through the ages. *Economic Botany*, *20*, pp. 223–243.

Park, J.C., 2003. Inhibitory effects of Korean plant resources on human immunodeficiency virus type 1 protease activity. *Oriental Pharmacy and Experimental Medicine*, *3*(1), pp. 1–7.

Ranalli, A., Contento, S., Lucera, L., Di Febo, M., Marchegiani, D. and Di Fonzo, V., 2006. Factors affecting the contents of iridoid oleuropein in olive leaves (*Olea europaea* L.). *Journal of Agricultural and Food Chemistry*, *54*(2), pp. 434–440.

Zhao, G., Yin, Z. and Dong, J., 2009. Antiviral efficacy against hepatitis B virus replication of oleuropein isolated from *Jasminum officinale* L. var. grandiflorum. *Journal of Ethnopharmacology*, *125*(2), pp. 265–268.

10.4.6 Family Orobanchaceae Vent. (1799)

The family Orobanchaceae consists of 15 genera and 150 species of parasitic herbs. The leaves are absent or very much reduced. The inflorescences are racemes, spikes, of solitary. The calyx is tubular and 4–6 lobed. The corolla is tubular, bilabiate, the lower lip 3-lobed. The androecium includes 4 stamens. The gynoecium consists of a unilocular ovary containing numerous ovules on parietal placentas. The fruit is a dehiscent capsule.

10.4.6.1 *Rehmannia glutinosa* (Gaertn.) Libosch. ex Fisch. & C.A. Mey.

Synonyms: *Digitalis glutinosa* Gaertn.; *Rehmannia chinensis* Libosch. ex Fisch. & C.A. Mey.

Common names: Ti huang (China)

Habitat: Roadsides

Distribution: China

Medicinal uses: This magnificent herb grows up to about 30 cm tall from an elongated rhizome with a repulsive aura, and somewhat worm-like. The stems are hairy. The leaves are simple, in basal rosette, and exstipulate. The blade is ovate, spathulate to elliptic, 2–10×1–5 cm, tapering at base, lobed (somewhat oak-leaves like), somewhat fleshy, dark green, marked with 5 pairs of secondary nerves sunken above, hairy, and round at apex. The racemes are basal, hairy, and somewhat purplish. The calyx is tubular, hairy, 1–1.5 cm long, and develops 5 lobes. The corolla is tubular, hairy, somewhat vaguely *Digitalis*-like, 3–4.5 cm long, of a peculiar type of pink to purple, about 1 cm long, and develops 5 bifid lobes. Four stamens adnate to the corolla are present. The ovary is minute and develops a slender style. The capsules are ovoid, dehiscent, and up to about 1.5 cm long.

Medicinal uses: Fever, leucorrhoea (China)

Commentaries

i. The roots are sun dried and steamed and used for the treatment of COVID-19 in China (Wang et al., 2020). Cycles of sun drying and steaming are characteristic of the Chinese traditional system of medicine which is among the oldest and most valuable written medicinal system in Asia and today taught in the medical universities of China. The oldest written herbal treatises are to be found in India, where Ayurveda is part of medical practice. Europe had a vast patrimony of medicinal knowledge and used numerous valuable medicinal plants and it is perhaps time for Western schools of medicine and pharmacy to provide students with deep knowledge in medicinal plants in a time where people are turning toward herbal remedies for their well-being and left dangerously so far with online products and other not pharmaceutically controlled herbals.
ii. The plant yields iridoid glycosides (Morota et al., 1989) including catalpol, leonuride (ajugol), aucubin, and melittoside (Oshio & Inouye, 1982).
iii. In the family Orobanchaceae, *Aeginetia indica* L. is given for sores in Cambodia, Laos, and Vietnam. Aqueous extract of *Aeginetia indica* L. inhibited the replication of the Hepatitis virus C (Lin et al., 2018).

References

Lin, C.W., Lo, C.W., Tsai, C.N., Pan, T.C., Chen, P.Y. and Yu, M.J., 2018. Aeginetia indica decoction inhibits hepatitis C virus life cycle. *International Journal of Molecular Sciences*, *19*(1), p. 208.

Morota, T., Sasaki, H., Nishimura, H., Sugama, K., Chin, M. and Mitsuhashi, H., 1989. Two iridoid glycosides from *Rehmannia glutinosa*. *Phytochemistry*, *28*(8), pp. 2149–2153.

Oshio, H. and Inouye, H., 1982. Iridoid glycosides of *Rehmannia glutinosa*. *Phytochemistry*, *21*(1), pp. 133–138.

Wang, S.X., Wang, Y., Lu, Y.B., Li, J.Y., Song, Y.J., Nyamgerelt, M. and Wang, X.X., 2020. Diagnosis and treatment of novel coronavirus pneumonia based on the theory of traditional Chinese medicine. *Journal of Integrative Medicine*, *18*(4), pp. 275–283

10.4.7 Family Plantaginaceae A.L. de Jussieu (1789)

The family Plantaginaceae consists of 3 genera and about 270 species of herbs or shrubs. The leaves are spiral, often in rosette, and exstipulate. The blade is often marked with secondary nerves somewhat parallel to the margins with a Monocots look. The inflorescences are spikes. The calyx, and the corolla are tubular and 4-lobed. The androecium includes 4 stamens inserted on the corolla tube and not uncommonly protruding from the corolla. The gynoecium consists of 2 carpels forming a 2-locular ovary with axile placentas, and a slender style. The fruits are tiny and dehiscent capsules.

10.4.7.1 *Plantago major* L.

Synonyms: *Plantago gigas* H. Lév.; *Plantago intermedia* Gilib.; *Plantago jehohlensis* Koidz.; *Plantago latifolia* Salisb.; *Plantago macronipponica* Yamam.; *Plantago pauciflora* Gilib.; *Plantago sawadai* (Yamam.) Yamam.; *Plantago sinuata* Lam.; *Plantago villifera* Kitag.

Common names: Common Plantain, great plantain; daun sendok (Indonesia); ekor anjing (Malaysia); ch'e ch'ien, da che qian (China); a kyaw ta htaung (Myanmar); isabgol (Nepal) ilanten (the Philippines)

Habitat: Meadows, roadsides

Distribution: From Turkey to the Philippines

Botanical observation: This herb grows up to about 30cm tall. The leaves are simple in basal rosettes and exstipulate. The petiole form a sheath at base, fibrous, grooved, and 3–5cm long. The blade is broadly elliptic, fleshy, hairy, 7–12 cm × 4.2–6cm and shows up to 10 longitudinal secondary nerves. The flowers are grouped in compact spikes which are 9–14cm long. The calyx presents 4 triangular lobes which are about 2mm long. The corolla is light pink to purple, to white, to 2mm long, and produce 4 tiny triangular lobes. The 4 anthers protrude out of the corolla as well as a linear stigma. The capsules are minute, papery, circumscissile, with an apical tip, and containing about 15 tiny seeds.

Plantago major L.

Medicinal uses: Boils (Nepal); dysentery (India, Vietnam, Cambodia, Laos, Indonesia); fever, gonorrhea, sores (Myanmar)

Strong antiviral (enveloped monopartite linear double-stranded DNA) hydrophilic hydroxycinnamic acid derivative: Caffeic acid (LogD –1.7 at pH 7.4; molecular mass = 180.1 g/mol) inhibited the replication of the Herpes simplex virus type-1 (strain KOS) and type-2 (strain 196) with the EC_{50} values of 15.3 and 87.3 µg/mL and the selectivity indexes of 671 and 118, respectively, in BCC-1/KMC cells (Chiang et al., 2002).

Strong antiviral (enveloped monopartite linear double-stranded DNA) amphiphilic flavonol: Baicalein (also known as 5,6,7-trihydroxyflavone; LogD 1.6 at pH 7.4; molecular mass = 270.2 g/mol) inhibited the replication of the Herpes simplex virus type-1 (strain KOS) with the EC_{50} value of 4.7 and the selectivity index of 4.2 in BCC-1/KMC cells (Chiang et al., 2002).

Strong antiviral (enveloped monopartite linear double-stranded DNA) hydrophilic hydroxycinnamic acid derivative: p-Coumaric acid (LogD –1.7 at pH 7.4; molecular mass = 180.1 g/mol) inhibited the replication of the Herpes simplex virus type-2 (strain 196) with the EC_{50} value of 32.8 and the selectivity index of 14.9 in BCC-1/KMC cells (Chiang et al., 2002).

Strong antiviral (non-enveloped monopartite linear double-stranded DNA) hydrophilic hydroxycinnamic acid derivative: Caffeic acid inhibited the replication of the Adenovirus-3 with the EC_{50} of 14.2 µg/mL and the selectivity index of 727 in BCC-1/KMC cells (Chiang et al., 2002).

Strong antiviral (non-enveloped monopartite linear double-stranded DNA) hydroxycinnamic acid derivative: Chlorogenic acid (LogD –3.9 at pH 7.4; molecular mass = 354.3 g/mol) inhibited the replication of the Adenovirus-3 with the EC_{50} of 13.3 µg/mL and the selectivity index of 301 in BCC-1/KMC cells (Chiang et al., 2002).

Strong antiviral (non-enveloped monopartite linear double-stranded DNA) hydrophilic hydroxycinnamic acid derivative: p-Coumaric acid inhibited the replication of the Adenovirus-11 with the EC_{50} values of 43.8 and the selectivity index of 11.2 in BCC-1/KMC cells (Chiang et al., 2002).

Commentaries

i. *Plantago asiatica* L. is used in Korea (local name: *Jilgyeongi*) for diarrhea, jaundice, and urinary tract infection, and in China for gonorrhea. Aqueous extract of this herb inhibited the replication of the Respiratory syncytial virus with IC_{50} value of 103.1 µg/mL and the selectivity index of 5.3 (Li et al., 2004).
ii. *Plantago ovata* Forsk. (local name: *Isapghol*) is used for dysentery and leprosy in India as well as *Plantago ciliata* Desf.
iii. In Pakistan, *Plantago lagocephala* Bunge is used for flu and boils (local name: *Danish*)
iv. Methanol extract (50 mL per well/ 1 mg/5 mL) of *Plantago lanceolata* L. inhibited the growth of *Escherichia coli*, *Klebsiella pneumoniae*, *Pseudomonas aeruginosa*, *Proteus mirabilis*, *Salmonella typhi*, *Salmonella paratyphi* A, *Salmonella paratyphi* B, *Salmonella typhimurium*, *Shigella boydii*, *Shigella flexneri*, *Shigella sonnei*, *Staphylococcus aureus*, and *Streptococcus faecalis* with inhibition zone diameter of 11.2, 13.4, 12, 11, 11.7, 10.4, 10.6, 10.2, 9.4, 12.3, 13.4, 16.4, and 15.6 mm, respectively (Satish et al., 2009).

References

Chiang, L.C., Chiang, W., Chang, M.Y., Ng, L.T. and Lin, C.C., 2002. Antiviral activity of Plantago major extracts and related compounds in vitro. *Antiviral Research*, *55*(1), pp. 53–62.

Li, Y., Ooi, L.S., Wang, H., But, P.P. and Ooi, V.E., 2004. Antiviral activities of medicinal herbs traditionally used in southern mainland China. *Phytotherapy Research*, *18*(9), pp. 718–722.

Satish, S., Mohana, D.C., Raghavendra, M.P., Babu, S., Raveesha, K.A. (2009) Antibacterial evaluation of some Iranian medicinal plants against some human pathogenic bacteria. *Asian Journal of Microbiology, Biotechnology and Environmental Sciences* 11(4), pp. 735–738.

10.4.8 Family Scrophulariaceae A.L. de Jussieu (1789)

The family Scrophulariaceae consists of about 220 genera and 4500 species of herbs (often gracile), shrubs, and rarely trees. The leaves are simple, alternate, opposite, and exstipulate. The inflorescences are racemes, spikes, cymes, or axillary. The calyx is tubular and 4- to 5-lobed. The corolla is 2-lipped. The androecium includes 4 stamens. The gynoecium includes a 2-locular ovary sheltering numerous ovules on axile placentas. The fruit is a capsule.

10.4.8.1 *Curanga fel-terrae* (Lour.) Merr.

Synonyms: *Curanga amara* Juss.; *Gratiola amara* Roxb.; *Picria felterrae* Lour.

Common names: Longri (India); pugun tano (Indonesia); pun (Laos); hempedu tanah (Malaysia); sagai-uak (Philippines); tan (Vietnam)

Curanga Fel-Terrae (Lour.) Merr.

Habitat: Forest edges, meadows, lawn

Distribution: India to the Philippines

Botanical observation: This discrete and creeping herb grows to 1 m long. The stem is slender, quadrangular, and rooting at nodes. The leaves are simple, opposite, sessile, and without stipules. The petiole is up to 1.3 cm long. The blade is broadly lanceolate, about 5×3 mm, serrate with a very few hairs underneath. The midrib is sunken above and secondary nerves are not visible above. The racemes are flowered, about 2.5 cm long, and axillary. The 4 sepals are about 6 mm long, oblong-ovate and 1.4 cm×1 cm in fruit. The corolla is bilabiate, white, 5 mm long. The lower lip is 6 mm long and 3-lobed. Four stamens are present. The capsules are ovoid, 5 mm long in the accrescent calyx.

Medicinal uses: Dysentery (India); fever (India; China); herpes (China)

Commentary: The plant has apparently not been studied for antimicrobial properties. It is said to be very bitter and yields cucurbitacin glycosides such as picfeltarraenin IB (Huang et al., 1998). It is given to expel worms for intestines which suggest the presence of neuroactive principles. As discussed earlier, neuroactive principles are often able to inhibit bacterial and/or fungal efflux pumps. Thus, antibiotic and/or antifungal potentiators could be produced by this common weed.

Reference

Huang, Y., De Bruyne, T., Apers, S., Ma, Y., Claeys, M., Vanden Berghe, D., Pieters, L. and Vlietinck, A., 1998. Complement-inhibiting cucurbitacin glycosides from Picria fel-terrae. *Journal of Natural Products*, *61*(6), pp. 757–761.

10.4.8.2 *Herpestis monnieri* (L.) Kunth.

Synonyms: *Bacopa monnieri* (L.) Pennell; *Bramia monnieri* (L.) Drake; *Gratiola monniera* L., *Lymachia monnieri* L.,

Common names: Thyme leaved gratiola, herb of grace, water hyssop; snao (Cambodia); jia ma chi xian, pa chi t'ien (China); brahmishak, brahmmi (India); phomi (Laos, Thailand); ulasimang-aso (Philippines); berimi (Malaysia), brahmi (India); ulasimang-aso (Philippines); ruot ga (Vietnam)

Habitat: Watery meadows

Distribution: India, China, Taiwan

Botanical description: This aquatic herb has a positive aura of life and grows up to about 25cm long. The roots are fibrous. The stem is creeping, rooting at nodes, terete, somewhat swollen and articulate, fleshy, glabrous, and orangish-brown. The leaves are simple, sessile, edible, opposite, without stipules, and somewhat decussate. The blade is spathulate, fleshy, somewhat like *Portulaca oleracea* L. and up to 3 mm×1.8 cm, the margin entire or dentate, and rounded at apex. The flower is solitary and axillary. The flower pedicel is up to 3 cm long and slender. The 5 sepals are 5 mm long, and lanceolate. The corolla is pure white pinkish to pale bluish, 0.8–1 cm long, obscurely 2-lipped with 5 lobes. The 4 stamens are somewhat purplish and of different length. The capsules are narrowly ovoid, enveloped in persistent calyx, and acute at apex. The seeds are yellow-brown, ellipsoid, truncate at one end, and longitudinally striated.

Medicinal uses: Bronchitis, diarrhea, leprosy (India)

Broad-spectrum antibacterial polar extract: Methanol extract inhibited the growth of *Bacillus subtilis* (ATCC 6051), *Proteus vulgaris* (ATCC 6380), *Salmonella typhimurium* (ATCC 23564), *Pseudomonas aeruginosa* (ATCC 25619), *Escherichia coli* (K-12), and *Staphylococcus aureus* (Mathur et al., 2010).

Broad-spectrum antifungal polar extract: Methanol extract inhibited the growth of *Candida albicans*, *Aspergillus niger*, *Saccharomyces cerevisiae,* and *Penicillium notatum* (Mathur et al., 2010).

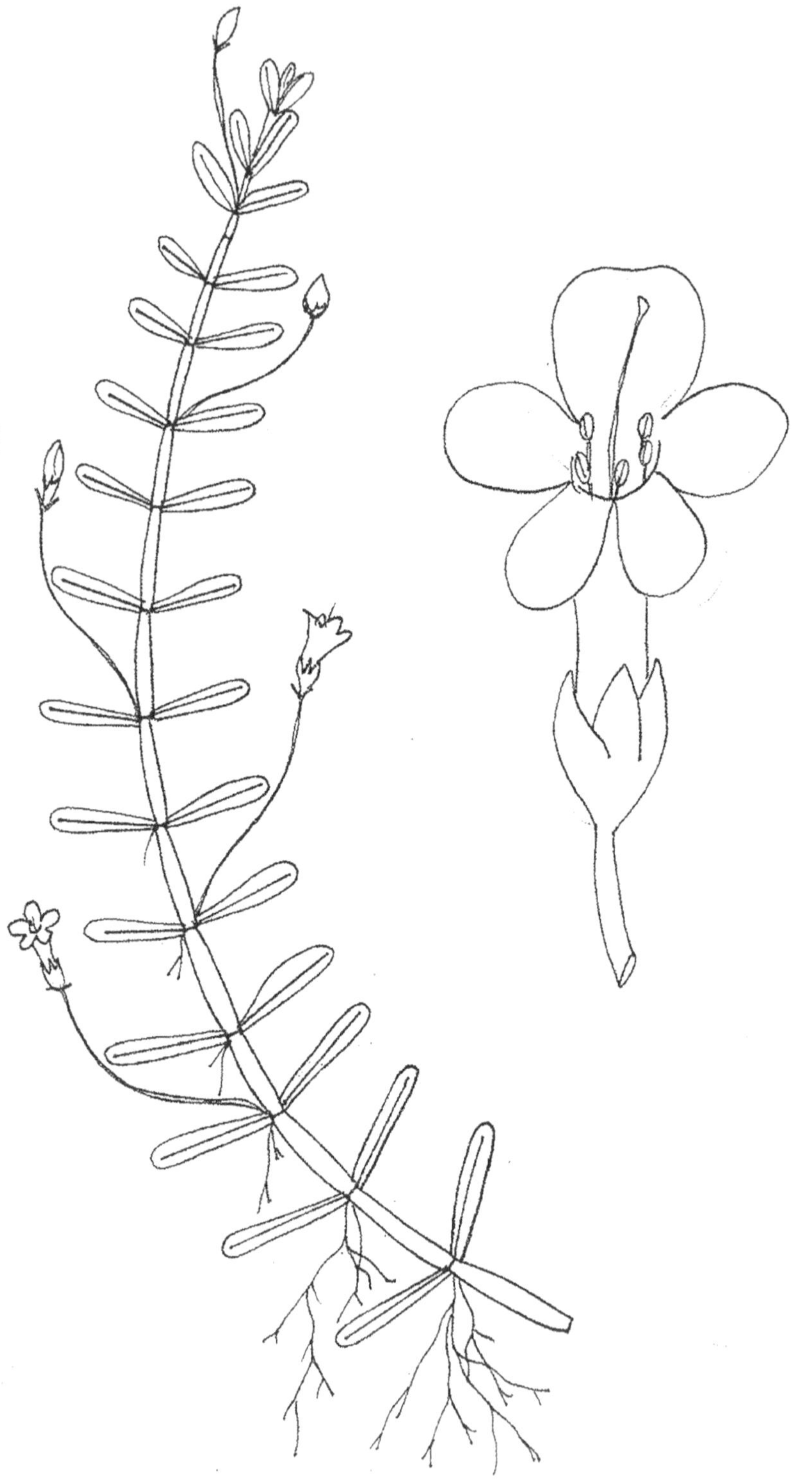

Herpestis monnieri (L.) Kunth.

Commentary

i. The plant is used for anxiety in Ayurvedic medicine and as may produce antibiotic and/or antifungal potentiators inhibiting efflux pumps.
ii. The plant yields triterpene saponins (Sivaramakrishna et al., 2005).

References

Mathur, A., Verma, S.K., Purohit, R., Singh, S.K., Mathur, D., Prasad, G.B.K.S. and Dua, V.K., 2010. Pharmacological investigation of *Bacopa monnieri* on the basis of antioxidant, antimicrobial and anti-inflammatory properties. *Journal of Chemical and Pharmaceutical Research*, *2*(6), pp. 191–198.

Sivaramakrishna, C., Rao, C.V., Trimurtulu, G., Vanisree, M. and Subbaraju, G.V., 2005. Triterpenoid glycosides from Bacopa monnieri. *Phytochemistry*, *66*(23), pp. 2719–2728.

10.4.8.3 *Limnophila repens* (Benth.) Benth.

Synonym: *Stemodia repens* Benth.

Common name: Creeping marshweed; maytia anos (Bangladesh); pu fu shi long wei (China); amraganda (India)

Habitat: Rivers, ponds

Distribution: India to Australia

Botanical description: This aquatic herb grows up to about 50 cm tall. The stems are quadrangular and hairy. The leaves are simple, decussate, sessile, and exstipulate. The blade is narrowly elliptic, linear-lanceolate, or lanceolate-elliptic, 0.5–3 cm × 3–10 mm, hairy below, serrate, and acute at margin. The flowers are magnificent, solitary, and axillary or arranged in axillary racemes. The calyx is tubular, about 5 mm long, hairy, and develops 5 linear lobes. The corolla is white, purple-red, or blue, 1 cm long, bilabiate, the lower lip 3-lobed. The fruit is an ovoid and septicidal capsule which is about 5 mm long.

Medicinal use: Dysentery (Bangladesh)

Commentaries

i. In India, *Limnophila gratissima* Bl. (local name: *Ambuja*) is given for fever and used as antiseptic and *Limnophila indica* R.Br. (synonym: *Limnophila gratioloides* R. Br.; local names: *Amragandhaka*, *Gulabi*) is given for dysentery, fever, and used as an antiseptic. *Limnophila indica* (L.) Druce (local name: *Karpur*) is used for fever and infections in Bangladesh.

   From *Limnophila indica* R.Br., the oleanane-type triteperne 3-oxo-olean-12(13),18(19)-dien-29α-carboxylic acid inhibited the growth of *Escherichia coli* (MTCC1667), *Salmonella typhimurium* (MTCC98), *Staphylococcus aureus* (MTCC96), *Bacillus subtilis* (MTCC121), *Pseudomonas aeruginosa* (MTCC741), *Pantoea ananatis* (MTCC2307), and *Listeria monocytogenes* (MTCC657), with the MIC values of 40, 100, 30, 25, 30, 75, and 30 μg/mL, respectively (Brahmachari et al., 2013). Plants in the genus *Limnophila* R. Br. produce antibacterial and antifungal flavones (Brahmachari et al., 2011).
ii. In Thailand, a plant in the genus *Limnophila* R.Br. is used for cough and fever. From this herb, the flavones nevadensin and isothymusin inhibited the growth of *Mycobacterium tuberculosis* (H37Ra) with the MIC values of 200 μg/mL (Suksamrarn et al., 2003).
iii. In the Philippines, *Limnophila roxburghii* G. Don (local name: *Kalao, Tala*) is used for food. It is not impossible that decrease in crop yields owed to global warming and resistant phytopathogenic microbes will compel the use of other plants for food. In such a possible dreadful situation, people living in remote and self-sufficient farming and plant collection will survive.

References

Brahmachari, G., Mandal, N.C., Roy, R., Ghosh, R., Barman, S., Sarkar, S., Jash, S.K. and Mondal, S., 2013. A new pentacyclic triterpene with potent antibacterial activity from Limnophila indica Linn.(Druce). *Fitoterapia*, *90*, pp. 104–111.

Brahmachari, G., Mandal, N.C., Jash, S.K., Roy, R., Mandal, L.C., Mukhopadhyay, A., Behera, B., Majhi, S., Mondal, A. and Gangopadhyay, A., 2011. Evaluation of the antimicrobial potential of two flavonoids isolated from Limnophila plants. *Chemistry & Biodiversity*, *8*(6), pp. 1139–1151.

Suksamrarn, A., Poomsing, P., Aroonrerk, N., Punjanon, T., Suksamrarn, S. and Kongkun, S., 2003. Antimycobacterial and antioxidant flavones from *Limnophila geoffrayi*. *Archives of Pharmacal Research*, *26*(10), pp. 816–820.

10.4.8.4 *Lindernia crustacea* (L.) F. Muell.

Common name: Malaysian false pimpernel

Synonyms: *Capraria crustacea* L., *Gratiola lucida* Vahl., *Pyxidaria crustacea* (L.) Kuntze, *Pyxidaria crustacea* (L.) Cham. & Schltdl., *Torenia flaccida* R. Br., *Torenia scabra* R. Br., *Vandellia alba* Benth., *Vandellia bodinieri* H. Lev, *Vandellia crustacea* (L.) Benth.

Habitat: Moist areas, paddy fields, grasslands, trail sides, open disturbed forests

Distribution: Tropical Asia

Botanical observation: This herb grows up to about 20 cm tall. The stem is brownish red, glabrous, angular, with internodes of about 3 cm long. The leaves are simple, opposite, and without stipules. The petiole is 2 mm long. The lade is triangular-ovate to broadly elliptic, 1.2–2 cm × 0.9–1.2 cm,

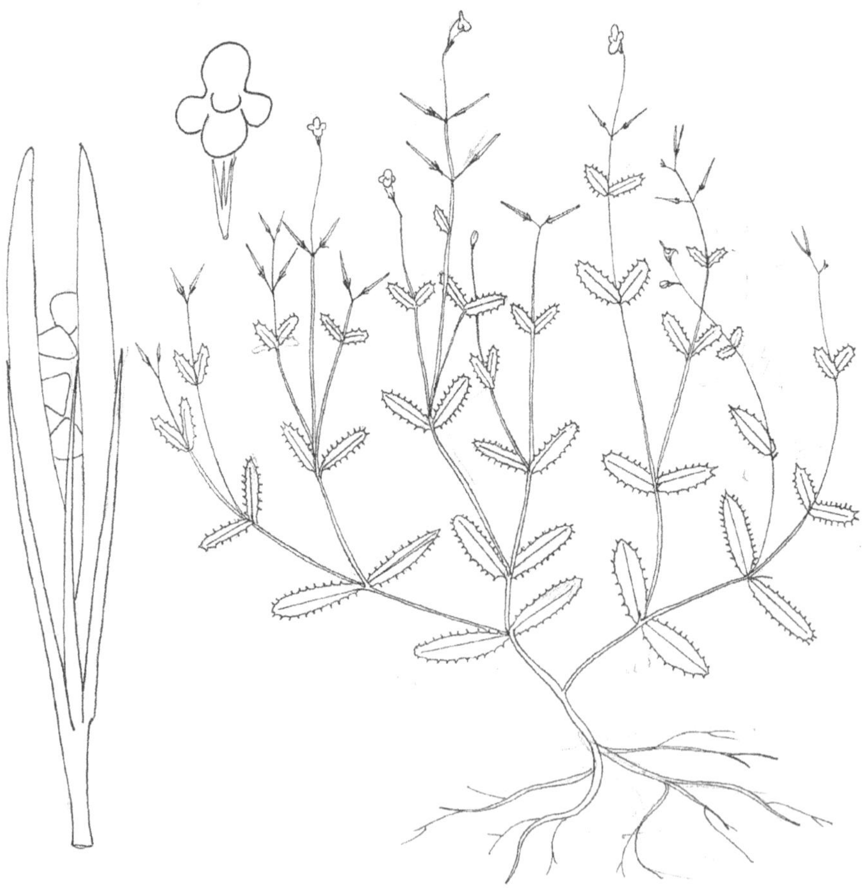

Lindernia crustacea (L.) F. Muell.

crenate or serrate, and with a few hairs. The flower is axillary and solitary or in short apical racemes. The calyx is urn like, 5 mm long, 5-lobed, the lobes triangular-ovate. The corolla is purplish at base, bilabiate, the tube slightly longer than the calyx, and the lower lip 3-lobed. The androecium includes 4 stamens. The style is slender. The capsules are broadly ellipsoid.

Medicinal uses: Dysentery (the Philippines); herpes (Taiwan)

Antiviral (enveloped monopartite linear double-stranded DNA) linear diterpene: Phytol inhibited the replication of the Epstein-Barr virus at the concentration of 20 µg/mL (Tsai et al., 2020).

Antiviral (enveloped monopartite linear double-stranded DNA) linear anthraquinone: Aloe-emodin inhibited the replication of the Epstein-Barr virus at the concentration of 20 µg/mL (Tsai et al., 2020).

Antiviral (enveloped monopartite linear double-stranded DNA) linear megastigmane glycoside: Byzantionoside B inhibited the replication of the Epstein-Barr virus at the concentration of 20 µg/mL (Tsai et al., 2020).

Antiviral (enveloped monopartite linear double-stranded DNA) linear flavone glycosides: Luteolin-7-*O*-β-D-glucopyranoside and apigenin-7-*O*-[β-D-apiofuranosyl (1 → 6)-β-D-glucopyranoside] inhibited the replication of the Epstein-Barr virus at the concentration of 20 µg/mL (Tsai et al., 2020).

Commentaries: In Thailand, *Lindernia ciliata* (Colsm.) Pennell (local name: *Ya kra tai jam*) and *Lindernia crustacea* (L.) F. Muell. (local name: *Ya kled hoi*) are used for fever.

Reference

Tsai, Y.C., Hohmann, J., El-Shazly, M., Chang, L.K., Dankó, B., Kúsz, N., Hsieh, C.T., Hunyadi, A. and Chang, F.R., 2020. Bioactive constituents of *Lindernia crustacea* and its anti-EBV effect via Rta expression inhibition in the viral lytic cycle. *Journal of Ethnopharmacology*, *250*, p. 112493.

10.4.8.5 *Scoparia dulcis* L.

Synonyms: *Scoparia grandiflora* Nash., *Capraria dulcis* Kuntze, *Gratiola micrantha* Nutt.

Common names: Sweet broomweed; ye gan cao (China); bandhonia, mithapata (Bangladesh); mithi pati (India); mala anis (the Philippines)

Habitat: Roadsides, drains, parkings, and waste lands

Distribution: Tropical Asia and Pacific

Botanical observation: This herb grows up to about 70 cm tall and has somewhat a positive aura. The stems are angled and somewhat coriaceous. The leaves are simple, decussate, subsessile, or whorled, and exstipulate. The blade is narrowly elliptic-lanceolate, up to about 2 cm × 5 mm, cuneate at base, serrate, and obtuse at apex. The flowers are arranged in axillary fascicles. The calyx is 5-lobed. The corolla is tubular, hairy at throat, white, and produces 4 lobes which are about 3 mm. The androecium includes 4 stamens. The capsules are about 3 mm across.

Medicinal uses: Boils, lung infection, leucorrhea, gonorrhea (Bangladesh); Dysentery (India)

Broad-spectrum antibacterial non-polar extract: Hexane extract (250 µg/6 mm disc) inhibited the growth of *Staphylococcus aureus* (ATCC 6538), *Bacillus cereus* (ATCC 13061), *Escherichia coli* (ATCC 25922), *Proteus mirabilis* (ATCC 35659), and *Salmonella typhimurium* (ATCC 13311) with the inhibition zone diameters of 9, 22, 12, 9, and 15 mm, respectively (Coulibaly et al., 2012).

Broad-spectrum nonpolar extract: Hexane extract (250 µg/6 mm disc) inhibited the growth of *Aspergillus niger* (DSM737), *Mucor rouxii* (DSM1151), *Penicillium roquefortii* (DSM1080), and *Fusarium oxysporum* (NCIM 1008) with the inhibition zone diameters of 23, 13, 43, and 13 mm, respectively (Coulibaly et al., 2012).

Very weak antibacterial (Gram-positive) antibacterial scopadulane-type diterpene: 4-Epi-scopadulcic acid B isolated from the aerial parts inhibited the growth of *Staphylococcus aureus*, *Staphylococcus aureus* (B26 strain), and methicillin-resistant *Staphylococcus aureus* (clinical isolate) (Phan et al., 2006).

Strong antimycobacterial mid-polar extract: Ethylacetate extract inhibited the growth of *Mycobacterium tuberculosis* with the MIC value of 12 μg/mL (Adebiyi et al., 2016).

Strong antiviral (enveloped monopartite linear double-stranded DNA) amphiphilic scopadulane-type diterpenes: Scopadulin (LogD 4.6 at pH 7.4; molecular mass = 442.5 g/mol) inhibited the replication of the Herpes simplex virus type-1 with the ED_{50} of 38 μg/mL (Hayashi et al., 1990). Scopadulcic acid B inhibited the replication of the Herpes simplex virus type-1 via a mechanism involving, at least in part, early host intracellular phase (Hayashi et al., 1998).

In vivo antiviral (enveloped monopartite linear double-stranded DNA) scopadulane-type diterpene: Scopadulcic acid B given orally at 200 mg/Kg/day to Hamsters afforded some levels of protection against the Herpes simplex virus (Hayashi et al., 1998).

Commentaries

i. Diterpenes are often able to inhibit the replication of Herpes simplex viruses *in vitro*. Scopadulane-type diterpenes could be examined for their antibiotic and or antifungal potentiating properties.
ii. *Paulownia tomentosa* (Thunb.) Steud. yields the prenylated flavanols tomentin A, tomentin B, and tomentin E which inhibited the Severe acute respiratory syndrome-associated coronavirus papain-like protease with the IC_{50} values of 6.2, 6.3, and 5 μM (Cho et al., 2013). Methyl-5-hydroxy-dinaphthol[1,2–2′,3′]furan-7,12-dione-6-carboxylate inhibited the replication of the Poliovirus type 1 (strain Brunhilde) and type 3 (strain Leon) with the ICSO value of 300 and 600 μg/mL, respectively, in HeLa cells (Kang et al., 1999). The prenylated flavones diplacone, 3′-*O*-methyl-5′-hydroxydiplacone,3′-*O*-methyl-5′-*O*-methyldiplacone, and 3′-*O*-methyldiplacol inhibited the replication of the Influenza virus A/Puerto Rico/8/34 (A/PR/8/34) of Madin-Darby canine kidney (Moon et al., 2016).
iii. *Scrophularia ningpoensis* Hemsl. is used for the treatment of COVID-19 in China (local name: *Xuan shen*) (Wang et al., 2020). Extract of this plant inhibited the replication of the Dengue virus serotype-2 in C6/36 cells (Maryam et al., 2020).
iv. *Veronica officinalis* L. is given for tuberculosis is Bangladesh (local name: *Chapta-pata*). Methanol extract of *Veronica persica* Poir. inhibited the replication of the Herpes simplex virus type 1 and type with the IC_{50} values of 0.6 and 0.7 μg/mL and the selectivity indexes of 451.6 and 383.5 in Vero cells (Sharifi-Rad et al., 2018).
v. In India, *Verbascum thapsus* L. (local name: *Bantamaku*) is given for diarrhea and in Nepal (local name: *Yugisingh*) is used for wounds. Methanol extract of this beautiful herb (the great mullein) inhibited the replication of the Bseudorabies virus (strain RC/79) with the IC_{50} value of 35 μg/mL and the selectivity index of 31.4 (Escobar et al., 2012) as well as Influenza A virus with an IC_{50} below 6.2 μg/mL (Rajbhandari et al., 2009). Methanol extract of leaves of *Verbascum sinaiticum* Benth. (25 mg/mL/11 mm well) inhibited the growth of *staphylococcus aureus* (ATCC 6538) and *Pseudomonas aeruginosa* (ATCC 27853) with inhibition zones of 19 and 16 mm, respectively (). This extract was inactive for *Escherichia coli* (ATCC 25922) (Tadeg et al., 2005).
vi. In India, the sinister herb *Picrorhiza kurrooa* Royle ex Benth. (local name: *Katukaa*) is used for cough, diarrhea, dysentery, fever, and jaundice. Methanol extract of rhizome of inhibited the replication of the Dengue virus serotype-1, -2, -3, and -4 with the IC_{50} values equal to 12.5, 16, 27.1, and 25 μg/mL, respectively (Sood et al., 2015).

References

Adebiyi, A., Bassey, E.E., Ayo, R., Bello, I., Habila, J. and Ishaku, G.A., 2016. Anti-mycobacterial, antimicrobial and phytochemical evaluation of Pulicaria crispa and Scoparia dulcis plant extracts. *Journal of Advances in Medical and Pharmaceutical Sciences*, pp. 1–11.

Cho, J.K., Curtis-Long, M.J., Lee, K.H., Kim, D.W., Ryu, H.W., Yuk, H.J. and Park, K.H., 2013. Geranylated flavonoids displaying SARS-CoV papain-like protease inhibition from the fruits of *Paulownia tomentosa*. *Bioorganic & Medicinal Chemistry*, *21*(11), pp. 3051–3057.

Coulibaly, A.Y., Konate, K., Youl, E.N.H., Sombie, P.A., Kiendrebeogo, M., Meda, N.R., Lamien, A., Zeba, B. and Nacoulma, O.G., 2012. Anti-proliferative effect of *Scoparia dulcis* L. against bacterial and fungal strains. *International Journal of Biological and Chemical Sciences*, *6*(6), pp. 3055–3063.

Escobar, F.M., Sabini, M.C., Zanon, S.M., Tonn, C.E. and Sabini, L.I., 2012. Antiviral effect and mode of action of methanolic extract of *Verbascum thapsus* L. on pseudorabies virus (strain RC/79). *Natural Product Research*, *26*(17), pp. 1621–1625.

Hayashi, K., Niwayama, S., Hayashi, T., Nago, R., Ochiai, H. and Morita, N., 1988. In vitro and in vivo antiviral activity of scopadulcic acid B from *Scoparia dulcis*, Scrophulariaceae, against herpes simplex virus type 1. *Antiviral Research*, *9*(6), pp. 345–354.

Hayashi, T., Kawasaki, M., Miwa, Y., Taga, T. and Morita, N., 1990. Antiviral agents of plant origin. III.: Scopadulin, a novel tetracyclic diterpene from *Scoparia dulcis* L. *Chemical and Pharmaceutical Bulletin*, *38*(4), pp. 945–947.

Kang, K.H., Huh, H., Kim, B.K. and Lee, C.K., 1999. An antiviral furanoquinone from *Paulownia tomentosa* Steud. *Phytotherapy Research: An International Journal Devoted to Pharmacological and Toxicological Evaluation of Natural Product Derivatives*, *13*(7), pp. 624–626.

Moon, H.I., Kim, Y.B. and Kim, S.K., 2016. Anti-influenza A activity of C-geranyl flavonoids isolated from *Paulownia tomentosa* and *Maclura pomifera*. *Bulgarian Chemical Communications*, *48*(2), pp. 303–305.

Phan, M.G., Phan, T.S., Matsunami, K. and Otsuka, H., 2006. Chemical and biological evaluation on scopadulane-type diterpenoids from *Scoparia dulcis* of *Vietnamese origin*. *Chemical and Pharmaceutical Bulletin*, *54*(4), pp. 546–549.

Rajbhandari, M., Mentel, R., Jha, P.K., Chaudhary, R.P., Bhattarai, S., Gewali, M.B., Karmacharya, N., Hipper, M. and Lindequist, U., 2009. Antiviral activity of some plants used in Nepalese traditional medicine. *Evidence-Based Complementary and Alternative Medicine*, *6*(4), pp. 517–522.

Sood, R., Raut, R., Tyagi, P., Pareek, P.K., Barman, T.K., Singhal, S., Shirumalla, R.K., Kanoje, V., Subbarayan, R., Rajerethinam, R. and Sharma, N., 2015. Cissampelos pareira Linn: Natural source of potent antiviral activity against all four dengue virus serotypes. *PLoS Neglected Tropical Diseases*, *9*(12), p. e0004255.

10.4.9 Family Verbenaceae J. Saint-Hilaire (1805)

The family Verbenaceae consists of about 90 genera and 2000 species of handsome trees, shrubs, or herbs. The stems are not uncommonly somewhat angled. The leaves are simple or compound, decussate, and extipulate. The inflorescences are cymes or racemes. The calyx is tubular and 4- to 5-lobed. The corolla is tubular and 4- to 5-lobed. The androecium includes 4 stamens. The gynoecium consists of 2 carpels merged in a 2- to 4-locular ovary, each locule sheltering a single ovule on axile placentas. The style is slender. The fruits are capsules or drupes. Plants in family produce antimicrobial flavonols, diterpenes, and iridoid glycosides. The family is being included in the Lamiaceae by some botanist and remains here treated as a distinctive family.

10.4.9.1 *Callicarpa macrophylla* Vahl

Synonyms: *Callicarpa dunniana* H. Lév.; *Callicarpa incana* Roxb.

Common names: Beauty berry; kojagaicho, dubhosa (Bangladesh); da ye zi zhu (China); daungsatpya I (Myanmar)

Habitat: Forests edges

Distribution: India, Sri Lanka, Nepal, Bhutan, Bangladesh, Myanmar, Vietnam, China, Thailand, Malaysia

Botanical observation: This graceful shrub grows to about 2 m tall. The stems are hairy. The leaves are simple, opposite, and exstipulate. The petiole is hairy and up to 3 cm long. The blade is 10–25×2.5–10.5 cm, elliptic to lanceolate, tapering or cuneate at base, serrate, acute at apex, hairy below, and marked with about 5–8 pairs of secondary nerves. The cymes are axillary, up to 8.5 cm long, somewhat rounded, and present numerous minute pinkish to purplish flowers. The calyx is tubular and hairy. The corolla is tubular, about 5 mm across, and develops 4 rounded lobes. The androecium includes 4 stamens protruding out of the corolla and with slender filaments. The ovary develops a linear style and a bifid stigma. The berries are globose, white or of a unique and almost supernatural type of purple, glossy, and minute.

Medicinal uses: Fever (Bangladesh); influenza (China); skin diseases (Myanmar); sores (Malaysia)

Antibacterial (Gram-positive) polar extract: Methanol extract of leaves (350 μg/disc) inhibited the growth of *Bacillus cereus* with the inhibition zone diameter of 22 mm (Mona, 2016).

Broad-spectrum antifungal polar extract: Ethanol extract of stems (200 μg/disc) inhibited the growth of *Cryptococcus neoformans*, *Candida albicans*, and *Aspergillus niger* with the inhibition zone diameters of 12, 13, and 14 mm, respectively (Yadav et al., 2012).

Commentaries

i. The plants build isopimarane-type diterpenes (Wang et al., 2017) as well as kaurane-type diterpenes such as the quinone calliterpenone (Chatterjee et al., 1972).
ii. In China, *Callicarpa bodinieri* H. Lév. is a remedy for blenorhea (China). *Callicarpa formosana* Rolf. (local name: *Du hong hua*) and *Callicarpa longifolia* Lam. are used for wounds in Taiwan.
iii. *Callicarpa tomentosa* (L.) Murray is a remedy for sores in Malaysia.
iv. In Indonesia, *Callicarpa pedunculata* R. Br. is used to treat foul sores and *Callicarpa candidans* (Burm.f.) Hochr. is used for boils in Indonesia.
v. *Callicarpa pentandra* Roxb. var *paloensis* (Elm.) Bakh. f *furfuracea* Bakh. is a remedy for cold in the Solomon Islands.
vi. Methanol extract of stems of *Callicarpa japonica* Thunb. at the concentration of 100 μg/mL inhibited RNA-dependent DNA polymerase and ribonuclease H activities of the Human immunodeficiency-1 reverse transcriptase (Min et al., 2001). From this plant, 5,6,7-trimethoxyflavone inhibited the replication of the Herpes simplex virus type-1 (HF strain), Herpes simplex virus type-1 (KOS strain), and Herpes simplex virus type-1 (A4-3 strain) in Vero cells with the IC_{50} values of 3.2, 5.8, and 9.4 μg/mL, respectively, and the selectivity indexes of 82, 46, and 29, respectively (Hayashi et al., 1997). 5,6,7-trimethoxyflavone inhibited the replication of the Human cytomegalovirus (Towne strain) in MRC-5 cells with the IC_{50} value of 8 and selectivity index of 4.9 as well as the Poliovirus (P1 type) with the IC_{50} value of 32 and a selectivity index of 8.3 (Hayashi et al., 1997).
vi. Aqueous extract of *Callicarpa nudiflora* Hook. et Arn. inhibited the Respiratory syncytial virus (Long strain) with the IC_{50} value of 37.5 μg/mL and selectivity index of 21.3 in Hep-2 cells (Ma et al., 2002).

References

Chatterjee, A., Desmukh, S.K. and Chandrasekharan, S., 1972. Diterpenoid constituents of *Callicarpa macrophylla* vahl: The sturctures and stereochemistry of calliterpenone and calliterpenone monoacetate. *Tetrahedron*, *28*(16), pp. 4319–4323.

Hayashi, K., Hayashi, T., Otsuka, H. and Takeda, Y., 1997. Antiviral activity of 5, 6, 7-trimethoxyflavone and its potentiation of the antiherpes activity of acyclovir. *The Journal of Antimicrobial Chemotherapy*, *39*(6), pp. 821–824.

Ma, S.C., Du, J., But, P.P.H., Deng, X.L., Zhang, Y.W., Ooi, V.E.C., Xu, H.X., Lee, S.H.S. and Lee, S.F., 2002. Antiviral Chinese medicinal herbs against respiratory syncytial virus. *Journal of Ethnopharmacology*, *79*(2), pp. 205–211.

Min, B.S., Kim, Y.H., Tomiyama, M., Nakamura, N., Miyashiro, H., Otake, T. and Hattori, M., 2001. Inhibitory effects of Korean plants on HIV-1 activities. *Phytotherapy Research*, *15*(6), pp. 481–486.

Mona, R.I., 2016. Investigation of cytotoxic and antibacterial activity of methanol extract of callicarpa macrophylla leaves. BRAC University.

Wang, Z.H., Huang, J., Niu, C., Wang, J.H. and Wu, B., 2017. Isopimarane-type diterpenoids from *Callicarpa macrophylla* Vahl. *Phytochemistry Letters*, *22*, pp. 224–228.
Yadav, V., Jayalakshmi, S., Singla, R.K. and Patra, A., 2012. Ex vivo screening of stem extracts of *Callicarpa macrophylla* Vahl. for antifungal activity. *Indo Global Journal of Pharmaceutical Sciences*, *2*(2), pp. 103–107.

10.4.9.2 *Clerodendrum infortunatum* L.

Common names: Hill glory bower; ghetu flower, Indian bhat tree; bhant (Bangladesh); barni, bhanti, bhatar, bockada, perugilai, perivelam (India), hkaung kri (Myanmar); paw kwaw (Thailand)

Habitat: Forest edges, roadsides

Distribution: India, Bangladesh, Myanmar, Thailand, and Malaysia

Botanical observation: This shrub grows to a height of 2.5 m. The stem is quadrangular and hairy at apex. The leaves are simple, decussate, and without stipules. The petiole is up to about 9.5 cm. The blade is dull green above, cordate to broadly lanceolate, thinly coriaceous, 5.2–11.5 cm × 15–20 cm, wavy, somewhat hairy on both surface, and presents about 6–10 pairs of secondary nerves. The cymes are about 20 cm long, terminal, and dull maroon or pinkish. The flowers are fragrant. The calyx is tubular, 1.5 cm × 1 cm, and develops 5 lanceolate lobes which are greenish to reddish. The corolla tube is pure white, up to about 2 cm long, with 5 whitish lobes, and reddish at throat. The 4 stamens and style are white and protrude out of the corolla. The berries are globose, about 5 mm across, black, glossy in persistent reddish calyx, the whole structure somewhat frightening.

Clerodendrum infortunatum L.

Medicinal uses: Dysentery (Bangladesh); fever, leucorrhoea, sores (Myanmar); fever, gonorrhea, wounds (India)

Moderate broad-spectrum antibacterial nonpolar extract: Benzene extract of leaves inhibited the growth of *Staphylococcus aureus* (ATCC 25923), *Bacillus subtilis* (UC564), *Bacillus polymixa*, *Streptococcus faecalis* (ATCC 29212), *Pseudomonas aeruginosa* (ATCC 25619), *Salmonella typhi, Vibrio cholerae, Shigella dysenteriae*, and *Escherichia coli* (NCTC 8196) (Prusty et al., 2008).

Moderate broad-spectrum antifungal non-polar extract: Benzene extract of leaves inhibited the growth of *Penicillium notatum* (ATCC 11625), *Aspergillus niger*, and *Candida albicans* (ATCC 18804), with the MIC values of 225, 175, and 150 µg/mL, respectively (Prusty et al., 2008).

Commentaries

i. The plant yields lupane-type triterpenes such as clerodolone, clerodone, clerodol, and the phytosterol clerosterol (Manzoor-Khuda & Sarela, 1965) as well as glycosides of hydroxycinnamic acid derivatives (Uddin et al., 2020). Extract of the plant exhibited anthelmintic effects suggesting the presence of neuroactive principles (Siju et al., 2011) which could inhibit bacterial and/or fungal efflux pumps.

ii. Plants in the genus *Clerodendrum* L. are often used for the treatment of microbial infections in Asia. *Clerodendrum indicum* (L.) Kuntze. is used in India (local name: *Kavalai*) for cough and fever, in Bangladesh (local name: *Ekdaira gach*) for carbuncles, and in Myanmar (local name: *Ngayant patu*) is used for bronchitis, fever, gonorrhea, leprosy, pneumoniae, and syphilis. In India, *Clerodendrum inerme* (L.) Gaertn. (local name: *Putigandhaa*) is used for fever and in Myanmar for venereal diseases. In India, *Clerodendrum phlomoides* L. (local name: *Agnimantha*) is used for dysentery and fever. *Clerodendrum serratum* (L.) Moon. (local name: *Bhaargi*) is used for cough and fever in India and syphilis in Malaysia. *Clerodendron japonicum* (Thunb.) Sweet is used to treat gonorrhea in China. *Clerodendrum buchananii* (Roxb.) ex Wall. is given for wounds in Indonesia. In the Solomon Islands, *Clerodendrum confusum* Hall. f. is used for sores. *Clerodendrum fragrans* Willd is given for smallpox and furuncles in Cambodia, Laos, and Vietnam. *Clerodendrum paniculatum* L. is given for gonorrhea in Taiwan. *Clerodendrum quadriloculare* (Blanco) Merr. is used for wounds in the Philippines. *Clerodendrum trichotomum* Thunb. is a remedy for gangrene in Korea, and from this plant, the glycoside of hydroxycinnamic acid derivative verbascoside inhibited Human immunodeficiency type-1 integrase with the IC_{50} value of 7.8 µM (Kim et al., 2001).

iii. Benzene and carbon tetrachloride are extremely hazardous solvents that must never be used for extraction.

References

Kim, H.J., Woo, E.R., Shin, C.G., Hwang, D.J., Park, H. and Lee, Y.S., 2001. HIV-1 integrase inhibitory phenylpropanoid glycosides from *Clerodendron trichotomum*. *Archives of Pharmacal Research*, *24*(4), pp. 286–291.

Manzoor-Khuda, M. and Sarela, S., 1965. Constituents of *Clerodendrum infortunatum* (bhat)—I: Isolation of clerodolone, clerodone, clerodol and clerosterol. *Tetrahedron*, *21*(4), pp. 797–802.

Prusty, A.K., Ghosh, T. and Sahu, S.K., 2008. Anthelmintic, antimicrobial and antipyretic activity of various extracts of *Clerodendrum infortunatum* Linn. leaves. *Oriental Pharmacy and Experimental Medicine*, *8*(4), pp. 374–379.

Siju, E.N., Rajalakshmi, G.R., Sreejith, K.R., Hariraj, N., Venugopal, V.P. and Sneha, V.N., 2011. Anthelmintic activity of leaves of Clerodendrum viscosum Vent. *International Journal of Pharmacology and Biological Sciences*, *5*(1), p. 25.

Uddin, M.J., Çiçek, S.S., Willer, J., Shulha, O., Abdalla, M.A., Sönnichsen, F., Girreser, U. and Zidorn, C., 2020. Phenylpropanoid and flavonoid glycosides from the leaves of *Clerodendrum infortunatum* (Lamiaceae). *Biochemical Systematics and Ecology*, *92*, p. 104131.

10.4.9.3 *Duranta repens* L.

Synonyms: *Duranta erecta* L.; *Duranta microphylla* Desf.; *Duranta plumieri* Jacq.

Common names: Kantamehedi (Bangladesh); jia lian qiao (China); barangi (India); nilkanda (Nepal); tan yod (Thailand); tan kwan (Vietnam)

Habitat: Cultivated

Distribution: Asia and the Pacific

Botanical description: This shrub grows up to about 3 m tall. The stems are lenticelled, thorny, squarish, and pubescent. The leaves are simple, spiral in groups of 3, and exstipulate. The petiole is none or up to 1 cm long. The blade is 2–6.5 cm × 1.5–3.5 cm, cuneate and tapering at base, serrate, acute at apex, and often light green. The racemes are slender, pendulous, axillary, and terminal spikes. The calyx is tubular, hairy, 4 mm long, and somewhat 5-lobed. The corolla is tubular, 8 mm long, bluish to pure white, bilabiate, the lower lip 3-lobed, and the upper lip 2-lobed. The androecium includes 4 stamens. The drupes are orange, globose, poisonous, and 5 mm across.

Medicinal uses: Pneumonia, ringworms (India); cold, cough, fever (Nepal)

Broad-spectrum antibacterial polar extract: Ethanol extract of fruits (100 μg/disc) inhibited the growth of *Shigella boydii, Shigella shiga*, *Shigella dysenteriae*, *Shigella flexneri*, *Shigella sonnei, Bacillus subtilis, Staphylococcus aureus, Streptococcus β-haemoliticus, Escherichia coli, Pseudomonas aeruginosa*, and *Klebsiella* sp. with the inhibition zone diameter of 20, 19, 22, 18, 20, 18, 16, 17, 20, 19, and 21 mm, respectively (Sharma et al., 2012).

Strong antibacterial (Gram-negative) polar extract: Ethanol extract of fruits inhibited the growth of *Shigella dysenteriae*, *Shigella sonnei*, *Shigella flexneri*, *Escherichia coli*, and *Klebsiella* sp. with the MIC values of 32, 64, 32, 32, and 32 μg/mL, respectively (Sharma et al., 2012).

Broad-spectrum antifungal polar extract: Ethanol extract of fruits (100 μg/disc) inhibited the growth of *Aspergillus niger*, *Aspergillus flavus*, *Aspergillus fumigatus*, *Candida albicans*, *Trichoderma* sp., and *Fusarium* sp. with the inhibition zone diameters of 18, 16, 17, 14, 12, and 11 mm, respectively (Sharma et al., 2012).

Commentary: The plant yields oleanane-type saponins (Hiradate et al., 1999).

References

Hiradate, S., Yada, H., Ishii, T., Nakajima, N., Ohnishi-Kameyama, M., Sugie, H., Zungsontiporn, S. and Fujii, Y., 1999. Three plant growth inhibiting saponins from *Duranta repens*. *Phytochemistry*, *52*(7), pp. 1223–1228.

Sharma, P., Khandelwal, S., Singh, T. and Vijayvergia, R., 2012. Phytochemical analysis and antifungal potential of *Duranta erecta* against some phytopatogenic fungi. *International Journal of Pharmaceutical Sciences and Research*, *3*(8), p. 2686.

10.4.9.4 *Gmelina arborea* Roxb. ex Sm.

Synonyms: *Gmelina rheedii* Hook.; *Premna arborea* Farw.

Common names: Gamarigach; Gamari, gamari guloh (Bangladesh); ashveta (India); yemene (Myanmar); gambari (Nepal); gemilina (the Philippines); so (Laoe); lo'i tho (Vietnam)

Habitat: Forests, cultivated

Distribution: Pakistan, India, Nepal, Sri Lanka, Bangladesh, Myanmar, Cambodia, Laos, Vietnam, Thailand, China, Malaysia, the Philippines

Botanical observation: This timber tree grows to about 10 m tall. The bark is grayish and somewhat smooth and the wood whitish. The stems are hairy and quadrangular at apex. The leaves are simple, opposite, and exstipulate. The petiole is hairy, slender, and up to 10 cm long. The blade is broadly cordate, cuneate to tapering at base with a pair of glands below, wavy, 8–20 × 4.5–15 cm, dull green, somewhat soft, acuminate at apex, and marked with 3–7 pairs of secondary nerves. The panicles are terminal, lax, and about 30 cm long. The calyx is tubular, about 5 mm long and 5-lobed. The corolla is of a unique yellowish-orangish to brown, wavy, 3–4 cm long, bilabiate, the lower lip peltate and somewhat orchidaceous, and the upper lip entire or 2-lobed. Four stamens are present. The ovary is minute and develops a slender style and a bilobed stigma. The drupes are glossy, yellowish-green, ovoid to oblong, and up to about 2.5 cm long. Beautiful specimens are growing in the Philippines.

Medicinal uses: Boils, diarrhea, smallpox, cholera, syphilis, gonorrhea, cuts sores, wounds (Bangladesh); AIDS, cough, fever, gonorrhea, meningitis, typhoid, ulcers (India); fever (Laos, Vietnam)

Broad-spectrum antibacterial halo developed by polar extract: Methanol extract of roots inhibited the growth *of Enterococcus faecalis, Staphylococcus aureus, Acinetobacter baumannii, Citrobacter freundii, Escherichia coli, Klebsiella pneumoniae,* and *Pseudomonas aeruginosa* (Mishra & Padhy, 2013).

Antifungal furofuran lignans: (+)-7′-*O*-Ethyl arboreol, (+)-paulownin, (+)-gmelinol, (+)-epieudesmin from the wood inhibited the growth of *Trametes versicolor* (Kawamura et al., 2004).

Commentaries

i. The plants yields furofuran lignans, hydroxycinnamic acid glycoside derivatives, as well as 2,6-dimethoxy-p-benzoquinone and 3,4,5-trimethoxyphenol (Falah et al., 2008). These are most probably antimicrobial *in vitro*. There is the gut feeling that interesting antimicrobial findings await discovery from this tree.
ii. Kawamura et al. (2004) observed antifungal synergy among lignans present in the wood, confirming the contention that plants modulate the production of phytoalexins according to the pathophytogenic challenge they faces.
iii. *Gmelina asiatica* L. (local name: *Gopabhadra*) is used in India for cough and gonorrhea and for syphilis in Cambodia (local name: *Anhchan*). In the Philippines, *Gmelina philippinensis* Cham. is a remedy for athlete's foot.

References

Falah, S., Katayama, T. and Suzuki, T., 2008. Chemical constituents from *Gmelina arborea* bark and their antioxidant activity. *Journal of Wood Science*, *54*(6), pp. 483–489.

Kawamura, F., Ohara, S. and Nishida, A., 2004. Antifungal activity of constituents from the heartwood of Gmelina arborea: Part 1. Sensitive antifungal assay against Basidiomycetes.

Mishra, M.P. and Padhy, R.N., 2013. In vitro antibacterial efficacy of 21 Indian timber-yielding plants against multidrug-resistant bacteria causing urinary tract infection. *Osong Public Health and Research Perspectives*, *4*(6), pp. 347–357.

10.4.9.5 *Lantana camara* L.

Synonyms: *Camara vulgaris* Benth.; *Lantana aculeata* L.

Common names: Lantana, large leaf lantana; chaturaangi (Bangladesh); unichedi (India); kembang telek (Indonesia); bunga pagar (Malaysia); baho baho the (Philippines); yi sun (Thailand); cay tram oi (Vietnam)

Habitat: Roadside, cultivated

Distribution: Pakistan to Hawaii

Botanical observation: It is a poisonous herb that grows up to about 1 m tall in Tropical Asia. It cultivated as ornamental plant. The stems are spiny. The leaves are simple, decussate, and exstipulate. The petiole is up to about 1 cm long. The blade is ovate to ovate-oblong, 5–10 cm × 2–5 cm, serrate, and acute at apex. The inflorescences are terminal umbels of showy flowers. The calyx is tubular, and up to about 3 mm long. The corolla is tubular, up to 1 cm long, 4-lobed, and orange to red. The androecium includes 4 stamens. The fruit is a glossy drupe which is up to 5 mm in diameter, globose; and black.

Medicinal uses:Small pox, tuberculosis (Bangladesh); leprosy (China)

Broad-spectrum antibacterial halo developed by polar extract: Ethanol extract of leaves of inhibited the growth of *Staphylococcus aureus, Staphylococcus epidermidis, Streptococcus pyogenes, Shigella dysenteriae, Shigella boydii, Shigella flexnerii, Shigella sonnei, Salmonella typhi, Escherichia coli, Enterococcus aerogenes*, and *Pseudomonas aeruginosa* with inhibition zone diameter of 10, 13, 11, 15, 12, 12, 13, 10, 10, 12, and 15 mm, respectively (0.5 mg/disc, type (Uddin et al, 2008).

Antimycobacterial polar extract: The methanol extract of leaves inhibited the growth of *Mycobacterium tuberculosis* (H37Rv) with the MIC value of 20 µg/mL (Kirimuhuzya et al., 2009).

Antifungal (filamentous) essential oil: Essential oil at 1 mg/mL inhibited the mycelial growth of *Corynespora cassiicola* by 25% (Passos et al., 2012).

Broad-spectrum antibacterial triterpene: 22β-Acetoxylantic acid inhibited the growth of *Staphylococcus aureus* and *Salmonella typhi* (Barre et al., 1997).

Commentaries

i. The plant yields verbascoside (Herbert et al., 1991).
ii. 5-Hydroxy-6,7,3′,4′,5′-pentamethoxyflavone from a member of the genus *Lantana* L. inhibited the growth of *Staphylococcus aureus, Salmonella typhimurium, Candida tropicalis, Aspergillus niger, Aspergillus fumigatus, Trichophyton mentagrophytes*, and *Microsporium canis* with the MIC values of 20, 5, 20, 200, 200, 50, and 50 µg/mL, respectively (Rwangabo et al., 1988).
iii. In the Tribe Lantaneae, *Lippia nodiflora* (L.) Michx. is given for gonorrhea in Taiwan. Essential oil of *Lippia graveolens* Kunth inhibited the growth of *Escherichia coli, Staphylococcus aureus*, and *Streptococcus mutans* with MIC values of 0.3 µg/mL (Miller et al., 2015).

References

Barre, J.T., Bowden, B.F., Coll, J.C., De Jesus, J., Victoria, E., Janairo, G.C. and Ragasa, C.Y., 1997. A bioactive triterpene from *Lantana camara. Phytochemistry, 45*(2), pp. 321–324.

Herbert, J.M., Maffrand, J.P., Taoubi, K., Augereau, J.M., Fouraste, I. and Gleye, J., 1991. Verbascoside isolated from *Lantana camara*, an inhibitor of protein kinase C. *Journal of Natural Products, 54*(6), pp. 1595–1600.

Kirimuhuzya, C., Waako, P., Joloba, M. and Odyek, O., 2009. The anti-mycobacterial activity of *Lantana camara* a plant traditionally used to treat symptoms of tuberculosis in South-western Uganda. *African Health Sciences, 9*(1), pp. 40–45.

Miller, A.B., Cates, R.G., Lawrence, M., Soria, J.A.F., Espinoza, L.V., Martinez, J.V. and Arbizú, D.A., 2015. The antibacterial and antifungal activity of essential oils extracted from Guatemalan medicinal plants. *Pharmaceutical Biology, 53*(4), pp. 548–554.

Oyourou, J.N., Combrinck, S., Regnier, T. and Marston, A., 2013. Purification, stability and antifungal activity of verbascoside from *Lippia javanica* and *Lantana camara* leaf extracts. *Industrial Crops and Products, 43*, pp. 820–826.

Passos, J.L., Barbosa, L.C.A., Demuner, A.J., Alvarenga, E.S., Silva, C.M.D. and Barreto, R.W., 2012. Chemical characterization of volatile compounds of *Lantana camara* L. and L. radula Sw. and their antifungal activity. *Molecules, 17*(10), pp. 11447–11455.

Rwangabo, P.C., Claeys, M., Pieters, L., Corthout, J., Vanden Berghe, D.A. and Vlietinck, A.J., 1988. Umuhengerin, a new antimicrobially active flavonoid from *Lantana trifolia. Journal of Natural Products, 51*(5), pp. 966–968.

10.4.9.6 *Premna corymbosa* Rottler & Willd.

Synonyms: *Cornutia corymbosa* Burm.f., *Premna cordifolia* Wight, *Premna foetida* Reinw. ex Blume; *Premna integrifolia* L., *Premna obtusifolia* R. Br.; *Premna scandens* Bojer, *Premna serratifolia* L., *Premna spinosa* Roxb.

Common names: Muney kerai (India); kaar (Marshall); agnibijaka (India); buas buas (Indonesia; Malaysia); kywe-thwe, taung tan kri, taung taunggi (Myanmar); osum (Palau); nigpr, alowalo (Papua New Guinea); aal (Yap)

Habitat: Forest edges, seashores, beaches, mangroves, roadsides

Distribution: Tropical Asia and Pacific

Botanical description: This tree grows to a height of about 12 m. The bole is curved and the bark is lenticelled. The stem is terete, glabrous, and longitudinally striated. The nodes are minutely marked with a transversal ring. The leaves are simple, opposite, and without stipules. The petiole is glabrous, channeled above, and 1.8–2.1 cm long. The blade is 4.5–6.8 cm × 11.5 × 14 cm, broadly elliptic to broadly lanceolate or cordate, glossy, somewhat coriaceous, wavy, obtuse at apex, acute, round or cordate at base, and marked with about 4 to 6 pairs of secondary nerves. The corymbs are axillary or terminal, with a characteristic architecture, many flowered, and about 6–10 cm in diameter. The calyx is cupular and hairy. The corolla is tubular with 4 minute round lobes, 3 mm across, greenish-white, and hairy at throat. Four stamens form the androecium. The style is linear and bifid. The berries are glossy, globose, 3 mm in diameter, 3–4 seeded on a persistent calyx.

Medicinal use: Bronchitis (India); ear infection, eye infection, syphilis (Palau); cough, pneumonia (Papua New Guinea); fever (India; Myanmar; Papua New Guinea); cold (India; Palau)

Broad-spectrum antibacterial halo developed by polar extract: Ethanol extract of roots inhibited the growth of *Staphylococcus aureus* ATCC-25923), *Klebsiella pneumoniae* (ATCC-25922), *Pseudomonas aeruginosa* (ATCC-25921), *Salmonella typhi*, and *Vibrio cholerae* (ATCC-25926) (Rajendran & Basha, 2010).

Moderate broad-spectrum antibacterial essential oil: Essential oil of leaves inhibited the growth of *Sarcina lutea* (IFO 3232), *Bacillus subtilis* (IFO 3026), *Escherichia coli* (IFO 3007), *Pseudomonas* sp. (ATCC 13867), *Klebsiella pneumoniae* (ATCC 10031), *Xanthomonas campestris* (IAM 1671), and *Xanthomonas campestris* (IAM 1671) (Rahman et al., 2016).

Broad-spectrum antifungal halo developed by polar extract: Ethanol extract of roots inhibited the growth of *Candida albicans* (ATCC 25928), *Aspergillus niger* (ATCC 2091), *Pencillium chrysogenum* (ATCC 2021), *Microsporum gypseum* (ATCC-2952), and *Epidermatophyton flocossum* (ATCC 2950) (Rajendran & Basha, 2010).

Commentaries

i. The plant yields abietane-type diterpenes (Habtemariam et al., 2015) as well as hydroxycinnamic acid glycoside derivatives such as verbascoside (see earlier) (Bose et al., 2013) and furofuran lignans (Woo et al., 2019).

ii. *Premna esculenta* Roxb. (local names: *Silazra; Lelom pada; Lamur; Tatui*) is a remedy for and fungal infection and tuberculosis in Bangladesh. In Myanmar, *Premna mollissima* Roth is used in India for gonorrhea, mouth ulcers, syphilis, and ringworms and is given after childbirth in Myanmar (local name: *Kyetyo*).

iii. In the Philippines, *Premna odorata* Blanco (local name: *Adgao, Alagaw*) is used for cough, tuberculosis, and wounds. Ethanol extract of this plant inhibited the replication of the Vesicular stomatitis virus with the EC_{50} value of 50 μg/mL in Vero cells (Hamidi et al., 1996). From the leaves of this plant, 1-heneicosyl formate (LogD 9.5 at pH 7.4; molecular mass = 340.5 g/mL) inhibited the growth of *Mycobacterium tuberculosis* ($H_{37}Rv$) with the MIC value of 8 μg/mL (Lirio et al., 2014). Natural products from medicinal plants with long alkyl chains and ketone moieties are often antimycobacterial. This is the case for fatty acids (Kondo & Kanai, 1976). Untreatable strains of *Mycobacterium tuberculosis* are emerging (Boire et al., 2013).

iv. Methanol extract of bark of *Premna mucronata* Roxb. inhibited the replication of the Dengue virus serotype-1, -2, -3, and -4 with the IC_{50} values of 12.5, 27.9, 21.3, and 12.5 μg/mL, respectively (Sood et al., 2015).

v. Plants in the genus *Premna* L. yield antibacterial and antifungal diterpenes such as 7α-hydroxy-6,1 l-cyclofarnes-3(15)-en-2-one inhibiting the growth of *Staphylococcus aureus*, *Salmonella typhimurium*, *Candida tropicalis*, *Aspergillus niger*, *Aspergillus fumigatus*, *Trichophyton mentagrophytes*, and *Microsporium canis* with the MIC values of

20, 5, 20, 200, 200, 50, and 50 μg/mL, respectively (Rahman et al., 2016). (5*R*,8*R*,9*S*,10*R*)-12-oxo-*ent*-3,13(16)-clerodien-15-oic acid isolated from a member of the genus *Premna* L. inhibited the growth of *Staphylococcus aureus* and *Bacillus subtilis* at the concentration of about 25 μg/mL (Habtemariam et al., 1990). The quinonemethide diterpene bharangin (molecular mass = 328.4 g/mol) from *Premna herbacea* Roxb. inhibited the growth of *Arthrobacter citrius* (NCIM 2320), *Bacillus cereus* (MTCC 430), *Bacillus pumilus* (NCIM 2327), *Staphylococcus aureus* (NCIM 2079), *Streptococcus epidermidis* (NCIM 2611), *Nocardia* sp., NCIM 2386), *Escherichia coli* (NCIM 2574), and *Clostridium* sp. (NCIM 2337) with the MIC values of 10, 10, 10, 6.2, 6.2, 10, 10, and 5 μg/mL, respectively (Murthy et al., 2006).

vi. 5, 7-Dihydroxy-8,4′-dimethoxyflavone 7-*O*-β-D-glucoside (also known as cirsitakaoside) isolated from *Premna szemaoensis* C. P'ei given intraperitoneally to mice at the dose of 30 mg/Kg increased immune response to the Vesicular stomatitis virus (Di et al., 2020).

References

Boire, N.A., Riedel, V.A.A., Parrish, N.M. and Riedel, S., 2013. Tuberculosis: from an untreatable disease in antiquity to an untreatable disease in modern times. *Journal of Ancient Diseases & Preventive Remedies*, *1*(2), pp. 1–11.

Bose, L.V., Varghese, G.K. and Habtemariam, S., 2013. Identification of acteoside as the active antioxidant principle of *Premna serratifolia* root wood tissues. *Phytopharmacology*, *4*(2), pp. 228–236.

Di, Q., Zhu, H., Pu, D., Zhao, X., Li, X., Ma, X., Xiao, W. and Chen, W., 2020. The natural compound *Cirsitakaoside enhances* antiviral innate responses against vesicular stomatitis virus in vitro and in vivo. *International Immunopharmacology*, *86*, p. 106783.

Habtemariam, S. and Varghese, G.K., 2015. A Novel diterpene skeleton: Identification of a highly aromatic, cytotoxic and antioxidant 5-methyl-10-demethyl-abietane-type diterpene from *Premna serratifolia*. *Phytotherapy Research*, *29*(1), pp. 80–85.

Habtemariam, S., Gray, A.I. and Waterman, P.G., 1993. A new antibacterial sesquiterpene from *Premna oligotricha*. *Journal of Natural Products*, *56*(1), pp. 140–143.

Habtemariam, S., Gray, A.I., Halbert, G.W. and Waterman, P.G., 1990. A novel antibacterial diterpene from *Premna schimperi*. *Planta Medica*, *56*(2), pp. 187–189.

Hamidi, J.A., Ismaili, N.H., Ahmadi, F.B. and Lajisi, N.H., 1996. Antiviral and cytotoxic activities of some plants used in Malaysian indigenous medicine. *Pertanika Journal of Tropical Agricultural Science*, *19*(2/3), pp. 129–136.

Kondo, E. and Kanai, K., 1976. Further studies on the lethal effect of long-chain fatty acids on mycobacteria. *Japanese Journal of Medical Science and Biology*, *29*(1), pp. 25–37.

Lirio, S.B., Macabeo, A.P.G., Paragas, E.M., Knorn, M., Kohls, P., Franzblau, S.G., Wang, Y. and Aguinaldo, M.A.M., 2014. Antitubercular constituents from *Premna odorata* Blanco. *Journal of Ethnopharmacology*, *154*(2), pp. 471–474.

Murthy, M.M., Subramanyam, M., Giridhar, K.V. and Jetty, A., 2006. Antimicrobial activities of bharangin from *Premna herbaceae* Roxb. and bharangin monoacetate. *Journal of Ethnopharmacology*, *104*(1–2), pp. 290–292.

Rahman, A., Shanta, Z.S., Rashid, M.A., Parvin, T., Afrin, S., Khatun, M.K. and Sattar, M.A., 2016. In vitro antibacterial properties of essential oil and organic extracts of *Premna integrifolia* Linn. *Arabian Journal of Chemistry*, *9*, pp. S475–S479.

Rajendran, R. and Basha, N.S., 2010. Antimicrobial activity of crude extracts and fractions of *Premna serratifolia* Lin. root. *Medicinal Plants-International Journal of Phytomedicines and Related Industries*, *2*(1), pp. 33–38.

Sood, R., Raut, R., Tyagi, P., Pareek, P.K., Barman, T.K., Singhal, S., Shirumalla, R.K., Kanoje, V., Subbarayan, R., Rajerethinam, R. and Sharma, N., 2015. *Cissampelos pareira* Linn: Natural source of potent antiviral activity against all four dengue virus serotypes. *PLoS Neglected Tropical Diseases*, *9*(12), p. e0004255.

Woo, S.Y., Wong, C.P., Win, N.N., Hoshino, S., Prema, Ngwe, H., Abe, I. and Morita, H., 2019. A new tetrahydrofuran lignan from *Premna serratifolia* wood. *Natural Product Communications*, *14*(1), p. 1934578X1901400130.

10.4.9.7 *Tectona grandis* L.f.

Synonyms: *Tectona theka* Lour.; *Theka grandis* (L. f.) Lam.

Common names: Teak; you mu (China); anila, bhuumisah (India); jati (Indonesia); kyun (Myanmar); sagwan (Pakistan); dalanang, teca (the Philippines); sak (Laos, Thailand); cay tek (Vietnam)

Habitat: Cultivated

Distribution: Pakistan to Southeast Asia

Botanical observation: This timber tree grows up to about 45 m. The wood is hard (teak wood) and the bark is light gray and smooth. The stems are grayish brown, quadrangular, and hairy. The leaves are simple, massive, decussate, and exstipulate. The petiole is stout and up to about 4 cm long. The blade is ovate-elliptic to ovate, 15–45 × 8–30 cm, hairy below, cuneate at base, wavy, dull pale green, obtuse at apex, and with 7 to 12 pairs of secondary nerves. The panicles are conspicuous, lax, many-flowered, and up to 30 cm long. The calyx is tubular, about 3 mm, with white stellate hairs and develops 5 rounded lobes. The corolla is whitish, tubular, about 6 mm long, and 5-lobed. The androecium includes 5 stamens protruding from the corolla tube. The ovary is hairy and develops a 5 mm long style. The drupes are somewhat oblate, light green, ribbed, and up to about 1.5 cm across. Specimens are growing in Malaysia by the sides of highways.

Medicinal uses: Bronchitis, cough, sores (India); cholera (Indonesia); dysentery (India; Pakistan); herpes, gonorrhea, urinary tract infection (Myanmar); fever (Thailand)

Moderate antibacterial (Gram-negative) polar extract: Ethanol extract of fruits inhibited the growth of *Escherichia coli* (ATCC 8739), *Pseudomonas aeruginosa*, *Klebsiella pneumoniae* (ATCC 1148), and *Klebsiella aerogenes* (ATCC 13048) with the MIC values of 64, 256, 128, and 64 µg/mL, respectively (Bitchagno et al., 2015). Aqueous ethanol extract of stem bark inhibited the growth *Micrococcus luteus* (MTCC (106)) and *Bacillus subtilis* (MTCC (121)) with the MIC values of 600 and 600 µg/mL (Singh et al., 2010).

Strong antibacterial (Gram-negative) anthraquinones: Tectograndone from the fruits inhibited the growth of *Escherichia coli* (ATCC 8739) with the MIC/MBC values of 32/128 µg/mL (Bitchagno et al., 2015). 6-Methyl-1,4-dihydroxyanthraquinone inhibited the growth of *Pseudomonas aeruginosa*, and *Klebsiella aerogenes* (ATCC 13048) with the MIC/MBC values of 128/256 and 16/128 µg/mL, (Bitchagno et al., 2015).

Strong antifungal (filamentous) amphiphilic naphthoquinone: Deoxylapachol (LogD 3.2 at pH 7.4; molecular mass = 226.2 g/mol) inhibited the growth of *Aspergillus niger* with the MIC value of 12.5 ppm, (Sumthong et al., 2006).

Strong antifungal (filamentous) amphiphilic anthraquinone: Tectoquinone (LogD 3.7 at pH 7.4; molecular mass = 226.2 g/mol) inhibited the growth of *Aspergillus niger* with the MIC value of 25 ppm (Sumthong et al., 2006).

Moderate antibacterial (Gram-negative) ursane-type triperpene: 2β-Hydroxyursolic acid inhibited the growth of *Escherichia coli* (ATCC 8739), *Pseudomonas aeruginosa*, *Klebsiella pneumoniae* (ATCC 1148), and *Klebsiella aerogenes* (ATCC 13048) with the MIC values of 128/256, 64/128, 64/128, and 64/128 µg/mL, respectively (Bitchagno et al., 2015).

Commentaries: The plants produce apocarotenoids (megastigmanes) (Macías et al., 2008) as well as quinones (Lacret et al., 2011).

References

Bitchagno, G.T.M., Fonkeng, L.S., Kopa, T.K., Tala, M.F., Wabo, H.K., Tume, C.B., Tane, P. and Kuiate, J.R., 2015. Antibacterial activity of ethanolic extract and compounds from fruits of *Tectona grandis* (Verbenaceae). *BMC Complementary and Alternative Medicine*, *15*(1), pp. 1–6.

Lacret, R., Varela, R.M., Molinillo, J.M., Nogueiras, C. and Macías, F.A., 2011. Anthratectone and naphthotectone, two quinones from bioactive extracts of *Tectona grandis*. *Journal of Chemical Ecology*, *37*(12), pp. 1341–1348.

Macías, F.A., Lacret, R., Varela, R.M., Nogueiras, C. and Molinillo, J.M., 2008. Bioactive apocarotenoids from *Tectona grandis*. *Phytochemistry*, *69*(15), pp. 2708–2715.

Singh, M., Khatoon, S., Singh, S., Kumar, V., Rawat, A.K.S. and Mehrotra, S., 2010. Antimicrobial screening of ethnobotanically important stem bark of medicinal plants. *Pharmacognosy Research*, *2*(4), p. 254.
Sumthong, P., Damveld, R.A., Choi, Y.H., Arentshorst, M., Ram, A.F., Van Den Hondel, C.A. and Verpoorte, R., 2006. Activity of quinones from teak (*Tectona grandis*) on fungal cell wall stress. *Planta Medica*, *72*(10), pp. 943–944.

10.4.9.8 *Vitex negundo* L.

Synonym: *Vitex paniculata* Lam.

Common names: Indian privet, 5-leaved chaste tree; bara, nishinda (Bangladesh); king tze, mu ching, shān pŭ jiāng (China); bana, nirgundi, nochi, suvaha (India); bangela (Iran); lenggundi (Malaysia); kiyubantin, kyaungban-gyi (Myanmar); simali (Nepal); nikka (Sri Lanka); lagundi (the Philippines)

Habitat: Forest ridge, roadsides

Distribution: Asia and Pacific

Botanical observation: This graceful treelet grows up to 6 m tall. The bark is gray and the stems are quadrangular and velvet. The leaves are simple, decussate, exstipulate, and 3–5 foliolate and without stipules. The petiole is 3.7–5 cm long, channeled, and velvety. The folioles are lanceolate, asymmetrical, 5.5–8.5 cm × 2–5 cm, velvety, and glaucous below. The blade shows 9–13 pairs of secondary nerves. The inflorescences are terminal, conical, and velvety panicles which are about 20 cm long. The calyx is 3 mm long, white, velvety, and produces 5 triangular lobes which are about 1 mm long. The corolla is 1 cm long, bluish purple, velvety outside, and hairy inside. The upper corolla lip is 2 mm long and bifid. The lower lip is 5 mm long and 3-lobed. The 4 stamens are hairy at base. The ovary is glabrous. The stigma is forked. The berries are 6 mm diameter andripen glossy and black.

Medicinal uses: Abscess, bronchitis, cholera, fever, cough, gonorrhea, typhus (India); cold, flu, cold, fever, sores, typhoid (Bangladesh); chronic bronchitis cold, cough, dysentery, gonorrhea, leucorrhea (China); cold, cough (the Philippines); dysentery (China; India); cough, cold, fever (Nepal); leprosy (India); carries (Iran); cough, cold, fever (Nepal)

Antibacterial (Gram-positive) non-polar extract: Hexane extract of leaves inhibited the growth of *Staphylococcus aureus* (ATCC 43300) (Triveni et al., 2016).

Broad-spectrum antibacterial polar extract: Methanol extract of leaves inhibited the growth of *Bacillus subtilis*, *Staphylococcus aureus*, *Micrococcus pyogenes*, *Pseudomonas aeruginosa,* and *Escherichia coli* (Gautam et al., 2008).

Moderate broad-spectrum antibacterial polar extract: Ethanol extract of leaves inhibited the growth of *Streptococcus faecalis*, *Klebsiella pneumoniae*; *Escherichia coli*, *Pseudomonas aeruginosa*, and *Staphylococcus aureus* with the MIC values of 125, 250, 250, 500, and 250 µg/mL, respectively (Sathiamoorthy et al., 2007).

Strong antifungal (yeast) polar extract: Ethanol extract of leaves inhibited the growth of *Cryptococcus neoformans* and *Trichophyton mentagrophytes* with the MIC values of 62.5 and 125 µg/mL, respectively (Sathiamoorthy et al., 2007).

Strong antibacterial (Gram-positive) amphiphilic flavone: Penduletin (LogD 1.4 at pH 7.4; molecular mass = 344.3 g/mol) isolated from the leaves inhibited the growth of methicillin-resistant *Staphylococcus aureus* with the MIC value of 10 µg/mL (Sichaem et al., 2021).

Strong broad-spectrum antibacterial amphiphilic flavones: Luteolin (LogD 1.1 at pH 7.4; molecular mass = 286.2 g/mol) isolated from the leaves inhibited the growth of *Streptococcus faecalis*, *Klebsiella pneumoniae*, *Escherichia coli*, *Pseudomonas aeruginosa*, and *Staphylococcus aureus* with MIC values of 25, 25, 50, 25, and 25 µg/mL (Sathiamoorthy et al., 2007). 5′-Hydroxy-3′, 4′, 3, 6, 7-pentamethoxyflavone isolated from the leaves inhibited the growth of *Klebsiella pneumoniae*, *Escherichia coli*, *Pseudomonas aeruginosa*, and *Staphylococcus aureus* with MIC values of 25, 50, 50, and 25 µg/mL (Sathiamoorthy et al., 2007).

Strong broad-spectrum antifungal flavones: Luteolin isolated from the leaves inhibited the growth of *Cryptococcus neoformans*, *Trichophyton mentagrophytes*, and *Candida parapsilosis*

(ATCC-22019) of MIC values of 25, 50, and 25 µg/mL, respectively (Sathiamoorthy et al., 2007). 5′-Hydroxy-3′, 4′, 3, 6, 7-pentamethoxyflavone inhibited the growth of *Cryptococcus neoformans*, *Trichophyton mentagrophytes*, *Aspergillus fumigatus*, and *Candida parapsilosis* (ATCC-22019) with the MIC values of 12.5, 25, 50, and 25 µg/mL, respectively (Sathiamoorthy et al., 2007).

Strong broad-spectrum antifungal iridoid glycoside: Negundoside (molecular mass = 496.5 g/mol) inhibited the growth of *Cryptococcus neoformans*, *Trichophyton mentagrophytes*, and *Candida parapsilosis* (ATCC-22019) of MIC values of 6.2, 6.2, and 50 µg/mL, respectively (Sathiamoorthy et al., 2007).

Moderate antiviral (enveloped monopartite linear single-stranded (+)-RNA) polar extract: Aqueous extract of leaves inhibited the replication of the Zika virus (MR766) with the IC_{50} value of 466.7 µg/mL in Vero cells (Vista et al., 2020).

Commentaries

i. The plant yields labdane-type and abietane-type diterpenes (Zheng et al., 2010).
ii. The plant is used against COVID-19 in Bangladesh (Azam et al., 2020). In the Philippines, a cough mixture is made from the leaves and sold in pharmacies. Thus, there is a need to examine the use of this plant as phytomedication against COVID-19. Note that the Severe acute respiratory syndrome-associated coronavirus 2 is like the Zika virus and enveloped monopartite linear and single-stranded virus.
iii. *Vitex peduncularis* Wall. ex Schauer in A. DC. is (local name: *Ash muli gach*) is a remedy for jaundice in Bangladesh, and in India is used for and eye infection (local name: *Charaigorwa*). In India, *Vitex agnus-castus* L. is given for cold, *Vitex leucoxylon* L. f. is used for fever (local name: *Kattunochi*), *Vitex agnus-castus* L. is used for eye infections and chills, *Vitex altissima* L.f. (local; name: *Nemaliadugu*) is used for leprosy, and *Vitex trifolia* L. is used for bronchitis and fever (local name: *Indranika*). *Vitex trifolia* L. is used for cold, cough, sores, and dysentery in Myanmar (local name: *Kyaung*-pan) and for cough and dysentery in Papua New Guinea (local name: *Dala*). *Vitex glabrata* R. Br. is a remedy for diarrhea in Thailand (local name: *Khai-nao*). In Malaysia, *Vitex pubescens* Vahl. is used for wounds and *Vitex trifolia* L. is a remedy for tuberculosis. In China, *Vitex quinata* (Lour.)Wall. is used for cough (local name: *Feng liu*).
iv. 5,4′-Dihydroxy-7,3′-dimethoxyflavone from *Vitex limonifolia* Wall. inhibited the replication of the Coxsackie virus 3B with the IC_{50} values of 1.8 µM (Ban et al., 2018).
v. 3-(4-hydroxyphenyl)-1-(2,4,6-trimethoxyphenyl)-2-propen-1-one isolated from a member of the genus *Vitex* L. inhibited the replication of the Human immunodeficiency virus type-1 by 77% at the concentration of 15.9 µM (Pan et al., 2014).
vi. *Stachytarpheta jamaicensis* (L.) Vahl. is given for syphilis in Cambodia, Laos, and Vietnam.

References

Azam, M.N.K., Al Mahamud, R., Hasan, A., Jahan, R. and Rahmatullah, M., 2020. Some home remedies used for treatment of COVID-19 in Bangladesh. *Journal of Medicinal Plants Studies*, *8*(4), pp. 27–32.

Ban, N.K., Thoa, N.T.K., Linh, T.M., Trang, D.T., Van Kiem, P., Nhiem, N.X., Tai, B.H., Van Minh, C., Song, J.H., Ko, H.J. and Kim, S.H., 2018. Labdane-type diterpenoids from *Vitex limonifolia* and their antivirus activities. *Journal of Natural Medicines*, *72*(1), pp. 290–297.

Gautam, L.M., Shrestha, S.L., Wagle, P. and Tamrakar, B.M., 2008. Chemical constituents from *Vitex negundo* (Linn.) of nepalese origin. *Scientific World*, *6*(6), pp. 27–32.

Pan, W., Liu, K., Guan, Y., Tan, G.T., Hung, N.V., Cuong, N.M., Soejarto, D.D., Pezzuto, J.M., Fong, H.H. and Zhang, H., 2014. Bioactive compounds from *Vitex leptobotrys*. *Journal of Natural Products*, *77*(3), pp. 663–667.

Sathiamoorthy, B., Gupta, P., Kumar, M., Chaturvedi, A.K., Shukla, P.K. and Maurya, R., 2007. New antifungal flavonoid glycoside from *Vitex negundo*. *Bioorganic & Medicinal Chemistry Letters*, *17*(1), pp. 239–242.

Sichaem, J., Nguyen, H.H., Nguyen, V.H., Mac, D.H., Mai, D.T., Nguyen, H.C., Tran, T.N.M., Pham, N.K.T., Nguyen, H.H., Niamnont, N. and Duong, T.H., 2021. A new labdane-type diterpenoid from the leaves of *Vitex negundo* L. *Natural Product Research*, *35*(14), pp. 2329–2334.

Triveni, A.G., Mendem, S.K., Shivannavar, C.T. and Gaddad, S.M., 2016. Antibacterial activity of *Vitex negundo* leave extract against methicillin resistant Staphylococcus aureus (MRSA). *International Journal of Pharmacy and Biological Sciences*, *6*(3), pp. 55–59.

Vista, F.E.S., Dalmacio, L.M.M., Corales, L.G.M., Salem, G.M., Galula, J.U. and Chao, D.Y., 2020. Antiviral effect of crude aqueous extracts from ten Philippine medicinal plants against Zika virus. *Acta Medica Philippina*, *54*(2).

Zheng, C.J., Huang, B.K., Wu, Y.B., Han, T., Zhang, Q.Y., Zhang, H. and Qin, L.P., 2010. Terpenoids from *Vitex negundo* seeds. *Biochemical Systematics and Ecology*, *38*(2), pp. 247–2

10.5 ORDER SOLANALES JUSS. EX BERCHT. & J. PRESL (1820)

10.5.1 Family Convolvulaceae A.L. de Jussieu (1789)

The family Convolvulaceae consists of about 40 genera and 1200 species of herbs, shrubs, or climbers. Upon incision, the stems not uncommonly yield some acrid sap. The leaves are simple, spiral, and exstipulate. The inflorescences are terminal or axillary cymes. The calyx includes 5 sepals. The corolla is funnel-shaped, more or less 5-lobed, ephemeral, and membranaceous. The androecium comprises 5 stamens. The gynoecium consists of 2 carpels fused into a 2-locular ovary, each locule sheltering 2 ovules. The fruits are papery capsules or berries.

10.5.1.1 *Evolvulus alsinoides* L.

Synonyms: *Convolvulus alsinoides* L.; *Evolvulus chinensis* Choisy; *Evolvulus debilis* Kunth

Common names: Dwarf morning glory; tu ding dui; yuan bian zhong (China); vishnukranta, shankhpushpi (India); sankha holi (Pakistan); bai to kan (Thailand); bat giao (Vietnam)

Habitat: Grassy areas, roadsides, open grounds

Distribution: Tropical Asia-Pacific

Botanical description: This beautiful herb grows to a length of 30 cm. The stems are terete and pilose. The leaves are simple, spiral, without stipules, and subsessile. The blade is linear-oblong, subglabrous, 0.8–1 cm × 2–3 mm, and slightly acuminate at apex. The inflorescences are axillary and hairy cymes which are 2–2.7 cm long. The calyx comprises 5 sepals which are linear and hairy. The corolla is of a peculiar blue and develops 5 round lobes which are about 3 mm long. The ovary develops 4 conspicuous styles. The fruit is a capsule.

Medicinal uses: Dysentery, hepatitis (India, Pakistan); fever (India; Pakistan; the Philippines)

Antibacterial (Gram-negative) polar extract: Ethanol extract (20 mg/mL/well) inhibited the growth of *Pseudomonas aeruginosa* and *Escherichia coli* with the inhibition zone diameters of 16 and 19 mm, respectively (Dash et al., 2002).

Moderate antifungal (filamentous) polar extract: Ethanol extract (200 µg/mL/well) inhibited the growth of *Malassezia furfur* (Deviha & Pavithram, 2015).

Strong antiviral (enveloped monopartite linear single-stranded (+) RNA) polar extract: Methanol extract inhibited the replication of the mouse coronavirus with the MIC value of 1 µg/mL (Vimalanathan et al., 2009).

Strong antiviral (non-enveloped monopartite linear single-stranded (+) RNA) polar extract: Methanol extract inhibited the replication of the Feline calicivirus with the MIC value of about 20 µg/mL (Vimalanathan et al., 2009).

Commentaries

i. What is (are) the anti-coronavirus principles in this plant?
ii. *Evolvulus nummularius* L. is a remedy for tooth infections in India.

References

Dash, G.K., Suresh, P., Sahu, S.K., Kar, D.M., Ganapaty, S. and Panda, S.B., 2002. Evaluation of *Evolvulus alsinoides* Linn. For anthelmintic and antimicrobial activities. *Journal of Natural Remedies*, *2*(2), pp. 182–185.

Deviha, M.S. and Pavithram, K.S., 2015. Antifungal activity by ethanolic extracts of medicinal plants against Malasseziafurfur: A potential application in the treatment of Dandruff. *international Journal of PharmTech Research*, *8*(3), pp. 440–444.

Vimalanathan, S., Ignacimuthu, S. and Hudson, J.B., 2009. Medicinal plants of Tamil Nadu (Southern India) are a rich source of antiviral activities. *Pharmaceutical Biology*, *47*(5), pp. 422–429.

10.5.1.2 *Ipomoea batatas* (L.) Lam.

Synonyms: *Batatas edulis* (Thunb.) Choisy; *Convolvulus batatas* L.; *Convolvulus edulis* Thunb.

Common names: Sweet potato; chinealu (Bangladesh); dam long (Cambodia); fan chu (China); kandagranthi (India); ubi jalak (Indonesia); lardaklahori (Iran); man kew (Laos); kau kau (Papua New Guinea); keledek (Malaysia); kaswan (Myanmar); kamote (Philippines); velkeleng (Sri Lanka); man tet (Thailand); koa lang (Vietnam)

Habitat: Cultivated

Distribution: Tropical Asia and the Pacific

Botanical observation: This vegetable native of tropical America grows up to 4m long. The tubers are red, white, or yellow, edible, and elongated. The stem is green or purplish, slightly angled, soft, fleshy, and rooting at nodes. The leaves are simple, alternate, and without stipules. The petiole is slender, channeled, slightly enlarged at base, and 5.5–10.5cm long. The blade is ovate-cordate, acute angular or more or less lobed, fleshy, soft, yellowish light green, 4–13cm × 3–13cm, and presents 5–10 pairs of secondary nerves. The inflorescences are axillary, 1–7-flowered, 2–10.5cm long, and angular. The flower pedicel is 0.2–1 cm long. The calyx includes 5 sepals which are elliptic, shortly acuminate, hairy, nerved, glabrous, and 0.7–1.3cm long. The corolla is pink, white, or pale purple to purple, with a darker center, infundibuliform, membranous, and 3–4cm long. The ovary is 4-celled. The capsules are globose.

Medicinal uses: Dysentery (Indonesia); fever (Malaysia)

Moderate antifungal (filamentous) ursane-type triterpene: Urs-13(18)-ene-3β-yl acetate inhibited the growth of *Sporothrix schenckii* and *Trichophyton mentagrophytes* with an MIC value of 50 µg/mL (Oluyori et al., 2016).

Commentaries

i. The Indian traditional systems of medicine uses the following plants in the genus *Ipomoea* L. to treat microbial infections: *Ipomoea fistulosa* Mart. ex Choisy is used for cold, fever, cuts and wounds, *Ipomoea mauritiana* Jacq. is a remedy for tuberculosis and leprosy, *Ipomoea aquatica* Forssk. is used for the treatment of ringworm and chickenpox (local name: *Kalaka*), *Ipomoea hederacea* Jacq. is used for fever (local name: *Krishnabija*), *Ipomooa digitata* L. is a remedy for leprosy, syphilis and gonorrhea (local name: *Gajeshta*), *Ipomoea pes-tigrides* L. used for boils, carbuncles, dog bites (local name: *Pulichovadi*), *Ipomea hispida* Roem. and Schult is used for leprosy and ulcers (local name: *Kalman*), *Ipomea reniformis* Chois. is used for fever bronchitis, rat bites (local name: *Adibhu*), *Ipomoea obscura* Ker·Gawl is given for mouth ulcers (local name: *Vachagandh*), and *Ipomea sepiaria* Koenig is utilized for ulcers (local name: *Nagini*).

ii. *Ipomoea aquatica* Forssk. is used for the treatment of fever, urinary tract infection, and sores in Myanmar and is given for and puerperal infection in China.

iii. *Ipomoea pes-caprae* (L.) R. Br. is used for boils in Indonesia and taken for diarrhea and blennorrhea in Cambodia (local name: *Trakuon kantek*). In Indonesia, *Ipomoea cairica* (L.) Sweet is a remedy for skin infections.

iv. In Myanmar, *Ipomoea carnea* Jacq. is used for cuts and wounds (local name: *Ajamari*), *Ipomoea nil* (L.) Roth is used for bronchitis, jaundice, and diarrhea (local name: *Siundelaharo*), and *Ipomoea purpurea* (L.) Roth. (local name: *Syudelaharo*) is given for bronchitis, jaundice, and diarrhea.

v. In Palau, *Ipomoea indica* (Burm.) Merr. (local name: *Oriemad*) is used to heal wounds and *Ipomoea littoralis* Bl. (local name: *Tohruh*) is given for abscesses. In Yap, *Ipomoea littoralis* Bl. (local name: *Gelew*) is given for cuts and leucorrhea.

Reference

Oluyori, A.P., Shaw, A.K., Preeti, R., Reddy, S., Atolani, O., Olatunji, G.A. and Fabiyi, O.A., 2016. Natural antifungal compounds from the peels of *Ipomoea batatas* Lam. *Natural Product Research*, *30*(18), pp. 2125–2129.

10.5.1.3 *Merremia peltata* (L.) Merr.

Synonym: *Convolvulus peltatus* L.

Common names: Big leaf rope; veliyana (Fiji); akar sambaing, balandete (Indonesia); akar balaan (Malaysia); kebeas (Palau); valeara (Papua New Guinea); boracan, bulakan (the Philippines), fue lautetele (Samoa); en luen (Thailand); fue mea (Tonga)

Quamoclit pinnata

Habitat: Roadsides, open forests

Distribution: Asia and Pacific.

Botanical observation: This climber grows up to 50 m long on trees and fences from tuberous roots. The stems are glabrous, lenticelled, fibrous, and yield white sap when incised. The internodes are of variable length but around 5 cm long. The leaves are simple, alternate, and without stipules. The petiole is about 8 cm long. The blade is dull light green to glossy, peltate, cordate; wavy, 13–20 cm × 10–15 cm, acuminate at apex, and shows 7–10 pairs of beautiful secondary nerves as well as scalariform tertiary nerves. The secondary nerves are curving at margins and reddish underneath. The cymes are large, many flowered, fleshy, axillary, and on 5–8 cm long and fleshy peduncles. The calyx comprises 5 fleshy, lanceolate subequal, sepals which are 1.8–2.5 cm long. The corolla is infundibuliform, membranous, pure white, 4–6 cm long, and discretely 5-lobed. The 5 stamens have elongated anthers. The stigma presents 2 broad lobes. The capsules are 2.5–3 cm long, in persistent sepals, dehiscent, and contain 4 trigonous seeds.

Medicinal uses: Boil, cough, cuts, sores, ulcers (Papua New Guinea); wounds (Indonesia; Papua New Guinea)

Viral enzyme inhibition by polar extract: Methanol extract inhibited the replication of the Human immunodeficiency virus type-1 reverse transcriptase and gpl20-CD4 binding (Yamamoto et al., 1997).

Commentaries

i. *Merremia mammosa* (Lour.) Hallier f. is used to treat tuberculosis and wounds in Indonesia. Ethyl acetate extract of this plant decreased the titer of Influenza A H1N1 hemagglutinin virus by 97.3% at concentration 1000 μg/mL (Purwitasari et al., 2020).

ii. In India, *Merremia emarginata* Burm. f. Hall. f. is used for pimples (local name: *Undirkani*), and methanol extract of leaves of this plant (100 μg/mL solution per disc) inhibited the growth of *Staphylococcus aureus*, *Bacillus cereus*, *Escherichia coli*, and *Pseudomonas aeruginosa* with the inhibition zone diameters of 12, 13.5, 12, and 9 mm, respectively (Elumalai et al., 2011). In India, *Merremia vitifolia* Hallier F. is used for blennorrhea, *Merremia tridentata* Hallier f. (local name: *Neyi kulovu*; *Prasarini*) is a remedy for herpes and urinary tract infections, and *Merremia hastata* Hall. (local name: *Paaran-kodi*) is used for carries.

iii. *Merremia umbellata* (L.) Hall. f. is used in Palau (local name: *Kebias*) for fever.

iv. In India, the following members of the family Convolvulaceae are used to treat microbial infection. The beautiful climber *Quamoclit pinnata* (Desr.) Bojer is used for carbuncles and dysentery (local name: *Tarulata*), *Lettsomia aggregata* Roxb. is a remedy for cough. *Argyreia speciosa* (L. f.) Sweet is used for ulcers (local name: *Ajantri*) and ethyl acetate extract of this plant inhibited the growth of *Mycobacterium tuberculosis* (H37Rv) with the MIC value of 50 μg/mL (Habbu et al., 2009). *Rivea ornata* (Roxb.) Choisy is given for bronchitis (local name: *Ajara*), *Erycibe paniculata* Roxb. is a remedy for cholera (local name: *Irimpiyatali*), *Operculina turpethum* (L.) Silva Manso is used for bronchitis and fever (local name: *Ardhachandra*), and *Cressa cretica* L (local name: *Amritasrava*) is a remedy for leprosy and part of an Ayurvedic formulation known as *Swasari Ras* recommended for COVID-19 (Balkrishna, 2020).

v. In Indonesia, *Argyreia paivae* A.R.Simões & P.Silveira is used for gonorrhea (local name: *Mau hale*). *Dichondra repens* J.R. Forst. & G. Forst. is used for boils in China. *Erycibe malaccensis* C.B. Clarke is used for sores in Malaysia.

vi. In the closely related family Cuscutaceae Bercht. & J. Presl. (1820), *Cuscuta chinensis* Lam. is used for blennorrhea in Cambodia, Laos, and Vietnam, and eye infections in India. In India, *Cuscuta reflexa* Roxb. (which looks from a distance very much like *Cassytha filiformis* L.) is a remedy for jaundice (local name: *Khavalli*), and in Thailand, it is a remedy for wounds (local name: *Foithong*). Methanol fraction of seeds of *Cuscuta reflexa* Roxb. (75 μg/mL/6 mm wells) inhibited the growth of *Staphylococcus aureus*, *Bacillus pumilus*, *Salmonella typhi*, *Salmonella typhimurium*, *Shigella boydii*, *Shigella sonnei*, *Shigella dysenteriae*, *Pseudomonas aeruginosa*, *Escherichia coli*, and *Vibrio cholerae* with inhibition zone diameter of 10, 11, 8, 7, 13, 14, 17, 14, 13, and 9 mm, respectively (Pal et al., 2006). Methanol extract of aerial part of *Cusuta japonica* Choisy inhibited the Human immunodeficiency virus protease (Park, 2003).

References

Balkrishna, A., 2020. Indian traditional ayurvedic treatment regime for Novel coronavirus, COVID-19. *Patanjali Research Institute*, *13*, pp. 1–28.

Elumalai, E.K., Ramachandran, M., Thirumalai, T. and Vinothkumar, P., 2011. Antibacterial activity of various leaf extracts of *Merremia emarginata*. *Asian Pacific Journal of Tropical Biomedicine*, *1*(5), pp. 406–408.

Habbu, P.V., Mahadevan, K.M., Shastry, R.A. and Manjunatha, H., 2009. Antimicrobial activity of flavanoid sulphates and other fractions of *Argyreia speciosa* (Burm. f) B.

Pal, D.K., Mandal, M., Senthilkumar, G.P. and Padhiari, A., 2006. Antibacterial activity of *Cuscuta reflexa* stem and *Corchorus olitorius* seed. *Fitoterapia*, *77*(7–8), pp. 589–591.

Park, J.C., 2003. Inhibitory effects of Korean plant resources on human immunodeficiency virus type 1 protease activity. *Oriental Pharmacy and Experimental Medicine*, *3*(1), pp. 1–7.
Purwitasari, N., Agil, M. and Studiawan, H., 2020. Activity of ethyl acetate fraction of *Merremia mammosa* hall as anti-influenza A (H1N1). *Indian Journal of Forensic Medicine & Toxicology*, 14(3).
Yamamoto, T., Takahashi, H., Sakai, K., Kowithayakorn, T. and Koyano, T., 1997. Screening of Thai plants for anti-HIV-1 activity. *Natural Medicines= 生薬學雜誌*, *51*(6), pp. 541–546.

10.5.2 Family Solanaceae A.L de Jussieu (1789)

The family Solanaceae consists of about 95 genera with 2300 species of herbs, shrubs, small trees. The leaves are simple, alternate, and exstipulate. The inflorescences are terminal or axillary cymes, racemes, or panicles. The calyx is tubular and 5-lobed. The corolla is tubular, star-shaped, or funnel-shaped, membranous, and 5-lobed. The androecium includes 5 stamens, the anthers characteristically united into somewhat and narrow conical mass. The gynoecium consists of two carpels fused in a bilocular ovary, each locule sheltering numerous ovules on axile placentas. The fruits are berries or capsules.

10.5.2.1 *Capsicum frutescens* L.

Synonyms: *Capsicum annuum* var. *frutescens* (L.) Kuntze

Common names: Chili pepper, tabasco pepper; marish (Bangladesh); lal mirch, mullaga (India); filfil-i-surkh (Iran); ngayok (Myanmar); meringel (Palau); lombo, seraimbsik (Papua New Guinea); sili (the Philippines); türk biberi (Turkey)

Habitat: Cultivated

Distribution: Asia and the Pacific

Botanical observation: This magnificent herb grows to about 1 m tall. The leaves are simple, spiral, and exstipulate. The petiole is 4–6 cm long, terete, smooth, and fleshy. The blade is lanceolate to elliptic, wavy, fleshy, pendulous, and 4–12 × 1.5–5 cm. The flowers are solitary, axillary, bending, and solitary or arranged in few-flowered axillary clusters. The calyx is tubular, minute, and 5-lobed. The corolla is tubular, about 1 cm across, 5-lobed, white, the lobes broadly lanceolate. The 5 stamens are bluish. The ovary is minute and develops a slender style. The berries are irregularly fusiform, erect, glossy, yellow, red, or purple, pungent to burning, up to about 10 cm long, and contain numerous minute yellowish seeds.

Medicinal uses: Abscesses, bronchitis, cuts, sores (Bangladesh); cholera, diarrhea, typhus, ulcers (India); cold (China); gonorrhea (Malaysia); pneumonia (Papua New Guinea); cholera (India; Malaysia, the Philippines); dengue (the Philippines)

Moderate broad-spectrum antibacterial polar extract: Aqueous extract of fruits inhibited the growth of *Salmonella typhimurium* (ATCC 13311), *Staphylococcus aureus*, and *Vibrio cholerae* (clinical isolate) with the MIC values of 500, 500, and 250 µg/mL, respectively (Koffi-Nevry et al., 2012).

Weak broad-spectrum antibacterial polar extract: Methanol extract inhibited the growth of *Escherichia coli* (ATCC 10536) and *Escherichia coli* (AG100 *Atet*: AG100 sur-expressing *AcrAB* pumps, containing TET^R gene actif) with MIC of 512 and 512 µg/mL, inhibited the growth of *Enterobacter aerogene* (ATCC 13048) and *Enterobacter aerogene* (EA-3: clinical MDR isolate) with MIC of 1024 and 512 µg/mL, respectively, and inhibited the growth of *Klebsiella pneumoniae* (ATCC 11296) and *Klebsiella pneumoniae* (KP63) with MIC of 1024 and 512 µg/mL, respectively, and inhibited the growth of *Providencia stuartii* (ATCC 29916) and *Providencia stuartii* (NAE16: MRD clinical isolate) with MIC of 1024 and 512 µg/mL, respectively (Touani et al., 2014).

Very weak antifungal (filamentous) polar extract: Aqueous extract of fruits inhibited the growth of *Aspergillus flavus*, *Penicillium* sp., and *Rhizopus* sp. with MIC values of 5000, 1250, and 5000 µg/mL, respectively (Soumya & Nair, 2002).

Strong antiviral (enveloped monopartite linear double-stranded DNA) polar extract: Aqueous extract of fruits at the concentration of 100 µg/mL inhibited the replication of the Herpes simplex virus type-1 (7401 H strain) by about 50% and with the IC_{50} value of 93 µg/mL (Hafiz et al., 2017).

Moderate antiviral (non-enveloped monopartite linear single-stranded (+) RNA) polar extract: Aqueous extract of fruits at the concentration of 100 µg/mL inhibited the replication of the Poliovirus type-1 (Sabin strain) by about 30% (Hafiz et al., 2017).

Moderate antiviral (enveloped monopartite linear single-stranded (−)-RNA) polar extract: Aqueous extract of fruits at the concentration of 100 µg/mL inhibited the replication of the Measles virus (Tanabe strain) by about 25% in Vero cellsHafiz et al., 2017.

Moderate antiviral (enveloped monopartite linear double-stranded DNA) polar extract: Methanol extract of fruits inhibited the replication of the Herpes simplex virus type-1 and type-2 with the MIC value of 100 µg/mL in Vero cells (Hafiz et al., 2017). Ethanol inhibited the replication of the Herpes simplex virus type-1 (KOS) with the IC_{50} value of 500 µg/mL and the selectivity index of 17.5 in Vero cells (Ordaz-Trinidad et al., 2018).

Strong antiviral (enveloped segmented linear single-stranded (−) RNA linear) amide alkaloid: Capsaicin inhibited the entry of Lassa virus in host cells with the EC_{50} of about 10 µM (Tang et al., 2020).

Commentaries

i. The plant yields the pungent amide alkaloids capsaicin and dihydrocapsaicin (Estrada et al., 2002).
ii. In India, *Capsicum annum* L. (local name: *Lodachina*) is used for sore throat and yellow fever and *Capsicum minimum* Roxb. is used for fever (local name: *Lalmirich*). In Bangladesh, some tribes take *Capsicum frutescens* L. as a means to prevent COVID-19 in rural area (Mohammed Rahmatullah personal communication). The severe acute respiratory syndrome-associated coronavirus like the Poliovirus is a monopartite linear single-stranded (+) RNA virus and thus chilies might have protective or curative effects against COVID-19 and other forthcoming viral pandemics.
iii. The Lassa virus appeared in Nigeria in 1969 and accounts since then for about 5000 deaths per year in Western Africa (Reuben et al., 2020). This zoonotic virus transmitted from rats (*Mastomys natalensis*) incurred for about 20% of patients' fatal hemorrhages (Reuben et al., 2020). The current treatment for this viral infection is ribavirin. The natural habitat of *Mastomys natalensis* is forest, and because of deforestation, this rat came in contact with humans whence zoonotic infection. The COVID-19 pandemic should hopefully end by the end of 2022 according to WHO experts, and it must be understood that if the current abuses on life forms (forests, animals) for material benefits do not stop we are bound to face other pandemic zoonoses similar to or even far more lethal than the Lassa virus on a global scale. Again, COVID-19 is a sinister and tragic warning to humans that the boundaries with the forces of life have been crossed. Now humans have a choice, will they choose to disappear?

References

Akanaca, P., Kurokawa, M., Tewtrakul, S., Pramyon, P., Sripandkulcha, B., Shirak, K. and Hattor, M., 2002. Inhibitory activities of Thai medicinal plants against herpes simplex type 1, poliovirus type 1, and measles virus. *Journal of Traditional and Complementary Medicine 19*, pp. 174–180.

Estrada, B., Bernal, M.A., Díaz, J., Pomar, F. and Merino, F., 2002. Capsaicinoids in vegetative organs of *Capsicum annuum* L. in relation to fruiting. *Journal of Agricultural and Food Chemistry*, *50*(5), pp. 1188–1191.

Hafiz, T.A., Mubaraki, M., Dkhil, M. and Al-Quraishy, S., 2017. Antiviral activities of *Capsicum annuum* methanolic extract against herpes simplex virus 1 and 2. *Pakistan Journal of Zoology*, *49*(1), p. 251.

Koffi-Nevry, R., Kouassi, K.C., Nanga, Z.Y., Koussémon, M. and Loukou, G.Y., 2012. Antibacterial activity of two bell pepper extracts: *Capsicum annuum* L. and *Capsicum frutescens. International Journal of Food Properties*, *15*(5), pp. 961–971.

Ordaz-Trinidad, N., Dorantes-Álvarez, L., Salas-Benito, J., Barrón-Romero, B.L., Salas-Benito, M. and Nova-Ocampo, M.D., 2018. Cytotoxicity and antiviral activity of pepper extracts (Capsicum spp). *Polibotánica*, *46*, pp. 273–285.

Reuben, R.C., Danladi, M.M. and Pennap, G.R., 2020. Is the COVID-19 pandemic masking the deadlier Lassa fever epidemic in Nigeria? *Journal of Clinical Virology*, *128*, p. 104434.

Soumya, S.L. and Nair, B.R., 2012. Antifungal efficacy of *Capsicum frutescens* L. extracts against some prevalent fungal strains associated with groundnut storage. *Journal of Agricultural Technology*, *8*(2), pp. 739–750.

10.5.2.2 *Datura metel* L.

Synonyms: *Datura alba* Rumph. ex Nees; *Datura fastuosa* L.; *Datura hummatu* Bernh.; *Datura innoxia* Mill; *Datura nilhummatu* Dunal

Common names: Downy Thorn-apple, metel, dhutra, dothura, dhutura (Bangladesh); yang jin hua (China); dhatra, kalo dhtura, madulam, neelaummam, umatthai, sagol-hidak (India); nanulah (Iran); padaing (Myanmar); tatul (Turkey)

Habitat: Cultivated

Distribution: Tropical Asia and the Pacific

Botanical observation: This dreadfully poisonous herb grows up to 2 m tall and has a sinister aura. The stems are terete, green to light purple, hairy, and woody at base. The leaves are simple, spiral, and exstipulate. The petiole is up to about 15 cm long. The blade is wavy, broadly lanceolate, fleshy, dull dark green, 7–20 × 4.5–13 cm, somewhat asymmetrical at base, acute to acuminate at apex and with about 5–11 pairs of secondary nerves. The inflorescence is axillary and solitary and bear magnificent showy flowers. The flower peduncle is about 2 cm long. The calyx is tubular, cylindrical, 5-lobed, the lobes triangular, and up to about 10 cm long. The corolla tube is up to 20 cm long, white, purplish, or yellowish, infundibuliform, somewhat convolvulaceous, and develop 5–10 broadly triangular and acuminate lobes. The 4 stamens are slender with about 1 cm long anthers. The capsules are globose, with somewhat a repulsive aura, nodding, irregularly dehiscent, spiny, about 5 cm in diameter, and shelter numerous seeds which are about 5 mm long.

Medicinal uses: Abscesses (Bangladesh); boils, bronchitis, earache, diarrhea, eye infections, gonorrhea, jaundice, leprosy, sores, wounds (India); gonorrhea, otitis (Pakistan); fever (Bangladesh; India); rabies (India; Pakistan); herpes (Indonesia).

Broad-spectrum antibacterial polar extract: Ethanol extract of leaves (100 µL of a 100 mg/mL solution/6 mm well) inhibited the growth of *Escherichia coli*, *Salmonella typhi*, *Bacillus cereus*, *Bacillus subtilis*, and *Staphylococcus aureus* with the inhibition zone diameters of 17.1, 15.2, 19.4, 18.5, and 26.7 mm, respectively (Kaushik & Goyal, 2008).

Strong broad-spectrum antibacterial polar extract: Ethanol extract of aerial part inhibited the growth of *Staphylococcus* sp., *Klebsiella* sp., *Escherichia coli*, *Streptococcus* sp., *Shigella* sp., *Pseudomonas* sp., *Salmonella* sp. and *Vibrio* sp. with MIC of 20, 50, 40, 50, 60, 60, 50 and 40 µg/mL, respectively (Gnanamani et al., 2003).

Weak antifungal (filamentous) mid-polar extract: Chloroform extract inhibited the growth of *Aspergillus fumigatus*, *Aspergillus flavus*, and *Aspergillus niger* with the MIC value of 625 µg/mL (Sharma, 2002).

Moderate antiviral (enveloped monopartite linear double-stranded DNA) tropane alkaloid: Atropine at the concentration of 200 µg/mL inhibited the replication of the Herpes simplex virus type-1 in HeLa cells via a mechanism involving, at least in part, inhibition of glycosylation of viral proteins and formation of defective viruses (Alarcón et al., 1984).

Very weak antiviral (enveloped monopartite linear single-stranded (−) RNA) polar extract: Aqueous extract of seeds inhibited the replication of the Rabies virus in Vero cells by 50% at the concentration of 1250 µg/mL (Roy et al., 2016).

Moderate antiviral (enveloped linear double-stranded DNA) tropane alkaloid: Atropine at the concentration of 400 µM inhibited the replication of the Vaccinia (strain WR) in primary monkey kidney cells by 80% (Yamazaki & Tagaya, 1980).

Very weak antiviral (non-enveloped monopartite linear single-stranded (+) RNA) tropane alkaloid: Atropine at the concentration of 400 μM inhibited the replication of the Poliovirus type-1 (strain Mahoney) in primary monkey kidney cells by 90% (Yamazaki & Tagaya, 1980).

Very weak antiviral (enveloped monopartite linear single-stranded (+) RNA) tropane alkaloid: Atropine at the concentration of 400 μM inhibited the replication of the Japanese encephalitis virus (strain Jath 160) in primary chick embryo by 31% (Yamazaki & Tagaya, 1980).

Very weak antiviral (enveloped monopartite linear double-stranded DNA) tropane alkaloid: Atropine at the concentration of 400 μM inhibited the replication of the Herpes simplex virus in primary monkey kidney cells by 20% (Yamazaki & Tagaya, 1980).

In vivo antiviral (enveloped monopartite linear single-stranded (+) RNA) tropane alkaloid: Scopolamine inhibited the replication of the Japanese encephalitis virus (SH1 strain) (Bhattacharjee et al., 2021).

Commentaries

i. The plant produces the tropane alkaloids atropine and scopolamine and the maximum yield was observed to be at the end of the reproductive stage (Afsharypuor et al., 1995) showing that the amount of secondary metabolites in plants and therefore the antimicrobial activity of extract depends on the stages of development of a plant.
ii. Atropine and scopolamine being able to bind to acetylcholine receptors may be able to inhibit bacterial and or fungal efflux pumps.
iii. Atropine and scopolamine could be able to inhibit the replication of coronaviruses.
iv. A few medicinal plants in Asia-Pacific are used for rabies and plants in the genus *Datura* L. are often used for this purpose. *Datura stramonium* L. is given for fever and wounds in Turkey (local names: *Deli patpat suzgecotu*), for wounds and tonsilitis (local name: *Daturoo*) in India, for rabies in Nepal (local name: *padaing-nyo)*, and for gonorrhea, eye infection, and rabies in Myanmar. Rabies virus binds to acetylcholine neuronal receptors (Lentz et al., 1986) explaining the use of plants in the genus *Datura* L. to treat rabies.
v. Consider that testing MIC by broth microdilution technique is possible for diluted polar compounds or extracts. Liquid broth MIC is difficult for mid-polar and non-polar compounds or extracts and even in concentrated polar extract because they do not dissolve in agar creating solid particles and the use of dimethyl sulfoxide may corrupt the results since by itself has antimicrobial effects. In clinical settings and microbiology labs, liquid broth MIC is done on known antibiotics which are mainly water soluble. Thus, for non- or mid-polar substances, paper disc or agar wells are more suitable for MIC.

References

Afsharypuor, S., Mostajeran, A. and Mokhtary, R., 1995. Variation of scopolamine and atropine in different parts of *Datura metel* during development. *Planta Medica*, *61*(04), pp. 383–384.

Alarcón, B., González, M.E. and Carrasco, L., 1984. Antiherpesvirus action of atropine. *Antimicrobial Agents and Chemotherapy*, *26*(5), pp. 702–706.

Bhattacharjee, A., Chaudhuri, R., Dash, J.J., Saha, M., Choudhury, L. and Roy, S., 2021. Pre-treatment with scopolamine naturally suppresses Japanese encephalitis viral load in embryonated chick through regulation of multiple signaling pathways. *Applied Biochemistry and Biotechnology*, *193*(6), pp. 1654–1674.

Gnanamani, A., Priya, K.S., Radhakrishnan, N. and Babu, M., 2003. Antibacterial activity of two plant extracts on eight burn pathogens. *Journal of Ethnopharmacology*, *86*(1), pp. 59–61.

Kaushik, P. and Goyal, P., 2008. In vitro evaluation of *Datura innoxia* (thorn-apple) for potential antibacterial activity. *Indian Journal of Microbiology*, *48*(3), p. 353.

Lentz, T.L., Benson, R.J., Klimowicz, D., Wilson, P.T. and Hawrot, E., 1986. Binding of rabies virus to purified *Torpedo acetylcholine* receptor. *Molecular Brain Research*, *1*(3), pp. 211–219.

Roy, S., Mukherjee, S., Pawar, S. and Chowdhary, A., 2016. Evaluation of in vitro antiviral activity of *Datura metel* Linn. against rabies virus. *Pharmacognosy Research*, *8*(4), p. 265.

Sharma, G.L., 2002. Studies on antimycotic properties of *Datura metel*. *Journal of Ethnopharmacology*, *80*(2–3), pp. 193–197.

Yamazaki, Z. and Tagaya, I., 1980. Antiviral effects of atropine and caffeine. *Journal of General Virology*, *50*(2), pp. 429–431.

10.5.2.3 *Physalis angulata* L.

Synonyms: *Physalis esquirolii* H. Lév. & Vaniot

Common names: Cape gooseberry, wild tomato, winter cherry, Chinese lantern plant, cut-leaf ground-cherry, ku zhi, pai chiang (China); popti, potichedi (India); daun kopo kopi (Indonesia); leletup (Malaysia); camates (the Philippines); kospeli (Fiji); tamaru ha'ari (French Polynesia); tomate chaca (Guam-Chamorro); bubeubedul (Palau); vivao (Samoa); camates, sisio (the Philippines); ba tom tok (Thailand); polopa (Tonga); lu lu cai (Vietnam)

Habitat: Vacant lands of land, wastelands, roadside

Distribution: Tropical Asia and Pacific

Botanical observation: This herb grows to a height of 1 m. The stems are angled, purplish-green, and succulent. The leaves are simple, spiral, hairy, and without stipule. The petiole is 8 mm–2 cm long. The blade is membranous, lanceolate, asymmetrical, dentate, 3–6.5 cm×2–4.5 cm, and marked with 4–5 pairs of arching secondary nerves. The flowers are solitary and axillary. The calyx is tubular, 4–6 mm long, and 5-lobed, the lobes triangular. The corolla is whitish-yellow, 1 cm long, infundibuliform, 5-lobed, and shows brownish spots at base. The androecium consists of 5 stamens with 2 mm long blue anthers. The capsules are 5-angled, lantern-shaped, membranous, angular, 2.7 cm×2 cm and develop at apex 5 tiny calyx lobes. The seeds are tiny, flat, yellowish, and reniform.

Medicinal use: Apparently none for microbial infections.

Strong broad-spectrum antibacterial steroid-fractions: Mixture of physalins extracted from the capsules at the concentration of 200 µg/mL abrogated the growth of *Staphylococcus aureus* (ATCC 6538P), *Staphylococcus aureus* (ATCC 25923), *Staphylococcus aureus* (ATCC 29213), and *Neisseria gonorrhoeae* (ATCC 49226) (Silva et al., 2005).

Antibacterial (Gram-positive) halo developed by steroid: Physalin B isolated from the capsules (200 µg/disc) inhibited the growth of *Staphylococcus aureus* (ATCC 6538P) with the inhibition zone of 9.2 mm (Silva et al., 2005).

Strong antimycobacterial hydrophilic steroid: Physalin D (LogD −0.06 at pH 7.4; molecular mass = 544.5 g/mol) inhibited the growth of *Mycobacterium tuberculosis* ($H_{37}Rv$) with the MIC value of 32 µg/mL (Januário et al., 2002).

Virus enzyme inhibition by polar extract: Extract inhibited human immunodeficiency virus type-1 protease (Park, 2003).

Commentaries: *Physalis minima* L. is a remedy for jaundice, mouth ulcers, sore throat, and tonsilitis in Thailand (local name: *Thong theng*) and used for gonorrhea, smallpox, and ear infection in India (local names: *Pataka*, *Poptiya*). *Physalis alkekengi* L. is used in Iran (local names: *Arusak posht,* Kakanj) for jaundice and syphilis.

References

Januário, A.H., Filho, E.R., Pietro, R.C.L.R., Kashima, S., Sato, D.N. and França, S.C., 2002. Antimycobacterial physalins from *Physalis angulata* L. (Solanaceae). *Phytotherapy Research: An International Journal Devoted to Pharmacological and Toxicological Evaluation of Natural Product Derivatives*, *16*(5), pp. 445–448.

Park, J.C., 2003. Inhibitory effects of Korean plant resources on human immunodeficiency virus type 1 protease activity. *Oriental Pharmacy and Experimental Medicine*, *3*(1), pp. 1–7.

Silva, M.T., Simas, S.M., Batista, T.G., Cardarelli, P. and Tomassini, T.C., 2005. Studies on antimicrobial activity, in vitro, of *Physalis angulata* L. (Solanaceae) fraction and physalin B bringing out the importance of assay determination. *Memorias do Instituto Oswaldo Cruz*, *100*(7), pp. 779–782.

10.5.2.4 *Solanum nigrum* L.

Synonym: *Solanum humile* Lam.

Common names: Black Nightshade, wonderberry; long k'oui (China); ranti (Indonesia); bahuphala, leipung, makoi, (India); aj-i-rizi (Iran); kkamajung (Korea); terong meranti (Malaysia); behi (Nepal), kach mach, makku (Pakistan); handikundi (Pakistan); bolagtab, hierba mora (the Philippines); kalukung (Sri Lanka); ca nuit ao (Vietnam)

Habitat: Wastelands, roadsides

Distribution: Tropical Asia

Botanical description: This herb grows to a height of 60 cm and has somewhat a sinister aura. The stems are fleshy. The young shoots are purplish. The leaves are simple, spiral, and exstipulate, in groups of 2–3. The petiole is up to 4 cm long. The blade is membranaous, 2.5–10 × 2–5.5 cm, lanceolate, tapering at base, dull green, and marked with about 6 pairs of secondary nerves. The inflorescences consist of about 3 cm long axillary or cauliflorous clusters of about 6 bending flowers which are 5 mm across. The calyx is tubular and 5 lobed, the lobes 2 mm long. The corolla is tubular, membranous, 5-lobed, the lobes yellowish at base, triangular, and up to 6 mm long. The 5 anthers are joined in a yellow conical mass. The berries are globose, black, glossy, about 8 mm in diameter, and nightmarishly poisonous.

Medicinal use: Dysentery (Pakistan; India; Nepal; China); blister, bronchitis, cough, diarrhea, tuberculosis (India); fever (Pakistan; Nepal)

Broad-spectrum antifungal halo developed by polar extract: Ethanol extract of fruits (1:500) inhibited the growth of *Aspergillus flavus*, *Aspergillus fumigatus*, *Aspergillus glaucus*, *Aspergillus terreus*, *Candida albicans*, *Candida tropicalis*, *Trichophyton tonsurans*, and *Trichophyton violaceum* (Shamim et al., 2004).

Strong antiviral (enveloped monopartite linear single-stranded (+) RNA) mid-polar extract: Chloroform extract inhibited the replication of the Hepatitis C virus in Huh-7 cells by about 70% via a mechanism involving, at least in part, inhibition of viral NS3 protease (Javed et al., 2011).

Commentaries

i. The plant yields steroidal alkaloids (Gu et al., 2018) as well as steroidal saponins (Ke et al., 1999) which account for the poisonous effects (Alexander, 1948; Towers, 1953). This plant is described as non-toxic in documents available globally providing further evidence that the neoliberal dismantlement of medicinal plant teaching under global corporate agenda (Baltodano, 2012) places populations in danger of fatal poisoning by plants and mushrooms. The full teaching of pharmacognosy (5 years including botany and toxicology) must be reimplemented in schools of pharmacy.

ii. Plants in the genus *Solanum* L. are used for medicinal purposes throughout Asia and the Pacific despite being often poisonous (see earlier comments): In Turkey, the poisonous *Solanum dulcamara* L. (local name: *Aci ot*) is a remedy for burns and cough, and used in India for leprosy and tuberculosis, in China (local name: *K'e ka*) for abscesses and boils, and in Cambodia for fever and eye infection (local name: *Chham em*). Methanol extract (50 mL of a 1 mg/5 mL solution/well) inhibited the growth of *Escherichia coli*, *Klebsiella pneumoniae*, *Pseudomonas aeruginosa*, *Proteus mirabilis*, *Salmonella typhi*, *Salmonella paratyphi* A, *Salmonella paratyphi* B, *Salmonella typhimurium*, *Shigella boydii*, *Shigella flexneri*, *Shigella sonnei*, *Staphylococcus aureus*, and *Streptococcus faecalis* with inhibition zone diameter of 11.4, 13.1, 12, 10, 14, 15.9, 15.2, 11.7, 12.5, 12.2, 12.7, 18.3, and 17.3 mm, respectively (Limsuwan et al., 2009).

    In Iran, *Solanum xanthocarpum* Schrad. & Wendl. (local name: *Tukhm-i-badanjan*) is used for flu, and in India (local names: *Anakranta, Rengni*) it is a remedy for toothache, cough, bronchitis, and fever.

    In Pakistan, *Solanum incanum* L. (local name: *Janglee baingan*) is a remedy for pneumonia, toothache, and sore throat, and *Solanum surattense* Burm.f. (local name: *Mokri*) is used for cough and toothache.

Solanum indicum L. (synonym: *Solanum anguivi* Lam) is a remedy for fever (local name: *Akranta*), bronchitis, and toothache in India, for cough in Myanmar, and for syphilis in Malaysia. In India, *Solanum jacquini* Miq. is used for cough and fever (local name: *Nella mollunga*). *Solanum verbascifolium* L. is a remedy for burns (local name: *Gandira*) in India and given for diarrhea in the Philippines. In India, the strange-looking *Solanum ferox* L. is used for fever (local name: *Chandra Pushpa*), and in Malaysia, it is given for syphilis. In India, *Solanum trilobatum* L. (local names: *Achuda Toodavullay*) is a remedy for bronchitis and cough and is used for herpes, sore throat, and mouth ulcer (local name: *Ma waeng*) in Thailand. In India, *Solanum gracilipes* Dcne. is given for otitis (local name: *Halun*), *Solanum albicaule* Kotschy ex Dunal is used for ulcers, *Solanum khasiana* C.B. Clarke. is given for toothache, and *Solanum trifoliatum* L. (local name: *Zitkantikuri*) is a remedy for bronchitis, cold, cough, and pox. *Solanum melongena* L. is used for bronchitis, cholera, and fever in India (local name: *Angana*) and for cough in Myanmar (local name: *kayan*).

In Bangladesh, *Solanum barbisetum* Nees (local name: *Kontikari*) is used for chickenpox, influenza, and typhoid

In Nepal, *Solanum aculeatissimum* Jacq. (local name: *Kantakaari*) is a remedy for toothache. *Solanum lyratum* Thunb. ex Murray is a remedy for influenza in China.

In Yap, *Solanum torvum* Sw. (local name *Rawelganafi*) is used for leprosy and from the fruits (which are edible) the isoflavane sulfate torvanol A and the steroidal saponin torvoside H inhibited the replication of the Herpes simplex virus-type-1 with the IC_{50} values of 9.6 and 23.2 µg/mL, respectively (Arthan et al., 2002). From the fruits, methyl caffeate inhibited the growth of *Proteus vulgaris*, *Klebsiella pneumoniae* (ESBL-3971), *Mycobacterium tuberculosis* ($H^{37}Rv$), and *Mycobacterium tuberculosis* (Rif^R) with the MIC values of 50, 25, 8, and 8 µg/mL, respectively (Balachandran et al., 2012). The fruits being edible (Malaysia) could be examined for their *in vivo* antimycobacterial or antiviral effects and may have potentials as antitubercular phytomedication.

iii. In the family Solanaceae, *Withania coagulans* (Stocks) Dunal (local name: *Akri*) is used for eye infection and wounds in Pakistan. *Withania somnifera* (L.) Dunal is used in Pakistan (local name: *Asgand*) and Bangladesh (local name: *Ashwogondha*) for tuberculosis and in India (local name: *Ashvagandha*) for bronchitis, carbuncles, ulcers, and syphilis. Aqueous extract at concentrations ranging from 10 to 1000 µg/mL inhibited the growth of *Mycobacterium tuberculosis* (Adaikkappan, et al., 2012). *Withania somnifera* (L.) Dunal is neuroactive and as such (see earlier) probably yields secondary metabolites that inhibits bacterial and/or fungal efflux pumps. Methanol extract of leaves (5 mm paper disc impregnated with 4 mg/mL solution) inhibited the growth of *Salmonella typhimurium* (TA100) and *Escherichia coli* (DH5a) with the inhibition zone diameters of 13 and 8.7 mm, respectively, and increased the sensitivity of *Salmonella typhimurium* (TA100) and *Escherichia coli* (DH5a) to Rifampicin and Isoniazid (Arora et al., 2004). Given orally at the dose of 100 mg/Kg, aqueous extract of roots protected mice against *Salmonella typhimurium* (Pwais et al., 2005). The plant is used in India for the treatment of Chikungunya virus infection (Viswanathan et al., 2008). Aqueous extract of this plant inhibited the replication of the Herpes simplex virus type-1 (Kambizi et al., 2007) and hydroalcoholic extract of roots abrogated the replication of the infectious bursal disease virus (UABz-2 strain) at 25 µg/mL (Pant et al., 2012). Methanol extract of roots of this herb inhibited the replication of Human immunodeficiency virus type-1 (NL4.3 strain) in CD4+ T cells by 47.2% at the maximum non-cytotoxic concentration of 15 µg/mL (Sabde et al., 2011). From *Withania somnifera* (L.) Dunal, the ergostane-type steroid withanone efficiently inhibited the binding of angiotensin-converting enzyme 2 and severe acute respiratory syndrome-associated coronavirus with the IC_{50} value of 0.3 µg/mL (Balkrishna et al., 2021).

iv. *Lycium chinense* Mill. (synonym: *Lycium barbarum* var. *chinense* (Mill.) Aiton) is used for tuberculosis in China (local name: *Gou qi*). The fruits of *Lycium barbatum* L series of pyrrole alkaloids (Youn et al., 2016), Pyrrole alkaloids are often antimycobacterial.

References

Adaikkappan, P., Kannapiran, M. and Anthonisamy, A., 2012. Antimycobacterial activity of Withania somnifera and Pueraria tuberosa against Mycobacterium tuberculosis H37Rv. *J. Acad. Indus. Res*, *1*(4), pp. 153–156.

Arora, S., Dhillon, S., Rani, G. and Nagpal, A., 2004. The in vitro antibacterial/synergistic activities of *Withania somnifera* extracts. *Fitoterapia*, *75*(3–4), pp. 385–388.

Arthan, D., Svasti, J., Kittakoop, P., Pittayakhachonwut, D., Tanticharoen, M. and Thebtaranonth, Y., 2002. Antiviral isoflavonoid sulfate and steroidal glycosides from the fruits of *Solanum torvum*. *Phytochemistry*, *59*(4), pp. 459–463.

Balachandran, C., Duraipandiyan, V., Al-Dhabi, N.A., Balakrishna, K., Kalia, N.P., Rajput, V.S., Khan, I.A. and Ignacimuthu, S., 2012. Antimicrobial and antimycobacterial activities of methyl caffeate isolated from *Solanum torvum* Swartz. fruit. *Indian Journal of Microbiology*, *52*(4), pp. 676–681.

Balkrishna, A., Pokhrel, S., Singh, H., Joshi, M., Mulay, V.P., Haldar, S. and Varshney, A., 2021. Withanone from *Withania somnifera* attenuates SARS-CoV-2 RBD and host ACE2 interactions to rescue spike protein induced pathologies in humanized zebrafish model. *Drug Design, Development and Therapy*, *15*, p. 1111.

Gu, X.Y., Shen, X.F., Wang, L., Wu, Z.W., Li, F., Chen, B., Zhang, G.L. and Wang, M.K., 2018. Bioactive steroidal alkaloids from the fruits of *Solanum nigrum*. *Phytochemistry*, *147*, pp. 125–131.

Javed, T., Ashfaq, U.A., Riaz, S., Rehman, S. and Riazuddin, S., 2011. In-vitro antiviral activity of *Solanum nigrum* against hepatitis C virus. *Virology Journal*, *8*(1), pp. 1–7.

Kambizi, L.G.B.M., Goosen, B.M., Taylor, M.B. and Afolayan, A.J., 2007. Anti-viral effects of aqueous extracts of *Aloe ferox* and *Withania somnifera* on herpes simplex virus type 1 in cell culture: research in action. *South African Journal of Science*, *103*(9), pp. 359–360.

Ke, H., Hisayoshi, K., Aijun, D., Yonngkui, J., Shigeo, I. and Xinsheng, Y., 1999. Antineoplastic agents-III: Steroidal glycosides from *Solanum nigrum*. *Planta Medica* 65, pp. 35–38.

Limsuwan, S., Subhadhirasakul, S. and Voravuthikunchai, S.P., 2009. Medicinal plants with significant activity against important pathogenic bacteria. *Pharmaceutical Biology*, *47*(8), pp. 683–689.

Owais, M., Sharad, K.S., Shehbaz, A. and Saleemuddin, M., 2005. Antibacterial efficacy of *Withania somnifera* (ashwagandha) an indigenous medicinal plant against experimental murine salmonellosis. *Phytomedicine*, *12*(3), pp. 229–235.

Pant, M., Ambwani, T. and Umapathi, V., 2012. Antiviral activity of Ashwagandha extract on infectious bursal disease virus replication. *Indian Journal of Science and Technology*, *5*(5), pp. 2750–2751.

Ragno, R., Marshall, G.R., Di Santo, R., Costi, R., Massa, S., Rompei, R., and Artico, M., 2000. Antimycobacterial pyrroles: Synthesis, antiMycobacterium tuberculosis activity and QSAR studies. *Bioorganic & Medicinal Chemistry 8*, pp. 1423–1432

Sabde, S., Bodiwala, H.S., Karmase, A., Deshpande, P.J., Kaur, A., Ahmed, N., Chauthe, S.K., Brahmbhatt, K.G., Phadke, R.U., Mitra, D. and Bhutani, K.K., 2011. Anti-HIV activity of Indian medicinal plants. *Journal of Natural Medicines*, *65*(3–4), pp. 662–669.

Shamim, S., Ahmed, S.W. and Azhar, I., 2004. Antifungal activity of Allium, Aloe, and *Solanum species*. *Pharmaceutical Biology*, *42*(7), pp. 491–498.

Towers, R.P., 1953. A case of poisoning by *Solanum nigrum*. *Irish Journal of Medical Science (1926–1967)*, *28*(2), pp. 77–80.

Viswanathan, M.V., Raja, D.K. and Khanna, S.D., 2008. Siddha way to cure Chikungunya. *Indian Journal of Traditional Knowledge*, *7*(2), pp. 345–346.

Youn, U.J., Lee, J.Y., Kil, Y.S., Han, A.R., Chae, C.H., Ryu, S.Y. and Seo, E.K., 2016. Identification of new pyrrole alkaloids from the fruits of *Lycium chinense*. *Archives of Pharmacal Research*, *39*(3), pp. 321–327.

11 Medicinal Plants in Asia-Pacific for Microbial Infections

The Clade Campanulids

11.1 ORDER ASTERALES LINK (1829)

11.1.1 FAMILY ASTERACEAE MARTYNOV (1820)

The family Asteraceae consists of about 1600 genera and 24000 species of herbs, shrubs, or climbers. The stems are often hairy. The leaves are simple, spiral, and exstipulate. The blade is simple or dissected and often hairy. The inflorescence is a very characterisitic flower-like aggregate of small flowers known as capitulum. The calyx is tubular and 5-lobed. The corolla is tubular and 5-lobed. The androecium includes 5 stamens. The gynoecium consists of a unicelled ovary sheltering a single ovule growing on a basal placenta. The stigma is often bifid. The fruits are achenes. Members in this huge family produce antimicrobial sesquiterpene lactones, flavones, pyrrolizidine alkaloids, and polyacetylenic fatty acids.

11.1.1.1 *Ageratum conyzoides* L.

Synonyms: *Carelia conyzoides* (L.) Kuntze; *Eupatorium conyzoides* (L.) E.H.L. Krause

Common names: White weed; motmoti, uchunti (Bangladesh); ajgandha, kepha pha pa (India); babadotan (Indonesia); tahi anjing (Malaysia); kado-po (Myanmar); ganaune ghans (Nepal); agmak (Palau); akikikanabebe, mungrimb (Papua New Guinea); bulak-manok (the Philippines); ya saap raeng (Thailand); cây bông thúi (Vietnam)

Habitat: Roadsides, wasteland

Distribution: Tropical Asia and Pacific

Botanical observation: This poisonous herbs grow up to 50cm tall. The stem is hairy. The leaves are simple, opposite, and exstipulate. The petiole is 1–3cm long. The blade is hairy, broadly lanceolate, 3–8 cm×2–5cm, cuneate or acute at base, marked with 2–3 pairs of secondary nerves, serrate, and acute at apex. The capitula are somewhat globose and about 5mm long, somewhat bluish to almost fluorescent (at dawn and sunset), and arranged in few-flowered cymes. The calyx is minutely 5-lobed. The corolla is tubular, 5-lobed, minute, and purplish. The achenes are 5-angled and minute.

Medicinal uses: Abscess (Palau); boils (India, Nepal; the Philippines; Papua New Guinea); cuts (India, Nepal; Papua New Guinea); diarrhea (India; Papua New Guinea); sores (Papua New Guinea); dysentery (Bangladesh; India; Papua New Guinea); fever (India), leprosy (India; Myanmar); skin infections l (Myanmar); wounds (India; Nepal; Papua New Guinea)

Antibacterial (Gram-positive) polar extract: Methanol extract inhibited the growth of *Staphylococcus aureus* (ATCC 29213) (Hoffman et al., 2004).

Broad-antifungal non-polar extract: Hexane extract inhibited the growth of *Candida albicans* and *Aspergillus fumigatus* (Hoffman et al., 2004).

Strong antifungal (filamentous) essential oil: Essential oil of leaves at the concentration of 1 μg/mL inhibited the growth of *Aspergillus flavus* (MI 190) by 63% with alteration of cytoplasmic

DOI: 10.1201/9781003180340-3

membrane (Nogueira et al., 2010). Essential oil (500 ppm) inhibited the growth of *Aspergillus parasiticus*, *Aspergillus oryzae*, and *Aspergillus flavus* by 56.1, 45.2, and 65%, respectively (Patil et al., 2010).

Strong antifungal non-polar extract: Hexane extract of stems (500 ppm) inhibited the growth of *Rhizoctonia solani*, *Phomopsis theae*, *Fusarium* sp., *Botryodiplodia theobromae*, and *Sclerotium rolfsii* by 100%, 35.2%, 77%, 100%, and 40.1%, respectively (Patil et al., 2010).

Strong antifungal amphiphilic chromene: Precocene II (also known as 6,7-dimethoxy-2,2-dimethylchromene; LogD 3.2 at pH 7.4; molecular mass = 220.2 g/mol) at 100 ppm abrogated the mycelial growth of *Rhizoctonia solani* and *Sclerotium rolfsii* (Iqbal et al., 2004).

Strong antiviral (non-enveloped monopartite linear single-stranded (+) RNA) polar extract: Methanol extract inhibited the replication of the Echovirus E7 and E19 with the IC_{50} values of 2×10^{-3} and 5.6×10^{-4} µg/mL, respectively, and the selectivity indexes of 746.7 and 2.8×10^{4}, respectively, in human rhabdomyosarcoma cells (Ogbole et al., 2018).

Commentaries

i. The plant produces series of chromenes (González et al., 1991) as well as hepatotoxic pyrrolidine alkaloids (Wiedenfeld & Röder, 1991).
ii. Methanol extract being able to inhibit the replication of Echoviruses (monopartite linear single-stranded (+) RNA) may be able to be active against the Severe acute respiratory syndrome-associated coronavirus. Severe acute respiratory syndrome-associated coronavirus 2-positive patients given after meal, daily, 50 mL of an ethanol extract of a mixture of plants including *Moringa oleifera* Lam, *Zingiber officinale* Roscoe, *Allium sativa* L., *Jatropha curcas* L., *Ageratum conyzoides*, L. *Citrus aurantiifolia* (Christm.) Swingle juice, as well as honey tested negative after 10 days (Adejuwon et al., 2020). A member of the genus *Ageratum* L. is used for the treatment of COVID-19 in China (Yang et al., 2020)
iii. *Ageratum houstonianum* Miller is used for cuts and wounds in Nepal (local name: *Buki phool*) and is also considered sacred.
iv. From the Subfamily Asteroideae; Tribe Heliantheae, *Eupatorium ayapana* Vent. (synonym: *Eupatorium triplinerve* Vahl; *Ayapana triplinervis* (Vahl) R.M. King & H. Rob.) is used in India (local name: *Ayapan*) to treat dysentery and in Bangladesh (local name: *Ayapanan*) for fever, cholera, tetanus, and putrefied ulcers. Essential oil of this herb inhibited the replication of the Zika virus with the IC_{50} value of 38 µg/mL and from this oil thymohydroquinone (LogD 1.8 at pH 7.4; molecular mass = 166 g/mol) inhibited the replication of the Zika virus with the IC_{50} value of 45 µg/mL and the selectivity index of 9.1 (Haddad et al., 2019). *Eupatorium odoratum* L. (local name: *Sing-hi-haar*) is given for cold and flu in Bangladesh. *Eupatorium fortunei* Turcz. (local name: *Pei lan*) is given for COVID-19 in China. Aqueous extract of *Eupatorium fortunei* Turcz. at the concentration of 100 µg/mL protected RAW 264.7 cells against the Newcastle disease virus, the Vesicular stomatitis virus, and the Influenza virus A/PuertoRico/8/34 H1N1 (Choi et al., 2017). Active principles here could be lignans such as the benzofuran lignan euparin (LogD 3.6 at pH 7.4; molecular mass = 216.2 g/mol) which inhibited the replication of the Respiratory syncytial virus (LONG strain) and Respiratory syncytial virus (A2 strain) with the IC_{50} values of 2.3 and 2.8 µM, respectively (Wang et al., 2016). Euparin inhibited the replication of the Poliovirus types 1, 2, and 3 with the EC_{50} values of 0.4, 0.1, and 0.1 µg/mL, respectively, and the selectivity indexes of 284.9, 1068, and 854.7, respectively (Visintini Jaime et al., 2013). Euparin is most probably active against the Severe acute respiratory syndrome-associated coronavirus and can be added to the list of first-line molecules to be tested for the forthcoming zoonotic viral pandemics.

v. In the Tribe Heliantheae, *Mikania cordata* (Burm. f.) B.L. Rob. is given in Bangladesh (local name: *Boina lota*) for cuts, dysentery, dengue, and wounds. The flavonol mikanin sulfate isolated from *Mikania micrantha* Kunth inhibited the replication of the Parainfluenza type 3 virus with IC_{50} values of 19.7 μM and selectivity index of 24 (But et al., 2009).

vi. In the Tribe Heliantheae, *Sigesbeckia orientalis* L. is traditionally used in China (local name: *Chu kao mu*) for abscesses, boils, dog bites, ringworm, and ulcers and in India (local name: *Kadambu*) for fever, ulcers, and ringworms. Extract of the whole herb inhibited the replication of the Newcastle diseases virus (Babbar et al., 1982) and here sesquiterpene lactones are probably involved. Methanol extract of *Sigesbeckia pubescens* (Makino) Makino (synonym: *Sigesbeckia orientalis* var. *pubescens* Makino) inhibited the Human immunodeficiency virus-1 protease (Park, 2003).

vii. In the Tribe Heliantheae, *Arnica montana* L. has been used by physicians of Europe up to about the 19th century for fever and tincture of the roots of this beautiful herb is used in China for fractures. In Korea, *Arnica palmata* Thunb. (*Syneilesis palmata* (Thunb.) Maxim.; local name: *Usannamul*) is grown around Buddhist temples for food and is used for bruises. Methanol extract of *Arnica palmata* Thunb. (*Syneilesis palmata* (Thunb.) Maxim. inhibited the Human immunodeficiency type-1 protease (Park, 2003).

viii. In the Tribe Heliantheae, *Adenostemma lavenia* (L.) O. Ktze. is used for wounds in India (local name: *Jangli-jira*), and in China (local name: *Fa ha mi*) it is given for flu, toothache, hepatitis, and pneumonia. The herb generates a series of kaurene-type diterpenes (Shimizu et al., 1990) which could be examined for their antimicrobial properties.

ix. In the Tribe Heliantheae, *Bidens pilosa* L. is used in India (local name: *Rekkala rani*) for cough and ulcers, in Nepal (local name: *Kalokuro*) for cuts and wounds, in Myanmar (local name: *K'sik nu geh*) for diarrhea, and in the Philippines it is used as a remedy for boils. Aqueous extract of this herb inhibited the replication of the Herpes simplex virus type-1 and type-2 (Chiang et al., 2003). In a subsequent study, aqueous extract at the concentration of 250 μg/mL was virucidal for the Herpes simplex virus type-1 (strain HF) and the Herpes simplex virus type-2 (strain Savage) and given orally at the dose of 1g/Kg twice daily to mice afforded some levels of protection against the Herpes simplex virus (401H strain) (Nakama et al., 2012). Aqueous extract of inhibited the replication of the Human immunodeficiency virus with the IC_{50} value of 54 μg/mL and a selectivity index of 7.7 in MT-4 cells (Matsuse et al., 1998).

x. In the Tribe Heliantheae, aqueous extract of *Bidens bipinnata* L. (synonym: *Bidens pilosa* var. *bipinnata* (L.) Hook. f.) inhibited the replication of the Respiratory syncytial virus with the IC_{50} value of 250 μg/mL and a selectivity index above 4 (Li et al., 2004). In Bangladesh, *Bidens sulphurea* (Cav.) Sch. Bip. local name: *Shumo phul*) is given for boils. From *Bidens cernua* L., the polyacetylene phenylheptatriyne (LogD 4.6 at pH 7.4; molecular mass = 164.2 g/mol) inhibited the growth of a broad array of fungi of which *Cryptococcus neoformans* with the MIC value of 12.5 μg/mL and MFC of 50 μg/mL and *Candida albicans* (ATCC 885-653) with the MIC value of 50 μg/mL and MFC of 100 μg/mL (Rybalchenko et al., 2010).

xi. In the Tribe Heliantheae, *Tagetes erecta* L. (synonym: *Tagetes patula* L.) is a remedy for abscesses, boils, colds, eye infections, sore throat, mumps, skin infection, and ulcers in India, in Bangladesh (local name:*Gada*) for cuts and wounds, and for cold in China (local name: *Wan shou ju*). Essential oil of flowers inhibited the growth of *Candida albicans* with the MIC value of 0.04 μL/mL, respectively (Safar et al., 2020). The flavonol glycoside patulitrin isolated from the flowers inhibited the growth of *Klebsiella pneumoniae* (Rhama & Madhavan, 2011). Ethanol extract of a member of the genus *Tagetes* L. inhibited the replication of the Herpes simplex virus type-1, West Nile virus, and Saint Louis encephalitis viruses (Martinez et al., 2020).

xi. In the Tribe Heliantheae, *Wedelia biflora* (L.) DC. in India is used for cuts, diarrhea, dysentery, sores, and wounds, and in Malaysia, it is a remedy for sores. The leaves yield 3′,7-di-*O*-methylquercetin which inhibited the growth of phytopathogenic fungi (Miles et al., 1993). In Bangladesh, *Wedelia chinensis* (Osbeck) Merr. (synonym: *Wedelia calendulacea* (L.) Less.; local name: *Bhringoraj*) is a remedy for tuberculosis and in India for jaundice. Hexane extract of *Wedelia trilobata* (L.) Hitchc. inhibited the growth of *Bacillus subtilis, Mycobacterium smegmatis, Staphylococcus aureus, Staphylococcus epidermidis, Proteus vulgaris, Pseudomonas aeruginosa, Salmonella paratyphi*, and *Shigella sonnei* (Taddei. & Rosas-Romero, 1999). From this plant, 3α-angeloyloxy-9β-hydroxy-*ent*-kaur-16-en-19-oic inhibited the replication of the Tomato spotted wilt virus (enveloped segmented linear (-) RNA virus) *in vivo* by 62.4% at a concentration of 100 µg/mL (Zhao et al., 2019). 3α-Angeloyloxy-9β-hydroxy-*ent*-kaur-16-en-19-oic inhibited the growth of *Candida albicans* (ATCC Y0109) with the MIC value of 125 µg/mL (Li et al., 2016).

xii. In the Tribe Heliantheae, *Xanthium spinosum* L. is a remedy for fever in Turkey (local name: *Dikenli sırça otu*). *Xanthium indicum* J. Koenig ex Roxb. is given for skin diseases in Bangladesh (local name: *Hagra, ghaghra*) and for fever in India (local name: *Otta*). In Pakistan, *Xanthium strumarium* L. (local name: *Chota dhatoora*) is given for dysentery, fever, leprosy, smallpox, and dental caries; in Cambodia, Laos, and Vietnam for abscesses; and in India (local name: *Shankheshwara*) for abscesses and ulcers. Methanol extract of leaves of this sinister herb (50 mg/mL/5 mm disc) inhibited the growth of *Klebsiella pneumoniae* (ATCC 15380), *Enterococcus faecalis* (ATCC 29212), *Bacillus subtilis* (ATCC 6633), *Proteus mirabilis* (ATCC 49132), and *Staphylococcus aureus* (ATCC 25923) with the inhibition zone diameters of 25, 9, 25, 6, and 25 mm, respectively (Devkota & Das, 2015). From this herb, the sesquiterpene 4-oxo-1 (5),2,11, (13)-xanthatriene-12,8-olide (also known as deacetylxanthumin) inhibited the mycelial growth of *Phytophthora drechsleri* with the MIC value of 12.5 µg/mL (Kim et al., 2002). Note the presence of a highly reactive furanolactone pharmacophore (see earlier) suggesting deacetylxanthumin to be a broad-spectrum antimicrobial phytoanticipin. Aqueous extract at the concentration of 250 µg/mL inhibited the replication of the Transmissible gastroenteritis virus by 48.6% (Kwon et al., 2003). The Transmissible gastroenteritis virus belongs to the Coronaviridae whence the possible presence of anti-Severe acute respiratory syndrome-associated coronavirus 2 principles in this plant. The seeds have been consumed in Bangladesh in time of famine and incurred fatal poisoning (Gurley et al., 2010).

xiii. In the Tribe Heliantheae, *Enydra fluctuans* Lour. is given for chickenpox, boil, and rabies in Bangladesh (local name: *Helancha*).

xiv. *In the Tribe* Heliantheae, the ornamental *Rudbeckia laciniata* L. yields the bisabolene-type sesquiterpene 2,10-bisaboladiene-1,4-endoperoxide which inhibited the growth of *Mycobacterium tuberculosis* (H37Rv) with the MIC value of 32 µg/mL (Pauli et al., 2005). One can observe that amphiphilic principles with endoperoxide moieties are often antimycobacterial.

References

Adejuwon, A.O., Obayemi, O.S., Odeleye, O.D. and Anatolyivna, V., 2020. Inhibitory actions of a medicinal plants' extract on SARS-CoV-2 and COVID-19. *Cancer Biology, 10*(3): 1–3.

Babbar, O.P., Joshi, M.N. and Madan, A.R., 1982. Evaluation of plants for antiviral activity. *Indian J Med Res, 76*(S), pp. 54–65.

But, P.P.H., He, Z.D., Ma, S.C., Chan, Y.M., Shaw, P.C., Ye, W.C. and Jiang, R.W., 2009. Antiviral constituents against respiratory viruses from Mikania micrantha. *Journal of Natural Products, 72*(5), pp. 925–928.

Choi, J.G., Lee, H., Hwang, Y.H., Lee, J.S., Cho, W.K. and Ma, J.Y., 2017. Eupatorium fortunei and its components increase antiviral immune responses against RNA viruses. *Frontiers in Pharmacology, 8*, p. 511.

Chiang, L.C., Chang, J.S., Chen, C.C., Ng, L.T. and Lin, C.C., 2003. Anti-Herpes simplex virus activity of Bidens pilosa and Houttuynia cordata. *The American Journal of Chinese Medicine*, *31*(03), pp. 355–362.

Devkota, A. and Das, R.K., 2015. Antibacterial activities of Xanthium strumarium L. *Journal of Natural History Museum*, *29*, pp. 70–77.

González, A.G., Aguiar, Z.E., Grillo, T.A., Luis, J.G., Rivera, A. and Calle, J., 1991. Chromenes from Ageratum conyzoides. *Phytochemistry*, *30*(4), pp. 1137–1139.

Gurley, E.S., Rahman, M., Hossain, M.J., Nahar, N., Faiz, M.A., Islam, N., Sultana, R., Khatun, S., Uddin, M.Z., Haider, M.S. and Islam, M.S., 2010. Fatal outbreak from consuming Xanthium strumarium seedlings during time of food scarcity in northeastern Bangladesh. *PloS One*, *5*(3), p. e9756.

Haddad, J.G., Picard, M., Bénard, S., Desvignes, C., Desprès, P., Diotel, N. and El Kalamouni, C., 2019. Ayapana triplinervis essential oil and its main component thymohydroquinone dimethyl ether inhibit Zika virus at doses devoid of toxicity in zebrafish. *Molecules*, *24*(19), p. 3447.

Hoffman, B.R., DelasAlas, H., Blanco, K., Wiederhold, N., Lewis, R.E. and Williams, L., 2004. Screening of antibacterial and antifungal activities of ten medicinal plants from Ghana. *Pharmaceutical Biology*, *42*(1), pp. 13–17.

Iqbal, M.C.M., Jayasinghe, U.L.B., Herath, H.M.T.B., Wijesekara, K.B. and Fujimoto, Y., 2004. A fungistatic chromene from Ageratum conyzoides. *Phytoparasitica*, *32*(2), pp. 119–126.

Kim, D.K., Shim, C.K., Bae, D.W., Kawk, Y.S., Yang, M.S. and Kim, H.K., 2002. Identification and biological characteristics of an antifungal compound extracted from Cocklebur (Xanthium Strumarium) against Phytophthora drechsleri. *The Plant Pathology Journal*, *18*(5), pp. 288–292.

Kwon, D.H., Kim, M.B., Yoon, D.Y., Lee, Y.H., Kim, J.W., Lee, H.G., Choi, I.S., Lim, J.S. and Choe, Y.K., 2003. Screening of plant resources of anti-viral activity. *Korean Journal of Medicinal Crop Science*, *11*(1), pp. 24–30.

Li, Y., Ooi, L.S., Wang, H., But, P.P. and Ooi, V.E., 2004. Antiviral activities of medicinal herbs traditionally used in southern mainland China. *Phytotherapy Research*, *18*(9), pp. 718–722.

Li, S.F., Ding, J.Y., Li, Y.T., Hao, X.J. and Li, S.L., 2016. Antimicrobial diterpenoids of Wedelia trilobata (L.) Hitchc. *Molecules*, *21*(4), p. 457.

Martinez, F., Massuh, Y., Aguilar, J.J., Ojeda, M.S., Contigiani, M.S., Montoya, S.C.N. and Konigheim, B.S., 2020. Cultivars of Tagetes minuta L. (Asteraceae) as a source of potential natural products with antiviral activity. *Journal of Herbal Medicine*, *24*, p. 100397.

Matsuse, I.T., Lim, Y.A., Hattori, M., Correa, M. and Gupta, M.P., 1998. A search for anti-viral properties in Panamanian medicinal plants: The effects on HIV and its essential enzymes. *Journal of Ethnopharmacology*, *64*(1), pp. 15–22.

Miles, D.H., Chittawong, V., Hedin, P.A. and Kokpol, U., 1993. Potential agrochemicals from leaves of Wedelia biflora. *Phytochemistry*, *32*(6), pp. 1427–1429.

Nakama, S., Tamaki, K., Ishikawa, C., Tadano, M. and Mori, N., 2012. Efficacy of Bidens pilosa extract against herpes simplex virus infection in vitro and in vivo. *Evidence-Based Complementary and Alternative Medicine*, *2012*. doi: 10.1155/2012/413453.

Nogueira, J.H., Gonçalez, E., Galleti, S.R., Facanali, R., Marques, M.O. and Felício, J.D., 2010. Ageratum conyzoides essential oil as aflatoxin suppressor of Aspergillus flavus. *International Journal of Food Microbiology*, *137*(1), pp. 55–60.

Ogbole, O.O., Akinleye, T.E., Segun, P.A., Faleye, T.C. and Adeniji, A.J., 2018. In vitro antiviral activity of twenty-seven medicinal plant extracts from Southwest Nigeria against three serotypes of echoviruses. *Virology Journal*, *15*(1), pp. 1–8.

Patil, R.P., Nimbalkar, M.S., Jadhav, U.U., Dawkar, V.V. and Govindwar, S.P., 2010. Antiaflatoxigenic and antioxidant activity of an essential oil from Ageratum conyzoides L. *Journal of the Science of Food and Agriculture*, *90*(4), pp. 608–614.

Park, J.C., 2003. Inhibitory effects of Korean plant resources on human immunodeficiency virus type 1 protease activity. *Oriental Pharmacy and Experimental Medicine*, *3*(1), pp. 1–7.

Pauli, G.F., Case, R.J., Inui, T., Wang, Y., Cho, S., Fischer, N.H. and Franzblau, S.G., 2005. New perspectives on natural products in TB drug research. *Life Sciences*, *78*(5), pp. 485–494.

Rhama, S. and Madhavan, S., 2011. Antibacterial activity of the flavonoid, patulitrin isolated from the flowers of Tagetes erecta L. *International Journal of PharmTech Research*, *3*(3), pp. 1407–1409.

Rybalchenko, N.P., Prykhodko, V.A., Nagorna, S.S., Volynets, N.N., Ostapchuk, A.N., Klochko, V.V., Rybalchenko, T.V. and Avdeeva, L.V., 2010. In vitro antifungal activity of phenylheptatriyne from Bidens cernua L. against yeasts. *Fitoterapia*, *81*(5), pp. 336–338.

Safar, A.A., Ghafoor, A.O. and Dastan, D., 2020. Screening of chemical characterization, antifungal and cytotoxic activities of essential oil constituents of Tagetes erecta L. from Erbil, Kurdistan Region-Iraq. *Polish Journal of Environmental Studies*, *29*(3), pp. 2317–2326.

Shimizu, S., Miyase, T., Umehara, K. and UENO, A., 1990. Kaurane-type diterpenes from Adenostemma lavenia O. KUNTZE. *Chemical and Pharmaceutical Bulletin*, *38*(5), pp. 1308–1312.

Taddei, A. and Rosas-Romero, A.J., 1999. Antimicrobial activity of Wedelia trilobata crude extracts. *Phytomedicine*, *6*(2), pp. 133–134.

Visintini Jaime, M.F., Campos, R.H., Martino, V.S., Cavallaro, L.V. and Muschietti, L.V., 2013. Antipoliovirus activity of the organic extract of Eupatorium buniifolium: Isolation of euparin as an active compound. *Evidence-Based Complementary and Alternative Medicine*, *2013*. doi: 10.1155/2013/402364.

Yang, Q.X., Zhao, T.H., Sun, C.Z., Wu, L.M., Dai, Q., Wang, S.D. and Tian, H., 2020. New thinking in the treatment of 2019 novel coronavirus pneumonia. *Complementary Therapies in Clinical Practice*, *39*, p. 101131.

Wang, W.J., Wang, L., Liu, Z., Jiang, R.W., Liu, Z.W., Li, M.M., Zhang, Q.W., Dai, Y., Li, Y.L., Zhang, X.Q. and Ye, W.C., 2016. Antiviral benzofurans from Eupatorium chinense. *Phytochemistry*, *122*, pp. 238–245.

Wiedenfeld, H. and Röder, E., 1991. Pyrrolizidine alkaloids from Ageratum conyzoides. *Planta medica*, *57*(06), pp. 578–579.

Zhao, L., Hu, Z., Li, S., Zhou, X., Li, J., Su, X., Zhang, L., Zhang, Z. and Dong, J., 2019. Diterpenoid compounds from Wedelia trilobata induce resistance to Tomato spotted wilt virus via the JA signal pathway in tobacco plants. *Scientific Reports*, *9*(1), pp. 1–12.

11.1.1.2 *Artemisia vulgaris* L.

Common names: Sagebrush mugwort; bei ai (China); barha, ndgadamana pushpachaamara, tirunama (India); baru cina (Indonesia); afsantin (Iran); bunga ayam (Malaysia); gilbas (the Philippines); walkolondu (Sri Lanka); kot chulalamfwawan (Thailand); cuu ngai da (Vietnam)

Habitat: Roadsides, waste lands

Distribution: Tropical Asia and the Pacific

Botanical description: This discrete herb grows to height of 2 m. The young stems are terete and hairy and older stems are purplish. The leaves are simple, spiral, and exstipulate. The blade is 3–10 × 1.5–10 cm, dull green above, with a slight odor, somewhat whitish-green and pubescent below, pinnately lobed at base and trifid or linear at apex. The racemes are pendulous and terminal and present numerous small whitish capitula (looks like somewhat *Chenopodium album* L. from far). The corolla is greenish-yellow and minute. The achenes are oblong and ellipsoid.

Medicinal uses: Apparently none for microbial infections

Moderate broad-spectrum antibacterial polar extract: Ethanol extract of whole plant inhibited the growth *of Enterobacter faecalis* (ATCC 29212), *Serratia marcescens* (ATCC 9986), *Proteus vulgaris* (ATCC 6830), *Klebsiella pneumoniae*, and *Staphylococcus aureus* with the MIC values of 400, 400, 400, 3125, and 400 µg/mL, respectively. (Hrytsyk et al., 2021).

Broad-spectrum antifungal polar extract: Ethanol extract inhibited the growth of *Candida albicans*, *Candida tropicalis*, and *Aspergillus niger* (Hrytsyk et al., 2021).

Moderate antiviral (enveloped linear monopartite dimeric single-stranded (+) RNA) polar extract: Methanol extract of flowers inhibited the growth of the Human immunodeficiency virus type-1 with the IC_{50} value of 350 µg/mL in peripheral blood mononuclear cells (Mohabatkar et al., 2015).

Antiviral (enveloped monopartite linear double-stranded DNA) polar extract: Methanol extract of aerial parts at the concentration of 25 µg/mL protected HeLa cells against the Herpes simplex virus type-1 (KOS strain) by 63.9% (Karamoddini et al., 2011).

Antiviral (enveloped monopartite linear single-stranded (+) RNA) sesquiterpene: Artemisinin inhibited the replication of the Bovine diarrhea virus at concentration below 100 µM (Karamoddini et al., 2011).

Commentaries

xii. *Artemisia scoparia* Waldst & Kit (local name: *Choree saroch*) is a remedy for burn in Pakistan. In India, *Artemisia absinthium* L. (local name: *Afsanteen*) is given for fever, *Artemisia maritima* L. (local name: *Chauhaara*) as well as *Artemisia vestita* Wall.

ex DC. (local name: *Gangaa tulasi*) is used for fungal infections, and *Artemisia nilagirica* (C.B. Clarke) Pamp. (synonym: *Artemisia vulgaris* var. *nilagirica* C.B. Clarke; local name: *Damanaka*) is a remedy for fever. In Nepal, *Artemisia indica* Willd. (local names: *Aftunteen, Mashiputrie, Titepati*) is used for diarrhea, fever, and wounds. *Artemisia chinensis* L. (synonym: *Crossostephium chinense* (L.) Makino) is used for cough in Vietnam (local name: *Nguyit bach*) and China. In China, *Artemisia argyi* Levl.et Van. (local name: *Ai căo*) is used for cold and cough, *Artemisia japonica* Thunb. for vaginitis, *Artemisia capillaris* Thunb. for cystitis and aqueous extract of seeds at the concentration of 250 µg/mL inhibited the Human immunodeficiency virus protease by 59.9% (Xu et al., 1996). *Artemisia apiacea* Hance is given for blennorrhea in China and methanol extract of roots of this herb at 100 µg/mL inhibited the Human immunodeficiency type-1 protease by 28% (Park et al., 2002). *Artemisia annua* L. is used to treat tuberculosis in China and one could examine the antimycobacterial properties of artemisinin. Being an endoperoxide it is most probably antimycobacterial (see earlier).

xiii. *Artemisia vulgaris* L. essential oil contains 1,8-cineole, α-thujone, camphor, and isobomeol (Haider et al., 2003) and the plant produces the antimalarial sesquiterpene endoperoxide artemisinin which is also present in *Artemisia annua* L. (Ranjbar et al., 2015). Artemisinin being active against the Bovine diarrhea virus probably accounts for the fact that ethanol extract of *Artemisia annua* L. inhibited the replication of the Severe acute respiratory syndrome-associated coronavirus (BJ-001) and (BJ-006) with the IC_{50} values of 34.5 and 39.2 µg/mL and selectivity indexes of 31 and 27, respectively (Li et al., 2005). Artemisinin could be of clinical value against the Severe acute respiratory-associated coronavirus 2 and other pandemic coronaviruses. COVID-19 patients who were given artemisinin 125 mg and piperaquine 750 mg on the first day followed by artemisinin 62.5 mg and piperaquine 375 mg/day for 6 days evoked clearance of the virus and levels of pulmonary protection (Li et al., 2021). In Morocco, the plant is used to prevent and/or treat COVID-19 (Benkhaira et al., 2021) and there is the possibility to use *Artemisia vulgaris* L. as a preventive and/or mitigating phytomedication against COVID-19. The purpose of this volumes is not to say that vaccines are useless, in fact thanks to vaccines millions of life have been saved.

xiv. The antiviral potentials of artemisinin derivative mixed with other antimalarial agents are being investigated. Mefloquine-dihydroartemisinin mixture inhibited the replication of the Severe acute respiratory syndrome-associated coronavirus strain (IHUMI-3) in Vero cells (Gendro et al., 2020). The antimalarial drug Artesunate (semisynthetic derivative of artemisinin) has favorable oral bioavailability and is a broad-spectrum antiviral agent *in vitro*: at the concentration of 15 µM it inhibited the replication of the Human cytomegalovirus (Towne strain) and the Herpes simplex virus type-1 by more than 95% and 83%, respectively (Effert et al., 2008). This sesquiterpene inhibited the replication of the Human herpes virus 6A and Epstein-Barr virus with the IC_{50} value of 3.8 and 7.2 µM, respectively (Effert et al., 2008). This derivative inhibited Hepatitis B virus DNA secretion with the IC_{50} value of 0.5 µM and inhibited the replication of the Human immunodeficiency virus type-1 (M-tropic) and (strain Ba-L) with the IC_{50} value of 600 nM (Effert et al., 2008). Is Artesunate of clinical value against COVID-19?

xv. Aqueous extract of stem and leaf of *Artemisia vulgaris* var. *maximowiczii* Nakai (synonym: *Artemisia princeps* Pamp) inhibited the replication of the Adenovirus-8 with the EC_{50} of 38.6 µg/mL and a selectivity index of 91.1 (Chiang et al 2003) and aqueous extract of leaves at the concentration of 250 µg/mL inhibited the Human immunodeficiency type-1 protease by 34% (Xu et al., 1996).

xvi. Anthelmintic principles in members of the genus *Artemisia* L. could be examined for inhibition of bacterial and fungal efflux pumps as well as artemisinin.

xvii. In the Subfamily Asteroideae; Tribe Anthemideae, *Achillea sibirica* Ledeb. is a remedy for boils in China and *Achillea millefolium* L. is used to treat cold in Indonesia. Essential oil of *Achillea millefolium* L. inhibited the growth of *Streptococcus mutans* and *Lactobacillus acidophilus* with MIC values of 1.4 and 3.7 µL/mL (Miller et al., 2015) and at the concentration of 1% inhibited the spore germination by *Didymella bryoniae* of by 36.4% (Fiori et al., 2000). In the Tribe Anthemideae, aqueous extract of flowers of *Chrysanthemum coronarium* L. at the concentration of 250 µg/mL inhibited the Human immunodeficiency type-1 protease by 12.9% (Xu et al., 1996). Apigenin 7-*O*-β-*D*-(4″-caffeoyl)glucuronide isolated from *Chrysanthemom morifolium* Ramat inhibited the Human immunodeficiency virus type-1 integrase with the IC_{50} value of 7.2 µg/mL and the replication of Human immunodeficiency virus tupe-1 (Strain IIIB) with the EC_{50} value of 41.8 µg/mL in MT-4 cells (Lee et al., 2003).

xviii. In the Tribe Anthemideae, members of the genus *Anthemis* L. are often used for medicinal purposes in Northwest Asia: *Anthemis altissima* L. (local name: *Papatya*) is given for jaundice, and *Anthemis cotula* L. (local name: *Hozan çiçegi*) is administered for dysentery, *Anthemis tinctoria* L is used for cough, and *Anthemis tinctoria* L. var. *tinctoria* (local name: *Boyaci papatyasi*) is used for a cold, jaundice, and toothache. *Anthemis nobilis* L. also known as Roman chamomile (synonym: *Chamaemelum nobile* (L.) All.) is taken in Iraq for acnea, bronchitis, cold, sore throat, and wounds. Methanol extract of this plant inhibited the replication of the Sindbis virus and Herpes simplex virus at the concentration of 200 and 50 µg/mL, respectively (Mouhajir et al., 2001). Since the Sindbis virus is an enveloped, monopartite, linear, single-stranded (+)RNA virus, Roman Chamomile could potentially be of value in the prevention or attenuation of COVID-19.

References

Benkhaira, N., Koraichi, S.I. and Fikri-Benbrahim, K., 2021. Ethnobotanical survey on plants used by traditional healers to fight against COVID-19 in Fez city, Northern Morocco. *Ethnobotany Research and Applications*, *21*, pp. 1–18.

Chiang, L.C., Cheng, H.Y., Liu, M.C., Chiang, W. and Lin, C.C., 2003. In vitro anti-herpes simplex viruses and anti-adenoviruses activity of twelve traditionally used medicinal plants in Taiwan. *Biological and Pharmaceutical Bulletin*, *26*(11), pp. 1600–1604.

Efferth, T., Romero, M.R., Wolf, D.G., Stamminger, T., Marin, J.J. and Marschall, M., 2008. The antiviral activities of artemisinin and artesunate. *Clinical Infectious Diseases*, *47*(6), pp. 804–811.

Gendrot, M., Duflot, I., Boxberger, M., Delandre, O., Jardot, P., Le Bideau, M., Andreani, J., Fonta, I., Mosnier, J., Rolland, C. and Hutter, S., 2020. Antimalarial artemisinin-based combination therapies (ACT) and COVID-19 in Africa: In vitro inhibition of SARS-CoV-2 replication by mefloquine-artesunate. *International Journal of Infectious Diseases*, *99*, pp. 437–440.

Haider, F., Dwivedi, P.D., Naqvi, A.A. and Bagchi, G.D., 2003. Essential oil composition of *Artemisia vulgaris* harvested at different growth periods under Indo-Gangetic plain conditions. *Journal of Essential Oil Research*, *15*(6), pp. 376–378.

Hrytsyk, R.A., Kutsyk, R.V., Yurchyshyn, O.I., Struk, O.A., Kireev, I.V. and Grytsyk, A.R., 2021. The investigation of antimicrobial and antifungal activity of some *Artemisia L.* species. *Pharmacia*, *68*, p. 93.

Karamoddini, M.K., Emami, S.A., Ghannad, M.S., Sani, E.A. and Sahebkar, A., 2011. Antiviral activities of aerial subsets of Artemisia species against Herpes Simplex virus type 1 (HSV1) in vitro. *Asian Biomedicine*, *5*(1), pp. 63–68.

Lee, J.S., Kim, H.J. and Lee, Y.S., 2003. A new anti-HIV flavonoid glucuronide from *Chrysanthemum morifolium*. *Planta Medica*, *69*(9), pp. 859–861.

Li, G., Yuan, M., Li, H., Deng, C., Wang, Q., Tang, Y., Zhang, H., Yu, W., Xu, Q., Zou, Y. and Yuan, Y., 2021. Safety and efficacy of artemisinin-piperaquine for treatment of COVID-19: An open-label, non-randomised and controlled trial. *International Journal of Antimicrobial Agents*, *57*(1), p. 106216.

Li, S.Y., Chen, C., Zhang, H.Q., Guo, H.Y., Wang, H., Wang, L., Zhang, X., Hua, S.N., Yu, J., Xiao, P.G. and Li, R.S., 2005. Identification of natural compounds with antiviral activities against SARS-associated coronavirus. *Antiviral Research*, *67*(1), pp. 18–23.

Miller, A.B., Cates, R.G., Lawrence, M., Soria, J.A.F., Espinoza, L.V., Martinez, J.V. and Arbizú, D.A., 2015. The antibacterial and antifungal activity of essential oils extracted from Guatemalan medicinal plants. *Pharmaceutical Biology*, *53*(4), pp. 548–554.

Mohabatkar, H., Behbahani, M. and Nejad, M.R.R., 2015. Effects of root, shoot, leaf and seed extracts of seven *Artemisia species* on HIV-1 replication and CD4 expression. *Journal of Coastal Life Medicine, 3*(12), pp. 996–999.
Oyedemi, S.O., Coopoosamy, R.M., 2015. Preliminary studies on the antibacterial and antioxidative potentials of hydroalcoholic extract from the whole parts of *Artemisia vulgaris* L. *International Journal of Pharmaceutics, 11*, pp. 561–569.
Park, J.C., 2003. Inhibitory effects of Korean plant resources on human immunodeficiency virus type 1 protease activity. *Oriental Pharmacy and Experimental Medicine, 3*(1), pp. 1–7.
Park, J.C., Hur, J.M., Park, J.G., Hatano, T., Yoshida, T., Miyashiro, H., Min, B.S. and Hattori, M., 2002. Inhibitory effects of Korean medicinal plants and camelliatannin H from *Camellia japonica* on human immunodeficiency virus type 1 protease. *Phytotherapy Research, 16*(5), pp. 422–426.
Ranjbar, M., Naghavi, M.R., Alizadeh, H. and Soltanloo, H., 2015. Expression of artemisinin biosynthesis genes in eight Artemisia species at three developmental stages. *Industrial Crops and Products, 76*, pp. 836–843.
Xu, H.X., Wan, M., Loh, B.N., Kon, O.L., Chow, P.W. and Sim, K.Y., 1996. Screening of traditional medicines for their inhibitory activity against HIV-1 protease. *Phytotherapy Research, 10*(3), pp. 207–210.

11.1.1.3 *Eclipta alba* (L.) Hassk.

Synonyms: *Eclipta erecta* L.; *Eclipta prostrata* (L.) L.; *Verbesina alba* L.; *Verbesina prostrata* L.

Common names: False daisy; kalo-keshi, kalashona (Bangladesh); li chang, mo han lian (China); bhringaraj, maka, oochi (India); dau sipat (Indonesia); keremak jantan (Malaysia); kyate hman (Myanmar); deberebelela (Palau); higis manok (the Philippines); kameng, ya sap (Thailand); co muc (Vietnam)

Habitat: Watery and shady meadows, by rivers, ponds, lakes, drains, roadside

Distribution: Asia and the Pacific

Botanical observation: This herb grows up to about 30 cm long and has a positive aura. The roots are massively fibrous. The stems are purplish, fleshy, hairy, and characteristically swollen between internodes. The leaves are simple, opposite, and exstipulate. The blade is 2.5–7.5 cm × 5–7 mm, hairy, sessile, serrate, and opposite. The flowers are arranged in axillary capitula attached to 2.5–3 cm long, slender, and straight pedicels. The capitula are 5–9 mm diameter, whitish, and look like some sorts of dwarfy sunflowers. The involucres are campanulate, and the bracts are brown green and longer than the flowers. The thalamus is flat and discoid. The ray flowers show small entire or bifid white ligules. The achenes are flattish and triangular.

Medicinal uses: Abscess, ear putrefaction (India); AIDS (Thailand); fever (Bangladesh; India); boils (Bangladesh; Myanmar); cold (India; Sri Lanka); cough (China; India; Sri Lanka); jaundice (India; Myanmar; Sri Lanka); cough, cold (India; Myanmar); chicken pox, small pox, venereal diseases (Myanmar); sore throat (China)

Strong broad-spectrum antibacterial amphiphilic 3-acyl coumarin: Wedelolactone (LogD 1.9 at pH 7.4; molecular mass = 314.2 g/mol) inhibited the growth of *Staphylococcus aureus* (ATCC 9144), *Salmonella typhimurium* (ATCC 23564), and *Staphylococcus epidermidis* (ATCC 155) with the MIC values of 20, 25, and 15 µg/mL, respectively (Dalal et al., 2010).

Strong antiviral (enveloped monopartite linear double-stranded DNA) amphiphilic 3-acyl coumarin: Wedelolactone inhibited the replication of the Human cytomegalovirus (AD169-GFP) in human foreskin fibroblasts with the IC_{50} value of 8.1 µM (CC_{50}: 173.4 µM) via a mechanism that involves at least in part the inhibition of viral DNA replication (Svrlanska et al., 2020).

Strong antiviral (enveloped monopartite linear single-stranded (+) RNA) amphiphilic 3-acyl coumarin: Wedelolactone inhibited the replication of the Hepatitis C virus by 80% at the concentration of 50 µM (Manvar et al., 2012).

Viral enzyme inhibition by 3-acyl coumarin: Wedelolactone inhibited RNA-dependent RNA polymerase (NS5B polymerase) with the IC_{50} value of 7.7 µM. Wedelolactone inhibited the Human immunodeficiency virus type-1 integrase with the IC_{50} value of 4 µM (Tewtrakul et al., 2007).

Viral enzyme inhibition by isoflavone: Orobol inhibited the Human immunodeficiency virus type-1 integrase with the IC_{50} value of 8.1 µM (Tewtrakul et al., 2007).

Viral enzyme inhibition by thiophene derivative: 5-Hydroxymethyl-(2,2′:5′, 2″)-terthienyl tiglate inhibited the Human immunodeficiency virus type-1 protease with the IC_{50} value of 58.3 μM (Tewtrakul et al., 2007).

Commentaries

i. Being planar, wedelolactone probably intercalates within the enzymes responsible of microbial genetic material replication and it is in fact an inhibitor of topoisomerase II (Benes et al., 2011). Wedelolactone might be able to inhibit Severe acute respiratory syndrome-associated coronavirus 2 RNA-dependent RNA polymerase. This plant being traditionally used in Thailand for AIDS (Tewtrakul et al., 200) could be (if proven to be non-toxic) theorically of value as phytomedication against COVID-19 and other forthcoming coronavirus pandemic waves or multidrug and/or vaccine-resistant COVID-19 mutant.

References

Benes, P., Knopfova, L., Trcka, F., Nemajerova, A., Pinheiro, D., Soucek, K., Fojta, M. and Smarda, J., 2011. Inhibition of topoisomerase IIα: Novel function of wedelolactone. *Cancer Letters*, *303*(1), pp. 29–38.

Dalal, S., Kataria, S.K., Sastry, K.V. and Rana, S.V.S., 2010. Phytochemical screening of methanolic extract and antibacterial activity of active principles of hepatoprotective herb, *Eclipta alba*. *Ethnobotanical Leaflets*, *2010*(3), p. 3.

Manvar, D., Mishra, M., Kumar, S. and Pandey, V.N., 2012. Identification and evaluation of anti hepatitis C virus phytochemicals from *Eclipta alba*. *Journal of Ethnopharmacology*, *144*(3), pp. 545–554.

Svrlanska, A., Ruhland, A., Marschall, M., Reuter, N. and Stamminger, T., 2020. Wedelolactone inhibits human cytomegalovirus replication by targeting distinct steps of the viral replication cycle. *Antiviral Research*, *174*, p. 104677.

Tewtrakul, S., Subhadhirasakul, S., Cheenpracha, S. and Karalai, C., 2007. HIV-1 protease and HIV-1 integrase inhibitory substances from *Eclipta prostrata*. *Phytotherapy Research: An International Journal Devoted to Pharmacological and Toxicological Evaluation of Natural Product Derivatives*, *21*(11), pp. 1092–1095.

11.1.1.4 *Emilia sonchifolia* (L.) DC.

Synonyms: *Cacalia sonchifolia* L.; *Crassocephalum sonchifolium* (L.) Less.; *Senecio sonchifolius* Moench, *Emilia javanica* (Burm. f.) C. B. Rob

Common names: Cupiol's shaving brush, redtassel flower; dihong pei ts'ao (China); djombang (Indonesia); usubeni nigana (Japan); satunbak merah (Malaysia); chua le, la mat gioi (Vietnam); fua lele, fua lele lili'i (Samoa)

Habitat: Roadsides, abandoned lants, lawns, parkings

Distribution: Tropical Asia and the Pacific

Botanical description: This adorable yet poisonous herb grows to a height of 30 cm. The stems are slender and somewhat glaucous. The leaves are simple, somehow fleshy, and without stipules. The basal leaves are pinnatifid and 5–10 cm long. The cauline leaves are hastate or linear, lobed or pinnate, sessile, amplexicaul, auricular, and somewhat glaucous. The inflorescences are solitary, shaving brush-shaped capitula on top of 3 cm long and thin pedicels. The involucres are cylindrical, glaucous, with asymmetrical, linear oblong lobes. The flowers are tubular and whitish to pink. The stamens and styles are pink-tipped crimson. The achene are hairy, brownish, and with tiny series of white prickles between the ribs.

Medicinal uses: Dysentery (China); diarrhea, ear infection, eye infection, fever (Bangladesh)

Commentaries

i. The plant produces a series of toxic pyrrolizidine alkaloids, such as senkirkine and doronine (Cheng et al. 1986). Essential oil in this plant accumulates β-caryophyllene and γ-muurolene (Joshi, 2018). Other constituents are caffeoylquinic acids (Shen et al., 2014). Note that pyrrolozidine alkaloids are antimicrobial.

ii. In the Subfamily Asteroideae; Tribe Senecioneae, *Senecio scandens* Buch.-Ham. ex D. Don is used to heal boils, furuncle, and eye infection in China (local name: *Qian li guang*). Ethanol extract of this plant inhibited the growth of methicillin-sensitive *Staphylococcus aureus* and methicillin-resistant *Staphylococcus aureus* with the MIC values of 1690 and 1820 µg/mL, respectively (Zuo et al., 2008).

iii. In the Tribe Senecioneae, *Gynura pseudochina* (L.) DC. is used for fever and herpes in Thailand (local name: *Kham khok*), sore throat in Vietnam (local name: *Tam thast*), and pimples in Indonesia (local name: *Daun dewa*). *Gynura procumbens* (Lour.) Merr. is given for fever in Cambodia and dysentery in Malaysia. Ethanol extract of *Gynura procumbens* (Lour.) Merr. inhibited the replication of the Herpes simplex virus type-1 (strain KOS) and Herpes virus type-2 (strain Baylor 186) with the IC_{50} values of 584 and 568 µg/mL, respectively, in Vero cells and afforded some levels of protection against herpes labialis when used externally in the form of a 2% gel (Jarikasem et al., 2013).

References

Al-Jaber, H.I., Mosleh, I.M., Mallouh, A., Abu Salim, O.M. and Abu Zarga, M.H., 2010. Chemical constituents of *Osyris alba* and their antiparasitic activities. *Journal of Asian Natural Products Research*, *12*(9), pp. 814–820.

Cheng, D.L. and Roeder, E. 1986. Pyrrolizidine Alkaloids from *Emilia sonchifolia*. *Planta Medica 52*, pp. 484–486.

Jarikasem, S., Charuwichitratana, S., Siritantikorn, S., Chantratita, W., Iskander, M., Frahm, A.W. and Jiratchariyakul, W., 2013. Antiherpetic effects of *Gynura procumbens*. *Evidence-Based Complementary and Alternative Medicine*, *2013*, 394865.

Joshi, R.K., 2018. Volatile constituents of *Emilia sonchifolia* from India. *Natural Product Communications*, *13*(10), p.1934578X1801301030.

Shen, S., Shen, L., Zhang, J., Li, G., Zhang, Y., Lu, H. and Si, J., 2014. Study on HPLC fingerprints of *Emilia sonchifolia* and content determination of caffeoylquinic acids. *China Journal of Traditional Chinese Medicine and Pharmacy*, *29*(12), pp. 3790–3793.

Zuo, G.Y., Wang, G.C., Zhao, Y.B., Xu, G.L., Hao, X.Y., Han, J. and Zhao, Q., 2008. Screening of Chinese medicinal plants for inhibition against clinical isolates of methicillin-resistant *Staphylococcus aureus* (MRSA). *Journal of Ethnopharmacology*, *120*(2), pp. 287–290.

11.1.1.5 *Gnaphalium luteoalbum* L.

Synonyms: *Gnaphalium affine* D. Don; *Gnaphalium helichrysoides* Wedd.; *Gnaphalium nanum* Willd.; *Gnaphalium pallidum* Lam.; *Pseudognaphalium luteoalbum* (L.) Hilliard & B.L. Burtt

Common names: Jersey cudweed; jabra (Bangladesh); si mian cao (China); bal-raksha (India); byaingchepui (Myanmar); buke phul; kairo jhar (Nepal); badok (the Philippines)

Habitat: Sun exposed and rocky or sandy areas, roadsides

Distribution: Tropical Asia and Pacific

Botanical observation: This beautiful glaucous and hairy herb grows to a height of 45 cm. The leaves are simple, sessile, and spiral. The blade is spathulate, oblong, obtuse, 2.5–6.3 cm × 3–1.3 cm, woolly, and fleshy. The capitula are orange or white and arranged in terminal corymbs which are about 2 cm across. The involucral bracts are straw-colored. The achenes are oblong, papillate, with tufted pappuses.

Medicinal uses: Diarrhea (Nepal); fever (India)

Bacterial virulence inhibition by flavone: Calycopterin inhibited the production of pyocyanin by *Pseudomonas aeruginosa* (ATCC 27853) with the EC_{50} value of 32 µM and at 140 µM inhibited bacterium motility by 38.4% (Froes et al., 2020).

Broad-spectrum antifungal polar extract: Acetone extract of leaves inhibited the growth of *Aspergillus niger*, *Fusarium oxysporum*, *Phytophthora nicotiana*, and *Pythium ultimum* with the MIC values of 630, 160, 20, and 630 µg/mL, respectively (Aderogba et al., 2014)

Weak antiviral (non-enveloped monopartite linear single-stranded (+) RNA) flavone: Calycopterin (5,4-hydroxy-3,6,7,8-tetramethoxyflavone) inhibited Poliovirus with the EC_{50} of 58 μM and a selectivity index of 34 (Mayer, 1999).

Commentary: In the Subfamily Asteroideae; Tribe Gnaphalieae, *Helichrysum pallasii* Ledeb. is given in Turkey for cough (local name: *Altunbas otu*) and *Helichrysum plicatum* DC (local name: *Herdemtaze*) is a remedy for cough, diarrhea, and urinary tract infections, *Helichrysum plicatum* DC. subsp. *polyphyllum* P. H. Davis & Kupicha (local name: *Kaymak çiçegi*) is used for diarrhea. The flavone galangin isolated from a member of the genus *Helichrysum* Mill. inhibited the replication of the Herpes simplex virus type 1 and Coxsackie B virus type 1 with IC_{50} values ranging from 12 to 47 μg/mL (Meyer et al., 1997). 2 4′,6′-Trihydroxy-3′-prenylchalcone and 4′,6′,5″-trihydroxy-6″,6″-dimethyldihydropyrano (2″,3″-2′,3′) chalcone isolated from a member of the genus *Helichrysum* Mill. inhibited the replication of the Influenza A virus (strain Panama) with the IC_{50} value of 100 μg/mL and *Mycobacterium tuberculosis* (H37Rv) with the MIC value of 50 μg/mL (Meyer et al., 1997).

References

Aderogba, M.A., McGaw, L.J., Bagla, V.P., Eloff, J.N. and Abegaz, B.M., 2014. In vitro antifungal activity of the acetone extract and two isolated compounds from the weed, *Pseudognaphalium luteoalbum*. *South African Journal of Botany*, *94*, pp. 74–78.

Froes, T.Q., Nicastro, G.G., de Oliveira Pereira, T., de Oliveira Carneiro, K., Reis, I.M.A., Conceição, R.S., Branco, A., Ifa, D.R., Baldini, R.L. and Castilho, M.S., 2020. Calycopterin, a major flavonoid from *Marcetia latifolia*, modulates virulence-related traits in *Pseudomonas aeruginosa*. *Microbial Pathogenesis*, *144*, p. 104142.

Mayer, R., 1999. Calycopterones and Calyflorenones, Novel Biflavonoids from *Calycopteris f loribunda*. *Journal of Natural Products*, *62*(9), pp. 1274–1278.

Meyer, J.J.M., Afolayan, A.J., Taylor, M.B. and Erasmus, D., 1997. Antiviral activity of galangin isolated from the aerial parts of *Helichrysum aureonitens*. *Journal of Ethnopharmacology*, *56*(2), pp. 165–169.

11.1.1.6 *Pluchea indica* (L.) Less.

Synonyms: *Baccharis indica* L.; *Erigeron denticulatum* Burm. f.

Habitat: Seashores, mangroves, swamps, sea muds, vacant lands, roadsides, cultivated

Common names: Indian fleabane, Indian pluchea, Indian camphorweed; pras anbok (Cambodia); kuo bao ju (China); ban japi phul (Bangladesh); kukarondh (India); luntas (Indonesia); beluntas (Indonesia; Malaysia); kayu (Myanmar); khlu (Thailand); cuc tan, cay luc, tu bi (Vietnam); kalapini (the Philippines); khlue tale, khlu (Thailand)

Distribution: Tropical Asia and the Pacific

Botanical observation: This shrub grows to a height of 1.5 m. The stems are terete, woody, green or purplish, and glabrous. The leaves are simple, alternate, and exstipulate. The blade is up to about 3.5 cm × 2 cm, dentate, membranous, obovate, edible (Indonesia, Malaysia), acuminate and narrow at the base, marked with about three pairs of secondary nerves, and release a subtle fragrance when crushed. The capitula are up to about 7 mm across, and arranged in terminal corymbose cymes which are about 5 cm long. The corolla is about 3 mm long, tubular, 5-lobed, and white or light pinkish. The achenes are minute, cylindrical, ribbed, and develop 2 mm long and white pappuses.

Medicinal uses: Dysentery (Bangladesh; Hawaii; Indonesia; Vietnam); fever (Hawaii; India; Thailand); leucorrhoea (Indonesia)

Strong antiviral (enveloped linear monopartite dimeric single-stranded (+) RNA) polar extract: Aqueous extract of leaves inhibited the replication of the Human immunodeficiency virus type-1 (strain IIIb) with the IC_{50} value of 1.7 μg/mL and a selectivity index of 94 in MT-4 cells (Locher et al., 1996).

Commentaries

i. What is(are) the anti-Human immunodeficiency virus principle in this shrub?
ii. The anti-COVID-19 potentials of this plant could be examined.
iii. In the Subfamily Asteroideae; Tribe Inuleae, *Blumea balsamifera* (L.) DC. is given for cholera in China (local name: *Ai na xiang*). Ethyl acetate extract of leaves of *Blumea balsamifera* (L.) DC. inhibited the growth of *Streptococcus mutans* (DMST 26095), *Streptococcus mitis* (ATCC 49456T), and *Streptococcus pyogenes* (DNAT 17020) with MIC/MBC values of 312.5/625, 78.1/312.5, and 156/312.5 µg/mL, respectively (Meerungrueang and Panichayupakaranant, 2014). Aqueous extract of leaves of *Blumea balsamifera* (L.) DC. inhibited the replication of the Zika virus (strain MR766) with the EC_{50} value of 261.6 µg/mL in Vero cells (CC_{50} = 444 µg/mL) (Vista et al., 2020). Note that *Blumea balsamifera* (L.) DC. (local name: *Sembung*) is used in Sulawesi against COVID-19 in a phytomedication also including *Zingiber officinale* Roxb. (local name: *Jahe merah*), *Phyllanthus niruri* L. (local name: *Meniran*), and *Andrographis paniculata* (Burm.f.) Wall. (local name: *Sambitolo*). In Thailand the plant is used for COVID-19 (Phumthum et al., 2021). The plant yields sesquiterpenes with strong activity against the influenza A virus (H3N2) *in vitro* (Xiong et al., 2020), and these could be active against the Severe acute respiratory syndrome-associated coronavirus. In the genus *Blumea* DC. methanol extract of *Blumea chinensis* (L.) DC. inhibited the replication of the Herpes simplex virus type-1 with the MIC value of 5 µg/mL in HeLa (Hamidi et al., 1996). *Blumea lacera* (Burm.f.) DC. is used to treat bronchitis in the Philippines, and *Blumea lanceolaria* (Roxb.) Druce is a remedy for influenza in Taiwan. From *Blumea lacera* (Burm.f.) DC., α-pinene-7β-*O*-β-D-2,6-diacetylglucopyranoside, 5,4′-dihydroxy-6,7,3′-trimethoxyflavone, and 3,5,4′-trihydroxy-6,7,3′-trimethoxyflavone inhibited the growth of *Candida albicans* (Ragasa et al., 2007). From *Blumea obliqua* (L.) Druce the thiophene polyacetylenes 5′-methyl-5-[4-(3-methyl-l-oxobutoxy)-l-butynyl]-2,2′-bithiophene and 5′-hydroxymethyl-5-[butyl-3-en-1-yn]-2,2′-bithiophene isovaleroxy ester inhibited the growth of *Epidermophyton floccosum* and *Pleurotus ostreatus* with the MIC values of 200 µg/mL (Ahmad & Alam, 1995). Observe that Asteraceae produce thiophene polyacetylenes phytoanticipins. In the Subfamily Asteroideae; Tribe Inuleae, *Inula japonica* Thunb. is used to heal abscesses in China (local name: *Xuan fu hua*) and yields the flavone axillarin (Qin et al., 2010) (LogD 0.6 at pH 7.4; molecular mass = 346.2 g/mol) that inhibited the replication of the Rhinovirus type-2 in HeLa cells with the EC_{50} value of 0.6 µg/mL (Tsuchiya et al., 1985) and inhibited the replication of the Herpes simplex virus type-2 with the EC_{50} value of 42.7 µg/mL (Álvarez et al., 2011). *Inula helenium* L. is a remedy for cough in Turkey (local name: *Anduz otu*) and inhibited the replication of the Mouse hepatitis virus *in vitro* (신현수, 2007). The magnificent herb *Inula oculus-christi* L. is used in Turkey (local name: *Yol otu*) for the treatment of urinary tract infection. In the Subfamily Asteroideae, aqueous extract of *Laggera pterodonta* (DC.) Sch. Bip. ex Oliv. inhibited the replication of the Herpes simplex virus type-1 with the IC_{50} value of 250 µg/mL and a selectivity index of 3 and the replication of the Respiratory syncytial virus with an IC_{50} value between 12.5 and 32 µg/mL (Li et al., 2004).

References

신현수, 2007. Coronavirus replication inhibition by herbal extracts (Doctoral dissertation).

Ahmad, V.U. and Alam, N., 1995. New antifungal bithienylacetylenes from *Blumea obliqua*. *Journal of Natural Products*, *58*(9), pp. 1426–1429.

Álvarez, Á.L., Habtemariam, S., Juan-Badaturuge, M., Jackson, C. and Parra, F., 2011. In vitro anti HSV-1 and HSV-2 activity of *Tanacetum vulgare* extracts and isolated compounds: An approach to their mechanisms of action. *Phytotherapy Research*, *25*(2), pp. 296–301.

Biswas, R., Dutta, P.K., Achari, B., Bandyopadhyay, D., Mishra, M., Pramanik, K.C. and Chatterjee, T.K., 2007. Isolation of pure compound R/J/3 from *Pluchea indica* (L.) Less. and its anti-amoebic activities against *Entamoeba histolytica. Phytomedicine*, *14*(7–8), pp. 534–537.

Giang, P.M., Sim, H.T. and Chinh, T.T.T., 2019. Triterpenoids and phytosterols isolated from *Pluchea indica* L. leaves. *VNU Journal of Science: Natural Sciences and Technology*, *35*(2).

Hamidi, J.A., Ismaili, N.H., Ahmadi, F.B. and Lajisi, N.H., 1996. Antiviral and cytotoxic activities of some plants used in *Malaysian indigenous* medicine. *Pertanika Journal of Tropical Agricultural Science*, *19*(2/3), pp. 129–136.

Le, T.B., Beaufay, C., Nghiem, D.T., Pham, T.A., Mingeot-Leclercq, M.P. and Quetin-Leclercq, J., 2019. Evaluation of the anti-trypanosomal activity of vietnamese essential oils, with emphasis on *Curcuma longa* L. and its components. *Molecules*, *24*(6), p. 1158.

Li, Y., Ooi, L.S., Wang, H., But, P.P. and Ooi, V.E., 2004. Antiviral activities of medicinal herbs traditionally used in southern mainland China. *Phytotherapy Research*, *18*(9), pp. 718–722.

Locher, C.P., Witvrouw, M., De Béthune, M.P., Burch, M.T., Mower, H.F., Davis, H., Lasure, A., Pauwels, R., De Clercq, E. and Vlietinck, A.J., 1996. Antiviral activity of Hawaiian medicinal plants against human immunodeficiency virus type-1 (HIV-1). *Phytomedicine*, *2*(3), pp. 259–264.

Meerungrueang, W. and Panichayupakaranant, P., 2014. Antimicrobial activities of some Thai traditional medical longevity formulations from plants and antibacterial compounds from *Ficus foveolata*. *Pharmaceutical Biology*, *52*(9), pp. 1104–1109.

Mohanty, S., Srivastava, P., Maurya, A.K., Cheema, H.S., Shanker, K., Dhawan, S., Darokar, M.P. and Bawankule, D.U., 2013. Antimalarial and safety evaluation of *Pluchea lanceolata* (DC.) Oliv. & Hiern: in-vitro and in-vivo study. *Journal of Ethnopharmacology*, *149*(3), pp. 797–802.

Phumthum, M., Nguanchoo, V. and Balslev, H., 2021. Medicinal plants used for treating mild COVID-19 symptoms among Thai Karen and Hmong. *Frontiers in Pharmacology*, p. 1844.

Qin, J.J., Zhu, J.X., Zhu, Y., Jin, H., Lv, Y. and Zhang, W.D., 2010. Flavonoids from the aerial parts of *Inula japonica. Zhongguo Tianran Yaowu*, *8*, pp. 257–259.

Ragasa, C.Y., Wong, J. and Rideout, J.A., 2007. Monoterpene glycoside and flavonoids from *Blumea lacera*. *Journal of Natural Medicines*, *61*(4), pp. 474–475.

Tsuchiya, Y., Shimizu, M., Hiyama, Y., Itoh, K., Hashimoto, Y., Nakayama, M., Horie, T. and Morita, N., 1985. Antiviral activity of natural occurring flavonoids in vitro. *Chemical and Pharmaceutical Bulletin*, *33*(9), pp. 3881–3886.

Vista, F.E.S., Dalmacio, L.M.M., Corales, L.G.M., Salem, G.M., Galula, J.U. and Chao, D.Y., 2020. Antiviral effect of crude aqueous extracts from ten Philippine medicinal plants against Zika virus. *Acta Medica Philippina*, *54*(2).

Wardah, W. and Kuncari, E.S., 2020. Kajian Etnobotani Pakundalang (Blumea balsamifera (L.) DC.) sebagai Solusi Alternatif untuk Kemandirian Kesehatan Masyarakat Banggai Kepulauan, Sulawesi Tengah. *Journal of Tropical Ethnobiology*, *3*(2), pp. 139–148.

Xiong, Y., Yi, P., Li, Y., Gao, R., Chen, J., Hu, Z., Lou, H., Du, C., Zhang, J., Zhang, Y. and Yuan, C., 2020. New sesquiterpeniod esters form *Blumea balsamifera* (L.) DC. and their anti-influenza virus activity. *Natural Product Research*, pp. 1–10.

11.1.1.7 *Spilanthes paniculata* Wall. ex DC

Synonyms: *Acmella paniculata* (Wall. ex DC.) R.K. Jansen

Common names: Para cress, toothache plant; qihui (China); daun getang (Indonesia); kaad (Laos); pokok getang kerbau, kerabu, galang, gutang (Malaysia); heukala (Myanmar); biri (the Philippines); phak khrat (Thailand); cuc ao, ngo ao (Vietnam)

Habitat: Waste lands, roadsides, cultivated

Distribution: Tropical Asia

Botanical description: This herb grows up to 30 cm tall and has a positive aura. The stems are hairy, fleshy, and purplish. The leaves are simple, opposite, and exstipulate. The petiole is 0.2–1 cm long. The blade is deltoid to lanceolate, 1–2.5 × 1.5–4 cm long, laxly serrate, and marked with 1 to 3 pairs of secondary nerves. The inflorescences consist of conical and showy reddish-yellow capitula on 2.5–16 cm long pedicels. The corolla is tubular and 4- to 5-lobed. The achenes are trigonal and present a pappus.

Medicinal uses: Dysentery (Cambodia, Laos and Vietnam); ear infection (India)

Broad-spectrum antibacterial polar extract: Methanol extract (500 µg/disc) inhibited the growth of *Bacillus cereus*, *Bacillus megaterium*, *Bacillus subtilis*, *Sarcinia lutea*, *Staphylococcus aureus*, *Salmonella paratyphi*, *Salmonella typhi*, *Vibrio parahaemolyticus*, *Vibrio mimicus*, *Escherichia coli*, *Shigella dysenteriae*, *Pseudomonas aeruginosa*, and *Shigella boydii* with the inhibition zone diameter of 15, 12, 15, 11, 11, 14, 11, 12, 14, 17, 11, 15, and 15 mm, respectively (Morshed et al., 2011).

Antifungal (yeast) mid-polar extract: Chloroform extract (1 mg/6 mm disc) inhibited the growth of *Cryptococcus neoformans* with the inhibition zone diameter of 7.5 mm (Phongpaichit et al., 20).

Moderate antifungal (yeast) mid-polar extract: Chloroform extract inhibited the growth of *Cryptococcus neoformans* with the MIC value of 128 µg/mL. *Saccharomyces cerevisiae* (ATCC 2601) with the MIC value of 256 µg/mL (Phongpaichit et al., 2005).

Weak antifungal (yeast) alkylamine: Spilanthol inhibited the growth of *Candida albicans* (ATCC 10231) (Fabri et al., 2021).

Commentaries

i. Spilanthol is neuroactive (Barbosa et al., 2016) and therefore probably able to inhibit bacterial or fungal efflux pumps.
ii. *Spilanthes acmella* (L.) Murray (local name: *Akor-korh*) is a remedy for urinary tract infection.

References

Barbosa, A.F., Carvalho, M.G.D., Smith, R.E. and Sabaa-Srur, A.U., 2016. Spilanthol: Occurrence, extraction, chemistry and biological activities. *Revista Brasileira de Farmacognosia*, *26*, pp. 128–133.

Fabri, R.L., Freitas, J.C., Lemos, A.S., Campos, L.M., Diniz, I.O., Pinto, N.C., Silva, T.P., Palazzi, C., Marchesini, P., Monteiro, C. and Barbosa, A.F., 2021. Spilanthol as a promising antifungal alkylamide for the treatment of vulvovaginal candidiasis. *Medical Mycology*.

Morshed, M.A., Uddin, A., Saifur, R., Barua, A. and Haque, A., 2011. Evaluation of antimicrobial and cytotoxic properties of *Leucas aspera* and *Spilanthes paniculata*. *International Journal of Biosciences*, *1*(2), pp. 7–16.

Phongpaichit, S., Subhadhirasakul, S. and Wattanapiromsakul, C., 2005. Antifungal activities of extracts from Thai medicinal plants against opportunistic fungal pathogens associated with AIDS patients. *Mycoses*, *48*(5), pp. 333–338.

11.1.1.8 *Vernonia cinerea* (L.) Less.

Synonyms: *Blumea chinensis* (L.) DC.; *Blumea esquirolii* H. Lév. & Vaniot; *Cacalia cinerea* (L.) Kuntze; *Conyza chinensis* L.; *Cyanthillium cinereum* (L.) H. Rob.; *Vernonia abbreviata* DC.; *Vernonia exilis* Miq.; *Vernonia kroneana* Miq.; *Vernonia parviflora* Reinw. ex Blume

Common names: Little iron weed; kuksim (Bangladesh); kbal ruy (Cambodia); devida, sahadevi, naichotte poonde, sahdevi uttamkanyaka (India); lidah anjing (Indonesia); yambaru higotai (Japan); nya phaen din yen (Laos); cenderong hari, concong hari, bujong semalam, ekor kudah, rumput sabasi, rumput susor, tahi babi (Malaysia); byaing-chay-pin, kadu-pyan (Myanmar); agas moro (the Philippines); ya khrao maeo. yaa saam wan (Thailand); bach dau ong (Vietnam)

Habitat: Waste lands, roadsides, meadows, lawn, parkings

Distribution: Tropical Asia

Botanical description: This slender and erect herb grows to about 30 cm tall. The leaves are simple, few, alternate, and exstipulate. The petiole is 2.5 mm–1.2 cm long. The blade is 3–7.5 cm × 0.6–1.2 cm, membranous, hairy, toothed, and ovate or lanceolate. The capitula are tubular, about 5 mm long, and arranged in lax and terminal corymbs. The corolla is tubular, of a unique type of magnificent purple, and develops 5 narrow lobes. The achenes are ribbed, hairy, and develop white and tufted pappuses.

Medicinal uses: Dysentery (Vietnam); eye infection (India), fever (Cambodia, India); wounds (Thailand; the Philippines)

Broad-spectrum antibacterial halo developed by non-polar extract: Benzene extract inhibited a broad spectrum of bacteria (Gupta et al., 2003).

Anticandidal halo developed by polar extract: Methanol extract (1 mg/9 mm well) inhibited the growth of *Candida albicans*, *Candida parapsilosis*, and *Candida tropicalis* with the inhibition zone diameters of 14, 18, and 12 mm, respectively (Dhanalakshmi, et al., 2013).

Commentaries

i. Antimicrobial principles in this herb have apparently not been isolated. It is known to yield a series of antiplasmodial sesquiterpene lactones with epoxide moieties (Chea et al., 2004) which are most probably broadly antimicrobial *in vitro*. Note that antiplasmodial sesquiterpene with peroxide or epoxide moieties are often antiviral and antimycobacterial (see earlier).

ii. *Vernonia amygdalina* Delile. is commonly grown in Malaysia for medicinal purposes, and from this plant the sesquiterpene lactone vernolide inhibited the growth of *Penicillium notatum*, *Aspergillus flavus*, *Aspergillus niger*, and *Mucor hiemalis* (Erasto et al., 2006). From this plant the sesquiterpene lactone ernodalol inhibited the growth of *Penicillium notatum*, *Aspergillus flavus*, *Aspergillus niger*, and *Mucor hiemalis* (Erasto et al., 2006). Vernolide inhibited the growth of *Bacillus cereus*, *Staphylococcus epidermidis*, *Staphylococcus aureus*, *Micrococcus kristinae*, *Streptococcus pyogenes*, and *Salmonella poona* with the MIC values of 500, 500, 500, 250, 500, and 500 µg/mL, respectively . Vernodalol inhibited the growth of *Bacillus cereus*, *Staphylococcus epidermidis*, *Micrococcus kristinae*, *Streptococcus pyrogens*, and *Salmonella poona* with the MIC values of 500, 500, 250, 500, and 500 µg/mL, respectively, (Erasto et al., 2006).

iii. In Bangladesh, *Vernonia patula* (Aiton) Merr. (local name: *Dando upp*) is used for tonsillitis. In India, *Vernonia albicans* DC. (local name: *Siru sanga chedi*) is a remedy for burns.

iv. In the Subfamily Carduoideae; Tribe Cynareae, *Saussurea costus* (Falc.) Lipsch (synonym: *Aucklandia costus* Falc.; *Aucklandia lappa* Decne; *Saussurea lappa* (Decne.) Sch. Bip.) is used in China (local names: *Niu bang*; *Yun mu xiang*) for the treatment of COVID-19 (Luo, 2020; Wang et al., 2020). The plant yields the sesquiterpene lactones costunolide and dehydrocostus lactone which inhibited the secretion of the Hepatitis B virus surface antigen on Hep3B cells with the IC_{50} values of 1 and 2 µM, respectively (Chen et al., 1995). This plant is different from *Arctium lappa* L. or the great burdock common in European meadows and of repute in European herbalism. *Arctium lappa* L. was recommended for syphilis by our peer the Italian physician Girolamo Fracastoro (1476–1553) (Riddell, W.R., 1927) while our peer the German physician Leonardt Fusch (1501–1566) advised its use for wounds "*Arctio ictibus auxiliatur*" (quoting Dioscoride) in his "*De Historia Stirpium Commentarii Insignes*" published around 1542. In China, *Arctium lappa* L. (local name: *Niu bang*) has been used from a remote period of time for leprosy (Van Wettum et al., 1901) and is also used for COVID-19 (Luo et al., 2020). *Arctium lappa* L. produces the dibenzylbutyrolactone lignan arctigenin which at the concentration of 313 µg/mL inhibited the replication of the Porcine circovirus type-2 (WH strain) in PK-15 cells and administered in mice at the dose of 200 µg/Kg afforded some levels of protection (Chen et al., 2016). Arctigenin given intraperitoneally at the dose of 10 mg/Kg twice a day for 7 days protected mice against the Japanese encephalitis virus (Swarup et al., 2008). The Japanese encephalitis virus is an enveloped monopartite linear single-stranded (+) RNA virus which give some rationale of *Arctium lappa* L. against COVID-19. *Arctium lappa* L. is anti-inflammatory suggesting that it may be able to mitigate the inflammatory burst in COVID-19. Again, it must be grasped that zoonotic pandemics could be better managed if besides vaccination, phytomedications prepared under strict pharmaceutical control or certified herbalists control could be available. The private sector does not like medicinal plants because, at least in part, plants cannot be patented.

Thus, it is to the public sector (public universities and public research centers) to handle this beneficial task.

v. In the Subfamily Carduoideae; Tribe Cynareae, members of the genus *Atractylodes* DC. are used for the treatment of fever and AIDS in China. *Atractylodes macrocephala* Koidz. is a remedy for COVID-19 in China (local name: *Bai shu*) (Wang et al., 2020; Yang et al., 2020) as well as *Atractylodes japonica* Koidz. (synonyms: *Atractylodes lancea* (Thunb.) DC.; *Atractylodes* atractylon *ovata* Thunb.) (local name: *Cang shu*) (Wang et al., 2020). The furanosesquiterpene atractylon (LogD 4.9 at pH 7.4; molecular mass=216.1 g/mol) inhibited the replication of the Influenza A/PR/8/34 virus (H1N1 subtype) and Influenza A/shenzhen/203/2001 (H3N2 subtype) in Madin-Darby canine kidney cells with the IC_{50} values of 8.9 and 9.4 µg/mL, respectively, and the selectivity indexes of 29.4 and 27.8, respectively (Chen et al., 2016). Given orally to mice at a dose of 40 mg/Kg for 5 days atractylon mitigated lung insults and serum of pro-inflammatory cytokines (Chen et al., 2016). Note the presence of an α-methylene-γ-lactone moiety in atractylon (see earlier). Extract of *Atractylodes japonica* Koidz. at the concentration of 100 µg/mL inhibited ribonuclease H activity of Human immunodeficiency virus type-1 reverse transcriptase and Human immunodeficiency virus-1 protease by 20.2% and 43%, respectively (Min et al., 2001) and at the concentration of 250 µg/mL inhibited Human immunodeficiency virus type-1 protease by 56.1% (Xu et al., 1996). (Park, 2003).

vi. In the Subfamily Carduoideae; Tribe Cynareae, the gorgeous *Carthamus tinctorius* L. (also known as safflower) is given for jaundice in Myanmar (local name: *Hsu pan*), fever in India (local name: *Kusumbha*), and COVID-19 and pneumonia in China (local name: *Cao hong hua*) (Wang et al., 2020). The flowers of *Carthamus tinctorius* L. are used as a source of a red dye on account of series of quinochalcone glycosides such as carthamin (Khazuma et al., 2000) which could be examined for their antimicrobial properties. Extract of this plant inhibited the replication of the Kaposi's sarcoma herpesvirus (Lee et al., 2013). Ethanol extract of the concentration of inhibited the replication of the Foot and mouth disease virus type O (local strain KPS/005/2545) at the concentration of 12 µg/mL (Chungsamarnyart et al., 2007). *Carthamus tinctorius* L. when experimentally infected by *Phytophthora drechsleri* yields the polyacetylene *trans-trans*-3,11-tridecadiene-5,7,9-triyne-1,2-diol which inhibits the growth of *Phytophthora drechsleri* and as such is a phytoalexin (Allen & Thomas, 1971). The ability of plants to generate antimicrobial phytoalexin upon challenge by a given phytopathogenic microorganism is mind-blowing and suggests the presence of communication mechanisms between plants and microorganism.

In the Tribe Cynareae, the magnificent *Cirsium japonicum* DC. is used to heal wounds in China (local name: *Ji*) and from this plant, the flavone glycoside pectolinarin inhibited the replication of the Dengue virus type-2 in LLCMK2 cells with the EC_{50} value of 86.4 µg/mL and a selectivity index of 4.6 (Simões et al., 2011). The flavone glycoside pectolinarin (Ishida et al., 1987) inhibited the Severe acute respiratory syndrome-associated coronavirus 3-chymotrypsin-like protease with the IC_{50} value of 51.6 µM (Jo et al., 2020). Methanol extract of aerial parts of *Cirsium japonica* var. *ussuriense* (Regel) Kitam. at 100 µg/mL inhibited the Human immunodeficiency-protease by 26% (Park, 2003). Hexane extracts of *Cirsium hypoleucum* DC. inhibited the growth of the Herpes simplex virus type-1 with the minimum cytopathogenic effect concentration of 100 µg/mL (Orhan et al., 2009).

In the Tribe Cynareae, *Cynara scolymus* L. (also known as artichoke) is used for jaundice in Turkey (local name: *Kangar firangi*). Hexane extracts of *Cynara scolymus* L. with the minimum cytopathogenic effect concentration of 100 µg/mL displayed antiviral effects (Orhan et al., 2009). From this stout herb, the sesquiterpene lactones cynaropicrin (LogD 1.6 at pH 7.4; molecular mass=346.3 g/mol) and grosheimol inhibited the replication of the Hepatitis C virus with the EC_{50} values of 1 and 1.2 µM, respectively (Elsebai et al., 2017) whence possible activity against the Severe acute respiratory syndrome-associated coronavirus 2 and other pandemic coronaviruses. This sesquiterpene lactone inhibited *Pseudomonas aeruginosa* MurA with the IC_{50} value 12.1 μM (Bachelier et al., 2006).

vii. In the Subfamily Asteroideae, Tribe Astereae, the magnificent *Aster tataricus* L. is a remedy for COVID-19 in China (local name: *Zi wan*) (Lem et al., 2020; Yang et al., 2020). From this plant, the shionane-type triterpene shion-22-methoxy-20(21)-en-3-one and shion-22(30)-en3,21-dione inhibited the secretion of the Hepatitis B virus surface antigen with the IC_{50} value of 0.8 g/mL, and shion-22-methoxy-20(21)-en-3-one inhibited the secretion of the Hepatitis B virus e-antigen with the IC_{50} value of 0.8 μg/mL (Zhou et). Aqueous extract of roots at the concentration of 250 μg/mL inhibited the Human immunodeficiency virus type-1 protease by 23.7% (Xu et al., 1996). Ethanol extract of aerial parts of *Aster triplolium* L. at 100 μg/mL inhibited the Human immunodeficiency virus type-1 protease by 34% (Park et al., 2002). From this plant, the abietane diterpene alkaloid dehydroabietylamine (LogD 3.3 at pH 7.4; molecular mass=285.4 g/mol) inhibited the growth of *Staphylococcus aureus* and *Pseudomonas aeruginosa* with the MIC values of 12.5 and 6.2 μg/mL, respectively (Ar & Ck, 2020). In the Subfamily Asteroideae; Tribe Astereae, ethanol extract of flowers of *Erigeron annuus* (L.) Pers at 100 μg/mL inhibited the Human immunodeficiency virus type-1 protease by 35% (Park et al., 2002). *Erigeron breviscapus* (Vaniot) Hand.-Mazz. produces the flavone glycoside scutellarin (also known as scutellarein-7-*O*-β-glucuronide or breviscapin) (LogD −4.8 at pH 7.4; molecular mass=462.3 g/mol) which inactivated the Porcine circovirus type 2 with the IC_{50} value of 1 μg/mL and the selectivity index of 18.6 (Sun et al., 2015).

viii. In the Subfamily Cichorioideae; Tribe Vernonieae, *Elephantopus scaber* L. is used for cuts, dysentery, and wounds in India (local names: *Bantamakhu*, *Jangli gobhi*), boils in Bangladesh (local name: *Mejurjhuti*), fever in Myanmar (local name: *Ma-tu-pin*) and Indonesia (local name: *Kateba*), and to treat blennorrhea in Cambodia, Laos, and Vietnam. Aqueous extract of leaves at the concentration of 250 μg/mL inhibited the Human immunodeficiency virus type-1 protease by 48% (Xu et al., 1996) and the Human immunodeficiency virus type-1 reverse transcriptase by 96.9% at the concentration of 200 μg/mL (Wiwat & Kwantrairat, 2014). From this herb, 1α, 2β-*O*-dicaffeoylcyclopentan-3β-ol inhibited the replication of the Respiratory syncytial virus (Geng et al., 2011). In Thailand the plant is used for COVID-19 (Phumthum et al., 2021). In Malayia, this plant is known as "*Tutup bumi*" and eaten by locals suggesting that it might be edible (this remain to be examined by toxicological means) and as such usable as antiviral phytomediaton. In the Subfamily Cichorioideae; Tribe Cichorieae, *Lactuca indica* L. is a remedy for measles in China (local name: *Chi guo ju*) and ethanol extract of this plant inhibited the replication of the Porcine epidemic diarrhea virus with the MIC value of 19.5 μg/mL (Trinh et al., 2021) and as such could be able to inhibit the replication of the Severe acute respiratory syndrome-associated coronavirus.

In the Subfamily Cichorioideae, Tribe Cichorieae, Aqueous extract of rhizome of *Taraxacum mongolicum* Hand.-Mazz. at the concentration of 250 μg/mL inhibited the Human immunodeficiency virus type-1 protease by 96.8% (Xu et al., 1996).

References

Allen, E.H. and Thomas, C.A., 1971. Trans-trans-3, 11-tridecadiene-5, 7, 9-triyne-1, 2-diol, an antifungal polyacetylene from diseased safflower (*Carthamus tinctorius*). *Phytochemistry*, *10*(7), pp. 1579–1582.

Bachelier, A., Mayer, R. and Klein, C.D., 2006. Sesquiterpene lactones are potent and irreversible inhibitors of the antibacterial target enzyme MurA. *Bioorganic & Medicinal Chemistry Letters*, *16*(21), pp. 5605–5609.

Chea, A., Hout, S., Long, C., Marcourt, L., Faure, R., Azas, N. and Elias, R., 2006. Antimalarial activity of sesquiterpene lactones from *Vernonia cinerea. Chemical and Pharmaceutical Bulletin*, *54*(10), pp. 1437–1439.

Chen, H.C., Chou, C.K., Lee, S.D., Wang, J.C. and Yeh, S.F., 1995. Active compounds from *Saussurea lappa* Clarks that suppress hepatitis B virus surface antigen gene expression in human hepatoma cells. *Antiviral Research*, *27*(1–2), pp. 99–109.

Chen, J., Li, W., Jin, E., He, Q., Yan, W., Yang, H., Gong, S., Guo, Y., Fu, S., Chen, X. and Ye, S., 2016. The antiviral activity of arctigenin in traditional Chinese medicine on porcine circovirus type 2. *Research in Veterinary Science*, *106*, pp. 159–164.

Cheng, Y., Mai, J.Y., Hou, T.L., Ping, J. and Chen, J.J., 2016. Antiviral activities of atractylon from Atractylodis Rhizoma. *Molecular Medicine Reports*, *14*(4), pp. 3704–3710.

Chungsamarnyart, N., Sirinarumitr, T., Chumsing, W. and Wajjawalku, W., 2007. In vitro study of antiviral activity of plant crude-extracts against the foot and mouth disease virus. *Agriculture and Natural Resources*, *41*(5), pp. 97–103.

Dhanalakshmi, P., Priya, A.J.P., Sagadevan, E., Lakshmi, Y.S., Manimaran, A., Sindhu, S. and Arumugam, P., 2013. Evaluation of inhibitory effect of *Vernonia cinerea* L. leaf extracts on different fungal species. *International Journal of Pharmacy and Pharmaceutical Sciences*, *5*(2), pp. 414–416.

Elsebai, M.F., Hakkola, J., Mehiri, M. and Diez, J., 2017. Cynaropicrin: A promising natural agent with antitumor and antiviral activities. *Multidisciplinary Digital Publishing Institute Proceedings*, *1*(10), p. 974.

Erasto, P., Grierson, D.S. and Afolayan, A.J., 2006. Bioactive sesquiterpene lactones from the leaves of *Vernonia amygdalina. Journal of Ethnopharmacology*, *106*(1), pp. 117–120.

Geng, H.W., Zhang, X.L., Wang, G.C., Yang, X.X., Wu, X., Wang, Y.F., Ye, W.C. and Li, Y.L., 2011. Antiviral dicaffeoyl derivatives from *Elephantopus scaber. Journal of Asian Natural Products Research*, *13*(7), pp. 665–669.

Gupta, M., Mazumder, U.K., Manikandan, L., Haldar, P.K., Bhattacharya, S. and Kandar, C.C., 2003. Antibacterial activity of *Vernonia cinerea. Fitoterapia*, *74*(1–2), pp. 148–150.

Ishida, H., Umino, T., Tsuji, K. and KOSUGE, T., 1987. Studies on antihemorrhagic substances in herbs classified as hemostatics in Chinese medicine. VII. On the antihemorrhagic principle in *Cirsium japonicum* DC. *Chemical and Pharmaceutical Bulletin*, *35*(2), pp. 861–864.

Jo, S., Kim, S., Kim, D.Y., Kim, M.S. and Shin, D.H., 2020. Flavonoids with inhibitory activity against SARS-CoV-2 3CLpro. *Journal of Enzyme Inhibition and Medicinal Chemistry*, *35*(1), pp. 1539–1544.

Kazuma, K., Takahashi, T., Sato, K., Takeuchi, H., Matsumoto, T. and Okuno, T., 2000. Quinochalcones and flavonoids from fresh florets in different cultivars of *Carthamus tinctorius* L. *Bioscience, Biotechnology, and Biochemistry*, *64*(8), pp. 1588–1599.

Lee, H., Cho, H., Son, M., Sung, G.H., Lee, T., Lee, S.W., Jung, Y.W., Shin, Y.S. and Kang, H., 2013. Dysregulation of KSHV replication by extracts from *Carthamus tinctorius* L. *Journal of Microbiology*, *51*(4), pp. 490–498.

Lem, F.F., Opook, F., Herng, D.L.J., Na, C.S., Lawson, F.P. and Tyng, C.F., 2020. Molecular mechanism of action of repurposed drugs and traditional Chinese medicine used for the treatment of patients infected with COVID-19: A systematic review. *medRxiv*

Luo, E., Zhang, D., Luo, H., Liu, B., Zhao, K., Zhao, Y., Bian, Y. and Wang, Y., 2020. Treatment efficacy analysis of traditional Chinese medicine for novel coronavirus pneumonia (COVID-19): An empirical study from Wuhan, Hubei Province, China. *Chinese Medicine*, *15*(1), pp. 1–13.

Min, B.S., Kim, Y.H., Tomiyama, M., Nakamura, N., Miyashiro, H., Otake, T. and Hattori, M., 2001. Inhibitory effects of Korean plants on HIV-1 activities. *Phytotherapy Research*, *15*(6), pp. 481–486.

Mouhajir, F., Hudson, J.B., Rejdali, M. and Towers, G.H.N., 2001. Multiple antiviral activities of endemic medicinal plants used by Berber peoples of Morocco. *Pharmaceutical Biology*, *39*(5), pp. 364–374.

Orhan, I., Deliorman-Orhan, D. and Özçelik, B., 2009. Antiviral activity and cytotoxicity of the lipophilic extracts of various edible plants and their fatty acids. *Food Chemistry*, *115*(2), pp. 701–705.

Park, J.C., 2003. Inhibitory effects of Korean plant resources on human immunodeficiency virus type 1 protease activity. *Oriental Pharmacy and Experimental Medicine*, *3*(1), pp. 1–7.

Park, J.C., Hur, J.M., Park, J.G., Hatano, T., Yoshida, T., Miyashiro, H., Min, B.S. and Hattori, M., 2002. Inhibitory effects of Korean medicinal plants and camelliatannin H from *Camellia japonica* on human immunodeficiency virus type 1 protease. *Phytotherapy Research*, *16*(5), pp. 422–426.

Phumthum, M., Nguanchoo, V. and Balslev, H., 2021. Medicinal plants used for treating mild Covid-19 symptoms among Thai Karen and Hmong. *Frontiers in Pharmacology*, p. 1844.

Riddell, W.R., 1927. The antisyphilitic pharmacopoeia of fracastorius (Continued). *The Public Health Journal*, *18*(6), pp. 267–279.

Simões, L.R., Maciel, G.M., Brandão, G.C., Kroon, E.G., Castilho, R.O. and Oliveira, A.B., 2011. Antiviral activity of *Distictella elongata* (Vahl) Urb.(Bignoniaceae), a potentially useful source of anti-dengue drugs from the state of Minas Gerais, Brazil. *Letters in Applied Microbiology, 53*(6), pp. 602–607.

Sj, A.R. and Ck, R., 2020. Dehydroabietylamine, a diterpene from *Carthamus tinctorious* L. showing antibacterial and anthelmintic effects with computational evidence. *Current computer-Aided Drug Design, 16*(3), pp. 231–237.

Suksamrarn, A., Chotipong, A., Suavansri, T., Boongird, S., Timsuksai, P., Vimuttipong, S. and Chuaynugul, A., 2004. Antimycobacterial activity and cytotoxicity of flavonoids from the flowers of *Chromolaena odorata. Archives of Pharmacal Research, 27*(5), pp. 507–511.

Sun, N., Yu, T., Zhao, J.X., Sun, Y.G., Jiang, J.B., Duan, Z.B., Wang, W.K., Hu, Y.L., Lei, H.M. and Li, H.Q., 2015. Antiviral activities of natural compounds derived from traditional Chinese medicines against porcine circovirus type 2 (PCV2). *Biotechnology and Bioprocess Engineering, 20*(1), pp. 180–187.

Swarup, V., Ghosh, J., Mishra, M.K. and Basu, A., 2008. Novel strategy for treatment of Japanese encephalitis using arctigenin, a plant lignan. *Journal of Antimicrobial Chemotherapy, 61*(3), pp. 679–688.

Trinh, T.B.N., Le, D.H., Nguyen, T.T.K., Nguyen, V.T., Nguyen, M.H., Muller, M., Le, V.P. and Nguyen, T.K.N., 2021. In vitro antiviral activities of ethanol and aqueous extracts of Vietnamese traditional medicinal plants against Porcine Epidemic Diarrhea virus: A coronavirus family member. *VirusDisease*, pp. 1–7.

Van Wettum, B.A.J., 1901. A Chinese opinion on leprosy, being a translation of a chapter from the medical standard-work 御纂醫宗金鑑 imperial edition of the golden mirror for the medical class. T'oung Pao, pp. 256–268.

Wang, S.X., Wang, Y., Lu, Y.B., Li, J.Y., Song, Y.J., Nyamgerelt, M. and Wang, X.X., 2020. Diagnosis and treatment of novel coronavirus pneumonia based on the theory of traditional Chinese medicine. *Journal of Integrative Medicine, 18*(4), pp. 275–283.

Wiwat, C. and Kwantrairat, S., 2014. HIV-1 reverse transcriptase inhibitors fromthai medicinal plants and *Elephantopus scaber* Linn. *Mahidol University Journal of Pharmaceutical Sciences, 40*(3), pp. 35–44.

Xu, H.X., Wan, M., Loh, B.N., Kon, O.L., Chow, P.W. and Sim, K.Y., 1996. Screening of traditional medicines for their inhibitory activity against HIV-1 protease. *Phytotherapy Research, 10*(3), pp. 207–210.

Yang, Q.X., Zhao, T.H., Sun, C.Z., Wu, L.M., Dai, Q., Wang, S.D. and Tian, H., 2020. New thinking in the treatment of 2019 novel coronavirus pneumonia. *Complementary Therapies in Clinical Practice, 39*, p. 101131.

11.1.2 Family Campanulaceae A.L. de Jussieu (1789)

The family Campanulaceae (including plants in the family Lobeliaceae) consists of about 90 genera and 2000 species of herbs or climber. The leaves are simple, alternate, and exstipulate. The flowers are arranged in axillary or terminal fascicles. The calyx and corolla are tubular and 5-lobed. The corolla is tubular and five-lobed or the lobes gathered at one side. The androecium includes 5 stamens. The gynoecium comprises 3 to 5 locular ovary, each locule contains numerous ovules. The fruits are capsules or berries marked at apex with an opening and vestigial sepals. Plants in this family are often used for the treatment of COVID-19 in China and produce piperidine alkaloids, polyacetylenes, and triterpene saponins.

11.1.2.1 *Platycodon grandiflorus* (Jacq.) A. DC.

Synonyms: *Campanula glauca* Thunb.; *Campanula grandiflora* Jacq.; *Platycodon autumnalis* Decne.; *Platycodon chinensis* Lindl. & Paxton; *Platycodon glaucus* (Thunb.) Nakai; *Platycodon sinensis* Lem.

Common names: Balloon flower, Chinese bell-flower; jie geng (China), asagao (Japan), do ra ji (Korea)

Habitat: Forests, meadows

Distribution: China, Korea, Japan

Botanical description: This magnificent herb grows to about 1 m tall and has a positive aura of life force. The root is pure white and anthropomorphic. The plant is often grown as ornamental.

The stems are glabrous, erect, and branched at apex. The leaves are simple, alternate, and exstipulate. The blade is ovate, about 4–7 cm × 1.5–24 cm, dentate, and glaucous below. The flowers are solitary and showy. The flower buds are peculiarly balloon-shaped or bulbous. The corolla is broadly campanulate, plain dark blue, and 5 cm in diameter. The 5 stamens are white. The gynoecium includes 5 stigmas which are white. The fruit is a dehiscent capsule containing numerous tiny black seeds.

Medicinal uses: Cholera (China); colds (Korea)

Antibacterial (Gram-positive) polar extract: Ethanol extract at 1% inhibited the growth of *Streptococcus mutans* (KCTC 3065) by 38.5% (Kim, 2014).

Anticandidal polar extract: Ethanol extract at 1% inhibited the growth of *Candida albicans* (KCTC 796) by 74.1% (Kim, 2014).

Strong antiviral (enveloped monopartite linear single-stranded (−) RNA) polar extract: Aqueous extract of roots inhibited the replication of the Respiratory syncytial virus with the IC_{50} value of 44.1 μg/mL and the selectivity index of 8 (Ma et al., 2002).

Strong antiviral (enveloped monopartite linear single-stranded (−) RNA) oleanane-type triterpene saponins: Platycodin D, platycodin D2, and platycodin D3 inhibited the replication of Hepatitis C virus (CON1) with the IC_{50} values of 0.3, 1, and 2.4 μg/mL, respectively (Kim et al., 2013). Platycodin D, platycodin D2, and platycodin D3 inhibited the replication of Hepatitis C virus (JHF-1) with the EC_{50} value of 3, 12, and 27 μg/mL, respectively (Kim et al., 2013).

Viral enzyme inhibition by polar extract: Aqueous extract of root at the concentration of 250 μg/mL inhibited the Human immunodeficiency virus-1 protease by 94% (Xu et al., 1996).

Viral enzyme inhibition by oleanane-type triterpene saponins: Platycodin D, platycodin D2, and platycodin D3 inhibited the Hepatitis C virus NS5B with the IC_{50} values of 5, 6, and 8 μg/mL, respectively (Kim et al., 2013).

Viral interaction with host cell inhibition by oleanane-type triterpene saponin: Platycodin D inhibited the interaction of the Severe acute respiratory syndrome-associated coronavirus 2 spike protein with angiotensin-converting enzyme with the EC_{50} value of 0.6 μM (Kim et al., 2021).

Commentaries

i. *Platycodon grandiflorum* (Jacquin) DC accumulates series of oleananes saponins known as platycosides (Zhang et al., 2015) and is used for COVID-19 in China (Shi et al., 2020).

ii. In the Subfamily Lobelioideae, Tribe Lobelieae, *Lobelia angulata* G. Forst. is used for boils, sore throat, and fever in India and *Lobelia chinensis* Lour. is used for Herpes zoster in China. Methanol extract of *Lobelia chinensis* Lour. inhibited the replication of the Herpes simplex virus type-1 with the IC_{50} value of 139.2 μg/mL and afforded some levels of protection in mice against the Herpes simplex virus type-1 when given orally at the dose of 50 mg/Kg for 7 days (Kuo et al., 2008). Chloroform extract inhibited the replication of the Chikungunya virus in Vero cells with the EC_{50} value of 28.4 μg/mL, (Sit et al., 2017). Chloroform extract inhibited the growth of *Bacillus subtilis*, *Staphylococcus aureus*, and *Klebsiella pneumoniae* with the MIC/MBC values of 310/313, 1250/1250, and 1250/1250 μg/mL, respectively (Sit et al., 2017). Hexane extract inhibited the growth of *Candida albicans*, *Candida parapsilosis*, *Candida krusei*, and *Trichophyton mentagrophytes* (Sit et al., 2017).

iii. In the Subfamily Campanuloideae; Tribe Campanuleae, the magnificent *Adenophora verticillata* Fisch (synonym: *Adenophora tetraphylla* (Thunb.) Fisch.; *Adenophora triphylla* (Thunb.) A. DC.) is used to treat tuberculosis in China (local name: *Lun ye sha shen*) as well as COVID-19 (Zhong, 2020). Methanol extract of this plant at the concentration of 100 μg/mL inhibited the fusion of the Human immunodeficiency virus to host cell by 60.7% (Chang et al., 2003).

iv. In the Subfamily Campanuloideae; Tribe Campanuleae, *Campanula pallida* Wall. is used for diarrhea, dysentery in Nepal (local name: *Ganobuti*), and wounds in India, and

Campanula glauca Thunb. is used to treat syphilis in Japan. Ethanol extract of *Campanula lactiflora* M. Bieb. inhibited the growth of *Mycobacterium tuberculosis* (H37Ra) with the MIC value of 200 µg/mL (Tosun et al., 2004). Extract of *Campanula punctata* Lam inhibited the replication of the Tobacco mosaic virus *in vivo* (Jing et al., 2012).

v. In the Subfamily Cyananthoideae; Tribe Codonopsideae, *Codonopsis pilosula* (Franch.) Nannf. is used to prevent gonorrhea and for COVID-19 in China (Zhang et al., 2021). It produces immunomodulatory and antiviral polysaccharides (Chen et al., 2018). The plant produces polyacetylenes (Jiang et al., 2015) which are probably antifungal (see earlier). Ethanol extract at the concentration of 1 mg/mL inhibited the growth of *Aspergillus fumigatus* (ATCC 46645) and *Candida albicans* by 55% and 100%, respectively (Zhang et al., 2013). In India, *Codonopsis clematidea* (Schrenk) Cl., *Codonopsis ovata* Benth., and *Codonopsis rotundifolia* Benth. are used for skin infections. In Bhutan, *Codonopsis bhutanica* Ludlow (local name: *Klub dud-rdor-rjenag-po*) is used for leprosy. In Korea, *Codonopsis lanceolata* Trautv. (local name: *Deodeok*) is used to treat bronchitis, cough, fever, furuncles, leucorrhea, sore throat, and tuberculosis.

vi. In the closely family Pentaphtagmaceae J.Agardh (1858), *Pentaphragma ellipticum* Poulsen (synonym: *Pentaphragma begoniifolia* Wall. ex G. Don; *Pentaphragma ellipticum* Poulsen; *Pentaphragma ridleyi* King & Gamble) is used to treat gonorrhea in Indonesia. This plant and its possible antimicrobial principles are on their way to disparition if the palm oil and logging madness keeps going on.

References

Chang, Y.S. and Woo, E.R., 2003. Korean medicinal plants inhibiting to Human Immunodeficiency Virus type 1 (HIV-1) fusion. *Phytotherapy Research: An International Journal Devoted to Pharmacological and Toxicological Evaluation of Natural Product Derivatives*, *17*(4), pp. 426–429.

Chen, L. and Huang, G., 2018. The antiviral activity of polysaccharides and their derivatives. *International Journal of Biological Macromolecules*, *115*, pp. 77–82.

Jiang, Y., Liu, Y., Guo, Q., Jiang, Z., Xu, C., Zhu, C., Yang, Y., Lin, S. and Shi, J., 2015. Acetylenes and fatty acids from *Codonopsis pilosula*. *Acta Pharmaceutica Sinica B*, *5*(3), pp. 215–222.

Jing, B., Ma, Z., Feng, J., Liang, H., Li, C. and Zhang, X., 2012. Evaluation of the antiviral activity of extracts from plants grown in the qinling region of China against infection by Tobacco mosaic virus (TMV). *Journal of Phytopathology*, *160*(4), pp. 181–186.

Kim, J., 2014. Antibacterial and anti-inflammatory effects of *Platycodon grandiflorum* extracts. *Journal of Digital Convergence*, *12*(3), pp. 359–366.

Kim, J.W., Park, S.J., Lim, J.H., Yang, J.W., Shin, J.C., Lee, S.W., Suh, J.W. and Hwang, S.B., 2013. Triterpenoid saponins isolated from *Platycodon grandiflorum* inhibit hepatitis C virus replication. *Evidence-Based Complementary and Alternative Medicine*, *2013*.

Kim, T.Y., Jeon, S., Jang, Y., Gotina, L., Won, J., Ju, Y.H., Kim, S., Jang, M.W., Won, W., Park, M.G. and Pae, A.N., 2021. Platycodin D, a natural component of Platycodon grandiflorum, prevents both lysosome-and TMPRSS2-driven SARS-CoV-2 infection by hindering membrane fusion. *Experimental & Molecular Medicine*, *53*(5), pp. 956–972.

Ma, S.C., Du, J., But, P.P.H., Deng, X.L., Zhang, Y.W., Ooi, V.E.C., Xu, H.X., Lee, S.H.S. and Lee, S.F., 2002. Antiviral Chinese medicinal herbs against respiratory syncytial virus. *Journal of Ethnopharmacology*, *79*(2), pp. 205–211.

Shi, Y.H., Huang, Y.F., Wang, W.Y., Yang, L., Zhou, H. and Sang, Z., 2020. Analysis on the current quality standards of Chinese materia Medica used in COVID-19 prevention and treatment. *Pharmacological Research*, *160*, p. 105074.

Sit, N.W., Chan, Y.S., Chuah, B.L., Cheng, R.J., Leong, W.M. and Khoo, K.S., 2017. Antiviral, antifungal and antibacterial activities of the Chinese medicinal plants, Houttuynia cordata, Lobelia chinensis and Selaginella uncinata. *Southeast Asian Journal of Tropical Medicine and Public Health*, *48*(3), pp. 616–627.

Tosun, F., Kızılay, Ç., Şener, B., Vural, M. and Palittapongarnpim, P., 2004. Antimycobacterial activity of some Turkish plants. *Pharmaceutical Biology*, *42*(1), pp. 39–43.

Xu, H.X., Wan, M., Loh, B.N., Kon, O.L., Chow, P.W. and Sim, K.Y., 1996. Screening of traditional medicines for their inhibitory activity against HIV-1 protease. *Phytotherapy Research*, *10*(3), pp. 207–210.

Zhang, L., Ravipati, A.S., Koyyalamudi, S.R., Jeong, S.C., Reddy, N., Bartlett, J., Smith, P.T., de la Cruz, M., Monteiro, M.C., Melguizo, Á. and Jiménez, E., 2013. Anti-fungal and anti-bacterial activities of ethanol extracts of selected traditional Chinese medicinal herbs. *Asian Pacific Journal of Tropical Medicine*, *6*(9), pp. 673–681.

Zhang, L., Wang, Y., Yang, D., Zhang, C., Zhang, N., Li, M. and Liu, Y., 2015. Platycodon grandiflorus: An ethnopharmacological, phytochemical and pharmacological review. *Journal of Ethnopharmacology*, *164*, pp. 147–161.

Zhang, J.L., Li, W.X., Li, Y., Wong, M.S., Wang, Y.J. and Zhang, Y., 2021. Therapeutic options of TCM for organ injuries associated with COVID-19 and the underlying mechanism. *Phytomedicine*, *85*, p. 153297.

Zhong, S., A traditional Chinese medicine complementary strategy for COVID-19.

11.1.3 Family Goodeniaceae R. Brown (1810)

The family Goodeniaceae consists of about 12 genera and 400 species of herbs, shrubs, or treelets. The leaves are simple, alternate, and exstipulate. The inflorescences are cymes, racemes, or solitary. The calyx is tubular and 5-lobed. The corolla is tubular and 5-lobed, the 5 lobes on 1 side. The androecium includes 5 stamens. The gynoecium includes a 2 to 4 loculed ovary, each locule sheltering 1 to many ovules. The fruit is a capsule or a berry.

11.1.3.1 *Scaevola taccada* (Gaertn.) Roxb.

Synonyms: *Scaevola frutescens* Krause, *Scaevola koenigii* Vahl, *Scaevola sericea* Forst.; *Scaevola sericea* Vahl

Common names: Scaevola, half flower, beach naupaka; cao hai tong (China); naupaka (Hawaii); kwyae (India); kesembung laut (Indonesia); ambong-ambong (Malaysia), vella muttagam (India); orpinletan (Myanmar); akajok, pahop (Papua New Guinea); boto (the Philippines); bong bong (Thailand); hep (Vietnam)

Habitat: Seashores, sandy beaches

Distribution: Tropical Asia and the Pacific

Botanical observation: This handsome shrub grows to a height of 6 m and has a positive aura. The stems are stout, glabrous, and smooth. The leaves are simple, sessile, spiral, and without stipules. The blade is fleshy, glossy, glabrous, spathulate, coriaceous (often observed in seashore trees), and 20.5–24×6.5–8 cm. Tufts of silky white hairs are present at the base of the leaves. The cymes are axillary and 7–9 cm long. The flowers are showy, whitish, and 1.6–1.8 cm long. The calyx is tubular, about 7 mm long, hairy, and produces 5 lobes that have somewhat vaguely a look of some sorts of spades. The corolla is characteristically hemitubular, 1–1.3 cm long, whitish, and produces 5 oblong lobes. The androecium consists of 5 stamens. The style is 2 cm long. The berries are greenish–white to pure white, lobed, glossy, irregularly globose, up to about 2 cm long, and crowned with the 5 calyx lobes and contain a rough and bony stone (the whole having vaguely a look of a minute guava fruits)

Medicinal uses: Cough, tuberculosis, toothache (Papua New Guinea); cuts (Hawaii); dysentery (India, the Philippines, Tonga); fever (India); syphilis

Strong antifungal furanocoumarin: Scataccanol isolated from the berries inhibited the growth of *Pythium insidiosum* with the MIC value of 5 µg/mL (Suthiwong et al., 2017).

Strong antifungal simple phenol: 4-Formylsyringol isolated from the berries inhibited the growth of *Pythium insidiosum* with the MIC value of 10 µg/mL (Suthiwong et al., 2017).

Strong antiviral (enveloped linear monopartite dimeric single-stranded (+) RNA) polar extract: Aqueous extract of fruits inhibited the replication of the Human immunodeficiency virus type-1 (Strain IIIb) in MT-4 cells with the IC_{50} value of 8.2 µg/mL and a selectivity index of 31 (Locher et al., 1996).

Commentaries

i. Members of the genus *Scaevola* L. yield prenylated coumarins, iridoid glycosides, and flavones (Ghisalberti et al., 2004).
ii. In Papua New Guinea, *Scaevola sericea* Vahl (synonym: *Scaevola taccada* (Gaertn.) Roxb.) is a remedy for tuberculosis and is used for cough.
iii. *Scaevola spinescens* R.Br. is used medicinally by the Aborigines of Australia who have the oldest medicinal practice in Asia and Pacific and a supernatural connection with Nature and the medicinal knowledge of Australian abotigines needs to be deepley explored and documented. Ethyl acetate extract of *Scaevola spinescens* R.Br. inhibited the growth of *Staphylococcus aureus* (ATCC 25923) and *Streptococcus pyogenes* (ATCC 10389) with the MIC/MBC values of 1000/1000 and 500/1000 µg/mL, respectively (Menjn, 2009). Aqueous extract inhibited the growth of *Alcaligenes faecalis*, *Citrobacter freundi*, *Klebsiella pneumoniae*, *Proteus mirabilis*, *Serratia marcescens*, and *Staphylococcus epidermidis* with the MIC values of 85.3, 2501, 66.8, 25.4, 350.3, and 489.3 µg/mL, respectively (Cock & Kukkonen, 2011). Extract of *Scaevola spinescens* R.Br. inhibited the replication of the Human cytomegalovirus (Semple et al., 1998).

References

Cock, I.E. and Kukkonen, L., 2011. An examination of the medicinal potential of *Scaevola spinescens*: Toxicity, antibacterial, and antiviral activities. *Pharmacognosy Research*, *3*(2), p. 85.

Ghisalberti, E.L., 2004. The goodeniaceae. *Fitoterapia*, *75*(5), pp. 429–446.

Locher, C.P., Witvrouw, M., De Béthune, M.P., Burch, M.T., Mower, H.F., Davis, H., Lasure, A., Pauwels, R., De Clercq, E. and Vlietinck, A.J., 1996. Antiviral activity of Hawaiian medicinal plants against human immunodeficiency virus type-1 (HIV-1). *Phytomedicine*, *2*(3), pp. 259–264.

Mejin, M., 2009. Isolation, structural elucidation and antibacterial activity of the chemical constituents of *Scaevola spinescens* (Doctoral dissertation).

Semple, S.J., Reynolds, G.D., O'leary, M.C. and Flower, R.L.P., 1998. Screening of Australian medicinal plants for antiviral activity. *Journal of Ethnopharmacology*, *60*(2), pp. 163–172.

Suthiwong, J., Thongsri, Y. and Yenjai, C., 2017. A new furanocoumarin from the fruits of *Scaevola taccada* and antifungal activity against *Pythium insidiosum*. *Natural Product Research*, *31*(4), pp. 453–459.

11.1.4 Family Menyanthaceae Dumort (1829)

The family Menyanthaceae consists of about 5 genera and 33 species of aquatic herbs. The leaves are simple or compound, alternate, and exstipulate. The petiole forms a sheath at base. The inflorescences are racemes, solitary, or fascicles. The calyx and corolla are tubular and 5-lobed. The androecium includes 5 stamens. The gynoecium comprises a unilocular ovary sheltering ovules growing on a basal placenta. The fruits are capsules. Plants in this family yield iridoid glycosides.

11.1.4.1 *Nymphoides cristata* (Roxb.) Kuntze

Synonyms: *Limnanthemum cristatum* (Roxb.) Griseb.; *Menyanthes cristata* Roxb.; *Villarsia cristata* (Roxb.) Spreng.

Common names: Crested floating-heart; jai chirata (Bangladesh); shui pi lian (China); tagarmuula (India); pusong lutang (the Philippines)

Habitat: Ponds, lakes, slow rivers

Distribution: Tropical Asia

Botanical observation: This superb nuphar-like herb grows to a length of about 2 m. The leaves are simple, alternate, exstipulate, and develop from a rhizome. The petiole is up to about 5 cm long. The blades are cordate, orbicular, floating, and 25–7.5 cm × 2–8 mm. The inflorescences are axillary clusters. The calyx is tubular and develops 5 lobes which are about 5 mm long. The 5 petals are pure white, membranous at margin, about 5 mm long, and yellow at base. The 5 stamens are about 3 mm long. The capsules are ellipsoid, about 5 mm long, beaked, and contain numerous tiny seeds.

Medicinal uses: Furuncules (Cambodia, Laos, and Vietnam)

Commentaries

i. The plant has apparently not been examined for antimicrobial effects. In India, *Nymphoides indica* (L.) Kuntze (local name: *Kumudini*) is given for dysentery and fever. Methanol extract of *Nymphoides indica* (L.) Kuntze inhibited the growth of bacteria and fungi (Amin et al., 2014). Plants in the genus *Nymphoides* Ség. yield flavoness (Daniel & Sabnis, 1978). Note that aquatic plants are often equipped with water soluble phytoalexins (tannins, flavonoids, simple phenolic).
ii. As per the theory of signatures, one could anticipate *Nymphoides cristata* (Roxb.) Kuntze to contain antimycobacterial principles and these could be flavonol glycosides (see earlier).
iii. The notoriously bitter *Menyanthes trifoliata* L., also known as marsh trefoil, is used for diarrhea, dysentery, fever, and herpes in India. Ethanol extract of *Menyanthes trifoliata* L. inhibited the growth of *Staphylococcus aureus* (ATCC 25923), *Morganella morganii* (E 10679/93), *Pseudomonad maltophila* (D 141/92), and *Corynebacterium amycolatum* with the MIC/MBC values of 100/100, 100/100, 100/100, and 20/100 µg/mL, respectively (Weckesser et al., 2007). *Morganella morgani* is a Gram-negative bacillus that accounts for lethal nosocomial infections in intensive care units and gains resistance to more and more antibiotics (Singla et al., 2010). This bacterium like other problematic methicillin-resistant *Staphylococcus aureus*, vancomycin-resistant Enterococci, extended-spectrum β-lactamase Enterobacteriaceae, and metallo-β-lactamase *Pseudomonas aeruginosa* (Singla et al., 2010) may some days completely resistant to antibiotics

References

Amin, A., Tuenter, E., Exarchou, V., Upadhyay, A., Cos, P., Maes, L., Apers, S. and Pieters, L., 2016. Phytochemical and pharmacological investigations on *Nymphoides indica* leaf extracts. *Phytotherapy Research*, *30*(10), pp. 1624–1633.

Amin, A., Upadhyay, A., Zafar, M., Cos, P., Maes, L., Apers, S., Exarchou, V. and Pieters, L., 2014. Antibacterial, antifungal, cytotoxic, antioxidant and antidiabetic compounds from *Nymphoides indica*; the first comprehensive phytochemical and pharmacological study. *Planta Medica*, *80*(16), p. P1L115.

Daniel, M. and Sabnis, S.D., 1978. Chemical systematics of family Gentianaceae. *Current Science*, *47*(4), pp. 109–111.

Singla, N., Kaistha, N., Gulati, N. and Chander, J., 2010. Morganella morganii could be an important intensive care unit pathogen. *Indian Journal of Critical Care Medicine: Peer-Reviewed, Official Publication of Indian Society of Critical Care Medicine*, *14*(3), p. 154.

Weckesser, S., Engel, K., Simon-Haarhaus, B., Wittmer, A., Pelz, K. and Schempp, C.Á., 2007. Screening of plant extracts for antimicrobial activity against bacteria and yeasts with dermatological relevance. *Phytomedicine*, *14*(7–8), pp. 508–516.

11.2 ORDER DIPSACALES JUSS. EX BERCHT. & J. PRESL (1820)

11.2.1 Family Adoxaceae E. Mey (1839)

The family Adoxaceae consists of about 4 genera and 220 species of herbs, shrubs, or treelets. The leaves are compound, opposite, and exstipulate. The inflorescences are terminal panicles or cymes. The calyx and corolla are tubular and 5-lobed. The androecium includes 5 stamens. The gynoecium comprises a unilocular or 3–5 locular ovary The fruits are drupes.

11.2.1.1 *Viburnum dilatatum* Thunb.

Synonym: *Viburnum brevipes* Rehder; *Viburnum fulvotomentosum* P.S. Hsu

Common names: Linden Viburnum, linden arrow-wood; jia mi (China)

Habitat: Cultivated

Distribution: China, Korea, Japan

Botanical observation: This handsome shrub grows up to 5 m tall. The stems are velvety when young and lenticelled. The leaves are simple, opposite, and without stipules. The petiole is 1–1.2 cm long and velvety. The blade is pubescent, papery, 2.8–4.7 cm × 5.7–7.4 cm, and orbiculate to elliptic. The base of the blade is round to cordate and the apex is shortly acuminate. The blade is marked with 7–9 pairs of secondary nerves which are showy, straight, and sunken. The margin is serrate, the midrib sunken, and filled with hairs above. The inflorescences are terminal, showy, and velvety panicles that have vaguely an umbelliferous look. The flowers are white and very small. The calyx comprises 5 velvety sepals. The corolla is tubular, pubescent, and 5-lobed, the lobes round. The androecium comprises 5 stamens. The drupes are up to 8 mm long, ellipsoid, bright red, and glossy.

Medicinal uses: Dysentery, sores (China)

Commentaries

i. The medicinal uses could be to be due to the astringency of flavonoids including anthocyanins which abound in the fruits of this ornamental tree (Kim et al., 2003). Note that astringent fruits are often of value for the prevention and for management of Influenza.
ii. In Turkey, *Viburnum lantana* L. (local name: *Germeso*) is a remedy for diarrhea.
iii. In Nepal, *Viburnum erubescens* var. *erubescens* Wall. ex DC. (local name: *Ashaare*) is used for pneumonia.

Reference

Kim, M.Y., Iwai, K., Onodera, A. and Matsue, H., 2003. Identification and antiradical properties of anthocyanins in fruits of *Viburnum dilatatum* Thunb. *Journal of Agricultural and Food Chemistry, 51*(21), pp. 6173–6177.

11.2.2 Family Caprifoliaceae A.L. de Jussieu (1789)

The family Caprifoliaceae consists of about 12 genera and 450 species of scandent shrubs. The leaves are simple, opposite, and exstipulate. The inflorescences are cymes or panicles. The calyx is tubular and 4–5 lobed. The corolla is tubular, 4–5 lobed, and characteristically salver-shaped or bilabiate. The androecium includes 5 stamens. The gynoecium includes a 2–5-loculed ovary, each locule sheltering one to many ovules growing on axile placentas. The fruits are drupes or capsules.

11.2.2.1 *Lonicera japonica* Thunb.

Common names: Japanese honeysuckle; jīn yín huā, ren dong (China); kim ngam, day nhan dong (Vietnam)

Habitat: Open forests, roadsides, thickets

Distribution China, Korea, Japan, and Taiwan

Botanical observation: This climbing shrub grows up to about 10 m long and has a positive aura of life force. The stems are terete, reddish, and when young tomentose. The leaves are simple, opposite, and exstipulate. The petiole is 5 mm long and pubescent. The blade is elliptic, 3–8 cm × 2–3 cm, truncate or cordate at base, and obtuse at apex. The flowers are axillary and grouped in pairs. The calyx is tubular, minute, 5-lobed, and hairy. The corolla is tubular, up to about 5 cm long, of a unique type of light yellow, bilobed, the upper lobe with four teeth and the lower recurved. The five stamens are recurved and protrude with the style out of the corolla. The berries are globose, up to about 7 mm across, glossy, and black.

Medicinal uses: Boils (Cambodia, Laos, and Vietnam); colds, fever, toothache (China); dysentery (Vietnam)

Broad-spectrum antibacterial halo developed by hydroxycinnamic acid derivatives: 3-*O*-Caffeoylquinic acid, 3,5-di-*O*-caffeoylquinic acid, and 4,5-di-*O*-caffeoylquinic acid (2 mg/12 mm disc) isolated from the leaves inhibited the growth of *Escherichia coli* with inhibition zone diameters of 21.3, 23.2, and 23.4 mm, respectively. 3-*O*-Caffeoylquinic acid, 3,5-di-*O*-caffeoylquinic acid, and 4,5-di-*O*-caffeoylquinic acid (2 mg/12 mm disc) inhibited the growth of *Staphylococcus aureus* with inhibition zone diameters of 24.6, 27.1, and 26.9 mm, respectively (Xiong et al., 2013).

Broad-spectrum antibacterial halo developed by iridoid glycoside: Secoxyloganin (2 mg/12 mm disc) isolated from the leaves inhibited the growth of *Escherichia coli* and *Staphylococcus aureus* with inhibition zone diameters of 13.5 and 14.4 mm, respectively (Xiong et al., 2013).

Broad-spectrum antibacterial halo developed by isoflavonol glycoside: Lutoloside (2 mg/12 mm disc) isolated from the leaves inhibited the growth of *Escherichia coli* and *Staphylococcus aureus* with inhibition zone diameters of 22.9 and 27.4 mm, respectively (Xiong et al., 2013).

Strong antibacterial monoterpenes: 7-Acetyl-8,9-dihydroxy thymol and 7,8-dihydroxy-9-buyryl thymol inhibited the growth of bacteria with IC_{50} values ranging from 27.6 to 128.5 µg/mL, respectively (Yang et al., 2018).

Strong antifungal (filamentous) essential oil: Essential oil of leaves inhibited the growth of *Microsporum canis* (KCTC 6348), *Microsporum canis* (KCTC 6591), *Microsporum canis* (KCTC 6591), *Trichophyton rubrum* (KCTC 6345), *Trichophyton rubrum* (KCTC 6352), *Trichophyton rubrum* (KCTC 6375), and *Trichophyton mentagrophytes* (KCTC 6077) with the MIC values of 62.5, 62.5, 250, 125, 125, 500, and 500 µg/mL, respectively (Rahman et al., 2014).

Moderate antiviral (enveloped circular double-stranded DNA) hydroxycinnamic acid derivative: 4-*O*-Caffeoylquinic acid ethyl ester inhibited the secretion of the Hepatitis B virus surface antigen and e-antigen as well as Hepatitis B virus DNA replication (Ge et al., 2018).

Viral enzyme inhibition by polar extract: Aqueous extract of flowers at the concentration of 250 µg/mL inhibited the Human immunodeficiency virus type-1 protease by 40% (Xu et al., 1996).

Commentaries

i. The plant is used in China for the treatment of COVID-19 as part of an herbal formula known as *Lian Hua Qing* (Luo, 2020; Wang et al., 2020; Zheng et al., 2020). Flower buds accumulates hydroxycinnamic acid derivatives such as the broad-spectrum antiviral chlorogenic acid (also known as 5-*O*-caffeoylquinic acid; see earlier).
ii. In Myanmar, *Lonicera macrantha* DC. (local name: *Leihruisen*) is used for diarrhea.
iii. In the closely related family Valerianaceae, Tribe Patrinieae, aqueous extract of roots of *Patrinia villosa* (Thunb.) Juss. (looks like somewhat an umbellifer) at the concentration of 250 µg/mL inhibited the Human immunodeficiency virus type-1 protease by 64.1% (Xu et al., 1996). Aqueous extract of this plant inhibited the replication of the Respiratory syncytial virus (Long strain) in Hep-2 cells with a IC_{50} value below 13 µg/mL and a selectivity index above 30 (Ma et al., 2002) and inhibited the replication of the Duck Hepatitis B virus with the IC_{50} value of 780 µg/mL (Leung et al., 2006). *Patrinia scabiosifolia* Link. is used to treat abscesses in China (local name: *Bai jiang*). The sublime *Nardostachys jatamansi* (D. Don) DC. (synonyms: *Nardostachys chinensis* Batalin; *Patrinia jatamansi* D. Don) is a remedy for skin diseases in China (local name: *Gan song*) and has apparently so far not been studied for antimicrobial effect. Extracts of this plant are neuroactive (Dhingra & Goyal, 2008) suggesting the possible presence of bacterial and or fungal efflux pumps inhibitors. In the Tribe Diervilleae, *Weigela florida* (Bunge) A. DC. is used for sores in China (local name: *Jin dai hua*)
iv. In the family Dipsacaceae A.L. de Jussieu (1789), the magnificent *Dipsacus asper* Wall. is used for wounds in Turkey (local name: *Pukiç*) and for cuts in China (local name: *Huan xu duan*). *Dipsacus asperoides* C.Y. Cheng & T.M. Ai (synonym: *Dipsacus asper* Wall. ex C.B. Clarke) is used in China (local name: *Chuan xu duan*) for the treatment of COVID-19 in combination with oseltamivir (Luo, 2020). Note that China uses medicinal plants as part of clinical practice. In India, *Dipsacus mitis* D. Don (synonym: *Dipsacus inermis* var. *mitis* (D. Don) Y.J. Nasir) is used for diarrhea. Methanol extract of *Dipsacus mitis* D.

Don inhibited the replication of the Human immunodeficiency virus type-1 with the EC_{50} value of 82 µg/mL and the selectivity index of 25 in Vero cells (Rajbhandari et al., 2001). *Dipsacus inermis* Wall. (local name: *Wopal hakh*) is a remedy for cough in Pakistan.

References

Dhingra, D. and Goyal, P.K., 2008. Inhibition of MAO and GABA: Probable mechanisms for antidepressant-like activity of *Nardostachys jatamansi* DC. in mice.

Ge, L., Wan, H., Tang, S., Chen, H., Li, J., Zhang, K., Zhou, B., Fei, J., Wu, S. and Zeng, X., 2018. Novel caffeoylquinic acid derivatives from *Lonicera japonica* Thunb. flower buds exert pronounced anti-HBV activities. *RSC Advances*, *8*(62), pp. 35374–35385.

Leung, K.T., Chiu, L.C., Lam, W.S., Li, Y., Sun, S.S. and Ooi, V.E., 2006. In vitro antiviral activities of Chinese medicinal herbs against duck hepatitis B virus. *Phytotherapy Research: An International Journal Devoted to Pharmacological and Toxicological Evaluation of Natural Product Derivatives*, *20*(10), pp. 911–914.

Luo, A., 2020. Positive SARS-Cov-2 test in a woman with COVID-19 at 22 days after hospital discharge: A case report. *Journal of Traditional Chinese Medical Sciences.*

Ma, S.C., Du, J., But, P.P.H., Deng, X.L., Zhang, Y.W., Ooi, V.E.C., Xu, H.X., Lee, S.H.S. and Lee, S.F., 2002. Antiviral Chinese medicinal herbs against respiratory syncytial virus. *Journal of Ethnopharmacology*, *79*(2), pp. 205–211.

Rahman, A., Al-Reza, S.M., Siddiqui, S.A., Chang, T. and Kang, S.C., 2014. Antifungal potential of essential oil and ethanol extracts of *Lonicera japonica* Thunb. against dermatophytes. *EXCLI Journal*, *13*, p. 427.

Rajbhandari, M., Wegner, U., Jülich, M., Schoepke, T. and Mentel, R., 2001. Screening of Nepalese medicinal plants for antiviral activity. *Journal of Ethnopharmacology*, *74*(3), pp. 251–255.

Wang, S.X., Wang, Y., Lu, Y.B., Li, J.Y., Song, Y.J., Nyamgerelt, M. and Wang, X.X., 2020. Diagnosis and treatment of novel coronavirus pneumonia based on the theory of traditional Chinese medicine. *Journal of Integrative Medicine.*

Xiong, J., Li, S., Wang, W., Hong, Y., Tang, K. and Luo, Q., 2013. Screening and identification of the antibacterial bioactive compounds from *Lonicera japonica* Thunb. leaves. *Food Chemistry*, *138*(1), pp. 327–333.

Xu, H.X., Wan, M., Loh, B.N., Kon, O.L., Chow, P.W. and Sim, K.Y., 1996. Screening of traditional medicines for their inhibitory activity against HIV-1 protease. *Phytotherapy Research*, *10*(3), pp. 207–210.

Yang, J., Li, Y.C., Zhou, X.R., Xu, X.J., Fu, Q.Y. and Liu, C.Z., 2018. Two thymol derivatives from the flower buds of *Lonicera japonica* and their antibacterial activity. *Natural Product Research*, *32*(18), pp. 2238–2243.

Zheng, S., Baak, J.P., Li, S., Xiao, W., Ren, H., Yang, H., Gan, Y. and Wen, C., 2020. Network pharmacology analysis of the therapeutic mechanisms of the traditional Chinese herbal formula Lian Hua Qing Wen in Corona virus disease 2019 (COVID-19), gives fundamental support to the clinical use of LHQW. *Phytomedicine*, *79*, p. 153336.

11.3 ORDER APIALES NAKAI (1930)

11.3.1 Family Apiaceae Lindley (1836)

The family Apiaceae consists of about 250 genera and 4000 species of herbs or shrubs. The stems are very characteristically often hollowed, ribbed, and equipped with a unique type of internal secretory structures. The leaves are compound, spiral, and exstipulate. The petiole forms a characteristic sheath at base. The inflorescences are characteristic umbels. The calyx and corolla are tubular and 5 lobed. The corolla is often asymmetrical and pure white. The androecium comprises 5 stamens. The ovary is bilocular. The fruits are schizocarps. Antimicrobial principles in this family include polyacetylenes, furanocoumarins, sesquiterpenes, and triterpene saponins.

11.3.1.1 *Angelica dahurica* (Fisch.) Benth. & Hook. f.

Synonym: *Callisace dahurica* Fisch.

Common names: Dahurian angelica; bai zhi (China); byakushi (Japan); gu rit dae (Korea)

Habitat: Meadows

Distribution: China, Taiwan, Japan, Korea, Russia

Botanical description: This magnificent umbellifer grows to 3 m tall and has a positive aura of life force. The taproot is brown, 5 cm thick, and strongly aromatic. The stems are stout, hollowed, ribbed, purplish green, and hairy. The petioles form oblong sheaths. The blade is triangular, 25–30 cm × 35–50 cm, and pinnate. The folioles are sessile, oblong, 1–4 cm × 1 4.5–10 cm, decurrent at base, and acute at apex. The umbels are up to 30 cm across. The calyx is minute. The corolla presents 5 pure white petals. The ovary develops a pair of styles. The schizocarps are globose and 7 mm in diameter.

Medicinal uses: Cold, fever, gonorrhea, herpes, pimples, sore throat (China)

Strong broad-spectrum antibacterial lipophilic polyacetylene: Falcarindiol (LogD 5.5 at pH 7.4; molecular mass = 260.3 g/mol) isolated from the roots inhibited the growth of *Staphylococcus aureus* (ATCC 25293) with MIC value of 32 µg/mL (Lechner et al., 2004). Falcarindiol inhibited the growth of *Micrococcus luteus* and *Bacillus cereus* with the MIC values of 50 µg/mL (Meot-Duros et al., 2010). Falcarindiol inhibited the growth of *Escherichia coli*, methicillin-sensitive *Staphylococcus aureus*, methicillin-resistant *Staphylococcus aureus*, and *Bacillus subtilis* (Matsuura et al., 1996).

Strong antimycobacterial lipophilic polyacetylene: Falcarindiol isolated from the roots inhibited the growth of *Mycobacterium fortuitum* (ATCC 6841) with the MIC value of 8 µg/mL (Lechner et al., 2004).

Strong antifungal lipophilic polyacetylene: Falcarindiol inhibited the germination of chlamydospores of *Mycocentrospora acerina* with the ED_{50} value of 31·8 µg/ml (Garrod et al., 1978).

Strong antiviral (enveloped linear segmented single-stranded (-) RNA) amphiphilic furanocoumarins: Isoimperatorin, oxypeucedanin (LogD 2.2 at pH 7.4; molecular mass = 286.2 g/mol), and imperatorin isolated from the roots inhibited the replication of the Swine influenza (H1N1, A/PR/8/34) with the EC_{50} value of 7.6, 5.9, and 11.3 µM, respectively, and the selectivity indexes of 14.8. 16.9, and 9.2, respectively, in Madin-Darby canine kidney cells (Lee et al., 2020). Isoimperatorin, oxypeucedanin and imperatorin isolated from the roots inhibited the replication of the Avian influenza (H9N2, A/Chicken/Korea/01210/2001) with the EC_{50} value of 6.2, 4.5, and 8.1 µM, respectively, and the selectivity indexes of 19.8, 22.4, and 12.9, respectively, in Madin-Darby canine kidney cells (Lee et al., 2020).

Strong antiviral (enveloped monopartite linear double-stranded DNA) amphiphilic furanocoumarin: Imperatorin (LogD 3.1 at pH 7.4; molecular mass = 270.2 g/mol) at the concentration of 15.6 µg/mL inhibited the replication of the Herpes simplex virus type-1 by 41.2%. In this experiment, xanthotoxin and isoimperatorin developed some levels of inhibition against Herpes simplex virus type-1 (Rajtar et al., 2017).

Commentaries

i. *Angelica dahurica* (Fisch.) Benth. & Hook. f. brings to being furanocoumarins of which xanthotoxol, 5-hydroxy-8-methoxypsoralen, psoralen, and xanthotoxin (Zhao et al., 2016). Note that plants mainly in the Clades Fabids, and Malvids, have acquired enzymes that prenylate flavonoids, coumarin, quinones, and xanthones. The attachment of isoprenic or geranyl moieties increase the lipophilicity of phenolics that become thus active against Gram-negative bacteria (see earlier). Further, isoprenyls and geranyls permit a stronger anchorage into the pockets of enzymes. *In Silico* papers are not reported here.

ii. *Angelica dahurica* (Fisch.) Benth. & Hook. f. is given for the prevention of COVID-19 in China (Deng, 2020).

iii. In the genus *Angelica* L., methanol extract of roots of *Angelica tenuissima* Nakai at the concentration of 100 µg/mL inhibited RNA-dependent DNA polymerase and ribonuclease H activities of the human immunodeficiency virus type-1 reverse transcriptase and protease by 25%, 1.6%, and 14.7%, respectively (Min et al., 2001). The chalcone isobavachalcone (LogD 4.3 at pH 7.4; molecular mass=324.3 g/mol) from *Angelica keiskei* Koidz. inhibited *Escherichia coli* (ATCC 10536), *Escherichia coli* (AG100A), *Enterobacter aerogenes* (ATCC 13048), *Enterobacter aerogenes* (EA298), *Klebsiella pneumoniae* (ATCC 11296), *Klebsiella pneumoniae* (KP63 MDR clinical isolates), *Pseudomonas aeruginosa* (PA124, MDR clinical isolates), and *Enterobacter cloacae* (Ec1194 clinical isolates) with MIC values of 128, 16, 256, 8, 32, 16, 64, and 64 µg/mL, respectively (Kuete et al., 2010). Isobavachalcone inhibited the replication of the Pseudorabies virus in PK15 cells at the concentration of 25.6 µM (Wang et al., 2020) and the replication of the Porcine reproductive and respiratory syndrome virus with the IC_{50} value of 3.1 µM and the selectivity index of 22 (Wang et al., 2018) and may be active against coronaviruses. *Angelica sinensis* (Oliv.) Diels is used for the treatment of COVID-19 in China (local name: *Dang hui*) (Wang et al., 2020) and for cold in Korea.

iv. Falcarindiol is a phytoalexin in members of the family Apiaceae (it abounds in carrot peel) and its mode of action involves, at least in part, bacterial and fungal cytoplasmic membrane damages with subsequent loss of cytoplasmic contents (Garrod et al., 1979).

v. In Pakistan, *Angelica glauca* Edgew. (local name: *Choora*) is used for cold and fever, in India (local name: *Choraka*) for cough and fever, and the Himalaya (local name: *Chura*).

vi. In India, *Angelica archangelica* L. var. *himalaica* (C. B. Clarke) Krishna and Badhwar (local name: *Chandaa*) is used for cough.

vii. During the Black Death, *Angelica archangelica* L. was used as protective remedy, and this plant had a mystical aura. Our peer, the French surgeon Ambroise Pare (1510–1590) recommends in his *Traicte de la peste, de la petite verole & rougeole: avec Une brefue defcription de la Lepre* published in 1568, recommends the use of *Angelica* (*cf. Angelica archangelica* L.) to prevent and cure plague. In 1551, the German theologian Philipp Melanchthon (1497–1560) and correspondent of our peer the German physician and botanist Leonardt Fusch (see earlier) prescribes a plant in the genus *Angelica* L. for protection against plague (Wickersheimer, 1923). In his 1578 edition *The Poore mans Ievvel, that is to say fay, a Treatife of the Peftilence*, Thomas Brasbridge (1537–1596) recommends the use of a plant in the genus *Angelica* L. and *Carduus Benedictus* (*cf, Cnicus benedictus* L. in the family Asteraceae) for protection against plague. The essential oil of *Angelica archangelica* L. inhibited the growth of *Yersinia enterocolitica* (ATCC 228) (Elgayyar et al., 2001) and as such could be able to inhibit the growth of *Yersinia pestis*, the zoonotic bacterium responsible for plague.

References

Deng, Y.J., 2020. Study on active compounds from Huoxiang Zhengqi Oral Liquid for prevention of coronavirus disease 2019 (COVID-19) based on network pharmacology and molecular docking. *Chinese Traditional and Herbal Drugs*, *5*, pp. 1113–1122.

Elgayyar, M., Draughon, F.A., Golden, D.A. and Mount, J.R., 2001. Antimicrobial activity of essential oils from plants against selected pathogenic and saprophytic microorganisms. Journal of Food Protection, 64(7), pp. 1019–1024.

Garrod, B., Lewis, B.G. and Coxon, D.T., 1978. Cis-heptadeca-1, 9-diene-4, 6-diyne-3, 8-diol, an antifungal polyacetylene from carrot root tissue. *Physiological Plant Pathology*, *13*(2), pp. 241–246.

Garrod, B., Lea, E.J.A. and Lewis, B.G., 1979. Studies on the mechanism of action of the antifungal compound falcarindiol. *New Phytologist*, *83*(2), pp. 463–471.

Lechner, D., Stavri, M., Oluwatuyi, M., Pereda-Miranda, R. and Gibbons, S., 2004. The anti-staphylococcal activity of Angelica dahurica (Bai Zhi). *Phytochemistry*, *65*(3), pp. 331–335.

Lee, B.W., Ha, T.K.Q., Cho, H.M., An, J.P., Kim, S.K., Kim, C.S., Kim, E. and Oh, W.K., 2020. Antiviral activity of furanocoumarins isolated from Angelica dahurica against influenza a viruses H1N1 and H9N2. *Journal of Ethnopharmacology*, *259*, p. 112945.

Matsuura, H., Saxena, G., Farmer, S.W., Hancock, R.E.W. and Towers, G.H.N., 1996. Antibacterial and antifungal polyine compounds from Glehnia littoralis ssp. leiocarpa. *Planta Medica*, *62*(03), pp. 256–259.
Meot-Duros, L., Cérantola, S., Talarmin, H., Le Meur, C., Le Floch, G. and Magné, C., 2010. New antibacterial and cytotoxic activities of falcarindiol isolated in Crithmum maritimum L. leaf extract. *Food and Chemical Toxicology*, *48*(2), pp. 553–557.
Rajtar, B., Skalicka-Woźniak, K., Świątek, Ł., Stec, A., Boguszewska, A. and Polz-Dacewicz, M., 2017. Antiviral effect of compounds derived from Angelica archangelica L. on Herpes simplex virus-1 and Coxsackievirus B3 infections. *Food and Chemical Toxicology*, *109*, pp. 1026–1031.
Wang, H.M., Liu, T.X., Wang, T.Y., Wang, G., Liu, Y.G., Liu, S.G., Tang, Y.D. and Cai, X.H., 2018. Isobavachalcone inhibits post-entry stages of the porcine reproductive and respiratory syndrome virus life cycle. *Archives of Virology*, *163*(5), pp. 1263–1270.
Wang, Y., Liu, T.X., Wang, T.Y., Tang, Y.D. and Wei, P., 2020. Isobavachalcone inhibits Pseudorabies virus by impairing virus-induced cell-to-cell fusion. *Virology Journal*, *17*(1), pp. 1–4.
Wickersheimer, E., 1923. Les recettes de Philippe Melanchthon contre la peste. *Revue d'Histoire et de Philosophie Religieuses*, *3*(2), pp. 169–173.
Zhao, A.H., Zhang, Y.B. and Yang, X.W., 2016. Simultaneous determination and pharmacokinetics of sixteen Angelicae dahurica coumarins in vivo by LC–ESI-MS/MS following oral delivery in rats. *Phytomedicine*, *23*(10), pp. 1029–1036.

11.3.1.2 *Bupleurum chinense* DC.

Synonyms: *Bupleurum chinense* var. *vanheurckii* (Müll. Arg.) Shan & Yin Li, *Bupleurum chinense* Franch., *Bupleurum flacutum* var. *ensifolium* H. Wolff, *Bupleurum togasii* Kitag., *Bupleurum vanheurckii* Müll. Arg.

Common names: Bupleurum; bee chia hu, chai hu (China), saiko (Japan), si ho (Korea)

Habitat: Meadows

Distribution: India, China, Japan, Korea, and Taiwan

Botanical description: This graceful umbellifer grows up to about 90 cm tall. The root is stout, elongate, brown, woody, and branched. The stems are zigzag-shaped, hollowed, and slender. The basal leaves are oblanceolate, 0.5–4 cm × 0.8–7 cm, tapering at base, and acuminate at apex. The middle leaves are elliptic, 1.5–4 cm × 1.8–10 cm, 7–9–nerved, glaucous below, and acute apex. The apical leaves are minute. The umbels are 5 cm across. The peduncles are slender and spreading. The umbellules are 5 mm in diameter and 5–10–flowered (the whole structure looking like a flower as with an asteraceous capitulum). The 5 bracts and the flowers are of a unique type of greenish-yellow and broadly lanceolate. The flowers are minute. The schizocarps are oblong, brown, and 3 mm in diameter.

Medicinal uses: Colds, hepatitis, pneumonia (China)

Strong antiviral (enveloped monopartite linear single-stranded (+) RNA) oleanane-type triterpene saponins: Saikosaponin A, saikosaponin B2, saikosaponin C, and saikosaponin D (Zhu et al., 2009) inhibited the replication of the Human coronavirus 229E (ATCC VR-740) with the EC_{50} values of 8.6, 1.7, 19.9, and 13.2 µM, respectively and the selectivity indexes of 26.6, 221.9, 19.2, and 13.3, respectively, in MRC-5 cells by a mechanism, involving at least in part, inhibition of viral attachment and penetration (Cheng et al., 2006). In a subsequent study, saikosaponin B2 inhibited the replication of the Human coronavirus 229E (ATCC VR-740) with the IC value of 1.7 µmol/L, respectively (Li et al., 2015).

Strong antiviral (enveloped circular double-stranded DNA) oleanane-type saponin: Saikosaponin D (Zhu et al., 2009) inhibited the replication of the Hepatitis B virus with the IC_{50} value of 0.3 µg/mL and selectivity index of 6.6 (Yin et al., 2008).

Strong antiviral (enveloped monopartite linear single-stranded (-) RNA) oleanane-type triterpene saponins: Saikosaponin A, saikosaponin B2, saikosaponin C, and saikosaponin D inhibited the replication of the Hepatitis C virus in HuH7 with the EC_{50} values of 1.1, 16.1, 86.1, and 2.8 µM and selectivity indexes of 29.6, 740.4, 716.7, and 23.3, respectively, via a mechanism involving, at least in part, virucidal effect, inhibition of virus attachment to host cell, and prevention of entry in host cell (Lin et al., 2015).

Commentaries

i. The plant is used for COVID-19 in China (Zhong et al., 2020). Oleanane-type triterpene saponins are often able to inhibit the replication of coronaviruses. One such saponin is glycyrrhizin (see earlier). Saponins could be virucidal since saikosaponin D at a concentration of about 5 µM inactivated the Measles virus and the Herpes simplex virus (Ushio & Abe, 1992). Saponins may alter the outer envelope of enveloped viruses and thus compromising anchoring of the virus to host cells. Further, saponins do not cross the cytoplasmic membrane of host cells (unless by transporters?) and thus cannot interfere *per se* with the intracellular viral replication process, and if they do, it may involve the stimulation of cellular membrane receptors. Fusion between the influenza viral envelope and host cell membranes is a required step in viral infection, as this releases the viral core into the host cytoplasm. Viral envelope contains cholesterol which promotes fusion of the virus with the host cell (Zawada et al., 2016), and cholesterol in host cell membrane promotes the fusion of the virion to the host cell (Ahn et al., 2002; Kielian & Helenius, 1984). Since triterpene saponins bind to membrane cholesterol (Wojciechowski et al., 2016), one could suggest that the mode of action of triterpene saponins involves at least in part binding to viral and host membrane cholesterol hampering virion-host cell fusion.

ii. In India, *Bupleurum flacutum* L. (local name: *Sipil*) is used for fever.

iii. In Pakistan, *Bupleurum longicaule* Wall. ex DC. (local name: *Gillo*) is used for fever.

References

Ahn, A., Gibbons, D.L. and Kielian, M., 2002. The fusion peptide of Semliki Forest virus associates with sterol-rich membrane domains. *Journal of Virology*, *76*(7), pp. 3267–3275.

Cheng, P.W., Ng, L.T., Chiang, L.C. and Lin, C.C., 2006. Antiviral effects of saikosaponins on human coronavirus 229E in vitro. *Clinical and Experimental Pharmacology and Physiology 33*(7), pp. 612–616.

Kielian, M.C. and Helenius, A., 1984. Role of cholesterol in fusion of Semliki Forest virus with membranes. *Journal of Virology*, *52*(1), pp. 281–283.

Li, D.Q., Wu, J., Liu, L.Y., Wu, Y.Y., Li, L.Z., Huang, X.X., Liu, Q.B., Yang, J.Y., Song, S.J. and Wu, C.F., 2015. Cytotoxic triterpenoid glycosides (saikosaponins) from the roots of Bupleurum chinense. *Bioorganic & Medicinal Chemistry Letters*, *25*(18), pp. 3887–3892.

Lin, L.T., Chung, C.Y., Hsu, W.C., Chang, S.P., Hung, T.C., Shields, J., Russell, R.S., Lin, C.C., Li, C.F., Yen, M.H. and Tyrrell, D.L.J., 2015. Saikosaponin b2 is a naturally occurring terpenoid that efficiently inhibits hepatitis C virus entry. *Journal of Hepatology*, *62*(3), pp. 541–548.

Ushio, Y. and Abe, H., 1992. Inactivation of measles virus and herpes simplex virus by saikosaponin d. *Planta Medica*, *58*(02), pp. 171–173.

Wojciechowski, K., Orczyk, M., Gutberlet, T. and Geue, T., 2016. Complexation of phospholipids and cholesterol by triterpenic saponins in bulk and in monolayers. *Biochimica et Biophysica Acta (BBA)-Biomembranes*, *1858*(2), pp. 363–373.

Yin, F., Pan, R., Chen, R. and Hu, L., 2008. Saikosaponins from Bupleurum chinense and inhibition of HBV DNA replication activity. *Natural Product Communications*, *3*(2), pp. 155–157.

Zawada, K.E., Wrona, D., Rawle, R.J. and Kasson, P.M., 2016. Influenza viral membrane fusion is sensitive to sterol concentration but surprisingly robust to sterol chemical identity. *Scientific Reports*, *6*(1), pp. 1–10.

Zhong, L.L.D., Lam, W.C., Yang, W., Chan, K.W., Sze, S.C.W., Miao, J., Yung, K.K.L., Bian, Z. and Wong, V.T., 2020. Potential targets for treatment of coronavirus disease 2019 (COVID-19): A review of Qing-Fei-Pai-du-tang and its major herbs. *The American Journal of Chinese Medicine*, *48*(05), pp. 1051–1071.

Zhu, Z., Liang, Z. and Han, R., 2009. Growth and saikosaponin production of the medicinal herb Bupleurum chinense DC. under different levels of nitrogen and phosphorus. *Industrial Crops and Products*, *29*(1), pp. 96–101.

11.3.1.3 *Cnidium monnieri* (L.) Cusson

Synonym: *Selinum monnieri* L.

Common names: Monnier's snowparsley; shay chwang tsze, she chuang (China), jashoshi (Japan), beol sa sang ja (Korea)

Habitat: Riparian grasslands

Distribution: India, Vietnam, Laos China, Korea, and Mongolia

Botanical description: This beautiful and gracile umbellifer grows up to 60 cm tall. The root is slender. The stems are striate and scabrous. The leaves are compound, spiral, and exstipulate. The lower petiole is 3–8 cm long. The blade is lanceolate, 2–5 cm × 3–8 cm, and pinnate. The folioles are deeply incised. The umbels are 2–3 cm across. The umbellules are densely flowered, and the flowers are minute and pure white. The schizocarps are ovoid and about 3 mm long.

Medicinal uses: Dysentery (China); leucorrhea (Korea)

Strong antifungal amphiphilic prenylated coumarins: Osthol (LogD 3.4 at pH 7.4; molecular mass = 244.2 g/mol) at the concentration of 30 µg/mL inhibited the growth of *Sclerotinia sclerotiorum*, *Thanatephorus cucumeris*, *Botrytis cinerea*, *Fusarium graminearum*, and *Colletotrichum capsici* by 30%, 63.8%, 57.6%, 53.7%, and 8.2%, respectively (Song et al., 2017). Bergapten at the concentration of 30 µg/mL inhibited the growth of *Sclerotinia sclerotiorum*, *Thanatephorus cucumeris*, *Botrytis cinerea*, *Fusarium graminearum*, and *Colletotrichum capsici* by 29%, 41.1%, 19.3%, 57.2%, and 8.2%, respectively (Song et al., 2017).

Strong antiviral (enveloped monopartite linear double-stranded DNA) coumarin derivatives: (+)-Cnidimonin A, (–)-cnidimonin A, (±) cnidimonin A, and (+)-cnidimonin B inhibited the replication of the Herpes simplex virus type-1 with the IC_{50} values of 19.2, 16, 2.1, and 23.1 µM, respectively (Su et al., 2017).

Antiviral (enveloped monopartite linear double-stranded DNA) prenylated coumarin: Bergapten inhibited the replication of the Herpes simplex virus at the concentration of 31.2 µg/mL (Rajtar et al., 2017).

Antiviral (non-enveloped monopartite linear single-stranded (+) RNA) prenylated coumarin: Bergapten inhibited the replication of the Coxsackievirus type B3 at the concentration of 31.2 µg/mL (Rajtar et al., 2017).

In vivo antiviral (non-enveloped linear monopartite single-stranded (+) RNA) prenylated coumarin: Osthol inhibited the replication of the Tobacco mosaic virus (Chen et al., 2020).

Commentaries

i. The plant produces a series of phytoanticipins coumarins such as bergapten, osthol, and imperatorin (Sun et al., 1990; Masuda et al., 1998). Note that bergapten inhibits efflux pumps in mammalian cells (Mirzaei et al., 2017) and as such could be able to inhibit bacterial and/or fungal efflux pumps.

ii. *Cnidium monnieri* (L.) Cusson belongs to the Subfamily Apioideae, Tribe Selineae. In the Subfamily Apioideae, Tribe Selineae, *Glehnia littoralis* F. Schmidt ex Miq. is used to treat tuberculosis in China. Methanol extract of *Glehnia littoralis* F. Schmidt ex Miq. (20 mg/disc) inhibited the growth of *Mycobacterium tuberculosis* (ATCC 35801) and *Mycobacterium avium* (ATCC 25291) (McCutcheon et al., 1997). In the Subfamily Apioideae, Tribe Selineae, methanol extract of roots *Ledebouriella seseloides* (Hoffm.) H. Wolff (synonym: *Saposhnikovia divaricata* (Turcz.) Schischk.) is used for the treatment of COVID-19 in China (Wang et al., 2020). Methanol extract of this herb inhibited the Human immunodeficiency type-1 protease (Park, 2003), and methanol extract of roots at the concentration of 100 µg/mL inhibited RNA-dependent DNA polymerase and ribonuclease H activities of Human immunodeficiency type-1 reverse transcriptase (Min et al., 2001). From this plant the coumarin derivative divaricoumarin A, divaricoumarin B, and divaricoumarin C at the concentration of 20 µM inhibited the replication of the Porcine epidemic diarrhea virus (Yang et al., 2015) which is a coronavirus.

References

Chen, Y.H., Guo, D.S., Lu, M.H., Yue, J.Y., Liu, Y., Shang, C.M., An, D.R. and Zhao, M.M., 2020. Inhibitory effect of osthole from Cnidium monnieri on tobacco mosaic virus (TMV) infection in Nicotiana glutinosa. *Molecules*, *25*(1), p. 65.

Masuda, T., Takasugi, M. and Anetai, M., 1998. Psoralen and other linear furanocoumarins as phytoalexins in Glehnia littoralis. *Phytochemistry*, *47*(1), pp. 13–16.

McCutcheon, A.R., Stokes, R.W., Thorson, L.M., Ellis, S.M., Hancock, R.E.W. and Towers, G.H.N., 1997. Anti-mycobacterial screening of British Columbian medicinal plants. International Journal of Pharmacognosy, 35(2), pp. 77–83.

Mirzaei, S.A., Dehkordi, N.G., Ghamghami, M., Amiri, A.H., Abdolahinia, E.D. and Elahian, F., 2017. ABC-transporter blockage mediated by xanthotoxin and bergapten is the major pathway for chemosensitization of multidrug-resistant cancer cells. *Toxicology and Applied Pharmacology*, *337*, pp. 22–29.

Rajtar, B., Skalicka-Woźniak, K., Świątek, Ł., Stec, A., Boguszewska, A. and Polz-Dacewicz, M., 2017. Antiviral effect of compounds derived from Angelica archangelica L. on Herpes simplex virus-1 and Coxsackievirus B3 infections. *Food and Chemical Toxicology*, *109*, pp. 1026–1031.

Song, P.P., Zhao, J., Liu, Z.L., Duan, Y.B., Hou, Y.P., Zhao, C.Q., Wu, M., Wei, M., Wang, N.H., Lv, Y. and Han, Z.J., 2017. Evaluation of antifungal activities and structure–activity relationships of coumarin derivatives. *Pest Management Science*, *73*(1), pp. 94–101.

Su, F., Zhao, Z., Ma, S., Wang, R., Li, Y., Liu, Y., Li, Y., Li, L., Qu, J. and Yu, S., 2017. Cnidimonins A–C, three types of hybrid dimer from Cnidium monnieri: Structural elucidation and semisynthesis. *Organic letters*, *19*(18), pp. 4920–4923.

Wang, S.X., Wang, Y., Lu, Y.B., Li, J.Y., Song, Y.J., Nyamgerelt, M. and Wang, X.X., 2020. Diagnosis and treatment of novel coronavirus pneumonia based on the theory of traditional Chinese medicine. *Journal of Integrative Medicinem*, *18*, pp. 275–283.

Yang, J.L., Dhodary, B., Ha, T.K.Q., Kim, J., Kim, E. and Oh, W.K., 2015. Three new coumarins from Saposhnikovia divaricata and their porcine epidemic diarrhea virus (PEDV) inhibitory activity. *Tetrahedron*, *71*(28), pp. 4651–4658.

11.3.1.4 *Torilis japonica* (Houtt.) DC.

Synonyms: *Caucalis japonica* Houtt.; *Caucalis praetermissa* (Hance) Franch., *Tordylium anthriscus* L., *Torilis anthriscus* (L.) C.C. Gmel., *Torilis anthriscus* var. *japonica* (Houtt.) H. Boissieu, *Torilis praetermissa* Hance

Common names: Upright hedge–parsley, Japanese hedge–parsley, xiao qie yi (China), yabujirami (Japan), sa sang ja (Korea)

Habitat: Forests, meadows

Distribution: Temperate Asia

Botanical observation: This gracile herb grows to 1 m tall. The basal and lower cauline leaves are petiolate. The petiole is 1.5–10 cm long. The blade is triangular, bipinnate, and about 20 cm × 15 cm. The pinnae are ovate and 1–2 cm ×2.5–5 cm. The peduncles are 2.5–25 cm long and hispid. The bracts are linear. The umbellules are 4–12–flowered and magnificent. The 5 sepals are small and deltoid. The 5 petals are pure white. The schizocarps are purple, globose, hairy, and about 5 mm long.

Medicinal uses: Dysentery (Korea)

Moderate broad-spectrum antibacterial essential oil: Essential oil of aerial parts inhibited the growth of *Klebsiella pneumoniae*, *Staphylococcus aureus*, *Salmonella enterolitica*, *Escherichia coli*, *Pseudomonas aeruginosa*, and *Proteus vulgaris* with the MIC values of 150, 175, 175, 150, 150, and 150 µL/mL, respectively (Kharkwala et al., 2017).

Moderate broad-spectrum anticandidal essential oil: Essential oil of aerial parts inhibited the growth of *Candida albicans* and *Candida guilliermondii* with the MIC values of 175 and 175µL/mL, respectively (Kharkwala et al., 2017).

Antibacterial (Gram-positive) sesquiterpene: Torilin at 0.1% abrogated the growth of *Bacillus subtilis* at a concentration of 50 ppm (Cho et al., 2008)

Strong antiviral (enveloped monopartite linear single-stranded (+)-RNA) polar extract: Methanol extract at the concentration of 50 µg/mL abrogated the replication of the Mouse hepatitis virus (strain MHV-A59), Mouse hepatitis virus (strain MHV-JHM), and Porcine epidemic diarrhea virus (Kim et al., 2010).

Strong antiviral (enveloped monopartite linear single-stranded (-) RNA) polar extract: Methanol extract at the concentration of 50 µg/mL abrogated the replication of the Vesicular stomatitis virus (Kim et al., 2010).

Commentary

i. The plant assembles the cytotoxic guaiane-type sesquiterpene torilin (Park et al., 2006) which could be able to inhibit bacterial and/or fungal efflux pumps because it decreases the resistance of mammalian cells to adriamycin, vinblastine, taxol, and colchicine (Kim et al., 1998). Torilin could be examined for activity against the Severe acute respiratory syndrome-associated coronavirus 2.

References

Cho, W.I., Choi, J.B., Lee, K., Chung, M.S. and Pyun, Y.R., 2008. Antimicrobial activity of torilin isolated from Torilis japonica fruit against Bacillus subtilis. *Journal of Food Science*, *73*(2), pp. M37–M46.

Kharkwala, G.C., Pandea, C., Tewari, G., Panwarb, A. and Pandeb, V., 2017. Terpenoid composition and antimicrobial activity of essential oil from Torilis japonica (Houtt.) DC. *Journal of the Indian Chemical Society*, *94*, pp. 191–194.

Kim, S.E., Kim, Y.H., Kim, Y.C. and Lee, J.J., 1998. Torilin, a sesquiterpene from Torilis japonica, reverses multidrug-resistance in cancer cells. *Planta Medica*, *64*(04), pp. 332–334.

Kim, H.Y., Eo, E.Y., Park, H., Kim, Y.C., Park, S., Shin, H.J. and Kim, K., 2010. Medicinal herbal extracts of Sophorae radix, Acanthopanacis cortex, Sanguisorbae radix and Torilis fructus inhibit coronavirus replication in vitro. *Antivir Therapy*, *15*(5), pp. 697–709.

Park, H.W., Choi, S.U., Baek, N.I., Kim, S.H., Eun, J.S., Yang, J.H. and Kim, D.K., 2006. Guaiane sesquiterpenoids fromTorilis japonica and their cytotoxic effects on human cancer cell lines. *Archives of Pharmacal Research*, *29*(2), p. 131.

11.3.1.5 *Trachyspermum ammi* (L.) Sprague

Synonyms: *Ammi copticum* L., *Bunium copticum* (L.) Spreng., *Carum copticum* (L.) C.B. Clarke, *Daucus copticus* (L.) Pers., *Ptychotis coptica* (L.) DC. *Sison ammi* L.; *Trachyspermum copticum* (L.) Link

Common names: Ajowan, ajwain, bishop's weed, xi ye cao guo qin (China); yavan (Bangladesh); ajma, ajwain, yamini, yavaanikaa (India); ziniyan (Iran); samone hpyu (Myanmar)

Habitat: Wastelands, roadsides, vacant lots, cultivated

Distribution: Iran, Afghanistan, Pakistan, India, Bangladesh, Myanmar, China, the Philippines

Botanical description: This gracile and elegant umbellifer grows up to about 40 cm tall. The petiole is 1–5 cm long. The blade is triangular, about 2 cm × 8 cm, and pinnate. The folioles are deeply incised. The umbels are 2.5–5 cm in diameter. The umbellules are 1 cm across and densely flowered. The 5 sepals are minute. The 5 petals are minute and white. The schizocarps are minute, ribbed, and aromatic.

Medicinal uses: AIDS (India); cholera (India, Iran); diarrhea (India, Myanmar, Pakistan); dysentery (Myanmar); post-partum (Bangladesh, India)

Broad-spectrum antibacterial halo developed by polar extract: Acetone extract of seeds inhibited the growth of *Enterococcus faecalis* (MTCC 439), *Staphylococcus aureus* (MTCC 96), *Escherichia coli* (MTCC 119), *Klebsiella pneumoniae* 1 (MTCC 109), *Klebsiella pneumoniae* (MTCC 530), *Pseudomonas aeruginosa* 1 (MTCC 647), *Pseudomonas aeruginosa* 2 (MTCC 741), *Salmonella typhi* (MTCC 531), *Salmonella typhimurium* 1 (MTCC 98), *Salmonella typhimurium* 2 (MTCC 1251), and *Shigella flexneri* (MTCC 1457) (Kaur & Arora, 2009).

Moderate antibacterial broad-spectrum polar extract: Methanol extract of seeds inhibited the growth of *Staphylococcus aureus* (ATCC 6538), *Staphylococcus aureus* (KCTC 1916), *Bacillus*

subtilis (ATCC 6633), *Pseudomonas aeruginosa* (KCTC 2004), *Salmonella typhimurium* (KCTC 2515), *Escherichia coli* (ATCC 8739), *Enterobacter aerogenes* (KCTC 2190), *Salmonella enteritidis* (KCTC 12021), and *Escherichia coli* (0157:H7 ATCC 43888) with the MIC values of 175, 175, 87.5, 112.5, 325, 500, 125, 375, and 450 µg/mL, respectively (Paul et al., 2011).

Strong antibacterial broad-spectrum essential oil: Essential oil of fruits inhibited the growth of *Staphylococcus aureus* (ATCC 6538), *Staphylococcus aureus* (KCTC 1916), *Bacillus subtilis* (ATCC 6633), *Pseudomonas aeruginosa* (KCTC 2004), *Salmonella typhimurium* (KCTC 2515), *Escherichia coli* (ATCC 8739), *Enterobacter aerogenes* (KCTC 2190), *Salmonella enteritidis* (KCTC 12021), and *Escherichia coli* (0157:H7 ATCC 43888) with the MIC values of 162.5, 175, 12.5, 12.5, 225, 462.5, 12.5, 250, and 450 µg/mL, respectively (Paul et al., 2011). In *Bacillus subtilis* (ATCC 6633), the essential oil induced cytoplasmic membrane pore formation and lysis (Paul et al., 2011). In a subsequent study, oil from fruits inhibited the growth of *Staphylococcus aureus*, *Bacillus subtilis*, *Enterococcus faecalis*, *Listeria monocytogenes*, *Escherichia coli*, *Pseudomonas aeruginosa*, *Salmonella typhi*, and *Shigella dysenteriae* (Aman et al., 2014).

Strong broad-spectrum antibacterial amphiphilic aromatic monoterpene: Thymol (LogD 3 at pH 7.4; molecular mass = 150.2 g/mol) inhibited the growth of *Escherichia coli*, *Pseudomonas aeruginosa*, *Proteus mirabilis*, *Klebsiella pneumoniae*, *Acinetobacter baumannii*, *Staphylococcus aureus*, *Enterococcus faecalis*, and *Bacillus subtilis* with the MIC values of 8, 4, 8, 8, 2, 16, 8, and 8 µg/mL, respectively (Orhan et al., 2012).

Strong anticandidal essential oil: Essential oil of fruits inhibited the growth of fluconazole-sensitive *Candida albicans* and fluconazole-resistant *Candida albicans* with the MIC values of 4.2 and 79.4 µg/mL, respectively (Sharifzadeh et al., 2015).

Weak antiviral (enveloped monopartite linear single-stranded (+) RNA) essential oil: Essential oil at the concentration of 500 µg/mL inhibited the replication of the Japanese encephalitis virus by 40% (Roy et al., 2015).

Strong anticandidal amphiphilic aromatic monoterpene: Thymol inhibited the growth of *Candida albicans* (ATCC 10231) and *Candida parapsilosis* (ATCC 22019) with the MIC values of 8 and 16 µg/mL, respectively (Orhan et al., 2012).

Strong antiviral (enveloped monopartite linear double-stranded DNA) amphiphilic aromatic monoterpene: Thymol (LogD 3 at pH 7.4; molecular mass = 150.2 g/mol) inhibited the replication of the Herpes simplex virus type-1 in Vero cells with the IC_{50} value of 0.002% (Sharifi-Rad et al., 2018). In a subsequent study, thymol at the concentration of 0.8 µg/mL inhibited the replication of the Herpes simplex virus type-1 in Madin-Darby canine kidney cells (Orhan et al., 2012).

Strong antiviral (enveloped monopartite linear negative stranded (-) RNA) amphiphilic aromatic monoterpene: Thymol at the concentration of 1.6 µg/mL inhibited the replication of the Parainfluenza virus type-3 in Vero cells (Orhan et al., 2012).

Commentaries

i. In India, *Trachyspermum roxburghianum* (DC.) Craib. (local name: *Karafs-e-Hindi*) is used to treat bronchitis.

ii. *Trachyspermum ammi* (L.) Sprague belongs to the Subfamily Apioideae, Tribe Pyramidoptereae. In the Subfamily Apioideae, Tribe Scandiceae, *Cuminum cyminum* L. also known as cumin is used in India (local name: *Shveta-jiraka*) for diarrhea. Aqueous extract of seeds (10 g/100 mL impregnated to 6 mm disc) inhibited the growth of *Escherichia coli*, *Klebsiella pneumoniae*, *Pseudomonas aeruginosa*, and *Staphylococcus aureus* with the inhibition zone diameters of 23.8, 15.9, 12.3, and 8.9 mm, respectively (Chaudhry and Tariq, 2008). Essential oil inhibited the growth of *Candida albicans* (clinical strain) with the MIC/MFC value of 11.7/23.3 µg/mL (Minooeianhaghighi et al., 2017). Methanol extract of fruits inhibited the replication of the Herpes simplex virus type-1 with the IC_{50} value of 180 µg/mL and a selectivity index of 2.5 in Vero cells (Motamedifar et al., 2010).

iii. The burning of resin of *Ferula assa-foetida* L. (Subfamily Apioideae, Tribe Scandiceae) (known as "*Hing*" in Arabic) was recommended by our peer the Persian physician Ibn Sina (see earlier) in time of pestilences. Ibn Sina called pestilences "*Amraz e Wabai*" in his monumental work (see earlier), describes air as mean of contamination, and recommends to quarantine people for 40 days, a method presumably later used by Venetians under the name of "*Quarantena*" (Farooqui et al., 2020)., During the Black Death, the burning of the resin of *Ferula assa-foetida* L. was recommended by our peer the Physician Ben Solomon Hen in his "*Tractatulus de pestilentia*" (see earlier) (Bos & Mensching, 2011). In Iran, the plant is used for wounds (local name: *Anghuzeh*), in Pakistan (local name: *Sup*) is used for cough, fever, and ulcers; in India, the plant is called "*Hingu*" and is used for cough. The coumarin derivatives 5′-hydroxyumbelliprenin, methyl galbanate, farnesiferol C (LogD 5.1 at pH 7.4; molecular mass = 282.4 g/mol), and *epi*-conferdione isolated from the resin inhibited the replication of the H1N1 in Madin-Darby canine kidney cells with the IC_{50} values of 0.5, 0.2, 0.2, and 0.3 µg/mL, respectively (Lee et al., 2009). The coumarin kellerin inhibited the replication of the Herpes simplex virus type-1 with the EC_{50} as 38 µg/mL in Vero cells (Ghannadi et al., 2014). *Ferula galbaniflua* Boiss. & Buhse also known as Galbanum is used for sores and wounds in Iraq and Iran (local name: *Gavshira*) and for cough in India (local name: *Gaosheer*). In *A Dialogue bothe pleasaunte and pietifull, wherein is a goodly regimente against the fever Pestilence, with a consolacion and comfort against death.* of our peer the English physician William Bullein (1515–1576) published in 1564, recommends Galbanum for plague.

iv. In India, *Ferula jaeschkeana* Vatke is used for dental carries. In Turkey, *Ferula orientalis* L. (local name: *Kintor*) is an antiseptic. The aliphatic bisulfide persicasulfide A isolated from the roots of *Ferula persica* Willd. inhibited the growth of *Aspergillus fumigatus* (PIM 112) and *Trichophyton mentagrophytes* with the MIC values of 62.5 and 62.5 µg/mL, respectively (Mirjani et al., 2005). The guaiane-type sesquiterpene diversolide F isolated from the roots of *Ferula diversivittata* Regel & Schmalh. inhibited the growth of *Staphylococcus aureus*, *Escherichia coli* (ATCC 8739), *Aspergillus niger* (ATCC 1624), and *Candida albicans* (ATCC 14053) (Iranshahi et al., 2008).

v. In the Subfamily Apioideae, Tribe Peucedaneae, *Anethum graveolens* L. also known as Dill is used for urinary tract infection in Iran (local name: *Shevīd*), bronchitis in India (local name: *Sowa*), and fever and sores in Myanmar (local name: *Sameik*). Essential oil of *Anethum graveolens* L. inhibited the growth of *Streptococcus mutans* and *Lactobacillus acidophilus* with MIC values of 2.5 and 5 µL/mL, respectively (Miller et al., 2015). Essential oil inhibited the growth of *Escherichia coli*, *Pseudomonas aeruginosa*, *Proteus mirabilis*, *Klebsiella pneumoniae*, *Acinetobacter baumannii*, *Staphylococcus aureus*, *Enterococcus faecalis*, *Bacillus subtilis*, *Candida albicans* (ATCC 10231), and *Candida parapsilosis* (ATCC 22019) (Orhan et al., 2012). Essential oil at the concentration of 0.8 µg/mL inhibited the replication of the Herpes simplex virus type-1 in Madin-Darby canine kidney cells and Parainfluenza virus type-3 in Vero cells (Orhan et al., 2012).

vi. In the Subfamily Mackinlayoidea, *Centella asiatica* (L.) Urban is used in the form of salad in Bangladesh (Mohammed Rahmatullah personal communication), in Malaysia (it is part of the plants eaten raw as therapeutic side dishes or *Ulams*), and Indonesia (where it is part of the *Jamus*). This water-loving herb is used for bronchitis, diarrhea, dysentery, and jaundice in India (local name: *Brahma-manduuki*) for fever and dysentery in Nepal (local name: *Ghortapre*), cold, cough, dysentery, fever, jaundice, ulcers, syphilis, and wounds in Bangladesh (local name: *Thankuni*), sores and urinary tract infection in Myanmar (local name: *Myin-hkwa*), cough in Sabah (local name: *Salapid*), dysentery in the Philippines (local name: *Jahog-jahong*), and cough and diarrhea in Papua New Guinea (local name: *Yotubukona*). The plant is used for leprosy (due to *Mycobacterium leprae*) in India, and ethanol extract inhibited the growth of *Mycobacterium tuberculosis* (Suresh et al., 2010). Hexane extract of leaves inhibited the growth of *Mycobacterium chelonae* with the MIC value of 78.1 µg/mL (Machado et al., 2015). Ethanol extract inhibited the replication of the Vesicular stomatitis virus with the MIC value of 100 µg/mL in Vero cells (Hamidi et al., 1996). The plant yields ursane-type triterpene saponins including asiaticoside and madecassoside (Aziz et al., 2007). Triterpene saponins often inhibit the anchoring and fusion of enveloped viruses to host cells (see earlier). Asiaticoside inhibited the secretion of the Hepatitis B virus surface antigen and e-antigen (Huang et al., 2013) and inhibited the replication of the Herpes simplex virus type-1 and type-2 (Yoosook et al., 2000). Asiaticoside and madecassoside inhibited the Human immunodeficiency virus type-1 viral protein R at the concentration of 10 µM (Jaisi et al., 2021). Asiaticoside and madecassoside are anti-inflammatory (Nurlaily et al., 2012) and account probably for the fact that an extract of the plant evoked some levels of pulmonary healing in COVID-19 convalescents (Cesarone et al., 2021). Thus, *Centella asiatica* (L.) Urban could be of value for the prevention or mitigation of COVID-19 and forthcoming coronavirus pandemics. There is the need to test these saponins *in vitro* against the anti-Severe acute respiratory syndrome-associated coronavirus.

vii. In the Subfamily Apioideae; Tribe Coriandreae, *Coriandrum sativum* L. also known as coriander (a constituent of curry powder) is used to treat jaundice in Iran (local name: *Gezhnis*), carbuncles, sores, ulcers, venereal diseases in Pakistan (local name: *Dhaniya*), jaundice in India (local name: *Dhania*), sores in Bangladesh (local name: *Donia*), bronchitis sore throat in Myanmar (local name: *Nannan*), and measles in China. Essential oil inhibited the growth of *Escherichia coli* (ATCC 25922), *Klebsiella pneumoniae* (ATCC 13883), *Salmonella typhimurium* (ATCC 13311), *Pseudomonas aeruginosa* (ATCC 27853), *Acinetobacter baumannii*, *Bacillus cereus* (ATCC 11778), *Staphylococcus aureus* (ATCC 25923), and *Enterococcus faecalis* (ATCC 29212) (Silva et al., 2011). Essential oil inhibited the growth of *Microsporum canis* and *Candida* sp. with the MIC/MFC values of 78/150 and 620/1250 µg/mL, respectively (Soares et al., 2012). Methanol extract inhibited the replication of the Hepatitis A virus (HM-175 strain) at the concentration of 50 µg/mL in FRhK-4 cells (Seo et al., 2017).

viii. In the Subfamily Apioideae, Tribe Oenantheae, methanol extract of aerial parts of *Cryptotaenia japonica* Hassk. inhibited the Human immunodeficiency virus type-1 protease by 37.6% at the concentration of 100 µg/mL (Min et al., 1999). *Oenanthe javanica* DC. is used as vegetable in many parts of Asia including Malaysia and a common sight in the wet markets of Malaysia and South Korea (local name: *Minari*) where it also affords a remedy for jaundice, and it is used for cough in Papua New Guinea (local name: *Igundaurautu*). From this vegetable, fraction enriched in flavones inhibited the replication of the Hepatitis B virus both *in vitro* and *in vivo* (Wang et al., 2005). Ethanol extract inhibited the replication of the Vesicular stomatitis virus with the MIC value of 20 µg/mL in Vero cells (Hamidi et al., 1996).

ix. In the Subfamily Apioideae, *Heracleum moellendorffii* Hance yields the polyacetylenes falcarindiol (see earlier) and panaxynol (Nakano et al., 1998) (also found in members of the closely related family Araliaceae which may eventually be incorporated in the Apiaceae (!)). Ethanol extract of this herb inhibited *in vivo* the replication of the Tobacco mosaic virus (strain U1) (Jing et al., 2012).

x. In the Subfamily Apioideae, Tribe Saniculae, *Eryngium foetidum* L. is a remedy for diarrhea and fever in India (local name: *Awaphadigom*), burns in Nepal (local name: *Bandhania*), and burns, diarrhea, fever, and pneumonia in Bangladesh (local name: *Bilati dhonia*). Ethyl acetate extract of leaves inhibited the growth of *Escherichia coli*, *Pseudomonas aeruginosa*, *Bacillus subtilis*, *Staphylococcus aureus*, and *Candida albicans* with the MIC values of 50, 3.1, 50, 3.1, and 1.5 µg/mL, respectively (Lingaraju et al., 2016). In Turkey, *Eryngium billardieri* Delile (local names: *Esekdikeni, Kerenge kera*) is used for abscesses and wounds and *Eryngium virens* Link (local name: *Bogadikeni*) is given for jaundice. In India, *Eryngium caeruleum* M. Bieb. (local name: *Pahaari gaajar*) is used for cough. Essential oil of aerial parts inhibited the growth of *Staphylococcus aureus*, *Listeria monocytogenes*, *Shewanella* sp., *Pseudomonas aeruginosa*, and *Escherichia coli* with the MIC/MBC values of 125/250, 250/500, 250/500, 500/1000, and 1000/1000 µg/mL, respectively (Mirahmadi et al., 2020). In the Subfamily Apioideae, Tribe Saniculeae, *Sanicula caerulescens* Franch. is used to heal sores in China.

xi. In the Subfamily Apioideae, Tribe Apieae, *Ammi visnaga* (L.) Lam. is used for cough in India. Methanol extract of fruits of this herb at the concentration of 100 µg/mL inhibited the Human immunodeficiency virus type-1 reverse transcriptase by 32.7% (Hussein et al., 1999). *Ammi visnaga* (L.) Lam. (local name: *Khella*) yields the vasorelaxant furanochromone khellin (Anrep et al., 1949) which has probably antibiotic or antifungal potentiating effects via efflux pump inhibition (see earlier).

References

Aman, S., Naim, A., Siddiqi, R. and Naz, S., 2014. Antimicrobial polyphenols from small tropical fruits, tea and spice oilseeds. *Revista de Agaroquimica y Tecnologia de Alimentos*, *20*(4), pp. 241–251.

Anrep, G.V., Kenawy, M.R. and Barsoum, G.S., 1949. The coronary vasodilator action of khellin. *American Heart Journal*, *37*(4), pp. 531–542.

Aziz, Z.A., Davey, M.R., Power, J.B., Anthony, P., Smith, R.M. and Lowe, K.C., 2007. Production of asiaticoside and madecassoside in Centella asiatica in vitro and in vivo. *Biologia Plantarum*, *51*(1), pp. 34–42.

Bos, G. and Mensching, G., 2011. The black death in hebrew literature: Abraham Ben Solomon Hen's "Tractatulus de pestilential". *Jewish Studies Quarterly*, *18*(1), pp. 32–63.

Cesarone, M.R., Hu, S., Belcaro, G., Cornelli, U., Feragalli, B., Corsi, M., Bombardelli, E., Cotellese, R., Hosoi, M. and Rosenkvist, L., 2021. Pycnogenol®-Centellicum® supplementation improves lung fibrosis and Post-Covid-19 lung healing. *Minerva Medica*. doi: 10.23736/S0026-4806.20.07225-0.

Chaudhry, N.M.A. and Tariq, P., 2008. In vitro antibacterial activities of kalonji, cumin and poppy seed. *Pakistan Journal of Botany*, *40*(1), p. 461.

Farooqui, S.P., Quamri, M.A., Basheer, R., Baig, S., Ahmed, K., Ahad, M. and Ansari, F., 2020. COVID-19 pandemic-perspective through Unani Medicine. *Journal of Research in Traditional Medicine*, *6*(2), pp. 37–46.

Ghannadi, A., Fattahian, K., Shokoohinia, Y., Behbahani, M. and Shahnoush, A., 2014. Anti-viral evaluation of sesquiterpene coumarins from Ferula assa-foetida against HSV-1. *Iranian Journal of Pharmaceutical Research: IJPR*, *13*(2), p. 523.

Hamidi, J.A., Ismaili, N.H., Ahmadi, F.B. and Lajisi, N.H., 1996. Antiviral and cytotoxic activities of some plants used in Malaysian indigenous medicine. *Pertanika Journal of Tropical Agricultural Science*, *19*(2/3), pp. 129–136.

Huang, Q., Zhang, S., Huang, R., Wei, L., Chen, Y., Lv, S., Liang, C., Tan, S., Zhuo, L. and Lin, X., 2013. Isolation and identification of an anti-hepatitis B virus compound from Hydrocotyle sibthorpioides Lam. *Journal of Ethnopharmacology*, *150*(2), pp. 568–575.

Hussein, G., Miyashiro, H., Nakamura, N., Hattori, M., Kawahata, T., Otake, T., Kakiuchi, N. and Shimotohno, K., 1999. Inhibitory effects of Sudanese plant extracts on HIV-1 replication and HIV-1 protease. *Phytotherapy Research: An International Journal Devoted to Pharmacological and Toxicological Evaluation of Natural Product Derivatives*, *13*(1), pp. 31–36.

Iranshahi, M., Hosseini, S.T., Shahverdi, A.R., Molazade, K., Khan, S.S. and Ahmad, V.U., 2008. Diversolides A–G, guaianolides from the roots of Ferula diversivittata. *Phytochemistry*, *69*(15), pp. 2753–2757.

Jaisi, A., Prema, P., Madla, S., Lee, Y.E., Septama, A.W. and Morita, H., 2021. Investigation of HIV-1 Viral protein R inhibitory activities of twelve Thai medicinal plants and their commercially available major constituents. *Chemistry & Biodiversity*. doi: 10.1002/cbdv.202100540.

Jing, B., Ma, Z., Feng, J., Liang, H., Li, C. and Zhang, X., 2012. Evaluation of the antiviral activity of extracts from plants grown in the qinling region of China against infection by Tobacco mosaic virus (TMV). *Journal of Phytopathology*, *160*(4), pp. 181–186.

Kaur, G.J. and Arora, D.S., 2009. Antibacterial and phytochemical screening of Anethum graveolens, Foeniculum vulgare and Trachyspermum ammi. *BMC Complementary and Alternative Medicine*, *9*(1), pp. 1–10.

Kuete, V., Ngameni, B., Tangmouo, J.G., Bolla, J.M., Alibert-Franco, S., Ngadjui, B.T. and Pagès, J.M., 2010. Efflux pumps are involved in the defense of Gram-negative bacteria against the natural products isobavachalcone and diospyrone. *Antimicrobial Agents and Chemotherapy*, *54*(5), pp. 1749–1752.

Lee, C.L., Chiang, L.C., Cheng, L.H., Liaw, C.C., Abd El-Razek, M.H., Chang, F.R. and Wu, Y.C., 2009. Influenza A (H1N1) antiviral and cytotoxic agents from Ferula assa-foetida. *Journal of Natural Products*, *72*(9), pp. 1568–1572.

Lingaraju, D.P., Sudarshana, M.S., Mahendra, C. and Rao, K.P., 2016. Phytochemical screening and antimicrobial activity of leaf extracts of Eryngium foetidum L.(Apiaceae). *Indo American Journal of Pharmaceutical Research*, *6*(2), pp. 4339–4344.

Machado, R.R.P., Dutra, R.C., Pittella, F., Raposo, N.R.B., Lesche, B., Duarte, R.S., Soares, G.L.G. and Kaplan, M.A.C., 2015. Screening antimycobacterial activity of Baccharis dracunculifolia, Centella asiatica, Lantana camara and Pterodon emarginatus. *Revista Brasileira de Plantas Medicinais*, *17*, pp.891–899.

Miller, A.B., Cates, R.G., Lawrence, M., Soria, J.A.F., Espinoza, L.V., Martinez, J.V. and Arbizú, D.A., 2015. The antibacterial and antifungal activity of essential oils extracted from Guatemalan medicinal plants. *Pharmaceutical Biology*, *53*(4), pp. 548–554.

Min, B.S., Bae, K.H., Kim, Y.H., Miyashiro, H., Hattori, M. and Shimotohno, K., 1999. Screening of Korean plants against human immunodeficiency virus type 1 protease. *Phytotherapy Research: An International Journal Devoted to Pharmacological and Toxicological Evaluation of Natural Product Derivatives*, *13*(8), pp.680–682.

Min, B.S., Kim, Y.H., Tomiyama, M., Nakamura, N., Miyashiro, H., Otake, T. and Hattori, M., 2001. Inhibitory effects of Korean plants on HIV-1 activities. *Phytotherapy Research*, *15*(6), pp. 481–486.

Minooeianhaghighi, M.H., Sepehrian, L. and Shokri, H., 2017. Antifungal effects of Lavandula binaludensis and Cuminum cyminum essential oils against Candida albicans strains isolated from patients with recurrent vulvovaginal candidiasis. *Journal de Mycologie Medicale*, *27*(1), pp. 65–71.

Mirahmadi, S.S., Aminzare, M., Azar, H.H. and Kamali, K., 2020. Effect of Eryngium caeruleum essential oil on microbial and sensory quality of minced fish and fate of Listeria monocytogenes during the storage at 4°C. *Journal of Food Safety*, *40*(2), p. e12745.

Mirjani, R., Shahverdi, A.R., Iranshahi, M., Amin, G. and Shafiee, A., 2005. Identification of antifungal compounds from Ferula persica. var. persica. *Pharmaceutical Biology*, *43*(4), pp. 293–295.

Motamedifar, M., Ghafari, N. and Talezadeh, S.P., 2010. The effect of Cumin seed extracts against herpes simplex virus type 1 in Vero cell culture. *Iranian Journal of Medical Sciences 35*(4), pp. 304–309.

Nakano, Y., Matsunaga, H., Saita, T., Mori, M., Katano, M. and Okabe, H., 1998. Antiproliferative Constituents in Umbelliferae Plants II.: Screening for Polyacetylenes in Some Umbelliferae Plants, and Isolation of Panaxynol and Falcarindiol from the Root of Heracleum moellendorffii. *Biological and Pharmaceutical Bulletin*, *21*(3), pp. 257–261.

Nurlaily, A., Noor Baitee, A.R. and Musalmah, M., 2012. Comparative antioxidant and anti-inflammatory activity of different extracts of Centella asiatica (L.) Urban and its active compounds, asiaticoside and madecassoside. *Med & Health*, *7*(2), pp. 62–72.

Orhan, İ.E., Özçelik, B., Kartal, M. and Kan, Y., 2012. Antimicrobial and antiviral effects of essential oils from selected Umbelliferae and Labiatae plants and individual essential oil components. *Turkish Journal of Biology*, *36*(3), pp. 239–246.

Park, J.C., 2003. Inhibitory effects of Korean plant resources on human immunodeficiency virus type 1 protease activity. *Oriental Pharmacy and Experimental Medicine*, *3*(1), pp. 1–7.

Paul, S., Dubey, R.C., Maheswari, D.K. and Kang, S.C., 2011. Trachyspermum ammi (L.) fruit essential oil influencing on membrane permeability and surface characteristics in inhibiting food-borne pathogens. Food Control, 22(5), pp. 725–731.

Roy, S., Chaurvedi, P. and Chowdhary, A., 2015. Evaluation of antiviral activity of essential oil of Trachyspermum Ammi against Japanese encephalitis virus. *Pharmacognosy Research*, *7*(3), p. 263.

Seo, D.J., Lee, M., Jeon, S.B., Park, H., Jeong, S., Lee, B.H. and Choi, C., 2017. Antiviral activity of herbal extracts against the hepatitis A virus. *Food Control*, *72*, pp. 9–13.

Sharifi-Rad, J., Salehi, B., Baghalpour, N., Kobarfard, F., Sharifi-Rad, M. and Mohammadizade, M., 2018. Antiviral activity of monoterpenes thymol, carvacrol and p-cymene against herpes simplex virus in vitro. *International Pharmacy Acta*, *1*(1), pp. 73–73.

Sharifzadeh, A., Khosravi, A.R., Shokri, H. and Sharafi, G., 2015. Antifungal effect of Trachyspermum ammi against susceptible and fluconazole-resistant strains of Candida albicans. *Journal de Mycologie Medicale*, *25*(2), pp. 143–150.

Silva, F., Ferreira, S., Duarte, A., Mendonca, D.I. and Domingues, F.C., 2011. Antifungal activity of Coriandrum sativum essential oil, its mode of action against Candida species and potential synergism with amphotericin B. Phytomedicine, 19(1), pp. 42–47.

Suresh, M., Rath, P.K., Panneerselvam, A., Dhanasekaran, D. and Thajuddin, N., 2010. Anti-mycobacterial effect of leaf extract of Centella asiatica (Mackinlayaceae). *Research Journal of Pharmacy and Technology*, *3*(3), pp. 872–876.

Wang, S.X., Wang, Y., Lu, Y.B., Li, J.Y., Song, Y.J., Nyamgerelt, M. and Wang, X.X., 2020. Diagnosis and treatment of novel coronavirus pneumonia based on the theory of traditional Chinese medicine. *Journal of Integrative Medicine*, 18, pp. 275–283.

Wang, W.N., Yang, X.B., Liu, H.Z., Huang, Z.M. and Wu, G.X., 2005. Effect of Oenanthe javanica flavone on human and duck hepatitis B virus infection 1. *Acta Pharmacologica Sinica*, *26*(5), pp. 587–592.

Yoosook, C., Bunyapraphatsara, N., Boonyakiat, Y. and Kantasuk, C., 2000. Anti-herpes simplex virus activities of crude water extracts of Thai medicinal plants. *Phytomedicine*, *6*(6), pp. 411–419.

11.3.2 Family Araliaceae A.L. de Jussieu (1789)

The family Araliaceae consists of about 90 genera and 1000 species of treelets shrubs, climbers, and herbs mainly abundant in temperate Asia or mountain tropical Asia. The stems armed with sharp prickles or not. The leaves are simple or compound (often palmately), spiral, and stipulate or not. The petiole is characteristically straight and swollen and clasping at base. The inflorescences are racemes or panicles of umbellules. The calyx is tubular. The corolla is made of 5 petals. A disc is present. Five stamens are present. The gynoecium includes 2–10 carpels fused into a plurilocular ovary, each locule sheltering a pendulous ovule. The fruits are drupes or berries. Plants in this family are often used to treat COVID-19 in China.

11.3.2.1 *Eleutherococcus nodiflorus* (Dunn) S.Y. Hu

Synonyms: *Acanthopanax gracilistylus* W.W. Sm.; *Acanthopanax hondae* Matsuda; *Acanthopanax nodiflorus* Dunn; *Acanthopanax pentaphyllus* (Thunb.) Witte; *Acanthopanax spinosus* var. *pubescens* Pamp.; *Acanthopanax villosulus* Harms; *Aralia palmata* Lour.; *Aralia pentaphylla* Thunb.; *Aralia scandens* Poir.; *Eleutherococcus gracilistylus* (W.W. Sm.) S.Y. Hu; *Eleutherococcus pentaphyllus* (Marchal) Nakai; *Eleutherococcus villosulus* (Harms) S.Y. Hu; *Panax spinosus* L. f.

Habitat: Meadows, riverbanks, roadsides

Distribution: China, Korea, and Taiwan

Botanical observation: This climbing shrub grows to about 3 m tall. The stems are prickly at nodes. The leaves are palmately compound, spiral, and exstipulate. The petiole is 3–8 cm long and prickly. The blade consists of 5 obovate to oblanceolate folioles which are 2.5–10 cm × 1.5–4.5 cm, dull green, cuneate at base, crenate, and acute to acuminate at apex. The panicles of umbels are terminal, and lax or axillary. The calyx is tubular and 5-lobed. The 5 petals are triangular, yellowish-green, and minute. Five stamens are present. A pair of style is present. The drupes are subglobose, darkish blue, about 5 mm across, and marked with persistent styles.

Medicinal uses: COVID-19 (China)

Commentaries

i. Plants in the genus *Eleutherococcus* Maxim. produce furofuranlignan and hydroxycinnamic acid glycosides called eleutherosides which stimulate the immune system (Steinmann et al., 2001). These immunostimulating principles may account for the use of *Eleutherococcus nodiflorus* (Dunn) S.Y. Hu against COVID-19 in China.
ii. *Eleutherococcus senticosus* (Rupr. ex Maxim.) Maxim. (also known as Siberian ginseng) is used for COVID-19 in China. Essential oil of fruits of this plant inhibited the growth of

Kocuria rhizophila (KACC 14744), *Micrococcus luteus* (KACC 14819), and *Escherichia coli* (KACC14818) with the MIC values of 125, 500, and 63 µg/mL, respectively (Jang et al., 2016). Ethanol extract inhibited the growth of *Nosema apis*, a lethal fungi to honey bees chronically exposed pesticides, global warming, and other environmental consequence of deforestation and other abuses on the fundamental "Rights of Nature" (Pettis et al., 2012). It must be grasped that once honey bees are gone, so will be our crops (Pettis et al., 2013). Extract of this plant inhibited the replication of the Human rhinovirus, Respiratory syncytial virus, and Influenza virus A virus (Glatthaar-Saalmüller et al., 2001). An extract at the concentration of 50 µg/mL inhibited the replication of the Hepatitis A virus (HM-175 strain) in fetal rhesus monkey kidney cells (Seo et al., 2017).

iii. *Eleutherococcus trifoliatus* (L.) S.Y. Hu is a remedy for leprosy in China. One may have noted that saponiferous plants are often used for leprosy in Asia and the Pacific.

iv. Methanol extract of stem bark of *Acanthopanax koreanum* Nakai inhibited the Human immunodeficiency virus type-1 protease (Park, 2003).

v. In the Subfamily Aralioideae, Tribe Aralieae, *Aralia chinensis* L. is given for abscesses in China (local name: *Huang mao cong mu*). Methanol extract of stems of *Aralia elata* (Miq.) Seem at the concentration of 100 µg/mL inhibited RNA-dependent DNA polymerase and ribonuclease H activities of the Human immunodeficiency virus type-1 reverse transcriptase by 58.9% and 24.3%, respectively (Min et al., 2001). The plant yields antiviral oleanane-type triterpene saponins (Li et al., 1994)

vi. In the Subfamily Aralioideae, Tribe Hedereae, *Hedera nepalensis* K. Koch var *sinensis* (Tobler) Rehder is used for abscesses in Cambodia, Laos, and Vietnam. The plant yield oleanane-type triterpenes (Kizu et al., 1985). Ethyl acetate extract inhibited the growth of *Escherichia coli*, *Klebsiella pneumoniae*, and *Bacillus subtilis* (Uddin et al., 2012).

vii. The resin of *Opopanax chironium* Koch was recommended by the our peer the physician Moses ben Maimon (1138–1204) for pestilence (Casal & Casal, 2004). The resin is part of of the medication for syphilis by our peer the Italian physician Girolamo Fracastoro (see earlier).

viii. *Tetrapanax papyrifer* (Hook.) K. Koch is used for the treatment of COVID-19 in China (Wang et al., 2020). It brings to being the oleanane-type triterpene, papyriogenin A (also known as 3,21-dioxooleana-11,13(18)-dien-28-oic acid; LogD 3.4 at pH 7.4; molecular mass = 466.6 g/mol) that inhibited the replication of the Human immunodeficiency virus (strain IIIB) with an EC_{50} value of 1.7 µM in H9 lymphocytes (Kuo et al., 2009)

ix. In India, *Macropanax undulatus* (Wall. ex G. Don) Seem. is a remedy for cold and fever. *Macropanax dispermus* (Blume) Kuntze is used in Myanmar for cough (local name: *M'büi an*) and is used for cough in Myanmar.

x. In Pakistan, *Schefflera arboricola* (Hayata) Merr. (local name: *Choti*) is given for cough and wounds. In Indonesia, *Schefflera elliptica* (Blume) Harms (local name: *Tangana*) is used for carries, cough, and wounds. From *Schefflera heptaphylla* (L.) Frodin, 3,4-di-*O*-caffeoylquinic acid and 3,5-di-*O*-caffeoylquinic acid inhibited the replication of the Respiratory syncytial virus with the IC_{50} value of 1.2 and 0.6 µg/mL, respectively (Li et al., 2005).

References

Casal, M.T. and Casal, M., 2004. Maimonides and the chemotherapy of infectious diseases. *Revista Española de Quimioterapia*, *17*(3), pp.289–294.

Glatthaar-Saalmüller, B., Sacher, F. and Esperester, A., 2001. Antiviral activity of an extract derived from roots of Eleutherococcus senticosus. *Antiviral Research*, *50*(3), pp. 223–228.

Jang, D., Lee, J., Eom, S.H., Lee, S.M., Gil, J., Lim, H.B. and Hyun, T.K., 2016. Composition, antioxidant and antimicrobial activities of Eleutherococcus senticosus fruit extracts. *Journal of Applied Pharmaceutical Science*, *6*(3), pp. 125–130.

Kizu, H., Kitayama, S., Nakatani, F., Tomimori, T. and Namba, T., 1985. Studies on nepalese crude drugs. III. On the saponins of Hedera nepalensis K. Koch. *Chemical and Pharmaceutical Bulletin*, *33*(8), pp. 3324–3329.
Kuo, R.Y., Qian, K., Morris-Natschke, S.L. and Lee, K.H., 2009. Plant-derived triterpenoids and analogues as antitumor and anti-HIV agents. *Natural Product Reports*, *26*(10), pp. 1321–1344.
Li, F., Tian, T.C. and Shi, C.Y., 1994. Study on antiviral saponins from Aralia elata. *China Journal of Chinese Medicinal Materials*, *19*, pp. 562–564.
Li, Y., But, P.P. and Ooi, V.E., 2005. Antiviral activity and mode of action of caffeoylquinic acids from Schefflera heptaphylla (L.) Frodin. *Antiviral Research*, *68*(1), pp. 1–9.
Min, B.S., Kim, Y.H., Tomiyama, M., Nakamura, N., Miyashiro, H., Otake, T. and Hattori, M., 2001. Inhibitory effects of Korean plants on HIV-1 activities. *Phytotherapy Research*, *15*(6), pp. 481–486.
Park, J.C., 2003. Inhibitory effects of Korean plant resources on human immunodeficiency virus type 1 protease activity. *Oriental Pharmacy and Experimental Medicine*, *3*(1), pp. 1–7.
Pettis, J.S., Vanengelsdorp, D., Johnson, J. and Dively, G., 2012. Pesticide exposure in honey bees results in increased levels of the gut pathogen Nosema. *Naturwissenschaften*, *99*(2), pp. 153–158.
Pettis, J.S., Lichtenberg, E.M., Andree, M., Stitzinger, J., Rose, R. and Vanengelsdorp, D., 2013. Crop pollination exposes honey bees to pesticides which alters their susceptibility to the gut pathogen Nosema ceranae. *PloS One*, *8*(7), p. e70182.
Seo, D.J., Lee, M., Jeon, S.B., Park, H., Jeong, S., Lee, B.H. and Choi, C., 2017. Antiviral activity of herbal extracts against the hepatitis A virus. *Food Control*, *72*, pp.9–13.
Steinmann, G.G., Esperester, A. and Joller, P., 2001. Immunopharmacological in vitro effects of Eleutherococcus senticosus extracts. *Arzneimittelforschung*, *51*(01), pp. 76–83.
Uddin, G., Khan, A.A., Alamzeb, M., Ali, S., Alam, M., Rauf, A. and Ullah, W., 2012. Biological screening of ethyl acetate extract of Hedera nepalensis stem. *African Journal of Pharmacy and Pharmacology*, *6*(42), pp. 2934–2937.
Wang, S.X., Wang, Y., Lu, Y.B., Li, J.Y., Song, Y.J., Nyamgerelt, M. and Wang, X.X., 2020. Diagnosis and treatment of novel coronavirus pneumonia based on the theory of traditional Chinese medicine. *Journal of Integrative Medicine*, 18, pp. 275–283.

11.3.2.2 *Panax notoginseng* (Burkill) F.H. Chen ex C.H. Chow

Synonym: *Aralia quinquefolia* var. *notoginseng* Burkill

Common name: San qi (China)

Habitat: Forests

Distribution: China, Korea

Botanical observation: This herb grows up to 60 cm tall from anthropomorphic roots. The blade is palmate, spiral, and exstipulate. The petiole is up to about 15 cm long. The blade comprises 5 folioles which are obovate or obovate-oblong, 3.5–13 cm × 1.5–7 cm, membranous, serrate, marked with about 5 pairs of secondary nerves, and acuminate at apex. The umbels are terminal on a 10–25 cm tall peduncle. The calyx is tubular. The corolla comprises 5 minute and oblong whitish petals. Five stamens are present. The berries are red, glossy, globose, about 1 cm across.

Medicinal use: COVID-19 (China)

Antiviral (enveloped monopartite linear negative-stranded (-) RNA) polar extract: Methanol extract of roots inhibited the replication of the Influenza A/PR/8/34-GFP virus in RAW 264.7 cells at the concentration of 100 µg/mL (Choi et al., 2017).

In vivo antiviral (enveloped monopartite linear negative-stranded (-) RNA) polar extract: Methanol extract of roots given to mice orally at a dose of 100 mg/Kg/day or 7 days protected rodent against Influenza A/PR/8/34 (Choi et al., 2017).

Commentaries

i. *Panax ginseng* C.A Meyer is given in China for the treatment of COVID-19 (Luo, 2020; Wang et al., 2020) on probable account of immunostimulating dammarane-type triterpene saponins such as ginsenoside Rg1 (Keranova et al., 1990). Extract of this herb inhibited the replication of the Respiratory syncytial virus (Lee et al., 2014). In India, *Panax ginseng* C.A Meyer is given for tuberculosis.

ii. Our peer, the American Naturalist E.O Wilson (1929 – 2021) said "Destroying the rainforest for economic gain is like burning a Renaissance painting to cook a meal" and one could add in the context of emerging global zoonotic pandemics, "to cook the last meal".

References

Choi, J.G., Jin, Y.H., Lee, H., Oh, T.W., Yim, N.H., Cho, W.K. and Ma, J.Y., 2017. Protective effect of Panax notoginseng root water extract against influenza A virus infection by enhancing antiviral interferon-mediated immune responses and natural killer cell activity. Frontiers in *Immunology*, 8, p. 1542.

Kenarova, B., Neychev, H., Hadjiivanova, C. and Petkov, V.D., 1990. Immunomodulating activity of ginsenoside Rg1 from Panax ginseng. *The Japanese Journal of Pharmacology*, *54*(4), pp. 447–454.

Luo, A., 2020. Positive SARS-Cov-2 test in a woman with COVID-19 at 22 days after hospital discharge: A case report. *Journal of Traditional Chinese Medical Sciences*, 7, pp. 413–417.

Lee, J.S., Ko, E.J., Hwang, H.S., Lee, Y.N., Kwon, Y.M., Kim, M.C. and Kang, S.M., 2014. Antiviral activity of ginseng extract against respiratory syncytial virus infection. *International Journal of Molecular Medicine*, *34*(1), pp. 183–190.

Wang, S.X., Wang, Y., Lu, Y.B., Li, J.Y., Song, Y.J., Nyamgerelt, M. and Wang, X.X., 2020. Diagnosis and treatment of novel coronavirus pneumonia based on the theory of traditional Chinese medicine. *Journal of Integrative Medicine*, *18*, pp. 275–283.

Bibliography

Abe, R. and Ohtani, K., 2013. An ethnobotanical study of medicinal plants and traditional therapies on Batan Island, the Philippines. *Journal of Ethnopharmacology, 145*, pp. 554–565.

Abu Bakar, F.I., Abu Bakar, M.F., Abdullah, N., Endrini, S. and Rahmat, A., 2018. A review of Malaysian medicinal plants with potential anti-inflammatory activity. *Hindawi*, 13 pages, Article ID 8603602.

Acharya, K.P. and Acharya, M., 2010. Traditional knowledge on medicinal plants used for the treatment of livestock diseases in Sardikhola VDC, Kaski, Nepal. *Journal of Medicinal Plants Research, 4*(2), pp. 235–239.

Acharya, E. (Siwakoti) and Pokhrel, B., 2006. Ethno-medicinal plants used by Bantar of Bhaudaha, Morang, Nepal. *Our Nature, 4*, pp. 96–103.

Adams, M., Alther, W., Kessler, M., Kluge, M. and Hamburger, M., 2011. Malaria in the renaissance: Remedies from European herbals from the 16th and 17th century. *Journal of Ethnopharmacology, 133*, pp. 278–288.

Ahmad, S.S., 2007. Medicinal wild plants from Lahore-Islamabad Motorway (M-2). *Pakistan Journal of Botany*, 39(2), pp. 355–375.

Ahmad, K., Weckerle, C.S. and Nazir, A. Ethnobotanical investigation of wild vegetables used among local communities in northwest Pakistan. *Acta Societatis Botanicorum Poloniae, 88*(1), p. 3616.

Ahmed, N., Mahmood, A., Mahmood, A., Sadeghi, Z. and Farman, M., 2015. Ethnopharmacological importance of medicinal flora from the district of Vehari, Punjab province, Pakistan. *Journal of Ethnopharmacology, 168*, pp. 66–78.

Ahmed, M.J. and Murtaza, G., 2015. A study of medicinal plants used as ethnoveterinary: Harnessing potential phytotherapy in Bheri, District Muzaffarabad (Pakistan). *Journal of Ethnopharmacology, 159*, pp. 209–214.

Ainslie, W., 1813. *Materia Medica of Hindoostan, and Artisan's and Agriculturist's Nomenclature*. Madras: Government Press.

Al-Adhroey, A.H., Nor, Z.M., Al-Mekhlafi, H.M. and Mahmud, R., 2010. Ethnobotanical study on some Malaysian anti-malarial plants: A community based survey. *Journal of Ethnopharmacology, 132*, pp. 362–364.

Altundag, E. and Ozturk, M., 2011. Ethnomedicinal studies on the plant resources of east Anatolia, Turkey. *Procedia Social and Behavioral Sciences, 19*, pp. 756–777.

Amber, R., Adnan, M., Tariq, A. and Mussarat, S., A review on antiviral activity of the Himalayan medicinal plants traditionally used to treat bronchitis and related symptoms. *Journal of Pharmacy and Pharmacology*. doi: 10.1111/jphp.12669

Amini, M.H. and Hamdam, S.M., 2017. Medicinal plants used traditionally in Guldara District of Kabul, Afghanistan. *International Journal of Pharmacognosy and Chinese Medicine, 1*(3), 000118.

Amiri, M.S., Joharchi, M.R. and TaghavizadehYazdi, M.E., 2014. Ethno-medicinal plants used to cure jaundice by traditional healers of Mashhad, Iran. *Iranian Journal of Pharmaceutical Research, 13*(1), pp. 157–162.

Asiatic Society of Bengal, 1812. Asiatic researches or transactions of the Society instituted in Bengal, for inquiring into the history and antiquities, the arts, sciences, and literature, of Asia. The Bavarian State Library.

Atigur Rahman, M., Uddin, S.B. and Wilcok, C.C., 2007. Medicinal plants used by Chakma tribe in Hill tracts Districts of Bangladesh. *Indian Journal of Traditional Knowledge, 6*(3), pp. 508–517.

Ayyanar, M. and Ignacimuthu, S., 2011. Ethnobotanical survey of medicinal plants commonly used by Kani tribals in Tirunelveli hills of Western Ghats, India. *Journal of Ethnopharmacology, 134*, pp. 851–864.

Azhar, M.F., Aziz, A., Haider, M.S., Nawaz, M.F. and Zulfiqar, M.A., 2015. Exploring the ethnobotany of *Haloxylon recurvum* (KHAR) and *Haloxylon salicornicum* (LANA) in Cholistan desert, Pakistan. *Pakistan Journal of Agricultural Sciences, 52*(4), pp. 1085–1090.

Balfour, E., 1857. *The Cyclopaedia of India and of Eastern and Southern Asia*. Madras: Scottish Press.

Bentley, R., 1872. *Dr. Pereira's Elements of Materia Medica and Therapeutics: Abridged and Adapted for the Use of Medical and Pharmaceutical Practitioners and Students*. London: Longmans, Green, and Co.

Bhagya, B. and Sridhar, K.R., 2009. Ethnobiology of coastal sand dune legumes of Southwest coast of India. *Indian Journal of Traditional Knowledge, 8*(4), pp. 611–620.

Bhandary, M.J. and Chandrashekar, K.R., 2011. Herbal therapy for herpes in the ethno-medicine of Coastal Karnataka. *Indian Journal of Traditional Knowledge, 10*(3), pp. 528–532.

Bhatia, H., Sharma, Y.P., Manhas, R.K. and Kumar, K., 2014. Ethnomedicinal plants used by the villagers of district Udhampur, J&K, India. *Journal of Ethnopharmacology, 151*, pp. 1005–1018.

Bose, D., Roy, J.G., Mahapatra, S.D. (Sarkar), Datta, T., Mahapatra, S.D. and Biswas, H., 2015. Medicinal plants used by tribals in Jalpaiguri district, West Bengal, India. *Journal of Medicinal Plants Studies, 3*(3), pp. 15–21.

Bulut, G., Zeki Haznedaroğlu, M., Doğan, A., Koyu, H. and Tuzlacı, E., 2017. An ethnobotanical study of medicinal plants in Acipayam (Denizli-Turkey). *Journal of Herbal Medicine, 10*, pp. 64–81.

Chakraborty, M.K. and Bhattacharjee, A., 2006. Some common ethnomedicinal uses for various diseases in Purulia District, West Bengal. *Indian Journal of Traditional Knowledge, 5*(4), pp. 554–558.

Chang, C.-S., Kim, H. and Chang, K.S., 2014. *Provisional Checklist of Vascular Plants for the Korea Peninsula flora (KPF)*. 660 pp.

Chowdhury, M. and Mukherjee, R., 2010. Ethno-medicinal survey of Santal tribe of Malda District of West Bengal, India. *Journal of Economic and Taxonomic Botany, 34*(3).

Chowdhury, M.S.H., Koike, M., Muhammed, N., Halim, M.A., Saha, N. and Kobayashi H., 2009. Use of plants in healthcare: A traditional ethno-medicinal practice in southeastern rural areas of Bangladesh. *International Journal of Biodiversity Science and Management, United Kingdom, 5*(1), pp. 41–51.

Chowlu, K., Mahar, K.S. and Das, A.K., 2017. Ethnobotanical studies on orchids among the *Khamti* Community of Arunachal Pradesh, India. *Indian Journal of Natural Products and Resources, 8*(1), pp. 89–93.

Colonel Heber Drury, 1873. *The Useful Plants of India*. London: William H. Allen & Co.

Cox, P.A., 1993. Saving the ethnopharmacological heritage of Samoa. *Journal of Ethnopharmacology, 38*, pp. 181–188.

Das, D. and P. Ghosh., 2017. Some important medicinal plants used widely in Southwest Bengal, India. *International Journal of Engineering Science Invention, 6*(6), pp. 28–50.

Das, T., Mishra, S.B., Saha, D. and Agarwal, S., 2012. Ethnobotanical survey of medicinal plants used by ethnic and rural people in Eastern Sikkim Himalayan Region. *African Journal of Basic & Applied Sciences, 4*(1), pp. 16–20.

Das, A.K. and Tongbram, Y., 2014. Study on medicinal plants used by Meitei community of Bishnupur district, Manipur. *International Journal of Current Research, 6*(2), pp. 5211–5219.

Debala Devi, A., Ibeton Devi, O., Chand Singh, T. and Singh, E.J., 2014. A study of aromatic plant species especially in Thoubal District, Manipur, North East India. *International Journal of Scientific and Research Publications, 4*(6).

Defilipps, R.A. and Krupnick, G.A., 2018. The medicinal plants of Myanmar. *PhytoKeys, 102*, pp. 1–341.

Defilipps, R.A., Maina, S.L. and Pray, L.A., 1988. The palauan and yap medicinal plant studies of masayoshi okabe, 1941–1943. Atoll Research Bulletin. National Museum of Natural History Smithsonian Institution, Washington, D.C., October 1988.

Dey, A. and De, J.N., 2010. A survey of ethnomedicinal plants used by the tribals of Ajoydha Hill region, Purulia District, India. *American-Eurasian Journal of Sustainable Agriculture, 4*(3), pp. 280–290.

Dey, A. and De, J.N., 2012. Ethnobotanical survey of Purulia district, West Bengal, India for medicinal plants used against gastrointestinal disorders. *Journal of Ethnopharmacology, 143*, pp. 68–80.

Durmuşkahya, C. and Öztürk, M., 2013. Ethnobotanical survey of medicinal plants used for the treatment of diabetes in Manisa, Turkey. *Sains Malaysiana, 42*(10), pp. 1431–1438.

Dutta, S.K., Vanlalhmangaiha, Akoijam, R.S., Lungmuana, Boopathi, T. and Saha, S., 2018. Bioactivity and traditional uses of 26 underutilized ethno-medicinal fruit species of North-East Himalaya, India. *Journal of Food Measurement and Characterization, 12*, pp. 2503–2514.

Elliott, E., Chassagne, F., Aubouy, A., Deharo, E., Souvanasy, O., Sythamala, P., Sydara, K., Lamxay, V., Manithip, C., Torres, J.A. and Bourdy, G., 2020. Forest Fevers: Traditional treatment of malaria in the southern lowlands of Laos. *Journal of Ethnopharmacology, 247.*

Flora of China Editorial Committee, 2018. *Flora of China*. Missouri Botanical Garden and Harvard University, Herbaria, USA.

Fluchiger, F.A. and Hangury, D., 1874. *A History of the Principal Drugs of Vegetable Origin, Met with in Great Britain and British India*. London: Macmillan and Co.

Frederick Porter Smith, 1871. *Contributions towards the Materia Medica and Natural History of China*. Shanghai: American Presbyterian Mission Press.

Gagnepain, F., 1934. Iridacées, Amaryllidacées et Liliacées nouvelles d'Asie, *Bulletin de la Société Botanique de France, 81*(1), pp. 66–74. doi: 10.1080/00378941.1934.10833934

Gao, X-M., Ji, B-K., Li, Y-K., Ye, Y-Q., Jiang, Z-Y., Yang, H-Y., Du, G., Zhou, M., Pan, X-X., Liu, W-X. and Hu, Q-F., 2016. New biphenyls from *Garcinia multiflora. The Journal of the Brazilian Chemical Society, 27*(1), pp. 10–14.

Gautam, R., Saklani, A. and Jachak, S.M., 2007. Indian medicinal plants as a source of antimycobacterial agents. *Journal of Ethnopharmacology, 110*, pp. 200–234.

Ghosh, C., 2017. Ethnobotanical survey in the Bamangola Block of Malda District, West Bengal (India): II. Medicinal and Aromatic plants. *Pleione, 11*(2), pp. 249–267.

Giesen, W., Wulffraat, S., Zieren, M. and Scholten, L., 2007. *Mangrove Guidebook for Southeast Asia.* Dharmasarn Co., Ltd., Bangkok, Thailand.

Girach, R.D., Brahmam, M., Misra, M.K. and Ahmed, M., 1998. Indigenous phytotherapy for filariasis from Orissa. *Ancient Science of Life, 17*(3), pp. 224–227.

Goswami, D., Mukherjee, P.K. and Kar, A., 2016. Screening of ethnomedicinal plants of diverse culture for antiviral potentials. *Indian Journal of Traditional Knowledge, 15*(3), pp. 474–481.

Grosvenor, P.W., Gothard, P.K., McWilliam, N.C., Suprlono, A. and Gray, D.O., 1995. Medicinal plants from Riau Province, Sumatra, Indonesia. Part 1: Uses. *Journal of Ethnopharmacology, 45*, pp. 75–95.

Gruyal, G.A. del Roasario, R. and Palmes, N.D., 2014. Ethnomedicinal plants used by residents in Northern Surigao del Sur, Philippines. *Natural Products Chemistry & Research, 2*, p. 4.

Gunawardana, S.L.A. and Jayasuriya, W.J.A.B.N. 2019. Medicinally important herbal flowers in Sri Lanka. *Hindawi,* 18 pages. Article ID 2321961.

Hanbury, D., 1862. Notes on Chinese Materia Medica. In: John E. Taylor (ed.), *The Pharmaceutical Journal and Transactions.*

Hasan, Md. M., Hossain, Sk. A., Ali, Md. A. and Alamgir, A.N.M., 2014. Medicinal plant diversity in Chittagong, Bangladesh: A database of 100 medicinal plants. *Journal of Scientific and Innovative Research, 3*(5), pp. 500–514.

Hooper, D., 1937. *Useful Plants and Drugs of Iran and Iraq.* The Library of the University of Illinois.

Hope, J. 1970. *Lectures on the Material Medica: Containing the Natural History of Drugs, Their Virtues and Doses: Also Directions for the Study of the Materia Medica; and An Appendix on the Method of Prescribing.* London: Edward and Charles Dilly.

Ijaz, F., Iqbal, Z., Rahman, I.U., Alam, J., Khan, S.M., Shah, G.M., Khan, K. and Afzal, A., 2016. Investigation of traditional medicinal floral knowledge of Sarban Hills, Abbottabad, KP, Pakistan. *Journal of Ethnopharmacology, 179*, pp. 208–233.

Jahan, I., Rahman, M.A. and Hossain, M.A., 2019. Medicinal species of Fabaceae occurring in Bangladesh and their conservation status. *Journal of Medicinal Plants Studies, 7*(4), pp. 189–195.

Jain, S.K. and Srivastava, S., 2005. Traditional uses of some Indian plants among islanders of the Indian Ocean. *Indian Journal of Traditional Knowledge, 4*(4), pp. 345–357.

Jan, G., Khan, M.A. and Gul, F., 2008. Ethnomedicinal plants used against diarrhea and dysentery in Dir Kohistan Valley (NWFP), Pakistan. *Ethnobotanical Leaflets, 12*, pp. 620–637.

Jazani, A.M., Maleki, R.F., Kazemi, A.h., Matankolaei, L.g., Targhi, S.T., kordi, S., Rahimi-Esboei, B. and Azgomi, R.N.D., 2018. Intestinal Helminths from the viewpoint of traditional Persian medicine versus modern medicine. *African Journal of Traditional, Complementary and Alternative Medicines, 15*(2), pp. 58–67.

Ji, H., Long, C. and Pei, S., 2004. An ethnobotanical study of medicinal plants used by the Lisu people in Nujiang, Northwest Yunnan, China. *Economic Botany, 58*, pp. S253–264. doi: 10.1663/0013-000

Kadir, M.F., Sayeed, M.S.B., Setu, N.I., Mostaf, A. and Mia, M.M.K., 2014. Ethnopharmacological survey of medicinal plants used by traditional health practitioners in Thanchi, Bandarban Hill Tracts, Bangladesh. *Journal of Ethnopharmacology, 155*, pp. 495–508.

Kar, A. and Borthakur, S.K., 2008. Medicinal plants used against dysentery, diarrhoea and cholera by the tribes of erstwhile Kameng District of Arunachal Pradesh. *Natural Product Radiance, 7*(2), pp. 176–181.

Kar, T., Mandal, K.K., Reddy, C.S. and Biswal, A.K., 2013. Ethnomedicinal plants used to cure diarrhoea, dysentery and cholera by some tribes of Mayurbhanj District, Odisha, India. *Life Sciences Leaflets, 2*, pp. 18–28.

Karki, M., Hill, R., Xue, D., Alangui, W., Ichikawa, K. and Bridgewater, P., 2017. Knowing our Lands and Resources: Indigenous and Local Knowledge and Practices related to Biodiversity and Ecosystem Services in Asia. Published in 2017 by the United Nations Educational, Scientific and Cultural Organization, 7, place de Fontenoy, 75352 Paris 07 SP, France.

Kayani, S., Ahmad, M., Sultana, S., Shinwari, Z.K., Zafar, M., Yaseen, G., Hussain, M. and Bibi, T., 2015. Ethnobotany of medicinal plants among the communities of Alpineand Sub-alpine regions of Pakistan. *Journal of Ethnopharmacology, 164*, pp. 186–202.

Ketpanyapong, W. and Itharat, A., 2016. Antibacterial activity of Thai medicinal plant extracts against microorganism isolated from post-weaning diarrhea in piglets. *Journal of the Medical Association of Thailand, 99*(Suppl. 4), pp. S203–S210.

Khadka, D., Dhamala, M.K., Li, F., Aryal, P.C., Magar, P.R., Bhatta, S., Bhatta, S., Basnet, A., Shi, S. and Cui, D., 2005. The use of medicinal plant to prevent COVID-19 in Nepal. Research Square. doi:10.21203/rs.3.rs–88908/v1

Khan, S.U., Khan, R.U., Mehmood, S., Khan, A., Ullah, I. and Bokhari, T.Z. Medicinal plants used to cure diarrhea and dysentery by the local inhabitants of District Bannu, Khyber PakhtoonKhwa, Pakistan. *Advances in Pharmaceutical and Ethnomedicines, 1*(1), pp. 15–18.

Khan, S.W. and Khatoon, S., 2008. Ethnobotanical studies on some useful herbs of Haramosh and Bugrote valleys in Gilgit, Northern areas of Pakistan. *Pakistan Journal of Botany, 40*(1), pp. 43–58.

Khare, C.P. *Indian Medicinal Plants: An Illustrated Dictionary.* New York: Springer.

Khisha, T., Karim, R., Chowdhury, S.R. and Banoo, R., 2012. Ethnomedical studies of Chakma communities of Chittagong Hill tracts, Bangladesh. *Bangladesh Pharmaceutical Journal, 15*(1), pp. 59–67.

Kirtikar, K.R., Basu, B.D., An, I.C.S. *Indian Medicinal Plants.* Delhi: Jayyed Press.

Koch, M., Kehop, D.A., Kinminja, B., Sabak, M., Wavimbukie, G., Barrows, K.M., Matainaho, T.K., Barrows, L.R. and Rai, P.P., 2015. An ethnobotanical survey of medicinal plants used in the East Sepik province of Papua New Guinea. *Journal of Ethnobiology and Ethnomedicine, 11*, p. 79.

Korkmaz, M. and Karakuş, S., 2015. Traditional uses of medicinal plants of Üzümlü district, Erzincan, Turkey. *The Pakistan Journal of Botany, 47*(1), pp. 125–134.

Koteswara Rao, J., Suneetha, J., Seetharami Reddi, T.V.V. and Aniel Kumar, O., 2011. Ethnomedicine of the Gadabas, a primitive tribe of Visakhapatnam district, Andhra Pradesh. *International Multidisciplinary Research Journal, 1/2*, pp. 10–14.

Kültür, Ş., 2007. Medicinal plants used in Kırklareli Province (Turkey). *Journal of Ethnopharmacology, 111*, pp. 341–364.

Lemmens, R.H.M.J. and Bunyapraphatsara, N., 2003. *Plant Resources of South-East Asia: Medicinal and Poisonous Plants.* Leiden: Backhuys Publishers.

Lemmens, R.H.M.J., Soerianegara, I. and Wong, W.C., 1995. *Plant Resources of South-East Asia: Timber Trees: Minor Commercial Timbers.* Leiden: Backhuys Publishers.

Li, D-l. and Xing, F-w., 2016. Ethnobotanical study on medicinal plants used by local Hoklos people on Hainan Island, China. *Journal of Ethnopharmacology, 194*, pp. 358–368.

Li, D.-l., Zheng, X.-l., Duan, L., Deng, S.-w., Ye, W., Wang, A.-h. and Xinga, F.-w., 2017. Ethnobotanical survey of herbal tea plants from the traditional markets in Chaoshan, China. *Journal of Ethnopharmacology, 205*, pp. 195–206.

Liu, B., Zhang, X., Bussmaan, R.W., Hart, R.H., Li, P., Bai, Y. and Long, C., 2017. *Garcinia* in Southern China: Ethnobotany, management, and niche modeling. *Economic Botany, 70*(4), pp. 416–430.

Lokho, K. and Narasimhan, D., 2013. Ethnobotany of Mao-Naga Tribe of Manipur, India. *Pleione, 7*(2), pp. 314–324.

Lone, P.A. and Bhardwaj, A.K., 2013. Traditional herbal based disease treatment in some rural areas of Bandipora district of Jammu and Kashmir, India. *Asian Journal of Pharmaceutical and Clinical Research, 6*(4).

Long, C-l. and Li, R., 2004. Ethnobotanical studies on medicinal plants used by the Red-headed Yao People in Jinping, Yunnan Province, China. *Journal of Ethnopharmacology, 90*, pp. 389–395.

Mahbubur Rahman, A.H.M. and Akter, M., 2013. Taxonomy and medicinal uses of Euphorbiaceae (Spurge) family of Rajshahi, Bangladesh. *Research in Plant Sciences, 1*(3), pp. 74–80.

Mahmood, A., Mahmood, A., Shaheen, H., Qureshi, R.A., Sangi, Y. and Gilani, S.A., 2011. Ethno medicinal survey of plants from district Bhimber Azad Jammu and Kashmir, Pakistan. *Journal of Medicinal Plants Research, 5*(11), pp. 2348–2360.

Malla, B., 2015. Ethnobotanical study on medicinal plants in Parbat district of western Nepal. A Dissertation submitted for the Partial Fulfilment of the Requirement for the Degree of Doctor of Philosophy in Environmental Science.

Manilal, K.S. and Sabu, T., 1985. Cyclea Barbata Miers (Menispermaceae): A new record of a medicinal plant from south India. *Ancient Science of Life, IV*(4), pp. 229–231.

Merrill, E.D., 1903. *A Dictionary of the Plant Names of the Philippine Islands.* Manila: Bureau of Public Printing.

Mikaili, P., Shayegh, J. and Asghari, M.H., 2012. Review on the indigenous use and ethnopharmacology of hot and cold natures of phytomedicines in the Iranian traditional medicine. *Asian Pacific Journal of Tropical Biomedicine,* pp. S1189–S1193.

Minh, V.V., Yen, N.T.K. and Thoa, P.T.K., 2014. Medicinal plants used by the Hre community in the Ba to district of central Vietnam. *Journal of Medicinal Plants Studies, 2*(3), pp. 64–71.

Mishra, D.N., 2009. Medicinal plants for the treatment of fever (*Jvaracikitsā*) in the *Madhavacikitsā* tradition of India. *Indian Journal of Traditional Knowledge, 8*(3), pp. 352–361.

Mitra, S. and Mukherjee, S.Kr., 2010. Ethnomedicinal usages of some wild plants of North Bengal plain for gastro-intestinal problems. *Indian Journal of Traditional Knowledge, 9*(4), pp. 705–712.

Mohammad, N.S., Milow, P. and Ong, H.C., 2012. Traditional medicinal plants used by the Kensiu Tribe of Lubuk Ulu Legong, Kedah, Malaysia. *Studies on Ethno Medicine, 6*(3), pp. 149–153.

Morteza-Semnani, K., 2015. A Review on Chenopodium botrys L.: Traditional uses, chemical composition and biological activities. *Pharmaceutical and Biomedical Research, 1*(2), pp. 1–9.

Moungsrimuangdee, B., 2016. A survery of Riparian species in the Bodhivijjalaya college's forest, Srinakharinwirot University, Sa Kaeo. *Thai Journal of Forestry, 35*(3), 15–29.

Movaliya, V. and Zaveri, M., 2014. A review on the Pashanbheda plant "Aerva javanica". *International Journal of Pharmaceutical Sciences Review and Research, 25*(2), pp. 268–275. Article No. 51.

Nazan Çömlekçđoğlu and Sengül Karaman, 2008. Kahramanmaras Sehir Merkezindeki Aktar'larda Bulunan Tıbbi Bitkiler. *KSU Journal of Science and Engineering, 11*(1).

Neamsuvan, O. and Ruangrit, T., 2017. A survey of herbal weeds that are used to treat gastrointestinal disorders from southern Thailand: Krabi and Songkhla provinces. *Journal of Ethnopharmacology, 196*, pp. 84–93.

Neamsuvan, O., Sengnon, N., Seemaphrik, N., Chouychoo, M., Rungrat, R. and Bunrasri, S., 2015. A survey of medicinal plants around upper Songkhla lake, Thailand. *African Journal of Traditional, Complementary and Alternative Medicines, 12*(2), pp. 133–143.

Neamsuvan, O., Singdam, P., Yingcharoen, K. and Sengnon, N., 2012. A survey of medicinal plants in mangrove and beach forests from sating Phra Peninsula, Songkhla Province, Thailand. *Journal of Medicinal Plants Research, 6*(12), pp. 2421–2437.

Nguyen-Pouplin, J., Tran, H., Tran, H., Phan, T.A., Dolecek, C., Farrar, J., Tran, T.H., Caron, P., Bodo, B. and Grellier, P., 2007. Antimalarial and cytotoxic activities of ethnopharmacologically selected medicinal plants from South Vietnam. *Journal of Ethnopharmacology, 109*, pp. 417–427.

Noor Hassan Sajib, S. Uddin, B. and Islam, Md. M., 2014. Angiospermic plant diversity of Subarnachar Upazila in Noakhali, Bangladesh. *Journal of the Asiatic Society Bangladesh, Science, 40*(1), pp. 39–60.

O'shaughnessy, W.B., 1842. *Bengal Dispensatory.* Calcutta: W. Thacker and Co., St. Andrew's Library.

Ong, H.G., Ling, S.M., Win, T.T.M., Kang, D-H., Lee, J-H. and Kim, Y-D., 2018. Ethnomedicinal plants and traditional knowledge among three Chin indigenous groups in Natma Taung National Park (Myanmar). *Journal of Ethnopharmacology, 225*, pp. 136–158.

Panda, S.K., Das, R., Leyssen, P., Neyts, J. and Luyten, W., 2018. Assessing medicinal plants traditionally used in the Chirang Reserve Forest, Northeast India for antimicrobial activity. *Journal of Ethnopharmacology, 225*, pp. 220–233.

Pandit, P.K. Inventory of Ethno Veterinary Medicinal Plants of Jhargram Division, West Bengal, India.

Patra, J.K. and Thatoi, H.N., 2011. Metabolic diversity and bioactivity screening of mangrove plants: A review. *Acta Physiol Plant, 33*, pp. 1051–1061.

Pereira, J., 1854. *The Elements of Materia Medica and Therapeutics.* Philadelphia: Blanchard and Lea.

Pereira, J., 1855. *Materia Medica and Therapeutics.* London: Longman, Brown, Green, and Longmans.

Perry, L.M. and Metzger J., 1980. *Medicinal Plants of Asia and Southeast Asia.* MIT Press, USA.

Pomet, P., Histoire générale des drogues. A Paris : Chez Jean-Baptiste Loyson, & Augustin Pillon, sur le Pont au Change, à la Prudence. Et au Palais, chez Estienne Ducastin, dans la Gallerie des Prisonniers, au bon Pasteur 1664.

Pradhan, B.K. and Badola, H.K., 2008. Ethnomedicinal plant use by Lepcha tribe of Dzongu valley, bordering Khangchendzonga Biosphere Reserve, in North Sikkim, India. *Journal of Ethnobiology and Ethnomedicine, 4*, p. 22.

Prakasa Rao, J. and Satish, K.V., 2016. *Persicaria perfoliata* (L.) H. Gross (Polygonaceae): A species new to Eastern Ghats of India. *Tropical Plant Research, 3*(2), pp. 249–252.

Qasim, M., Gulzar, S. and Khan, M.A., 2011. Halophytes as medicinal plants. *Conference Paper.*

Qureshi, R. and Raza Bhatti, G., 2008. Ethnobotany of plants used by the Thari people of Nara Desert, Pakistan. *Fitoterapia, 79*, pp. 468–473.

Rahaman, C.H. and Karmakar, S., 2014. Ethnomedicine of Santal tribe living around Susunia hill of Bankura District, West Bengal, India: The quantitative approach. *Journal of Applied Pharmaceutical Science, 5*(02), pp. 127–136.

Rahman, M.A., 2010. Indigenous knowledge of herbal medicines in Bangladesh. 3. Treatment of skin diseases by tribal communities of the hill tracts districts. *Bangladesh Journal of Botany, 39*(2), pp. 169–177.

Rahmatullah, M., Azam, Md. N.K., Malek, I., Nasrin, D., Jamal, F., Rahman, Md. A., Khatun, Z., Jahan, S., Seraj, S. and Jahan, R., 2012. An ethnomedicinal survey among the Marakh Sect of the Garo Tribe of Mymensingh District, Bangladesh. *International Journal of PharmTech Research, 4*(1), pp. 141–149.

Rahmatullah, M., Haque, M.E., Kabir Mondol, M.R., Hasan, M., Aziz, T., Jahan, R. and Seraj, S., 2014. Medicinal formulations of the Kuch Tribe of Bangladesh. *The Journal of Alternative and Complementary Medicine, 20*(6), pp. 428–440.

Rahmatullah, M., Mollik, Md. A.H., Ahmed, Md. N., Bhuiyan, Md. Z.A., Hossain, Md. M., Azam, Md. N.K., Seraj, S., Chowdhury, M.H., Jamal, F., Ahsan, S. and Jahan, R., 2010. A survey of medicinal plants used by folk medicinal practitioners in two villages of Tangail district, Bangladesh. *American-Eurasian Journal of Sustainable Agriculture, 4*(3), pp. 357–362.

Rajendran, A., Ravikumar, K. and Henry, A.N., 2000. Plant genetic resources and knowledge of traditional medicine in Tamil Nadu. *Ancient Science of Life, XX*(1&2), pp. 25–28.

Rama Chandra Prasad, P., Sudhakar Reddy, C., Raza, S.H. and Dutt, C.B.S., 2008. Folklore medicinal plants of North Andaman Islands, India. *Fitoterapia, 79*, pp. 458–464.

Rao, P.K., Hasan, S.S., Bhellum, B.L. and Manhas, R.K., 2015. Ethnomedicinal plants of Kathua district, J&K, India. *Journal of Ethnopharmacology, 171*, pp. 12–27.

Ray, T., 2014. Customary use of mangrove tree as a folk medicine among the sundarban resource collectors. *International Journal of Research in Humanities, Arts and Literature (IMPACT: IJRHAL), 2*(4), pp. 43–48.

Reed, C.F., 1977. Economically Important Foreign Weeds. Potential Problems in the United States. United States Department of Agriculture.

Rehman, H., Begum, W., Anjum, F. and Tabasum, H., 2014. Rheum emodi (Rhubarb): A Fascinating Herb. *Journal of Pharmacognosy and Phytochemistry, 3*(2), pp. 89–94.

Romulo, A., Zuhud, E.A.M., Rondevaldova, J. and Kokoska, L., 2018. Screening of in vitro antimicrobial activity of plants used in traditional Indonesian medicine. *Pharmaceutical Biology, 56*(1), pp. 287–293.

Roy, S., 2015. An ethnobotanical study on the medicinal plants used by Rajbanshis of Coochbehar district, West Bengal, India. *Journal of Medicinal Plants Studies, 3*(5), pp. 46–49.

Safa, O., Soltanipoor, M.A., Rastegar, S., Kazemi, M., Dehkordi, K.N. and Ghannadi, A., 2012. An ethnobotanical survey on hormozgan province, Iran. *Avicenna Journal of Phytomedicine, 3*(1), Winter 2013, pp. 64–81.

Saha, M.R., De Sarker, D. and Sen, A., 2014. Ethnoveterinary practices among the tribal community of Malda district of West Bengal, India. *Indian Journal of Traditional Knowledge, 13*(2), pp. 359–367.

Saikia, A.P., Ryakala, V.K., Sharma, P., Goswami, P. and Bora, U., 2006. Ethnobotany of medicinal plants used by Assamese people for various skin ailments and cosmetics. *Journal of Ethnopharmacology, 106*, pp. 149–157.

Sandakan, S., 2005. Preferred check-list of Sabah trees.

Sankaranarayanan, S., Bama, P., Ramachandran, J., Kalaichelvan, P.T., Deccaraman, M., Vijayalakshimi, M., Dhamotharan, R., Dananjeyan, B. and Sathya Bama, S., 2010. Ethnobotanical study of medicinal plants used by traditional users in Villupuram district of Tamil Nadu, India. *Journal of Medicinal Plants Research, 4*(12), pp. 1089–1101.

Sapma, S., 2017. Indigenous plant. *ECHO Asia Conference*. October 2–6, 2017. Chiang Mai.

Savithramma, N., Yugandhar, P. and Linga Rao, M., 2014. Ethnobotanical studies on Japali Hanuman Theertham-A sacred grove of Tirumala hills, Andhra Pradesh, India. *Journal of Pharmaceutical Sciences and Research, 6*(2), pp. 83–88.

Shah, A., Ahmad, M., Marwat, S.K. and Zafar, M., 2013. Ethnobotanical study of medicinal plants of semi-tribal area of Makerwal & Gulla Khel (lying between Khyber Pakhtunkhwa and Punjab Provinces). *American Journal of Plant Sciences, 4*, pp. 98–116.

Shang, X., Tao, C., Miao, X., Wang, D., Tangmuke, Dawa, Wang, Y., Yang, Y. and Pan, H., 2012. Ethnoveterinary survey of medicinal plants in Ruoergai region, Sichuan province, China. *Journal of Ethnopharmacology, 142*, pp. 390–400.

Sharma, J., Gairola, S., Sharma, Y.P. and Gaur, R.D., 2014. Ethnomedicinal plants used to treat skin diseases by Tharu community of District Udham Singh Nagar, Uttarakhand, India. *Journal of Ethnopharmacology, 158*, pp. 140–206.

Simpson, D.A. and Inglis, C.A., 2001. Cyperaceae of economic, ethnobotanical and horticultural importance: A checklist. *Kew Bulletin, 56*(2), pp. 257–360.

Singh, J.P., Rathore, V.S. and Roy, M.M., 2015. Notes about Haloxylon salicornicum (Moq.) Bunge ex Boiss., a promising shrub for arid regions. *An International Journal, 62*, pp. 451–463.

Singh, B., Sultan, P., Hassan, Q.P., Gairola, S. and Bedi, Y.S. Ethnobotany, traditional knowledge, and diversity of wild edible plants and fungi: A case study in the Bandipora District of Kashmir Himalaya, India. *Journal of Herbs, Spices & Medicinal Plants.* DOI: 10.1080/10496475.2016.1193833

Song, M-J. and Kim, H., 2011. Ethnomedicinal application of plants in the western plain region of North Jeolla Province in Korea. *Journal of Ethnopharmacology, 137*, pp. 167–175.

Song, P., Sekhon, H.S., Lu, A., et al., 2007. M3 Muscarinic receptor antagonists inhibit small cell lung carcinoma growth and mitogen-activated protein kinase phosphorylation induced by acetylcholine secretion. *Cancer Research, 67*, pp. 3936–3944.

Srivastava, R.C., 2014. Family polygonaceae in India. *Indian Journal of Plant Sciences, 3*(2), pp. 112–150.

Sur, P.R., Sen, R., Halper, A.C. and Bandyopadhyay, S., 1987. Observation on the ethnobotany of malda-west Dinajpur Districts West Bengal-I. *Journal of Economic and Taxonomic Botany, 10*(2).

Tabata, M., Sezik, E., Yeşilada, E., et al. Traditional medicine in Turkey III. Folk medicine in East Anatolia, Van and Bitlis Provinces. *Pharmaceutical Biology.* doi: 10.3109/13880209409082966

Thapa, L.B., Dhakal, T.M., Chaudhary, R. and Thapa, H., 2013. Medicinal plants used by Raji ethnic tribe of Nepal in treatment of gastrointestinal disorders. *Our Nature, 11*(2), pp. 177–186.

Tolken, H.R. The species of Arthrocnemum and Salicornia (Chenopodiaceae) in Southern Africa. *Bothalia, 9*(2), pp. 255–307.

Tribedi, G.N., Mudgal, V. and Pal, D.C., 1993. Some less known ethnomedicinal uses of plants in sunderbans, India. *Bulletin of the Botanical Survey of India, 35*(1–4), pp. 6–10.

Tripathi, Y.C., Prabhu, V.V., Pal, R.S. and Mishra, R.N., 1996. Medicinal plants of Rajasthan in Indian system of medicine. *Ancient Science of Life, XV*, pp. 190–212.

Tumpa, S.I., Hossain, Md. I. and Ishika, T., 2014. Ethnomedicinal uses of herbs by indigenous medicine practitioners of Jhenaidah District, Bangladesh. *Journal of Pharmacognosy and Phytochemistry, 3*(2), pp. 23–33.

Uddin, S.N. and Hassan, Md. A., 2012. Angiosperm flora of Rampahar reserve forest under Rangamati district in Bangladesh. I. Liliopsida (Monocots). *Bangladesh Journal of Plant Taxonomy, 19*(1), pp. 37–44.

Uddin, M.Z., Hassan, Md. A. and Sultana, M., 2006. Ethnobotanical survey of medicinal plants in Phulbari Upazila of Dinajpur district, Bangladesh. *Bangladesh Journal of Plant Taxonomy, 13*(1), pp. 63–68.

Uddin, K., Mahbubur Rahman, A.H.M. and Rafiul Islam, A.K.M., 2014. Taxonomy and traditional medicine practices of Polygonaceae (Smartweed) Family at Rajshahi, Bangladesh. *International Journal of Advanced Research, 2*(11), pp. 459–469.

Ullah, M., Khan, M.U., Mahmood, A., Malik, R.N., Hussain, M., Wazir, S.M., Daud, M. and Shinwari, Z.K., 2013. An ethnobotanical survey of indigenous medicinal plants in Wana district south Waziristan agency, Pakistan. *Journal of Ethnopharmacology, 150*, pp. 918–924.

Ullah, S., Khan, M.R., Shah, N.A., Shah, S.A., Majid, M. and Farooq, M.A., 2010. Ethnomedicinal plant use value in the Lakki Marwat District of Pakistan. *Journal of Ethnopharmacology.*

Upadhyay, O.P., Kumar, K. and Tiwari, R.K., 2008. Ethnobotanical study of skin treatment uses of medicinal plants of Bihar. *Pharmaceutical Biology, 36*(3), pp. 167–172.

Vaidya, B.N. and Joshee, N., Report on Nepalese Orchid Species with Medicinal Properties. https://www.researchgate.net/publication/312888202.

Van Sam, H., Nanthavong, K. and Kessler, P.J.A., 2004. Trees of Laos and Vietnam: A field guide to 100 economically or ecologically important species. *Blumea, 49*, pp. 201–349.

Veldkamp, J.F. and Flipphi, R.C.H., 1987. A revision of Leptonychia (Sterculiaceae) in Southeast Asia. *BLUMEA, 32*, pp. 443–457.

Wagh, V.V. and Jain, A.K. Ethnopharmacological survey of plants used by the *Bhil* and *Bhilala* ethnic community in dermatological disorders in Western Madhya Pradesh, India. *Journal of Herbal Medicine.*

Wangchuk, P., Keller, P.A. Pyne, S.G., Taweechotipatr, M., Tonsomboon, A., Rattanajak, R. and Kamchonwongpaisan, S., 2011. Evaluation of an ethnopharmacologically selected Bhutanese medicinal plants for their major classes of phytochemicals and biological activities. *Journal of Ethnopharmacology, 137*, pp. 730–742.

Warrier, K.C.S. and Warrier, R.R., 2019. *Sacred Groves: Repositories of Medicinal Plants.* Coimbatore: Institute of Forest Genetics and Tree Breeding.

Watt, G., 1889. *A Dictionary of the Economic Products of India.* Calcutta: Government Printing.

Wazir, A.R., Shah, S.M. and Razzaq, A. Ethno botanical studies on plant resources of Razmak, North Waziristan, Pakistan.

WHO, 2009. *Medicinal Plants in Papua New Guinea*. Western Pacific Region:WHO.

Yaseen, G., Ahmad, M., Shinwari, S., Potter, D., Zafar, M., Zhang, G., Shinwari, Z.K. and Sultana, S., 2019. Medicinal plant diversity used for livelihood of Public health in deserts and arid regions of Sindh-Pakistan. *The Pakistan Journal of Botany, 51*(2), pp. 657–679.

Yazan, L.S. and Armania, N., 2014. *Dillenia* species: A review of the traditional uses, active constituents and pharmacological properties from pre-clinical studies. *Pharmaceutical Biology*, *52*(7), pp. 890–897.

Zheng, X-l., Wei, J-h., Sun, W., Li, R-t., Liu, S-b. and Dai, H-f., 2013. Ethnobotanical study on medicinal plants around Limu Mountains of Hainan Island, China. *Journal of Ethnopharmacology, 148*, pp. 964–974.

Index

B

C

I

J

N

Q

R

T

U

V

W

X

Y

Z

www.ingramcontent.com/pod-product-compliance
Lightning Source LLC
LaVergne TN
LVHW080649011026
849160LV00014B/1364